Foundation Analysis

RONALD F. SCOTT

California Institute of Technology

PRENTICE-HALL, INC., Englewood Cliffs, NJ 07632

Library of Congress Cataloging in Publication Data

SCOTT, RONALD F
  Foundation analysis.

  (Civil engineering and engineering mechanics series)
  Includes index.
  1. Foundations.   I.   Title.
TA775.S36   624.1′5   80-23345
ISBN   0-13-329169-3

Civil Engineering and Engineering Mechanics Series

N. M. NEWMARK and W. J. HALL, *Editors*

Editorial/production supervision
      and interior design by Lori Opre
Manufacturing buyer: Anthony Caruso

Printed in the United States of America

10 9   8   7   6   5   4   3   2   1

PRENTICE-HALL INTERNATIONAL, INC., *London*
PRENTICE-HALL OF AUSTRALIA PTY. LIMITED, *Sydney*
PRENTICE-HALL OF CANADA, LTD., *Toronto*
PRENTICE-HALL OF INDIA PRIVATE LIMITED, *New Delhi*
PRENTICE-HALL OF JAPAN, INC., *Tokyo*
PRENTICE-HALL OF SOUTHEAST ASIA PTE. LTD., *Singapore*
WHITEHALL BOOKS LIMITED, *Wellington, New Zealand*

To *Sydney Böttøcks*

Contents

Appendix B Plasticity   493

Appendix C Figures and Tables   509

Preface

In the titles of most books in the foundation engineering field the word "design" appears. That is not the case with this work, and the omission is deliberate. Design is the most complicated process that an engineer is called upon to perform; it is an intricate amalgam of experience, judgment, measurement, and analysis, usually pursued on a trial-and-error basis. One can go through the motions of explaining design procedures to a class, but I am not sure that we can actually *teach* design. The best technique appears to lie in the case history approach, followed by examples of increasing difficulty. An engineer in a particular field begins to have a feel for the quantities appropriate to a given problem—"design"—after performing many analyses of similar situations. Give an engineer even a slightly different problem—design of an aircraft wing structure to a bridge engineer—and the learning process must begin all over again. The past experience is merely a guide.

Intimately related to design is analysis. Analysis cannot begin without some conjectured member or framework, which originates as a guess. Every foundation design process begins with a tentative selection of the foundation scheme (footings, mat, piles) based on general site conditions and structural constraints, followed by estimated member sizes and reinforcing amounts. To this arrangement the design loads are applied. Including the known range of soil response, analysis yields initial values of deflection, bending moments, and shears for comparison with allowable values for the structure and its components. Member sizes are thereupon modified to bring quantities computed in a second cycle into closer alignment with tolerable values. Several such iterations result in convergence of the two sets of quantities to any desired degree. In this process, analysis plays an important part; design in its conventional sense consists of the guesses made in between parts of the analysis.

A much subtler question relates to economy and the closeness of approach of the moments or stresses to the material limits at all points of the structure. In good economical design the material will approach its limits to the same degree at many disparate points in the structure simultaneously under the design loads. Developed to this state, design is the essence of engineering.

In foundations the interaction of variable and nonlinear soil response with the behavior of structural entities and components makes exact analyses impossible. The designer's ingenuity lies in the choice of an analytical method that most reasonably approximates real life but still permits a solution to be attained. There is an economic reward, and substantial satisfaction, in hitting upon an analysis technique that best matches the situation to be evaluated. Formerly, procedures that could be carried through by hand or with simple calculating engines (slide rule, mechanical calculator) were all that were available. Engineers (von Karman, Westergaard, Timoshenko) achieved eminence through the elegance of their analytical approaches or the invention of new mathematical tools. Hand calculations for some structures (bridges, space frames, concrete dams) required prodigious amounts of labor and were usually accomplished by dividing the overall task into smaller portions each negotiated by tabulation. Techniques appropriate to these limitations were developed. Now, with large digital computers, the computational limits are no longer a major difficulty. The machine's capabilities by the late 1960's had surpassed our abilities to characterize (linear, viscous, plastic, hysteretic, dissipative) the materials with which we had to deal. In addition to the pertinent quantities deriving from machine analysis, irrelevant numbers in vast amounts spew forth. Ingenuity in analysis can now be exercised in limiting the amount of calculation to be performed and in selecting the most appropriate analysis for each occasion. This process can always be aided by careful choice of simplified hand calculation at an early stage in the design operation.

It is to the last area that this book is directed. There are many techniques of analysis, often the precursors of machine computation, which do not seem to have been widely applied in foundation analysis. With the use of a programmable hand calculator or a small desk-top computer, to free the engineer from the drudgery of mechanical manipulations, many of these methods permit the analysis of surprisingly complex problems and avoid the complexity and inconvenience of programming a large computer with its single-case limitation. The simpler techniques frequently permit useful parametric studies to be made in the preliminary stages of design; when the choice has been narrowed down, the few remaining alternatives can be refined or eliminated by detailed examinations on a big machine.

In the foundation engineering field, I have not found a convenient source of codification of detailed hand or elementary computer calculational routines. Through the years I have collected those that have been useful to me in consulting practice and in teaching. Soil-structure interaction is so complex that I have found it very helpful in both these areas to discover analyses that elucidate the mechanism or effects in particular cases without being unduly complex. Here I have presented a number of these methods, with applications, in what seems to me a fairly logical order of increasing complexity of interaction effects. In order to smooth the exposition, and relate it more directly to foundation situations, I have emphasized the application of the methods, to the omission of mathematical details of proofs of convergence and the like. Wherever appropriate, I have tried to illustrate the latter requirement and to give confidence in the use of a particular method by comparing a numerical result to an exact or much more closely approximate solution, where it exists.

All the constitutive models used are relatively simple—linearly elastic and elasto-plastic—in order to keep the computations within the scope of hand or desk calculator operations. However, the basic techniques indicated will support the use of more advanced constitutive relations, such as those involving successive yield surfaces, of which a number of examples are now current. The penalty for the better representation is, of course, more computation at a higher level of complexity.

I hope that the book will be of value to practicing engineers, whom it may encourage back into the area of personal calculation, and also to educators and students of foundation engineering. For the assistance of the latter and entertainment of the former, I have included a number of problems and exercises with each chapter.

A solutions manual is available for these, although I must emphasize that in the design problems an "answer" is a highly personal thing. Accordingly, the solutions manual includes, whenever appropriate, some discussion of assumptions and results to help the instructor and student in the learning process.

I have many people to thank for their assistance during the writing of this book. The outline formed while I was on leave from Caltech at Cambridge University, partially supported by Guggenheim and Churchill Fellowships. Much of the material and many of the exercises were reviewed by my graduate students in a succession of courses and were refined by that contact. At different times portions of the manuscript were typed by Alice Gear, Jeanette Hannigan, Sharon Vedrode, Gloria Jackson, and Ruth Stratton; I am grateful for their patience and precision. My family provided the support that is necessary to an effort of this kind.

RONALD F. SCOTT

Soil Behavior, Elasticity, and Plasticity: Soil-Structure Interaction Classification 1

1.1 INTRODUCTION

The stresses generated in a structure resting on soil and the displacements the structure undergoes are determined by the loads acting in and on the structural framework and by the relationship of the stiffness of the structure to that of the material on which it rests. Within a limited degree of uncertainty arising from the behavior of structural concrete and riveted and other joints, the stiffness of a steel or concrete structure can be characterized relatively well in comparison with that of the underlying soil or rock.

This is the case because the properties of structural materials are well known, so that structural stiffness is related to the proportioning and geometrical arrangement of the elements of the structure. Up to fairly well-defined limits of load or relative deformation the behavior is essentially linear. Yielding begins beyond these limits in a generally stable and consistent fashion which has been extensively investigated for a variety of structural forms and materials. Within the requirements of various building codes, the properties of structural materials are circumscribed.

In contrast, except in rare cases in which special preparations are undertaken, the nature and distribution of the soil and rock underlying a structure cannot be described precisely, nor are the materials easily accessible to investigation and testing. Difficulties are generated because these underlying materials are neither homogeneous nor isotropic. In general, in comparison with structural fabrics, they are also soft, a property which renders samples difficult to retrieve in a pristine condition. Their properties are evanescent, so that behavior in a laboratory test may be widely different from that in the ground. Unlike those in the structure itself, stress conditions in the soil underlying a structure are difficult to evaluate, and even more difficult to simulate in tests. The fact that the materials' properties are dependent on both past and present stress conditions, and will change under the superimposed stresses, only emphasizes the problems of characterization and analysis. Generally, satisfactory engineering analyses of steel and concrete structures can be made by using linearly elastic or elastoplastic material models. On the other hand, the behavior of granular materials is more complex.

Although a great deal of work has been done on the stress-strain or constitutive relations of soils, results are, at present, still incomplete. Only under the simplest of boundary conditions, and under special conditions of material behavior, can satisfactory estimates of stresses and displacements be made. This situation is likely to obtain for some time because of the difficulty of the subject and the limited effort being devoted to it.

Before the introduction and widespread use of large digital computers, the computational difficulties associated with the solution of continuum mechanics problems under boundary conditions common to soil engineering situations were so great that gross simplifications of both the material properties and the geometry were made. Field problems were classified into two groups: those of displacement and stress distribution calculation on the one hand (elliptical equations) and problems involving computations of a failure load or bearing capacity (hyperbolic equations) on the other hand. For the first class, the material behavior was taken to be that of a linearly elastic solid. Even linearly elastic problems pose great analytical difficulties under boundary conditions encountered in geotechnical circumstances. For the second class, a rigid plastic model was adopted. With this representation, other analytical difficulties present themselves. Under these descriptions soil engineering calculations fell into one of the two classes: (a) settlement and stress distribution or (b) bearing capacity.

Research on the engineering properties of soils has concentrated (and still does) on areas relevant to these two concepts; in particular, emphasis has been placed on failure behavior. Generalized constitutive relations have seldom been examined; instead, the effect of numerous variables (water content, density, mineral type, grain size, grain shape, distribution of grain sizes, stress, stress history, pore pressure, time, temperature, and so on) on the strength of clay and silt, and on sand and gravel has been exhaustively studied.

For the bulk of boundary condition problems relevant to soil engineering, the material has been considered to occupy a half-space, with loading applied at the upper boundary. Stress and deformation problems have been solved (for linearly elastic materials) involving various complexities of loading at the upper boundary and loads inside the solid itself. One, two, and then three layers of solid of different, but still linearly elastic, properties have been treated. Some attempts at solutions for anisotropic materials have been made. Other geometrical arrangements (quarter-space, holes, and inclusions) have received less attention.

Because of the separation of general stress and displacement problems in nonlinear materials into these two compartments, a growing understanding of the complexities of the real soil behavior caused difficulties to appear. Since, for example, pore pressures play an important role in the determination of the stresses under which the soil would fail, it became necessary to calculate them in appropriate circumstances. This could be done, at a first level, by assigning to pore pressure the hydrostatic stress component in a linearly elastic incompressible medium of the appropriate geometry and approximate loading conditions. These pore pressures can be applied (albeit inconsistently) to a failure analysis. However, the pore pressures generated by the nonlinear behavior of soil have importance equal to or greater than those developed by the linear part of the behavior. How can these more realistic pore pressures be computed since the analysis is linear? They are obtained by using linear analysis to obtain stresses. The stresses are applied in laboratory tests on the real nonlinear material to determine the pore pressures which develop. Various empirical relations have been developed to describe the results of such tests. As another example, it is recognized that nonlinear stress-strain soil behavior gives rise to strains and thus displacements and surface settlements which are significantly different from those computed by the linear model. Since the nonlinear analysis cannot be performed, how can the corrected settlements be obtained? This is done in a fashion similar to that described above, by using a linear analysis to give stresses and by making subsequent laboratory tests under these stresses to give the nonlinear material strains. The vertical components of the measured strains are then applied at the appropriate coordinates in the geometrical model and summed to give surface displacements. The fact that the resultant stresses and strains are necessarily incompatible is ignored.

These difficulties of constructing tractable, analyzable geometrical models for soil problems have historically been such that the principal development in the field has been the study and analysis of case histories, in conjunction with a growing appreciation of the complexities of material behavior. Thus, for example, since it is difficult, if not impossible, to retrieve a sample of sand in a condition identical to that in which it exists in the ground, methods have been developed of estimating the state (relative density) of the material in its natural condition *without* obtaining samples. These have involved a variety of static and dynamic probes of the material, in which the static force or the number of blows of a standard weight required to advance a penetration device have been measured. By observation and comparison of, for example, the settlement of known loads of simple geometry at the ground surface, with the property of the material measured by such a penetration resistance, it has been possible to provide guides as to the amount by which a footing at a given pressure will settle in practice. Such a technique does not rely on methods of continuum mechanics for computation or justification.

The soil engineering literature and technology are replete with such approaches to the solution of practical problems. Earth pressures on retaining structures, permissible differential settlements of structures, thicknesses of highway and airfield pavements, and allowable loads on piles and pile structures are all examples relevant to this discussion. In many of these cases, mathematical analyses are highly simplified or even totally absent.

The development of a variety of computational methods involving the use of large digital computers in the last decade makes tractable many problems that could not be tackled hitherto. In the early stages of the development of finite difference and finite element computer techniques, only situations involving linear materials could be handled. Even with this limitation, the ability to solve problems with complex boundary and geometry conditions allowed analysis to replace estimation in many practical cases concerned with at first two, and later three, dimensions. The capability of performing dynamic analyses was developed. More recently, bilinear, elastic-plastic, cap model, and more complicated material representations have been incorporated into finite element computations, further extending the range of analysis.

The nature of the problem facing the soil engineer has changed. It has become of growing importance to achieve, firstly, better characterizations of the soil behavior, and secondly, better definition of material boundaries and in-place properties in the field. Our ability to analyze problems has, for the first time, outstripped our ability to describe the material and boundary conditions to a comparable degree. In addition, the lack of quantitative data on the performance of full-scale structures severely limits our ability to test the numerical methods of analysis. How can a 2-meter diameter, 100-meter long pile be tested? How can the postulated performance of an earth dam in an earthquake be checked? This will probably continue to remain a fundamental difficulty.

Because the application of the principles of soil mechanics to the behavior of structures in practice has explained the subsequent performance of the structures to a precision compatible with that to which the boundary conditions and structural properties are known, there is some question as to the incentives for further detailed research in soil mechanics along the lines indicated above. There are few conventional configurations in soil and structural engineering practice where existing, largely empirically based, methods of evaluation and analysis do not give displacements, settlements, or stresses to a generally adequate degree of accuracy for civil engineering purposes, that is to say, within a spread of 30 percent to 50 percent about the subsequently measured performance. Failures due directly to soil behavior have been rare in recent decades, and they result most usually from field investigations which are inadequate for various reasons. There may not be a sufficient density of bore holes or test samples, cavities in porous foundations may escape definition, or a critical layer or fracture surface may be so small as to avoid detection by the methods used.

Apart from the economy and reliability of existing techniques of investigation and analysis, there are other reasons why they are unlikely to be displaced entirely by computer methods. In every civil engineering situation the preliminary stages require various estimates to be made for design purposes. These involve loads, stresses, settlements, and displacements. For these estimates, hand or simple computer calculations provide the flexibility required to permit the effect of changes in quantities or assumptions to be assessed rapidly. Only when a design is essentially complete is it likely that a detailed computer analysis will be attempted. Such an analysis may still be fairly costly, in personnel time if not in computer costs, and is not in general well-adapted to a study of the effect of design variables. Instead, a few such analyses may be used either to pinpoint the occurrence of locally high stresses or large deformations or to assess with possibly greater accuracy the relative settlements and displacements that the structure will both cause and undergo as it is built. A knowledge of existing approaches to analysis, by largely elementary methods, of proposed designs including both structures and soils is therefore likely to be of value to the soil engineer for some time to come. Its application will enable the engineer to assess alternative designs relatively rapidly in given circumstances and to make parametric studies of the effects of certain variables under his control, and last, but not least, it will provide him with a necessary check on the subsequently more complex calculations performed on computers.

It is the purpose of this book to survey some of these practical methods of analysis for design. A number of well-known techniques will be reexamined and a variety of methods rarely used will be introduced and described. Many of the methods are adaptable to computation on small programmable calculators or desk-top computers.

1.2 SEQUENCE OF PROBLEM AREAS

Since, as described in the preliminary paragraphs of this section, the principal practical problems in soil engineering are generally those involved with soil-structure interaction, it is of interest to classify various typical situations on the basis of increasing complexity. At the lowest interaction level, analyses deal with those circumstances where the loads imposed on the soil are largely independent of the soils' response, although the stresses generated in the structure are affected by the soil behavior. Herein, we use the word *soil* to describe the foundation material which supports the *structure*. The structure is generally a steel or concrete construction. On occasion, as with an earth embankment or dam, soil is the material of both structure and foundation.

In order of increasing complexity of soil-structure interaction, therefore, we may list a variety of cases of foundation engineering interest.

1.2.1 Vertical loads

Vertical loads are ultimately due to the weight of the structure. Most soil engineering problems in practice fall into this category. The most simple are those in which the structure has extreme flexibility. For example, the one-dimensional loading of an underlying soil, uniform laterally, by a fill of horizontal extent great in comparison with the thickness of the foundation material, is a case widely treated in soil mechanics and engineering text books. The basic assumption here, for no interaction to occur between the fill and the soil, is that lateral variations in the soil occur over distances which are large, compared with the thickness of the fill.

These conditions would *not* be met, for example, if a 5-meter thick fill is placed over a former stream channel a few meters wide, filled with a compressible natural material. In this case, complicated stress patterns would be generated in the fill in the vicinity of the channel because of the greater compressibility of the material there compared to the soil on each side.

Another case of somewhat less practical interest in the general context of foundation engineering is that of a static wheel load on a level natural ground surface. The flexibility of the suspension systems of independently sprung vehicles is such that small variations in the level of the wheels do not substantially affect the load on the wheel. Thus individual differences in the displacement of the soil under each wheel do not significantly alter the load transmitted through the wheel. Spatial variations of the soil properties in dimensions of the order of the length or width of the vehicle have to be large to affect individual wheel loads. The ability of the vehicle to move slowly or to supply a particular draw-bar pull (so that the problem remains a static one) depends on the performance of a single wheel on the soil under its share

6

of the vehicle load. This situation occurs in relation to what is termed *off-the-road locomotion* when agricultural or military vehicles travel across natural ground. The question of the mobility of the vehicle under these circumstances is an interesting one.

Most vehicles, however, travel along artificially prepared surfaces, or pavements, placed on the natural soil. The stresses and deflections of the pavement depend on the relative flexibility of the structure (which, in this case, is the pavement and any prepared material underlying it) and foundation soil. If the wheels applying the loads are closely spaced, the stiffness of their springing system may also play a part.

Buildings transfer their loads, both dead and live, to the underlying soil through footings resting at ground surface or below. For ground of moderate to considerable stiffness and strength (the two properties are in proportion to one another in most materials), each column usually bears on a single footing, which is so proportioned as to prevent the possibility of either failure or harmful settlements occurring. In the case in which the building is relatively flexible, for example, if it is a single-story steel-frame structure, there is little or no interaction between the deflections in the underlying soil and the loads acting on the footings.

In the case of softer foundation soils, or as a consequence of circumstances imposed by property lines or architectural considerations, it may be necessary to combine the footings of two or more columns into one footing. Such an arrangement is called a *combined footing*. On occasion several columns in a row all apply their loads to a single long footing. This is frequently done, for example, in the case of a row of relatively closely spaced columns. The resulting foundation structure is called a *strip footing*. In the limiting case in which the structural loads are transferred to the soil by a bearing wall, a *wall footing* is required. These considerations of passing from individual or combined to strip or wall footings involve a growing degree of interaction between the structure and the underlying soil.

When the underlying soil is very weak and compressible, or when differential settlements must be severely limited for structural or functional reasons, several methods are available for the execution of the foundation work. If a calculation of the total area of individual or combined footings based on settlement or bearing capacity requirements shows it to be greater than one-half the area of the building itself, economy in the foundations will result if the entire building is placed, in effect, on a single slab, which is called the *mat*. The mat is a logical extension of the concept of individual and combined footings. It may be a slab, but more commonly it consists of an intersecting grillage of inverted T or other beams. On occasion a steel framework may be used to transfer the building's loads to the concrete foundation slab. The distribution of stresses and displacements in the mat and underlying soil is a three-dimensional problem in which there is a considerable degree of inter-action between structure (including the mat) and soil.

In difficult foundation conditions the selection of an appropriate foun-dation depends on a number of requirements. These include the nature and function of the building itself in terms of the loads it will transmit to the soil, the settlements it can tolerate, and the consequences of a transgression of established limits, as well as considerations of cost and special site condi-tions. Buildings especially sensitive to differential settlements include both conventional and nuclear power generating facilities, structures housing certain kinds of machine operations or assembly lines, and particle accel-erators. The latter present perhaps the most stringent requirements of all, in that lines and levels must be maintained with deviations less than a milli-meter along paths hundreds to thousands of meters in length.

Building sites in cities may offer certain difficulties associated with the removal of soil, in the case in which a basement or bored piles are considered for the foundations, and in connection with the protection of adjacent struc-tures against movements occasioned by construction operations. In compe-tition, in effect, with mat foundations, therefore, are foundations composed of piles, of which a wide variety is available, ranging from massive concrete caissons to prestressed concrete piles, steel H-piles, and steel tubes. In gen-eral, supporting piles are arranged in groups whose interaction with the sup-portive soil can be extremely complicated. Knowledge of single-pile behavior is basic to the understanding of the interaction in groups of piles.

In the circumstances considered up to now, loads have been applied vertically, generally as a consequence of the weight of a structure, to the soil. However loads are often applied to or by the soil laterally, and situations where this occurs constitute another class of interaction problems, which will be examined next.

1.2.2 Lateral loads

When lateral forces must be absorbed in a foundation soil, as in certain bridge piers, or in anchorages for tie-rods employed with sheet piling or for earthquake-resisting structures, a foundation must be constructed to provide for the transfer of the lateral loads into the soil. Commonly this employs piles, either individually or in groups. The interaction of pile and soil stresses and deflections under lateral and moment loadings has been studied in considerable detail.

In a variety of situations in civil engineering construction, soil must be constrained from moving laterally, as in quay and harbor walls, in excava-tions, and in highway and other cuts. The constraint is supplied by a wide range of supporting structures. Sheet pile walls with or without ties, cast-in-place walls or walls with tie-backs, strutted bulkheads, and cellular bulkheads or cofferdams are all techniques employed for this purpose. They form a group of foundation structures whose interaction with the soil they rest on and support is most complex. Studies and analyses of the performance of such structures have not been as extensive as those of other foundation types, but they still provide a substantial body of information.

On occasion, the structure imposing the load on the underlying soil or rock in itself consists of soil or is composed largely of soil. We will classify this as a problem area separate from those discussed previously.

1.2.3 Soil structures

The stresses and displacements developed in an earth dam or highway embankment and in the supporting soil or rock represent perhaps an extreme case of the interaction problem. This occurs because, as mentioned previously, the mechanical behavior of soil is complex and not well understood, and in an earth dam, soil constitutes both the structure and the foundation material. By comparison, the behavior of a conventional steel frame or concrete structure is well-defined. An additional difficulty arises because of the typically long construction time required for a large earth dam. Each increment of material added to the dam first imposes its gravitational load on the existing structure (soil) which transfers it to the natural supporting soil or rock. This incremental soil layer subsequently becomes a part of the structure, so that the geometry of the structure is continuously changing. The stresses and displacements developed in this construction sequence are different from those that would be generated were the earth dam to be in effect instantaneously built, in the absence of a gravitational field, and then gravity turned on. The latter procedure is the one conventionally assumed for analysis.

An earth dam (or other dam) may be pierced with openings for inspection, drainage, or measurement purposes. Highway embankments contain openings to permit the passage of small streams or highway drainage. These openings or culverts are sustained with rings or portions of rings of steel or concrete which are stressed by the superimposed burden of soil. The pipes or arch sections are usually placed early in a construction sequence and the embankment soil is built around and over them. Stresses and displacements occur both in the ring and in the soil medium in this process; the interaction is complex but has been frequently studied. Similar behavior occurs in the imbedment of sewage or drainage pipes in trenches in the ground. A special class of this problem was encountered by the builders of the pyramids of Egypt, when they incorporated shafts, tunnels, and chambers, sometimes of considerable dimensions, in the growing structure.

The most complex and least understood interaction problems at present are the dynamic stresses and displacements generated by shock loads or by the vibrations caused by machinery and earthquakes. Of these the circumstances of greatest human, economic, and engineering consequences are those attendant upon earthquakes. They constitute the last of the interaction problems to be listed here.

1.2.4 Dynamic loads

Although a great deal of public attention is paid to the possibility of ground displacement due to fault movements during an earthquake, by far the greatest human hazard is a consequence of the effect of widespread shaking upon civil engineering structures. The most direct concern is with the performance of structures occupied by people and structures performing vital functions such as the supply of water, gas, or electricity, along with the associated operating systems and generating stations. The transmission of earthquake vibrations through soil or rock into the structure and the consequent dynamic stresses and displacements in structure and ground present an extremely complex interaction problem. Much research needs still to be done in respect to the vibrations of structures and varied types, piled foundations, retaining walls, and dams.

Many of the greatest potential hazards in an earthquake are associated with the behavior of natural soil and rock slopes and earth dams. A growing number of investigations is being devoted to these subjects and several methods of analysis have been proposed.

In this book, various analytical and computational methods will be illustrated with specific and typical examples from the above listing. Individual, combined, strip, and mat foundation analysis methods are discussed first. Highway or airfield pavement slabs are examined from both the points of view of elastic stresses and displacements and of plastic or collapse behavior. Pile stresses and deflections under axial, torsional, and lateral loads when the soil response is either plastic or elastic-plastic are also treated. A limited amount of attention is given to pile group performance under axial and lateral loads. Unfortunately lack of space precludes treatment of retaining structures, earth dams and embankments, pipes and culverts, and dynamic soil behavior.

Some of the basics of the theories of elasticity and plasticity are summarized in the appendices, which may be useful for occasional reference from the body of the text. Appendix C presents a variety of detailed numerical results calculated for some of the problems discussed in various chapters.

PROBLEMS

1. Explain, with drawings of sample cases, the design problems presented by the interaction of soil and structural mechanical properties.
2. Assume that some calculational method, employing realistic soil properties, gave you all the information on stresses, displacements, etc., that you wanted to know in connection with the design of an earth dam or other structure. How would you use this information? Outline the steps you would take to design a suitably safe structure: (a) under static conditions (own weight, water load, seepage); (b) under given earthquake conditions (history of ground accelerations, or spectrum of ground motion).
3. You are given the task of designing a pressure cell to measure one normal component of stress at a point in a soil mass. Plot a diagram of the ratio

$$\frac{\text{stress indicated by the pressure cell}}{\text{stress in the soil if the cell were absent}}$$

versus the ratio

$$\frac{\text{stiffness of the cell}}{\text{stiffness of the soil}}$$

In what stiffness range would you design the cell? After reading through Chapter 5 of this book you may try a quantitative analysis. (It may be helpful to apply Baker's method to this problem.)
4. Is previous experience helpful in the design of structures where the interaction with soil is important? Explain.
5. Since, for many large civil engineering structures, such as dams, offshore production platforms, and nuclear power plant containment vessels, detailed field measurements of loads, stresses, and displacements under varied loading and soil conditions are scanty or even totally lacking, what basis do we have for confidence in the design calculations?
6. Real soils are anisotropic, inhomogeneous, nonlinear, and inscrutable. Considering the deficiencies of field investigations, the limited range of material models available, and the limitations of our calculational methods, how can we justify the engineering design computations that we make?

Methods of Analysis   2

2.1 INTRODUCTION

There are a number of stages in the analysis of a soil or foundation engineering problem. They may be classified in the following way:

1. It is necessary first to understand the field conditions and the critical elements to be obtained and assessed by subsequent analysis. This involves an understanding of the size, nature, and magnitude of loading associated with the structure. From these considerations, the region of foundation soil that will participate in and respond to the stresses imposed by the structure may be estimated. A site and laboratory investigational program is organized. The selection of field and laboratory tests will be predicated on the nature of the difficulties posed by the structure and site conditions. The problem may be principally one of stresses and deformations rather than failure, or, alternatively, failure or the possibility of collapse may be the overriding criterion. From the field and laboratory studies (which will not be discussed here) the boundary conditions of the problem are established, including the presence of layers of different soils, and the basic material properties are characterized. All the information is then present and leads to the next and most critical step.

2. The field situation is always too complex to be tackled directly as it stands. Instead, it must be idealized to a solvable problem. It is here that judgment, experience, or intuition, so heavily emphasized in soil engineering, plays the most important role. The problem must be simplified without loss of its essential features. In the best of such idealizations, the calculated important stresses, displacements, or bearing capacities will not differ by more than 30 percent to 50 percent from those that would be obtained were it possible to analyze the original conditions in all their complexity. Since the latter analysis cannot normally be done, this implies that the final arbiter of a satisfactory analysis is the behavior of the completed structure. The crucial stage in the idealization process is usually the rejection of unimportant data from the mass of information normally available. From the suitably idealized problem follows the next step.

3. The mathematical relations describing the idealization and the boundary conditions are formulated and solved. The solution is, of course, a necessary prerequisite to the next stage in the procedure, a stage second only in importance to that described in paragraph 2 above.

4. From the analysis, stresses and displacements are obtained usually at a variety of locations in the idealized model. These have to be interpreted in terms of structural performance, that is to say, total and differential settlements, bending moments, relation of stresses to yield values, and so on. The model is idealized and the calculated stresses and displacements are, to a greater or lesser extent, fictitious, but the actual structure is real and will undergo real stresses and movements, so that this step is usually difficult. Relating the computed values to the performance of the real structure is a complicated task, in which again experience is of assistance. In some circumstances, an additional step is indicated by the results of the calculations.

5. It may be desirable to test or develop confidence in the calculated results by performing field trials of portions or suitably scaled models of the prototype structure. A single pile or small group of piles may be driven and tested to check the estimated bearing capacity or load-deflection relations; a section of embankment may be raised until a failure occurs; the deformational properties of the foundation soil may be elucidated by a few plate-loading tests. On occasion, additional field investigations may be indicated to resolve critical items—the depth, thickness, or properties of a thin clay layer, the relative density of a sand zone, or the presence of cavities in limestone. Generally, when the construction gets underway, some check on the calculations is given by continuous or intermittent observations of settlements, pore pressures, tie-rod loads, or stresses in the soil. Sometimes, the results of these measurements permit or even require that changes in construction be made.

The function of this book is the treatment of a variety of analytical techniques that can be employed at stages 3 and 4 above in the solution of foundation engineering problems. A discussion of the principal methods of solutions available begins here. In subsequent chapters the techniques will be applied to a variety of problems of soil-structure interaction.

The emphasis throughout is placed on obtaining numerical answers, not on mathematical formalism. As a consequence, many of the techniques treated are numerical or approximate in nature. The methods selected are those that minimize the required calculational effort; where feasible, the results will be compared either with "exact" analytical solutions or with numbers obtained from more extensive numerical computations carried out on digital computers.

The approaches range from analytical in the classic mathematical sense to almost completely numerical where only simple arithmetic operations need be performed. They are all adaptable to computer coding, with programs ranging from simple to highly complex. A number of the methods employed in computer calculations, such as the *finite element* method, can be reduced in scope to a level at which hand calculations can give useful results and some examples of such computations are given.

It is still far from widely recognized that some level of understanding of the principles of solid mechanics is immensely useful in the handling of problems in soil mechanics and foundation engineering. Appendix A is a reference guide to this understanding. It is suggested that this appendix be examined before the rest of the book is tackled. It will be particularly helpful to review the material on strain energy, work, and potentials.

Much of soil engineering has to do with the interaction of structural elements such as beams, plates, or shells, generally straight or plane but occasionally curved, with soil. Pavements, individual and combined footings, mats, piles, walls, and pipes are all examples. Because the analytical treatment of problems in which the soil is considered to be a continuum is so difficult, a number of approximations to the soil reaction have been proposed from time to time. These foundation models are discussed later in the book, but they are used without detailed explanations in the earlier examples which are intended to illustrate numerical *methods*.

2.2 THE ANALYTICAL SOLUTION OF STRESS
AND DISPLACEMENT PROBLEMS
IN CONTINUA

The classical methods of solid mechanics follow the lines outlined in Section A.1 in Appendix A. The technique requires solution of the equilibrium equations (A.9) together with constitutive relations, such as equations (A.40) for the general linear elastic material or equations (A.41) for a Hookean elastic solid. At the same time, the compatibility relations, equations (A.28) and the boundary conditions must be satisfied. A number of techniques has been described for obtaining suitable solutions for given problems, and the available methods are discussed in solid mechanics text-books. The analytical techniques have led to equations for stresses, strains, and displacements in a number of problems of interest in soil mechanics, such as those involving point, line, and distributed loads on the surface of a semi-infinite medium, when the medium is homogeneous and isotropic.

The algebraic and numerical solutions to a variety of these problems have been listed and tabulated in a number of books. These results are generally useful directly in the analysis of many soil problems, but they may also be employed in the construction of influence functions for use with some of the numerical techniques that will be discussed herein. However, few solutions are available to problems in which soil layers of differing properties appear or in which the boundary geometries are complicated. It is fair to say that, for areas of soil mechanics interest, the majority of continuum boundary value problems of linear elasticity for which solutions can be achieved with the expenditure of even considerable amounts of mathematical ingenuity and manipulative labor has already been solved. The analytical solution of a new problem generated by a situation encountered in soil engineering would require an effort such that it is unlikely to be attempted by purely mathematical approaches.

Consequently, in most soil engineering calculations the technique employed has been one of idealizing the real-life problem to one for which a mathematical solution has been available (see paragraph 2 of Section 2.1). This process can be described as one of obtaining an exact solution to an approximate problem.

An alternative technique which will be described here, is to leave the original field problem as it stands, to the best of one's knowledge of the material properties, and to obtain an approximate numerical solution. Thus the exact problem is solved approximately. With this approach, generally referred to as *numerical analysis*, solutions as close to the correct solution as desirable can generally be achieved. The limitation is usually one of economics. There is a wide variety of numerical techniques available for analyzing problems of interest in foundation engineering. Among them are (1) a class of approaches called the *method of undetermined parameters*, which includes collocation and finite element techniques, and (2) methods which make direct use of the differential equations, such as finite difference techniques, including simultaneous equation solving or iteration or relaxation methods. Most of the computational techniques have been known for many years, but some of them have only gained widespread use recently with the development of digital computers. The increased calculational capacity of large computers has been such that, for many problems, only one or two methods have been used, usually finite element or finite difference, almost to the exclusion of other approximate techniques. For a large proportion of foundation engineering situations, particularly in the early stages of design calculations, it can be argued that these methods oversolve the problem in view of the scarcity or inadequacy of the data at hand. They supply much more information than the designer can make use of. The designer may be interested in displacement or normal stress at only one or two locations in the problem field, but a finite element calculation, for example, supplies all displacements and all stresses in every element of the modeled field. The superfluity of information frequently gives the analyst a spurious sense of the accuracy of the calculation.

The increased availability of small computers from pocket-sized to desk-top size makes a wide range of calculational methods available. These are methods too complex or too tedious for ready calculation by hand, but they are of such a scale that their use by a large computer is hardly worth the programming effort. They make accessible to the designer a surprisingly large range of linear and nonlinear solution techniques that would have been too formidable a few years ago. Using these methods and a calculator, the designer can readily check, in an approximate way, the effect of changes in many of the variables in typical foundation problems. Once the design choice has been narrowed down or even finalized, one or a few detailed solutions by the finite element method can refine the quantities of deflection, stress or bending moment. Once they are known the slab section size or reinforcing quantity are determined.

A number of numerical methods especially applicable to foundation engineering problems will be summarized below. The treatment is not intended to be rigorous or exhaustive; instead, it emphasizes the practical aspects of the computation. More detail and a wider range of examples can be found in textbooks on numerical analysis (2, 3, 5).[1]

2.3 METHOD OF UNDETERMINED PARAMETERS

In this approach a function is selected that contains arbitrary constants and that satisfies certain boundary conditions in the problem being studied. Various techniques can then be applied to evaluate the constants by satisfying different criteria of goodness of fit of the function to the equation describing the physics of the problem. Each criterion used will, in general, give rise to a different set of constants, solving the problem more or less precisely as compared to an exact solution. The labor involved in the solution varies with the criterion selected. The advantage of these approximate procedures is that a solution can nearly always be obtained, even for situations, including

---

[1] The numbers in parentheses throughout the book indicate the references at the end of each chapter.

nonlinear problems, for which exact solutions would be difficult if not impossible. In many cases, the results are perfectly satisfactory for engineering analyses, although occasionally the numerical values may be very inaccurate. Care must be taken to check the solutions on completion, and it is desirable to conduct running checks during the numerical procedures.

The various criteria of fit are described and illustrative examples are given in the following sections. It should be noted that the boundary condition requirements on the selected function vary with the solution technique. With collocation methods, for example, all the boundary conditions must be satisfied by the adopted function; in energy methods only certain conditions, called *essential*, must be met. These points will be discussed in more detail as the different approaches are examined.

2.3.1 Collocation

Generally, less labor is involved in collocation than in the other methods to be described. It proceeds as follows. The function satisfying the boundary conditions is substituted in the differential equation to give an expression termed the *residual*, which includes the space variable or variables and the arbitrary coefficients. If the function were the exact solution, it would satisfy the equation at all points in the domain. Since it is not, it can be made to satisfy the differential equation at as many points in the field of the problem as there are unknown coefficients. This is done by substituting a selected coordinate or coordinates in the residual and requiring the residual to be zero at the coordinate location. For each position at which this is done, an equation is obtained in the unknown constants. With equal numbers of equations and coefficients, the equations can be solved simultaneously to give the constants. Naturally, the more constants included in the original function, the more locations there are at which the residual can be equated to zero, and the better a solution obtained. When the original differential equation is linear, the simultaneous algebraic equations are also linear; when it is nonlinear, so are the simultaneous equations to be solved. Experience and judgment play a part in the selection of the trial function with which to approximate the problem.

The collocation technique can be best demonstrated by an example. Other examples appear throughout the book. In the study of the axial loading of piles (see Chapter 8) an approximate equation can be developed which is written in its dimensionless linear form as

$$\frac{d^2\psi}{ds^2} - \psi = 0 \qquad\qquad (2.1)[2]$$

---

[2]This equation also appears in Chapter 4 and in another context entirely describes one-dimensional flow in a leaky aquifer (11). It turns up in many other problems of physics.

where ψ describes the dimensionless axial displacement of the pile as a function of dimensionless distance s along the pile from the ground surface. There are two boundary conditions for this problem, at the top and bottom (tip) of the pile. The usual condition at the top is the applied load. In the dimensionless terms of the equation this condition is represented by the requirement

$$\left(\frac{d\psi}{ds}\right)_{s=0} = -1 \qquad (2.2)$$

A variety of boundary conditions can present themselves at the pile tip. If the pile extends to essentially rigid bedrock, the tip displacement may be zero. Alternatively, tip displacement may be unconstrained, but tip load may be controlled by soil conditions below the tip. For this example, the simplest condition to use is zero displacement. If the pile length (dimensionless) is described by s_l, then, when $s = s_l$

$$\psi = 0 \qquad (2.3)$$

in this case. The function we choose to represent the pile deflection as a function of length must be able to account for the two boundary conditions. Of the variety of functions available, an obvious first choice in many of these problems is a polynomial

$$\psi = a_0 + a_1 s + a_2 s^2 + \cdots a_n s^n \qquad (2.4)$$

We select $n = 2$ to give three undetermined parameters a_0, a_1, and a_2. More terms in equation (2.4) will give ultimately more simultaneous linear equations to solve and, possibly, a better solution. Equation (2.4) must satisfy the boundary conditions, (2.2) and (2.3); therefore, we have

$$\left(\frac{d\psi}{ds}\right)_{s=0} = a_1 = -1 \qquad (2.5)$$

and a_1 is determined immediately.

From the tip requirement, equation (2.3), we get

$$a_0 + a_1 s_l + a_2 s_l^2 = 0 \qquad (2.6)$$

Equations (2.5) and (2.6) ensure that the function (2.4) satisfies the boundary conditions. We therefore need one more equation to determine the two coefficients a_0 and a_2; it is obtained by collocation. The second derivative of the function, (2.4), is

$$\frac{d^2\psi}{ds^2} = 2a_2 \qquad (2.7)$$

and this, as well as the function, (2.4), itself can be substituted in the differential equation (2.1) to give the residual, R.

$$R = 2a_2 - (a_0 + a_1 s + a_2 s^2) \tag{2.8}$$

In the collocation method, R is made equal to zero at selected points, s on the pile. Since we only need one more equation here, we can only collocate (ensure that equation (2.1) is satisfied) at one point on the pile. The point must, of course, be in the range $0 < s < s_l$. Assuming that the pile length, s_l, is greater than unity (it need not be, of course; it can have any value), we might try a collocation at $s = 1$. Substituting this into equation (2.8) and equating R to zero give

$$-a_0 - a_1 + a_2 = 0 \tag{2.9}$$

Substituting for a_1 from equation (2.5) and combining equations (2.6) and (2.9) give

$$a_0 = \frac{s_l(s_l + 1)}{s_l^2 + 1}; \qquad a_2 = \frac{s_l - 1}{s_l^2 + 1} \tag{2.10}$$

If, for example, the pile length, $s_l = 2$, then

$$a_0 = 1.2; \qquad a_2 = 0.2 \tag{2.11}$$

and the function, equation (2.4), is finally

$$\psi = 1.2 - s + 0.2s^2 \tag{2.12}$$

A check of this expression indicates that the boundary conditions are satisfied.

How good equation (2.12) is, as a description of the pile deflection, can be measured in two ways: (a) the magnitude of the residual as a function of s along the pile and (b) the departure of the solution from the exact analytical result, which is known in this case (see Chapter 8). Both measures are shown in Table 2.1 for the case $s_l = 2.0$. It is seen that the maximum relative error in this case is almost one-third, with the calculated ground surface pile displacement being 25 percent too high. When the force in the pile, which is proportional to the gradient of the displacement function, is considered, the approximate solution gives

$$-\psi' = -\frac{d\psi}{ds} = 1 - 0.4s \tag{2.13}$$

which represents a straight line from the value 1.0 at the pile top to 0.2 at the base. This is also compared with the dimensionless force from the exact solution in Table 2.1. Although the approximate solution is, of course, exactly

TABLE 2.1

Collocation Error

s	0	0.2	0.4	0.6	0.8	1.0	1.2	1.4	1.6	1.8	2.0
R	-0.8	-0.608	-0.432	-0.272	-0.128	0	0.112	0.208	0.288	0.352	0.4
ψ (equation 2.12)	1.2	1.008	0.832	0.672	0.528	0.4	0.288	0.192	0.112	0.048	0
ψ (exact solution)	0.964	0.782	0.631	0.506	0.401	0.312	0.236	0.169	0.109	0.054	0
% Error	+24.5	+28.9	+31.8	+32.8	+31.6	+28.1	+22.0	+13.5	+2.6	-10.3	0

$-\psi'$ (approx. solution)	1.0	0.92	0.84	0.76	0.68	0.6	0.52	0.44	0.36	0.28	0.2
$-\psi'$ (exact solution)	1.0	0.826	0.685	0.572	0.481	0.410	0.355	0.315	0.287	0.271	0.266
% Error	0	+11.4	+22.6	+32.9	+41.3	+46.3	+46.3	+39.6	+25.3	+3.3	-24.7

correct at the pile top, errors of 46 percent occur in the reduced force level half-way down the pile. This might be expected, since the straight line variation given by the approximate theory cannot compare very well with the nonlinear force function of depth exhibited by the exact solution. A better result might be expected from an approximate solution containing four unknown coefficients. It can also be seen in Table 2.1 that the residual calculated as a function of depth along the pile does not give a good indication of either the magnitude or the sense of the error.

Another approach to collocation is to require that *integrals* of the residual be zero over some sections or subdomains of the problem region. The number of subdomains chosen corresponds to the number of additional equations required to obtain the desired coefficients. As in the case of point collocation, there appears to be no systematic method, in any problem, by which the points or the subdomain can be chosen to improve the accuracy of the approximation. They are selected essentially arbitrarily. Again, we illustrate by example.

2.3.2 Subdomain collocation

The same example equation (2.1) is employed as before, and the same function (2.4) is selected to approximate the solution. Equations (2.5) and (2.6) still hold, and we need one more equation. This time we will require that the integral of the residual, equation (2.8), be made equal to zero over the region of the pile $0 < s < 1.0$. Thus

$$\int_0^1 R \, ds = \int_0^1 [-a_0 - a_1 s + (2 - s^2)a_2] \, ds = 0$$

and we get

$$a_0 + \frac{a_1}{2} - \frac{5}{3}a_2 = 0. \tag{2.14}$$

Again, substituting for a_1 from equation (2.5) and combining equations (2.6) and (2.14) give

$$a_0 = \frac{1}{2} + \frac{5(s_l - 0.5)}{3s_l^2 + 5}; \qquad a_2 = \frac{3(s_l - 0.5)}{3s_l^2 + 5} \tag{2.15}$$

With a pile length, as before, of $s_l = 2.0$, we have

$$a_0 = \frac{16}{17} = 0.9412; \qquad a_2 = \frac{9}{34} = 0.2647 \tag{2.16}$$

so that the displacement function in this case is

$$\psi = 0.9412 - s + 0.2647s^2 \tag{2.17}$$

If this solution is compared with the exact solution of Table 2.1, it will be found that the relative error near the top of the pile is much reduced but that it is increased about two-thirds of the way down the pile where the absolute displacement is quite small. The same statement can be made for the force in the pile; the values at the pile top are better represented than before, but at the pile tip the force in the approximate solution has a small negative value which is clearly incorrect.

Another, and in this case, well-known method for minimizing the overall error in the approximate solution is to perform a least-squares calculation; this is described next.

2.3.3 Least squarcs

In this technique we minimize the integral of the square of the residual over the entire domain of the problem. The minimization in general is accomplished by taking derivatives of the integral with respect to each of the unknown coefficients and equating them to zero. In the example currently being examined only one additional equation is sought; therefore, only one coefficient derivative is required.

From equation (2.8) we have

$$\int_0^{s_l} R^2 \, ds = \int_0^{s_l} (-a_0 - a_1 s + (2 - s^2)a_2)^2 \, ds \qquad (2.18)$$

After performing the integration, it is differentiated with respect to either a_0 or a_2 and the result is equated to zero to give one more equation. If a_0 is selected, the operation eventually gives

$$2a_0 + s_l a_1 + \left(\frac{2}{3}s_l - 4\right)a_2 = 0 \qquad (2.19)$$

In conjunction with equations (2.5) and (2.6), this can be solved to yield

$$a_0 = \frac{s_l^3 + 12s_l}{4(s_l^2 + 3)}; \qquad a_2 = \frac{3s_l}{4(s_l^2 + 3)} \qquad (2.20)$$

For $s_l = 2$, we have

$$a_0 = \frac{8}{7} = 1.1429; \qquad a_2 = \frac{3}{14} = 0.2143 \qquad (2.21)$$

Now the displacement function is

$$\psi = 1.1429 - s + 0.2143s^2 \qquad (2.22)$$

which may be compared with the other results.

The last method to be discussed in the topic of general operations on a residual is somewhat similar to the least-squares technique, and in certain cases, it leads to the same result.

2.3.4 Galerkin method

With this technique, the residual is multiplied, in turn, by each of the individual functions (not including the coefficient) used to construct the function ψ, and the result is integrated over the region of the problem and equated to zero. This will give as many integrals and equations as there are components to ψ.

In the example being used, we only need one more equation, but we will show all the integrals that may be used in general.

$$\int_0^{s_l} R \cdot 1 \, ds = \int_0^{s_l} [-a_0 - a_1 s + (2 - s^2)a_2] \, ds = 0 \qquad (2.23)$$

$$\int_0^{s_l} R \cdot s \, ds = \int_0^{s_l} s[-a_0 - a_1 s + (2 - s^2)a_2] \, ds = 0 \qquad (2.24)$$

$$\int_0^{s_l} R \cdot s^2 \, ds = \int_0^{s_l} s^2[-a_0 - a_1 s + (2 - s^2)a_2] \, ds = 0 \qquad (2.25)$$

If the operation indicated by equation (2.23) is completed, it will be found that it will give an equation identical, in this case, to equation (2.19) so that the Galerkin method, for this selection, will give a result the same as obtained by the least-squares approach. The calculations indicated by the other two equations (2.24) and (2.25) will also have the same consequence as the second two minimizations performed by least squares, that is, minimizing with respect to a_1 and a_2, respectively. For a different result, we will do the latter computation. From equation (2.25) we have

$$a_0 + \frac{3}{4} s_l a_1 + 3\left(\frac{s_l^2}{5} - \frac{2}{3}\right)a_2 = 0 \qquad (2.26)$$

which, with the substitution of a_1 from equation (2.5) can be solved simultaneously with equation (2.6) to give

$$a_0 = \frac{3s_l^3 + 40s_l}{8s_l^2 + 5}; \qquad a_2 = \frac{5s_l}{8s_l^2 + 5} \qquad (2.27)$$

Once again using $s_l = 2$, we see that the values are

$$a_0 = \frac{13}{9} = 1.4444; \qquad a_2 = \frac{5}{36} = 0.1389 \qquad (2.28)$$

and the function is then

$$\psi = 1.4444 - s + 0.1389s^2 \tag{2.29}$$

On comparison with the analytically correct result, it will be seen that this function is worse than the others investigated so far.

In these residual methods it will almost always be found that simple collocation requires less labor than the subdomain, least-squares, and Galerkin methods, since the latter require integration. It is difficult to make definite statements about accuracy, which depends on the function selected, the number of terms, and on the collocation points or subdomains selected. In the above example collocation at other points than the one used will give results of very variable accuracy. As usual in textbooks, the example selected has deliberately been kept simple to emphasize the principles involved. It is still realistic, though, and the solutions could be used in practice for estimates of pile displacements or loads, especially when it is considered how poorly the numerical soil properties are generally known. With comparatively little extra effort, four terms and coefficients can be used with a corresponding increase in accuracy. There is another reason for selecting this particular problem; it lends itself well to an examination of the effect of nonlinear soil behavior on the pile response. This aspect can be handled in the same way as the linear approximate analysis demonstrated above. It requires only a little more work and will be discussed in detail in Chapter 9.

So far, the residual methods have required the use of functions which must satisfy *all* the boundary conditions. There is another, now very widely used, technique which only requires that certain boundary conditions be met, whereas the constraints imposed by others are relaxed. This is the *potential energy* or *Ritz* method (2) which will now be described. Although it also falls under the general category of methods of undetermined parameters, there are a sufficient number of variants in its usage that a separate section will be devoted to the technique.

2.4 ENERGY METHODS

As an elementary illustration of the use of energy principles, it is convenient to tackle the example of Appendix A, as illustrated in Figure A-1, and solved there by the straightforward solid mechanics method by a different procedure.

In equation (A.59) it is shown that the difference between the strain energy stored in a body and the work done on it by external (or body forces) forces is equal to zero during a virtual displacement. We now consider equation (A.48) as an expression of variation in a quantity π, which is the total potential energy of the body, as in equation (A.60)

$$\pi = U_v - W \tag{2.30}$$

where U_v is the total strain energy stored in the body and W is the work done by surface and body forces in moving through the displacements to the final strained state of the body. The principle of minimum potential energy states that of all possible displacement states of a loaded body that state of displacement which minimizes the potential energy is the correct one.

For Figure A-1, for example, the potential energy, according to equation (2.30), is

$$\pi = \frac{1}{2} R_2 d_2 + \frac{1}{2} R_3 d_3 - Pd \tag{2.31}$$

where d is the displacement of the point of application of the load P. Using equations (A.4) and (A.5) for the springs, equation (2.31) becomes

$$\pi = \frac{1}{2} k_1 d_2^2 + \frac{1}{2} k_2 d_3^2 - Pd \tag{2.32}$$

Now, let it be assumed that under the load P the beam AC rotates through a small angle θ, where

$$\theta = \frac{d_2}{a} = \frac{d_3}{b} = \frac{d}{b} \tag{2.33}$$

so that equation (2.32) becomes

$$\pi = \frac{1}{2} k_1 a^2 \theta^2 + \frac{1}{2} k_2 c^2 \theta^2 - Pb\theta \tag{2.34}$$

The assumed displacement function here is a straight line passing through zero at the left end of the beam; the undetermined coefficient is the slope θ of the line. We have to find a value of θ that minimizes the potential energy, π, of the system. This is obtained by differentiating π with respect to θ and making the result equal to zero

$$\frac{\partial \pi}{\partial \theta} = (k_1 a^2 + k_2 c^2)\theta - Pb = 0$$

or   (2.35)

$$\theta = \frac{Pb}{k_1 a^2 + k_2 c^2}$$

This solves the problem since equation (2.33) gives d, d_2, and d_3 in terms of θ, and equations (A.4) and (A.5) give R_2 and R_3 in terms of d_2 and d_3. It should be noted that the equilibrium equations were not used in the form of equations (A.1) and (A.2); they are included in the derivation of the potential energy expression. In equations (A.4) and (A.5) the parameters k_1 and k_2 (sometimes called *spring constants*) are correctly referred to as *spring stiff-*

nesses since their use gives the forces acting on the springs. Analogously, the minimum potential energy principle has enabled us to calculate the stiffness k of the beam/spring system if we write equation (2.35) in the form

$$P = \frac{k_1 a^2 + k_2 c^2}{b} \theta$$

$$= k\theta$$

(2.36)

since P is now related to a stiffness times a deflection.

2.4.1 Ritz method

In the simple example above the unknown function is the prescribed straight line of the deflected rigid beam of which slope is the coefficient to be determined. The only effective boundary requirement is that the deflection be zero at the rigid support. For the general problem, the function must approximate the elastic line of the deflected member, or the force or moment distribution in the complementary formulation (described in Sections 5.1.1. 8.2.5. and A.6), and boundary conditions must be considered in detail. In the energy approach the boundary conditions may be separated into two groups— essential and additional—for mathematical reasons, which are explained in numerical analysis treatises (2 and 3). Fortunately, the two classes may also be distinguished on a physical basis. The trial function in the Ritz method must satisfy only the *essential* boundary conditions. We are largely concerned in this book with second- and fourth-order differential equations expressed in terms of displacements, forces, or moments. The general rule is that if the order of the equation in ψ is $2n$, then the essential boundary conditions to be satisfied are ψ and its first $n - 1$ derivatives; the rest are nonessential or additional. In a fourth-order equation that means that there are two essential conditions. In a second-order equation only Ψ is essential. The two essential conditions for fourth-order equations are the requirements placed on the dependent variable and its gradient at the ends or other points of the beam. If displacement is the variable, then the required end (say) displacements and slopes are the essential conditions. When moment is the variable, the end moments and shears are the essential boundary conditions. These are the conditions to be satisfied by the assumed function in the energy method. The consequence in, for example, a beam-bending problem is that the displacements and, if specified, slopes at each end of the beam must be taken care of by the form of the function chosen, but when the coefficients are determined to fulfill the energy requirement, the moments at the beam ends will not, in general, be correct. In a plane problem in linear elasticity the displacement conditions along the boundaries will be correct, but the stresses there will not. They may not be zero, for example, along an unloaded portion of the perimeter.

The Ritz method consists in determining the coefficients of the selected function to minimize an expression termed a *functional*. The functional is related to the differential equation describing the physics of the problem, and formal mathematical methods exist for defining the functional of a given differential equation (2). For the circumstances involved in structural engineering, however, a simpler interpretation of the functional in terms of the potential energy of the configuration can be made. The potential energy or functional, π, is equal to the strain energy of the system, U_v, less the work done, W, by the applied load causing the strains, as in the elementary example earlier. Thus, in more complicated cases, the functional still follows equation (2.30), in which both U_v and W can be expressed in terms of the function assumed with the undetermined parameters. Minimization of π is accomplished now by differentiating it with respect to each of the unknown coefficients in turn and equating the result to zero. It is possible, therefore, to obtain as many equations as there are unknown coefficients if there are no other constraints on their magnitude. When some other conditions are present, the minimization can supply as many additional equations as are required. For approximate solutions obtained by any method, calculation of the numerical value of the functional for each approximation gives a measure of its relative merit. The lower the value of the potential energy, the better the solution.

We will need the functionals for pile axial loading and beam-bending problems when pile or beam is imbedded in an elastic foundation. In dimensionless terms the axial loading expression is

$$\pi = \frac{1}{2} \int_0^{s_t} \left(\frac{d\psi}{ds}\right)^2 ds + \frac{1}{2} \int_0^{s_t} \psi^2 \, ds - \psi_0 \tag{2.37}$$

where ψ and s are the same terms as before and ψ_0 is the value of ψ at the pile top, $s = 0$, where it is assumed the load acts. If in some case the axial load is applied at some other point in the pile below ground surface, the last term is the value of ψ at the load point.

For a beam on an elastic foundation (see Chapter 5), the equivalent functional is

$$\pi = \frac{1}{2} \int \left(\frac{d^2\psi}{ds^2}\right)^2 ds + \frac{1}{2} \int \psi^2 \, ds - \psi_a \tag{2.38}$$[3]

Here ψ is the dimensionless deflection at right angles to the beam axis, and the coordinate s is measured along the axis. The integrals are taken over

---

[3]This is the formulation when the parameter λ^4 is defined as k/EI in the dimensional analysis. In the analytical solution (see equation 5.12) λ^4 is taken as $k/4EI$, and with this definition, the coefficient outside the first integral is $\frac{1}{8}$ intead of $\frac{1}{2}$.

the whole length of the beam and are left unspecified here because symmetry is frequently employed in the formulation. An applied load at a point with displacement ψ_a is assumed here. If moment is applied instead, then the slope at the moment point, ψ'_a, is required instead. Alternatively, if a distributed load is placed on the beam, the potential energy due to it must be worked out *via* another integral. This will be discussed elsewhere.

When the complementary formulation is to be employed, with force or moment instead of displacement, the functional to be minimized is the complementary energy, expressed in terms of the complementary strain energy and the complementary energy of the required displacement [equation (A.61)].

The Ritz method will be illustrated by the same axially loaded pile example as before. Again, we choose the function

$$\psi = a_0 + a_1 s + a_2 s^2 \tag{2.39}$$

to represent the deflected pile shape, and the required bottom boundary condition is $\psi = 0$ at $s = s_l$, so that

$$a_0 + a_1 s_l + a_2 s_l^2 = 0 \tag{2.40}$$

as before.

Forming the square of the function and of its first derivative and calculating π give the result

$$\pi = \frac{1}{2}(a_0^2 + a_1^2)s_l + \frac{1}{2}(a_0 a_1 + 2a_1 a_2)s_l^2 + \frac{1}{6}(2a_0 a_2 + a_1^2 + 4a_2^2)s_l^3$$

$$+ \frac{a_1 a_2 s_l^4}{4} + \frac{a_2^2 s_l^5}{10} - a_0 \tag{2.41}$$

By differentiating this equation with respect to each of the coefficients a_0, a_1, and a_2 in turn and equating each result to zero, three more equations can be produced. Only two more are needed to go with the essential condition, equation (2.40), to give the three required for a solution. We take the first two derivatives:

$$\frac{\partial \pi}{\partial a_0} = s_l a_0 + \frac{s_l^2}{2} a_1 + \frac{s_l^3}{3} a_2 - 1 = 0 \tag{2.42}$$

$$\frac{\partial \pi}{\partial a_1} = \frac{1}{2} s_l^2 a_0 + \frac{1}{3}(3s_l + s_l^3)a_1 + \frac{1}{4}(4s_l^2 + s_l^4)a_2 = 0 \tag{2.43}$$

With s_l selected to be 2.0, as previously, the coefficients of the three equations become

a_0	a_1	a_2	Load Vector	
1	2	4	0	
2	2	$\dfrac{8}{3}$	1	(2.44)
2	$\dfrac{14}{3}$	8	0	

Simultaneous solution gives the result

$$a_0 = 0.75; \qquad a_1 = 0; \qquad a_2 = -0.1875 \tag{2.45}$$

so that

$$\psi = 0.75 - 0.1875s^2 \tag{2.46}$$

Here the solution is much worse than any obtained by the residual methods; in particular, the load at $s = 0$ is seen to be zero (gradient of ψ at $s = 0$) which is a physical impossibility. As pointed out before, this occurs because no control was placed on the deflected shape of the pile other than to require the displacement to be finite at the top and zero at the tip. Without the latter condition, no restraint at all would have been exercised on the pile. Ignoring the load, however, the pile top displacement is 0.75, which is about 25 percent lower than that of the exact solution and is therefore in error by an amount comparable to those in the previously employed residual techniques.

Another example will give a better opportunity to appraise the method. We choose a semi-infinite beam resting on an elastic foundation (see Chapter 5) and subjected to a point load at its free end. The equation describing the beam deflection is a fourth-order one, and an exact solution must satisfy two boundary conditions at each end. At the loaded end the two conditions are that the shear force equal the applied load and the moment equal zero. The other end of the beam is at infinity, where its displacement, slope, and all other quantities are equal to zero. We can choose an approximate function giving zero displacement (an essential condition) at infinity, but the displacement condition at the loaded end is not specified, and we cannot therefore choose a function to match it. However, an equivalent problem would consist of giving the free end of the beam a specified displacement, without an applied moment, and then determining the force needed to cause the displacement. In the latter circumstances, two essential conditions, displacement at each end of the beam would be specified, and the approximate solution adopted would have to suit those conditions. Thus, functions which permit either an end displacement or an end shear are equivalent in this case.

We assume the function

$$\psi = a_1 e^{-s} + a_2 e^{-2s} \tag{2.47}$$

for the deflected shape. More terms could be added for greater accuracy, but these are sufficient to illustrate the method. It can be seen that the function vanishes at infinity, as it should, and that the deflection at $s = 0$ is $(a_1 + a_2)$ which meets the requirement of a finite displacement at that point. In order to compare the final result with the known exact solution, we shall employ the variant of the functional equation (2.38), in which the fraction outside the first integral is $\frac{1}{8}$ rather than $\frac{1}{2}$. When the approximate solution is differentiated, squared, and so on, and substituted in the functional, the final result is

$$\pi = \frac{5}{16}a_1^2 + \frac{2}{3}a_1a_2 + \frac{5}{8}a_2^2 - a_1 - a_2 \qquad (2.48)$$

From this the derivatives are obtained:

$$\frac{\partial \pi}{\partial a_1} = \frac{5}{8}a_1 + \frac{2}{3}a_2 - 1 = 0 \qquad (2.49)$$

$$\frac{\partial \pi}{\partial a_2} = \frac{2}{3}a_1 + \frac{5}{4}a_2 - 1 = 0 \qquad (2.50)$$

and the solution is

$$a_1 = 1.7319; \qquad a_2 = -0.1237 \qquad (2.51)$$

so that

$$\psi = 1.7319e^{-s} - 0.1237e^{-2s} \qquad (2.52)$$

The displacement at the end $s = 0$ is

$$\psi_0 = 1.6082 \qquad (2.53)$$

versus 2.0 for the exact solution [see equation (5.46)]. However, the moment at the end $s = 0$ of the beam is proportional to $(\psi'')_0$, where

$$(\psi'')_0 = \left(\frac{d^2\psi}{ds^2}\right)_0 = 1.7319 - 0.4948 = 1.2371$$

and this is very far from the value zero which is the actual end condition for moment.

When the problem to be solved is a two- or three-dimensional continuum one, use of the Ritz method is still possible, though more difficult to apply in practice. Two- or three-dimensional functions satisfying the essential boundary conditions must be established, with coefficients to be obtained by minimization. However, if the problem region is subdivided into smaller domains, a systematic approach can be developed. The analysis is essentially simple, but for most practical problems the computational labor is so great that a computer is required. Occasionally, useful results can be obtained by a hand solution, and examples will be given below. The technique is called the *finite element* method.

2.4.2 Finite element method

In this approach the continuum of the real problem is approximated by a number of finite or discrete lumps whose mechanical behavior is selected to simulate that of the continuum (elastic, plastic, visco-elastic, etc.). The method was originally developed about 1955 in the aircraft industry and has been used and more frequently in structural analysis in general. In its original form, the *finite elements* represented real structural elements, beams, and columns, and the operational equations were developed on a physical basis. This approach is not convenient in the application of the technique to continuum mechanics, and, in consequence, the formulations will be established from the energy stored and work done point of view.

The method will be set out in terms of a plane strain problem, although there are no additional difficulties in considering a general three-dimensional situation. In the structural example of Section 2.4 the identification of the discrete elements with parts of the structure (springs) is obvious, and this visualization of a structure as an assemblage of elemental components is not difficult for many types of real structures. In soil mechanics we will, however, be largely involved with the representation of soil masses or continua, and therefore it is necessary to set out a technique for handling stress and displacement problems in continua.

The approach is to overlay the region of interest with an assemblage of polygons, in general, which have common vertices and sides. Although four-sided polygons may and frequently are selected, it is convenient to use triangles in many soil engineering applications. Triangles can easily be adapted to a variety of boundary and internal geometries. Examples are shown in Figures 2-1(a) and (b) for two soil mechanics problems. In Figure 2-1(a) it is appropriate, although not necessary, to use a regular triangular network, whereas in Figure 2-1(b), a regular rectangular pattern, adaptable to the boundary conditions, is employed. It is desirable to use smaller rectangles in the regions, for example, near the footing, where stress concentrations are expected to occur. In Figure 2-1(b) the footing is also represented by finite elements for the case in which its flexibility is of interest. Should it be assumed rigid in comparison with the soil, the elements are omitted. Instead, the appropriate mesh points at the ground surface are all given the same *displacement*.

The method of solving the problem is then to assume a displacement field for each triangular (or other shape) discrete element, equivalent to the assumption of a given rotation θ in the example problem of Section 2.4, such that a node point common to several triangles undergoes the same displacement when it is calculated from the displacement field for each of the con-

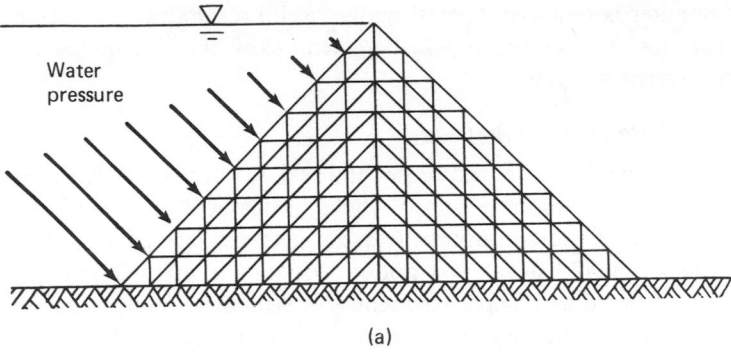

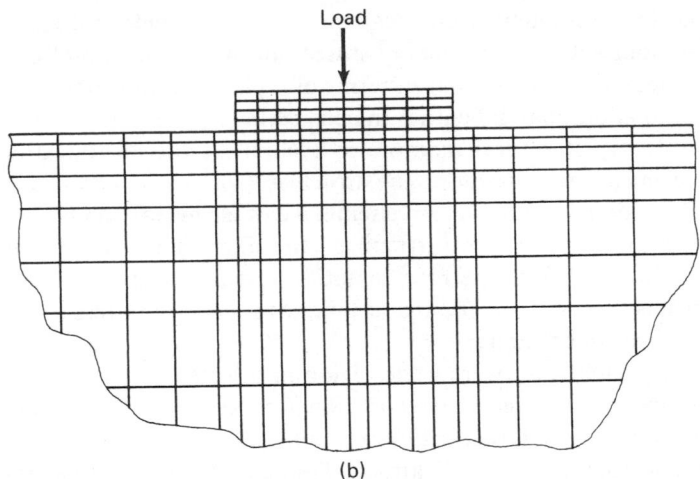

Figure 2-1. Finite element meshes: (a) dam subjected to water load and triangular elements; (b) footing on soil, or rock, rectangular elements.

necting triangles. It will be recognized that this is an application of the idea of consistent deformations. In line with this the displacement fields selected must match the known boundary displacements, as essential conditions. Next, the strains in each triangle are calculated from its displacement field, which contains unknown constants whose value will eventually be obtained. At this stage the material behavior enters the calculations, so that the stresses in each element can be derived from the strains and so that a potential energy expression, which can be written in a form independent of the constitutive relations of the material, can be specialized for the particular material properties of interest, in terms of strains or stresses. Substituting the calculated strains, for example, of one triangular element into the energy expression gives an equation for the energy of that element. Since the potential energy includes provision for (a) boundary forces and (b) body forces, these will be included for (a) any triangle that is subjected to a boundary load at a node point and (b) all triangles if body forces are important. When this is done for all triangles, their potential energies are added together to give the total potential energy for the whole region. It can be seen how this was arrived at in equation (2.34) in the preliminary example.

Now we have to find the values of the unknown constants that characterize the displacement fields. These values have to be such as to minimize the total potential energy; this minimum is obtained by differentiating the total potential energy expression with respect to each unknown constant, or, more conveniently, to each nodal displacement in turn and setting the resultant equation equal to zero, as was done in the Ritz method. However, in the previous examples the result of this operation was only a few equations in terms of force equaling stiffness times deflection. In the general finite element case, there are hundreds or thousands of node points, each of which has two components of displacement in the plane problem, and the differentiation process leads to one equation for each component at each node point. Thus, we will have a system of many algebraic equations that have to be solved simultaneously to give the nodal displacements. Once the displacements at each node of a triangle have been obtained, the constants in the displacement field for that triangle can be calculated and consequently the strains and stresses can then be computed. The evaluation of the system of simultaneous equations requires a large digital computer: A variety of programs is available for carrying out the required operations extremely rapidly.

As in the example problem, each of the equations obtained by the differentiation of the total potential involves forces and displacements. However, in general, each equation includes the displacement components at several node points and several forces. The coefficient multiplying θ in equation (2.36) is a stiffness. Consequently, in the full set of equations the *system* of coefficients multiplying the displacement components represents the stiffness of the entire system and is called the *stiffness matrix*. A large number of books has been devoted to the assembly and manipulation of stiffness matrices in connection with the finite element method (4, 12).

It will be recognized that each equation resulting from the differentiation of the total potential energy expression with respect to a displacement component is an equilibrium equation in the direction of the component. The system of equations therefore also represents the equilibrium of the problem region expressed through the equilibrium of force components at every node point. Consequently, the procedure outlined above has taken into account all the necessary elements of analysis: consistency of deformations, constitutive relations, and equilibrium.

It must be noted that the representation of the area of the problem by means of the network of triangles is a conceptual procedure only; it does *not* correspond to a model representation, for example, in which the real region might be represented by a number of real triangles pinned together at their node points. The actual displacements of a plane solid triangular element subjected to loads at its nodes are complicated and are not what is assumed to occur in the present approach. Rather, consistency of deformations requires that not only the node points of the triangles but also *the sides* of adjacent triangles remain attached. Consequently, the assumed displacement for any triangle must give displacements along its sides which are identical with the displacements along the contiguous sides of adjacent triangles, derived from *their* displacement fields. In selecting the displacement fields, this requirement must be satisfied.

The simplest displacement field that meets these requirements is one in which the displacements are assumed to be *linear* functions of the coordinates in each triangle, so that the triangle with the coordinates of Figure 2-2(a) is assumed to deform as shown in Figure 2-2(b). With this assumption, the displacements in the nth triangle become

$$u_n = a_n x + b_n y + c_n \tag{2.54}$$

$$v_n = d_n x + e_n y + f_n \tag{2.55}$$

in which the $a_n, b_n, \ldots, f_n$ are constants holding for triangle n only, and x and y are the coordinates of points in the triangle, including the vertices. The linear equations (2.54) and (2.55) allow the triangle to displace, rotate, stretch, and deform, but lines which were straight before straining remain straight afterward. Once the coefficients $a_n, b_n, \ldots, f_n$ have been determined, the following calculation gives the displacement of triangle n. Each corner of the triangle has two degrees of freedom, in the x- and y-directions, and so its total displacement requires six constants ($a_n, b_n \ldots$) for its description. If a rectangular element were to be employed with four corners, eight degrees of freedom are involved and eight constants must be used. In this case,

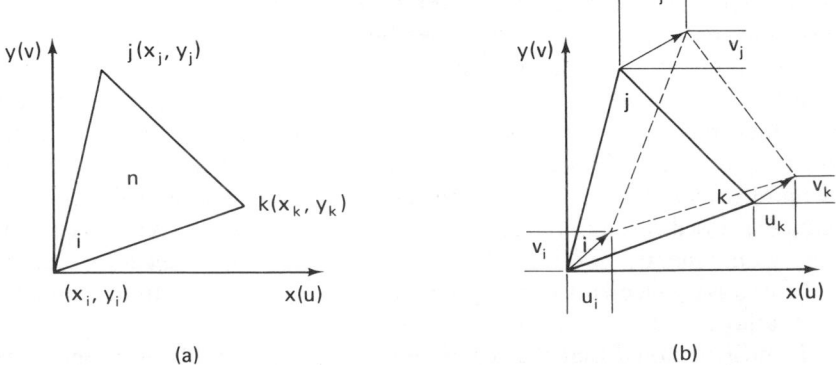

(a)   (b)

Figure 2-2. Displacements of triangular finite element: (a) coordinate scheme; (b) displaced element.

equations identical to (2.54) and (2.55), each with the addition of a term in xy multiplied by another constant, are needed.

Other displacement functions, such as parabolic, may also be employed. Since they give more degrees of freedom to the element, compatibility must be ensured by adding node points, for example, at the midpoints of the triangle sides.

With the displacements given by equations (2.54) and (2.55), the *strains in triangle n* are given by equation (A.26):

$$(\epsilon_x)_n = \frac{\partial u_n}{\partial x} = a_n; \qquad (\epsilon_y)_n = \frac{\partial v_n}{\partial y} = e_n;$$

$$(\gamma_{xy})_n = \frac{1}{2}\left[\frac{\partial u_n}{\partial y} + \frac{\partial v_n}{\partial x}\right] = \frac{1}{2}(b_n + d_n) \tag{2.56}$$

From equations (2.56) and (A.48) the *stresses in triangle n* can be computed. We assume here that the problem is plane strain, so that $\epsilon_z = 0$.

$$(\sigma_x)_n = \frac{E}{(1 + v)(1 - 2v)}[(1 - v)(\epsilon_x)_n + v(\epsilon_y)_n]$$

$$= \frac{E}{(1 + v)(1 - 2v)}[(1 - v)a_n + ve_n] \tag{2.57}$$

$$(\sigma_y)_n = \frac{E}{(1 + v)(1 - 2v)}[(1 - v)(\epsilon_y)_n + v(\epsilon_x)_n]$$

$$= \frac{E}{(1 + v)(1 - 2v)}[va_n + (1 - v)e_n] \tag{2.58}$$

$$(\tau_{xy})_n = G(\gamma_{xy})_n = \frac{E}{2(1 + v)}(b_n + d_n) \tag{2.59}$$

Thus, the strains and stresses are constant in each triangle but are different from one triangle to another. Specifically, the stresses are never zero at the external boundaries of the triangles, even where there is no boundary load. This occurs because stress is not an essential condition.

Considering the node points of triangle n, we see that their displacements can be written from equations (2.54) and (2.55) in terms of the constants a_n, b_n, since the coordinates of the vertices are known. This gives six equations:

$$u_i = a_n x_i + b_n y_i + c_n$$
$$v_i = d_n x_i + e_n y_i + f_n$$
$$u_j = a_n x_j + b_n y_j + c_n$$
$$v_j = d_n x_j + e_n y_j + f_n \tag{2.60}$$
$$u_k = a_n x_k + b_n y_k + c_n$$
$$v_k = d_n x_k + e_n y_k + f_n$$

At some of the node points of triangles in a given problem, such as shown in Figure 2-1(a) and (b), the displacements will be specified by the boundary conditions. For example, in Figure 2-1(a) we might require that the vertices occurring along the base of the dam have zero displacements in both the x- and y-directions. This would represent the effect of a rigid foundation to which the dam is bonded. In other situations, such as along the centerline (axis of symmetry) of Figure 2-1(b), the vertical displacement of each node point would be unknown, but, from symmetry, the lateral displacement would be required to be zero if the foundation and soil are homogeneous. These conditions are made use of, by solving equation (2.60) for the coefficients a_n, b_n, ..., f_n in terms of the displacements of the triangle vertices, some of which in general may be known and some not.

The solution of equation (2.60) can be carried out by determinants to give

$$a_n = \frac{\begin{vmatrix} u_i & y_i & 1 \\ u_j & y_j & 1 \\ u_k & y_k & 1 \end{vmatrix}}{2A_n} = \frac{[u_i(y_j - y_k) + u_j(y_k - y_i) + u_k(y_i - y_j)]}{2A_n}$$

$$b_n = \frac{\begin{vmatrix} x_i & u_i & 1 \\ x_j & u_j & 1 \\ x_k & u_k & 1 \end{vmatrix}}{2A_n} = \text{etc.} \tag{2.61}$$

$$c_n = \frac{\begin{vmatrix} x_i & y_i & u_i \\ x_j & y_j & u_j \\ x_k & y_k & u_k \end{vmatrix}}{2A_n} = \text{etc.}$$

and similar expressions for d_n, e_n, f_n, where A_n is the area of triangle n. Thus, for each triangle, the coefficients a, b, ..., f are expressed in terms of the displacements of the vertices. This corresponds to equation (2.33) in the beam or springs example, where the unknown θ was expressed in terms of the spring displacements, which were also unknown. At this stage, what was known of the displacement geometry was introduced in the same way as we will introduce it here in the general case for those boundary displacements which are known.

The next step, as in the example, is to write the expression for the potential energy of the system. In this case, we must develop the equation for the potential energy of one triangle, n, in plane strain and then sum the energy for all the triangles. For a linearly elastic, isotropic body, the strain energy density U from equation (A.53) can be written in terms of strains as follows (plane strain case):

$$
\begin{aligned}
U &= \frac{vE}{2(1+v)(1-2v)}(\epsilon_x + \epsilon_y)^2 + \frac{E}{2(1+v)}(\epsilon_x^2 + \epsilon_y^2) + \frac{2E}{2(1+v)}\gamma_{xy}^2 \\
&= \frac{E}{2(1+v)(1-2v)}[(1-v)(\epsilon_x^2 + \epsilon_y^2) + 2v\epsilon_x\epsilon_y + 2(1-2v)\gamma_{xy}^2] \\
&= \frac{E}{2(1+v)(1-2v)}\left\{(1-v)\left[\left(\frac{\partial u}{\partial x}\right)^2 + \left(\frac{\partial v}{\partial y}\right)^2\right] + 2v\frac{\partial u}{\partial x}\cdot\frac{\partial v}{\partial y}\right. \\
&\quad \left. + \frac{(1-2v)}{2}\left[\frac{\partial u}{\partial y} + \frac{\partial v}{\partial x}\right]^2\right\}
\end{aligned}
\tag{2.62}
$$

Considering that the state of strain is constant in any one triangle, n, for example, the total strain energy U_y for that triangle is equal to the strain energy density times the area of the triangle, A_n. (Unit thickness of the elements is understood.) The terms $\partial u/\partial x$, $\partial v/\partial y$, etc; appearing in equation (2.62) have been evaluated in equation (2.56) for the displacements of equations (2.54) and (2.55) in triangle n, so that equation (2.62), multiplied by the area, and with the replacement of displacement gradients from equation (2.56), gives U_{vn}, the strain energy in the triangle n:

$$
U_{vn} = \frac{EA_n}{2(1+v)(1-2v)}\left\{(1-v)[a_n^2 + e_n^2] + 2va_ne_n \right. \\
\left. + \frac{(1-2v)}{2}[b_n + d_n]^2\right\}
\tag{2.63}
$$

The total potential energy in equation (2.30) includes a term W in the work done by the surface and body forces as the strained region moves from its initial to its final displaced position. For a single triangular element, n, this quantity is given by the expression

$$
W_n = \int A_n(Xu_n + Yv_n)\,dx\,dy + \int S_n(T_xu_n + T_yv_n)\,ds
\tag{2.64}
$$

in which X and Y are the components of the body force per unit of area (in the two-dimensional problem) in the x- and y-directions and the first integral is carried out over the area of the element; T_x and T_y are the components of the surface force (applied load) per unit of length of the boundary and the integration is carried out over the length of the boundary S_n. Many triangles in a typical problem with an edge forming one of the boundaries of the region will have no surface load acting. In general, there are no surface loads on internal triangles. Consequently, the second term in equation (2.64) will only appear in the energy expression for a few triangles in a typical problem. In many problems body forces per unit of volume are neglected and in these cases the first integral in equation (2.64) will not appear in the formulation of the problem.

Referring to equation (2.30), we can put equations (2.63) and (2.64) together to get the total potential energy of the nth element in the following equation. In this equation, equations (2.54) and (2.55) are substituted for the displacements u_n and v_n in equation (2.64):

$$
\pi_n = \frac{EA_n}{2(1 + v)(1 - 2v)} \left\{ (1 - v)[a_n^2 + e_n^2] + 2va_ne_n + \frac{(1 - 2v)}{2}[b_n + d_n]^2 \right\}
$$

$$
- \int_{A_n} \{X(a_nx + b_ny + c_n) + Y(d_nx + e_ny + f_n)\} \, dx \, dy \qquad (2.65)
$$

$$
- \int_{S_n} \{T_x(a_nx + b_ny + c_n) + T_y(d_nx + e_ny + f_n)\} \, ds
$$

Since the constants a_n, b_n, ..., f_n have all been obtained [equation (2.61)] in terms of the generally unknown displacements u_i, v_i, ..., and known coordinates, x_i, y_i, ..., of the node points, equation (2.65) can be written in terms of u_i, v_i, etc.

Following the procedure in the simple example given before, we sum up the potential energies, π_n, of each triangle to give the total potential energy of the region under sutdy, π.

Next, we must take the derivative of the resulting potential energy expression with respect to each unknown nodal displacement, $\partial \pi / \partial u_i$, $\partial \pi / \partial v_i$, etc., and equate each derivative to zero. This is equivalent to obtaining the equation of equilibrium in each of the two coordinate directions at each node point. Since there is one equation for each unknown displacement component, we will now have a set of simultaneous algebraic equations in the unknown nodal displacements. The term for the strain energy [first term in brackets in equation (2.65)] includes the *squares* of the coefficients a_n, b_n, ..., f_n. These coefficients, it is seen from equation (2.61), are linear in the displacements u_i, v_i, ..., etc., so that the final form of the total potential energy expression [after substitution for a_n, b_n from equation (2.61)] includes the squares of the unknown displacements u_i, v_i, ..., etc., in its strain energy term, whereas the body and surface force terms are linear in these displacements. Consequently, differentiation with respect to the displacements produces a system of simultaneous algebraic equations, linear in the unknowns u_i, v_i, etc., and with constant coefficients as follows:

$$k_{11}u_1 + k_{12}v_1 + k_{13}u_2 + k_{14}v_2 + \cdots k_{1(2m)}v_m = C_1$$
$$k_{21}u_1 + k_{22}v_1 + k_{23}u_2 + k_{24}v_2 + \cdots \qquad\qquad = C_2$$
$$k_{31}u_1 + k_{32}v_1 \cdots \qquad\qquad\qquad\qquad = C_3$$
$$\cdots\cdots\cdots\cdots\cdots\cdots\cdots\cdots\cdots\cdots\cdots\cdots\cdots \qquad (2.66)$$
$$\cdots\cdots\cdots\cdots\cdots\cdots\cdots\cdots\cdots\cdots\cdots\cdots\cdots$$
$$k_{(2m)1}u_1 + k_{(2m)2}v_1 + \cdots + k_{(2m)(2m)}v_m = C_{(2m)}$$

In equation (2.66) the constants k_{11}, k_{12}, ..., etc., arise through the differentiation of the strain energy terms in the summation of equation (2.65) and therefore include the elastic moduli of the elements. These are therefore spring constants or stiffnesses in the same way that

$$k = \frac{(k_1 a^2 + k_2 c^2)}{b}$$

was a stiffness in equation (2.36) of the simple example. This system of coefficients k_{11}, k_{12}, ..., is the *stiffness matrix* referred to previously. The constants C_1, C_2, ... in equation (2.66) come about through the differentiation of the body and surface force terms in the total potential energy expression.

Although the system of equations obtained appears (and is) formidable when a number of elements comparable to those shown in Figure 2-1 is used to describe a problem region, there are two aspects of the formulation which help to simplify the solution of the equations. The first is that, although there appear to be great many terms in each equation, this is not, in fact, the case, since the total strain energy expression only contains a few terms in each u_i and v_i. Consequently, when the expression is differentiated with respect to a particular displacement, u_i, say, only those terms including u_i give rise to corresponding ku_i terms in the equation; the others give zero. Thus, each equation consists of a few terms and many zeros. The second consideration that leads to simplification is that, because the energy expression is quadratic in the displacements, the stiffness coefficients are not all different; each k_{12}, ..., k_{36}, ..., k_{ij}, etc., is equal to k_{21}, ..., k_{63}, ..., k_{ji}, respectively. The stiffness matrix is said to be *symmetric*.

In general, the set of simultaneous algebraic linear equations (2.66) is solved by means of a large computer. Efficient computer programs exist for the solution of such equations when there are only a few terms in each equation and the matrix is symmetric (4.12). The point will not be dealt with further here, but the solution is facilitated by careful numbering of the node points in a given problem, so that the nonzero terms in each equation in equation (2.66) are close together and the stiffness matrix is banded. When the solution is obtained in terms of the nodal displacements at all the nodes of the problem net, the coefficients a_n, b_n, etc., can be obtained from equation (2.61) and the stresses in each triangle from equations (2.57) through (2.59).

It can be seen that there is no reason why the E and v cannot be different for each triangle, so that layered or inhomogeneous media can be treated if desired. Should an anisotropic but linear medium be present, the elastic constants in the stress-strain version of equation (A.40) must be known and an alternative form of the strain energy expression, equation (2.62), set up, but no great increase in complexity is required. Elastic-plastic solutions and other nonlinear problems can be solved in a step-by-step manner by checking the stress conditions in each element against a yielding or other criterion and by changing the element's stress-strain behavior (e.g., to one with a much smaller modulus of elasticity) when the criterion is exceeded. It will be seen that the use of a yielding criterion expressed in terms of the strain energy density or strain energy per triangle would be convenient in the finite element approach.

It may be pointed out that the same technique of discrete element analysis described above can be used to solve Laplace's equation for flow in porous media (11), except that a different equation equivalent to the total potential energy expression, equation (2.65), for one triangle, must be minimized. For the solution of the two-dimensional Laplace equation in total head, h, the following function, due to Euler, must be obtained for each triangle:

$$\pi = \frac{A_n}{2}\left[k_x\left(\frac{\partial h}{\partial x}\right)^2 + k_y\left(\frac{\partial h}{\partial y}\right)^2\right] \qquad (2.67)$$

where k_x and k_y are the coefficients of permeability of the soil in the x- and y-directions. The function π must be summed for all the triangles before being differentiated with respect to each unknown total head at a nodepoint to give the system of equations to be solved.

In this problem it is the total head which is considered to vary linearly in each triangle. Thus, in triangle n

$$h_n = a_n x + b_n y + c_n \qquad (2.68)$$

and, consequently, a_n, b_n, and c_n can be evaluated in terms of the nodal values of h.

Some examples of finite element calculations follow.

(a) Brazilian test cylinder. A simple example is chosen: the so-called *Brazilian test cylinder* under plane strain. The test is performed because it provides a relatively uniform tensile stress across the vertical diameter of the cylinder. The configuration is shown in Figure 2-3(a), and body forces are considered to be absent. Although not directly pertinent to soil engineering problems, this example was selected because the boundary conditions

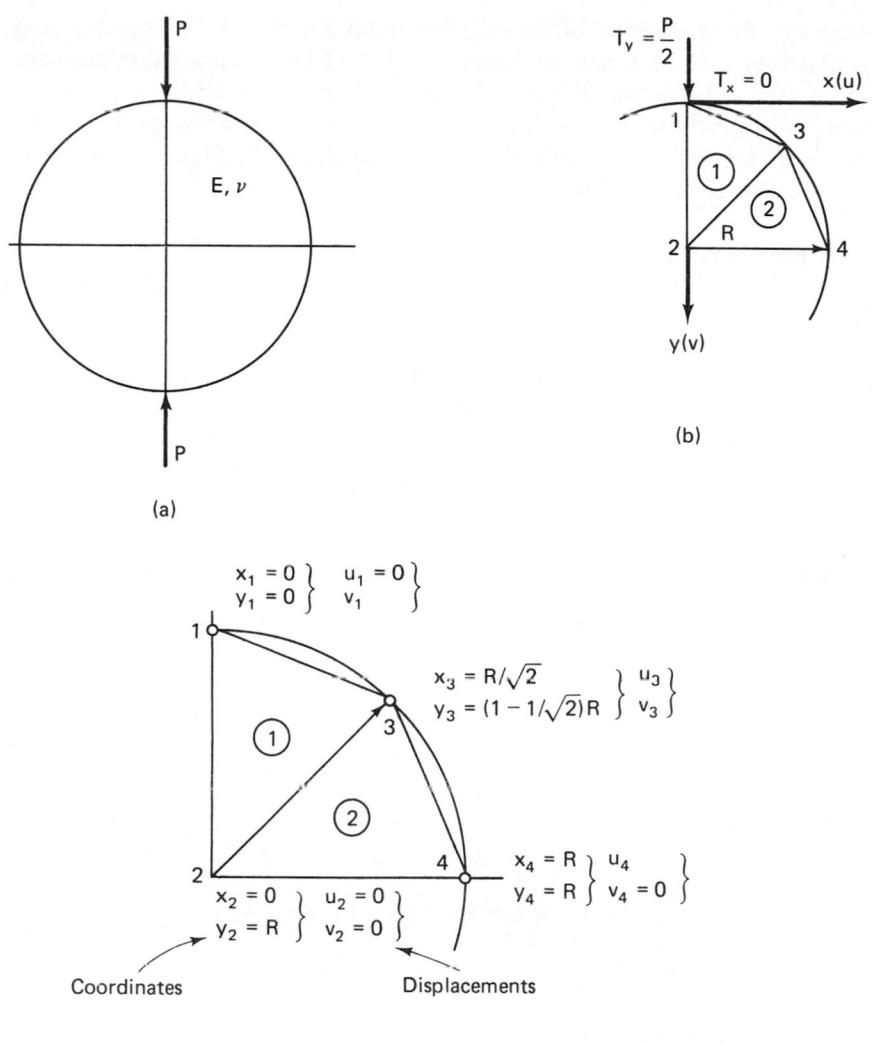

(a)

(b)

(c)

Figure 2-3. Brazilian cylinder sample problem: (a) test cylinder and loading; (b) coordinate system and mesh selected; (c) coordinates and displacement boundary conditions.

are definite and an analytical solution is available (8) for comparison with the numerical calculations. A more relevant illustration follows the cylinder example.

Since the cylinder has two axes of symmetry as shown in Figure 2-3(a), only one-quarter of the region needs to be analyzed. For the hand computations, with the simple triangular mesh of Figure 2-3(b), it can be seen from symmetry that the center of the cylinder, point 2, can be taken to have zero displacement, point 1 has an unknown vertical (y-direction) displacement and a zero x-displacement, point 4 has an unknown x-displacement and a zero y-displacement, and both displacements are unknown at point 3. The triangles, coordinates, and displacements are shown in Figure 2-3(c). The solution proceeds as follows.

Displacement Equations

In $\Delta\text{①}$

pt. 1 $\begin{cases} u_1 = a_1 x_1 + b_1 y_1 + c_1 = 0 \therefore c_1 = 0 \\ v_1 = d_1 x_1 + e_1 y_1 + f_1 \qquad \therefore f_1 = v_1 \end{cases}$

pt. 2 $\begin{cases} u_2 = a_1 x_2 + b_1 y_2 + c_1 = 0 \therefore b_1 = 0 \\ v_2 = d_1 x_2 + e_1 y_2 + f_1 = 0 \therefore e_1 R + f_1 = 0 \end{cases}$ $\qquad\qquad$ (2.69)

pt. 3 $\begin{cases} u_3 = a_1 x_3 + b_1 y_3 + c_1 \therefore a_1 = \dfrac{\sqrt{2}\,u_3}{R} \\[2mm] v_3 = d_1 x_3 + e_1 y_3 + f_1 \therefore d_1 \dfrac{R}{\sqrt{2}} + e_1\left(1 - \dfrac{1}{\sqrt{2}}\right)R + f_1 = v_3 \end{cases}$

Consequently,

$$d_1 = \frac{-v_1}{R} + \frac{\sqrt{2}\,v_3}{R}; \quad e_1 = -\frac{v_1}{R}$$

In $\Delta\text{②}$

pt. 2 $\begin{cases} u_2 = a_2 x_2 + b_2 y_2 + c_2 = 0 \therefore b_2 R + c_2 = 0 \\ v_2 = d_2 x_2 + e_2 y_2 + f_2 = 0 \therefore e_2 R + f_2 = 0 \end{cases}$

pt. 3 $\begin{cases} u_3 = a_2 x_3 + b_2 y_3 + c_2 \\ v_3 = d_2 x_3 + e_2 y_3 + f_2 \end{cases} \therefore$ etc. $\qquad\qquad$ (2.70)

pt. 4 $\begin{cases} u_4 = a_2 x_4 + b_2 y_4 + c_2 \\ v_4 = d_2 x_4 + e_2 y_4 + f_2 \end{cases} \therefore$ etc.

Consequently,

$$a_2 = \frac{u_4}{R}; \quad b_2 = -\frac{\sqrt{2}\,u_3}{R} + \frac{u_4}{R}; \quad c_2 = \sqrt{2}\,u_3 - u_4$$

$$d_2 = 0; \quad e_2 = -\frac{\sqrt{2}\,v_3}{R}; \quad f_2 = \sqrt{2}\,v_3$$

These have been solved directly, but, in general, equation (2.61) would be employed.

Equation (2.65) gives the total potential energy of one triangular element. It must be evaluated for each triangle.

The area of both triangles $= A_1 = A_2 = R^2/2\sqrt{2}$, so that

$$\pi_1 = \frac{ER^2}{4\sqrt{2}\,(1+v)(1-2v)} \left\{ (1-v)\left[\frac{2u_3^2}{R^2} + \frac{v_1^2}{R^2}\right] - \frac{2\sqrt{2}\,vu_3v_1}{R^2} \right.$$
$$\left. + \frac{(1-2v)}{2R^2}(-v_1 + \sqrt{2}\,v_3)^2 \right\} - \frac{P}{2}v_1 \tag{2.71}$$

The last term is due to the surface force $(= T_y)$ acting at point 1. For the second triangle,

$$\pi_2 = \frac{ER^2}{4\sqrt{2}\,(1+v)(1-2v)} \left\{ (1-v)\left[\frac{u_4^2}{R^2} + \frac{2}{R^2}v_3^2\right] - \frac{2\sqrt{2}\,vu_4v_3}{R^2} \right.$$
$$\left. + \frac{(1-2v)}{2R^2}(-\sqrt{2}\,u_3 + u_4)^2 \right\} \tag{2.72}$$

in which there is no contribution from surface loading. The total potential energy $\pi = \pi_1 + \pi_2$ is obtained by summing equations (2.71) and (2.72). It will be seen that there are four unknown displacements, v_1, u_3, v_3, and u_4, in this problem; so the system of linear equations describing the problem is obtained by differentiating π, the total potential energy, with respect to each of the unknowns in turn.

Thus, we get

$$\frac{\partial \pi}{\partial v_1} = B\{[2(1-v) + (1-2v)]v_1 - 2\sqrt{2}\,vu_3$$
$$- \sqrt{2}\,(1-2v)v_3 + 0\} - \frac{P}{2} = 0$$

$$\frac{\partial \pi}{\partial u_3} = B\{-2\sqrt{2}\,vv_1 \qquad\qquad + [4(1-v) + 2(1-2v)]u_3$$
$$+ 0 - \sqrt{2}\,(1-2v)u_4\} = 0 \tag{2.73}$$

$$\frac{\partial \pi}{\partial v_3} = B\{-\sqrt{2}\,(1-2v)v_1 \qquad\quad + 0$$
$$+ [4(1-v) + 2(1-2v)]v_3 - 2\sqrt{2}\,vu_4\} = 0$$

$$\frac{\partial \pi}{\partial u_4} = B\{0 \qquad\qquad - \sqrt{2}\,(1-2v)u_3$$
$$- 2\sqrt{2}\,vv_3 + [2(1-v) + (1-2v)u_4]\} = 0$$

in which

$$B = \frac{E}{4\sqrt{2}\,(1+v)(1-2v)}$$

and zeros have been entered to show that, in this case, one term has no effect on each equation in turn. The B's have been left in the last three equations to show that the coefficients of the u's and v's are spring constants. If these coefficients are examined, it will be seen that they are symmetrical (i.e., the coefficient of u_3 in the first equation is equal to the coefficient of v_1 in the second equation, etc) as discussed earlier.

Now the system of equations (2.73) has to be solved to give v_1, u_3, v_3, and u_4.

These are found to be as follows:

$$v_1 = \frac{\sqrt{2}\,(1 + v)(2 - v)(3 - 4v)}{8(1 - v)}\frac{P}{E}$$

$$u_3 = \frac{v(1 + v)(5 - 4v)}{8(1 - v)}\frac{P}{E}$$

$$v_3 = \frac{(1 + v)(2 - 5v + 4v^2)}{8(1 - v)}\frac{P}{E}$$
(2.74)

$$u_4 = \frac{\sqrt{2}\,v(1 + v)(3 - 4v)}{8(1 - v)}\frac{P}{E}$$

Now equations (2.57) through (2.59) can be used to give the stresses in triangles 1 and 2, since all the constants $a_1, b_1, \ldots, f_1$ and $a_2, b_2, \ldots, f_2$ have been determined following equations (2.69) and (2.70), respectively, in terms of v_1, u_3, v_3, and u_4.

For a material with $E = 3 \times 10^7$ psi, $(2.1 \times 10^5 \text{ MN/m}^2)$, $v = 0.2$, and a load of $P = 10^4$ lb (45 kN), the above displacements are: $v_1 = 0.350 \times 10^{-3}$ in. $(8.9 \times 10^{-3}$ mm), $u_3 = 0.525 \times 10^{-4}$ in. $(1.33 \times 10^{-3}$ mm), $v_3 = 0.725 \times 10^{-4}$ in. $(1.84 \times 10^{-3}$ mm), and $u_4 = 0.389 \times 10^{-4}$ in. $(0.99 \times 10^{-3}$ mm). A numerical finite element analysis by computer utilizing 221 triangular elements to represent the quadrant of Figure 2-3(c) gave the respective values 0.805×10^{-3} in., 0.27×10^{-4} in., 0.38×10^{-4} in., and 0.82×10^{-4} in. $(20.4 \times 10^{-3}$, 0.7×10^{-3}, 0.97×10^{-3}, 2.1×10^{-3} mm), at the same points. The analytical solution (8) gives $[\infty]$, 0.24×10^{-4}, 0.36×10^{-4}, 0.84×10^{-4} in. $(\infty, 0.61 \times 10^{-3}, 0.91 \times 10^{-3}, 2.1 \times 10^{-3}$ mm), respectively. The infinite value for the vertical deflection of point 1 in the analytical solution is a consequence of the line load application at that point, which is a singularity in the mathematical analysis. In reality, the load, even applied by a knife-edge, occupies a finite width, and the deflection is finite. In the numerical solution, infinity is represented by a displacement (v_1), an order of magnitude greater than the others.

(b) Triangular dam. A somewhat more realistic problem is selected for the second example; it is the triangular dam illustrated in Figure 2-4(a). To keep the calculations as simple as possible for hand evaluation, and to introduce the effect of body force loading, we will obtain a solution for the deflections of the dam under gravity loading. This implies, in effect, that

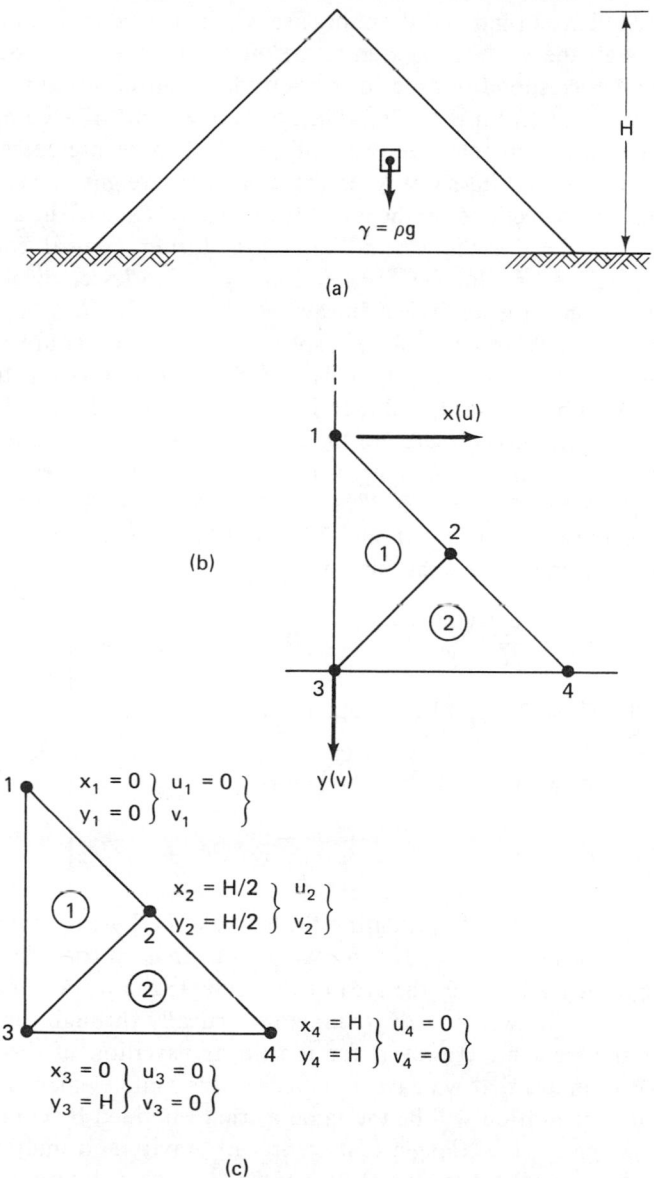

Figure 2-4. Earth dam under gravity loading: (a) dam and loading; (b) coordinate system and mesh selected; (c) coordinates and displacement boundary conditions.

the dam is to be built under zero gravity, and then gravity is to be turned on, so that the structure deforms, in this case, elastically. The foundation rock will be taken as rigid and the dam base will be considered to be bonded to it. Although the problem is seen to be an artificial one, the deflections obtained will correspond in order of magnitude to those resulting from the ultimate settlement (from the soil's effective unit weight) of a dam of these proportions built from relatively wet soil in which pore pressures develop during construction. Alternatively, if the total unit weight is used in the calculations, the deflections are the pseudostatic equivalents of the amplitudes that would develop in the vertical direction during vertical earthquake accelerations of ± 1 g. Smaller accelerations will give lesser displacements in proportion, since the analysis is linear.

The coordinates and boundary displacement conditions are shown in Figures 2-4(b) and (c), respectively. It is seen that only one symmetrical half of the dam need be analyzed, and that, in consequence of this condition and the boundary restrictions, there are three unknowns to be obtained: u_1, u_2, and v_2. We will therefore end up with three simultaneous equations for solution.

By applying the same equations (2.69) and (2.70) as before to triangles 1 and 2, we determine the constants in this case to be

$$\left. \begin{array}{lll} a_1 = \dfrac{2u_2}{H} & b_1 = 0 & c_1 = 0 \\[2ex] d_1 = \dfrac{2v_2 - v_1}{H} & e_1 = \dfrac{-v_1}{H} & f_1 = v_1 \end{array} \right\} \quad (2.75)$$

$$\left. \begin{array}{lll} a_2 = 0 & b_2 = \dfrac{-2u_2}{H} & c_2 = 2u_2 \\[2ex] d_2 = 0 & e_2 = \dfrac{-2v_2}{H} & f_2 = 2v_2 \end{array} \right\} \quad (2.76)$$

There are several ways of calculating the work done by the gravity forces in this case. In general, the body force work done in each triangle should be obtained by integrating over the area of the triangle the work done on each element $dx\,dy$ by its weight $\gamma\,dx\,dy$ moving vertically through the displacement v of the element. However, for the linear variation of displacement with the coordinates that we have assumed in this finite element model, the result of this integration will be the same as that obtained by concentrating the whole weight of the triangle at its center of gravity and multiplying the weight by the vertical movement of the center of gravity. Alternatively, one-third of the weight of the triangle can be considered to be attached to each of the three node points, and the work done can be calculated from the sum of each one-third weight moving through the vertical displacement of its associated node. If a parabolic or other distribution of displacement were to be assumed, the work done by the weight should be obtained by the integration method.

In the present case, all approaches give the same result:

$$\text{In triangle 1, work done} = \frac{\gamma H^2}{4}\frac{(v_1 + v_2)}{3}$$

$$\text{In triangle 2, work done} = \frac{\gamma H^2}{4}\frac{v_2}{3}$$

(2.77)

Including the constants, equations (2.75) and (2.76), and the work done for the two triangles, equation (2.77), in the total potential energy expression, equation (2.65), and summing for both triangles, we get, eventually

$$\pi = \frac{E}{8(1 + v)(1 - 2v)}\Big[(1 - v)(v_1^2 + 4u_2^2 + 4v_2^2) - 4vu_2v_1$$

$$+ \frac{(1 - 2v)}{2}(4v_2^2 + v_1^2 - 4v_1v_2 + 4u_2^2)\Big] - \frac{\gamma H^2}{12}(v_1 + 2v_2)$$

(2.78)

Since this exercise is too laborious for us to proceed further, still retaining v as an independent variable, we will assign it the value 0.3 at this stage. Equation (2.78) reduces to

$$\pi = \frac{E}{4.16}(0.9v_1^2 + 3.6u_2^2 + 3.6v_2^2 - 1.2v_1u_2 - 0.8v_1v_2)$$

$$- \frac{\gamma H^2}{12}(v_1 + 2v_2)$$

(2.79)

Differentiating π successively with respect to the three unknowns v_1, u_2, and v_2 and setting the results equal to zero give the three required equations:

$$\frac{E}{4.16}(\quad 1.8v_1 - 1.2u_2 - 0.8v_2) = \frac{\gamma H^2}{12}$$

$$\frac{E}{4.16}(-1.2v_1 + 7.2u_2 + \quad 0 \quad) = 0$$

(2.80)

$$\frac{E}{4.16}(-0.8v_1 + \quad 0 \quad + 7.2v_2) = \frac{\gamma H^2}{6}$$

Once again it is seen that the set of equations is symmetrical about the diagonal. This a useful check on the arithmetic. The numbers appearing in the equations are exact. Solving equation (2.80) gives

$$v_1 = \frac{22.88}{81.6}\frac{\gamma H^2}{E}; \qquad u_2 = \frac{v_1}{6} = \frac{22.88}{489.6}\frac{\gamma H^2}{E};$$

$$v_2 = \frac{v_1}{9} + \frac{0.52}{5.4}\frac{\gamma H^2}{E} = \Big(\frac{22.88}{734.4} + \frac{0.52}{5.4}\Big)\frac{\gamma H^2}{E}$$

(2.81)

Again, these numerical values are exact.

It is of interest to evaluate these displacements for some reasonable values of the material properties. Let us choose $H = 100$ m (328 ft), $\gamma = 20$ kN/m³ (127 pcf), and $E = 20$ MN/m² (2900 psi) as more-or-less soil-like properties. For these, we find $\gamma H^2/E = 10$ m (32.8 ft), and

$$v_1 = 2.804 \text{ m}; \qquad u_2 = 0.467 \text{ m};$$
$$v_2 = 1.275 \text{ m} \qquad (9.2, 1.5, 4.2 \text{ ft}) \tag{2.82}$$

That means that the dam's crest subsided almost 3 percent of its height, with the other displacements in proportion.

For comparison, the same configuration was established for a computer solution involving several hundred triangular elements; the vertical displacement of the crest was found to be 3.148 m(10.3 ft). No analytical solution exists for this case. Correspondence between the crude hand solution and the computed result is fortuitously good.

The stress components could be calculated in either of these two examples. The two triangles give different stresses, but they are the same throughout each triangle because of the assumption that displacement is linear in the distance coordinate. In consequence, the stresses do not satisfy the boundary condition of no normal stress along the free solid surface 1–2–4 in Figure 2-4. This is a consequence of the displacement assumption; it was not selected to match all the boundary conditions. The minimization process picks the set of coefficients or variables from among those made available to it, which minimizes the potential energy function. If the form of the approximate solution entered into the problem does not permit all the boundary conditions to be satisfied, then the solution naturally will not suit these requirements either. In this case, displacement is the essential boundary condition, not stress.

When the loading on a symmetrical structure is asymmetric, the finite element approach may include an undesirably large number of unknowns for hand solution. In this case, reduction in the number of equations to be solved at one time can be achieved by dividing the load into symmetric and antisymmetric components. Figure 2-5 illustrates an earth dam subjected to water load on one side. For the four-triangle solution arrangement shown, there are six unknown displacements to be determined; they evolve from six simultaneous equations. If the problem is split up, as in the illustration, into symmetric and antisymmetric components, the number of simultaneous equations to be solved at one time is reduced. The symmetric geometry has three unknown displacements as in Figure 2-4, and the antisymmetric has four. In the antisymmetric problem only one-half of the dam needs to be analyzed, as in the symmetric case. When the displacements are calculated for the two conditions, they are added together to give the compiete solution.

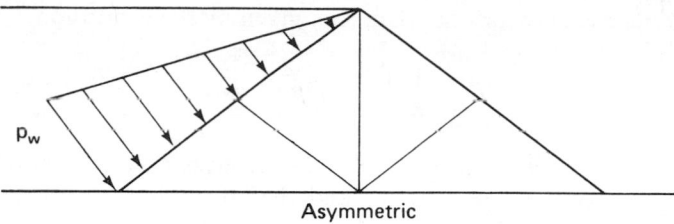

Asymmetric

=

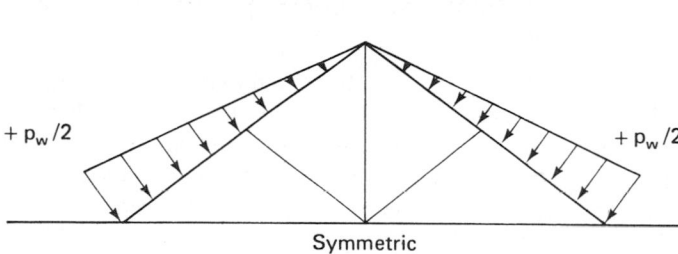

Symmetric

+

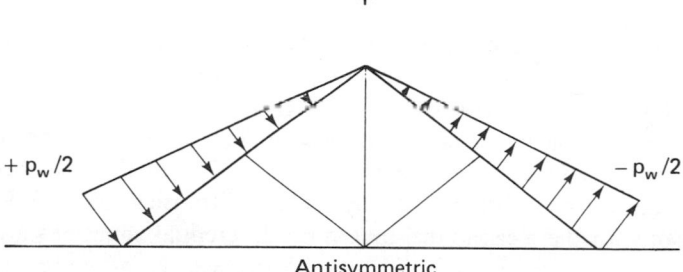

Antisymmetric

Figure 2-5. Decomposition of earth dam water loading into symmetric and antisymmetric components.

The division of a loading system into symmetric and antisymmetric components is a common device in structural analysis and will be encountered again later.

Finally, although this dam problem *is* artificial with respect to real static dam deflections, it does have a use in the solution of a dynamics problem. For vertical excitation of the dam, the deflections we have calculated represent the first mode of vibration (the *breathing mode*) of this limited three degree of freedom model. Utilization of Rayleigh's quotient (3) enables the modal frequency to be calculated by equating the kinetic and potential energies.

The kinetic energy for one triangle is given by the expression

$$E = \frac{1}{2} \int\int (\dot{s})^2 \rho \, dx \, dy \tag{2.83}$$

where $\dot{s}$ is the velocity vector of an element of mass $\rho \, dx \, dy$ at point (x, y), ρ is the density of the dam material, and the integration is performed over the area of the triangle. This equation can be rewritten in terms of the angular frequency of the dam in this mode, ω

$$E = \frac{1}{2}\omega^2 \int\int s^2 \rho \, dx \, dy = \frac{1}{2}\rho\omega^2 \int\int (u^2 + v^2) \, dx \, dy \tag{2.84}$$

when s is taken as the displacement vector equal to $(u^2 + v^2)^{1/2}$. The potential energy U_v is the strain energy stored in the structure in its displaced state and is given by the term $E/8(1 + v)(1 - 2v)\{\ldots\}$ in equation (2.78) for both triangles. Summing up the kinetic energy for the two triangles of the dam and equating it to the potential energy give the modal frequency

$$\omega^2 = \frac{2U_v}{\rho \overset{2}{\underset{}{\sum}} \int\int (u^2 + v^2) \, dx \, dy} \tag{2.85}$$

Substituting the values obtained for the displacements, and thus for the coefficients $a_n, b_n, \ldots$ etc. in U_v, u, and v in this equation, and evaluating gives the required dam frequency.

Another approach, which is also useful for cases with one or more dimensions, is the finite difference computation. This method has been used for some time in the solution of steady-state and transient water flow conditions in soils. In these cases, the governing differential equations are represented by *finite difference* equations, and the system of finite difference equations set up for the problem is solved simultaneously (9, 11).

For certain idealizations of the response of the soil foundation material, the finite difference technique is useful both for hand calculations and for simple computer codes.

2.5 FINITE DIFFERENCE METHOD

The finite difference method is described in many textbooks of numerical analysis (2, 3, and 7) and will be developed here again with respect to the beam on the elastic foundation (see Chapter 5).

$$EI\frac{d^4w}{dx^4} + kw = 0 \tag{2.86}$$

Here, E and I are the Young's modulus and moment of inertia of the beam, respectively, k is the spring constant (units of FL^{-2}) of the foundation, and w and x are the vertical deflection and distance coordinate along the beam, respectively, as shown in Figure 2-6.

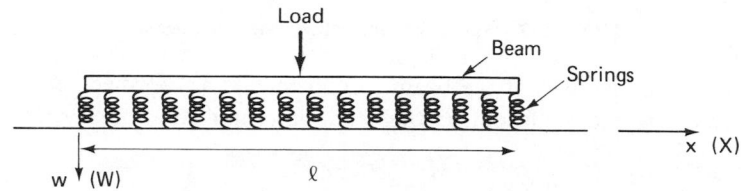

Figure 2-6. Flexible beam on elastic (Winkler) foundation.

The analytical solution of this equation is possible and will be discussed in Chapter 5, but it becomes increasingly difficult as various conditions, such as beam modulus or spring constant changing along the beam, are introduced. We will therefore look at the solution of the equation by finite differences. First, however, it is convenient to reduce the equation to a dimensionless form by introducing a parameter α with the dimensions of L^{-1},

$$\alpha = \sqrt[4]{\frac{k}{EI}} \qquad (2.87)^4$$

Substituting α in the beam equation gives the dimensionless equation

$$\frac{d^4 W}{dX^4} + W = 0 \qquad (2.88)$$

where $W = \alpha w$ and $X = \alpha x$. Once the solution to deflection w or W has been obtained, various engineering quantities of interest can be derived, such as slope w', moment EIw'', and shear EIw''', where the primes represent successive derivatives with respect to x.

Instead of the continuous distribution of w as a function of x, for the finite difference solution we choose to divide the beam up into sections of usually equal length H (dimensionless) as shown in Figure 2-7. There will then be $n = L/H$ such segments, when $L = \alpha l$ is the (dimensionless) length of the beam. The method then proceeds to calculate the values of deflection W_i at the points i, called *nodal points*, between the finite sections of the beam. The values of W_i are controlled by finite difference representations of the differential equation and boundary conditions and are therefore approximations to the correct values $W(X)$ at the same points. The finite difference

---

[4] Note the difference between α and the parameter λ of equation (5.12).

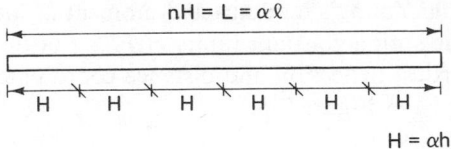

Figure 2-7. Finite difference subdivision of a beam.

forms of the derivatives of the beam displacements are formed successively as follows, with the coefficients in the usual binomial expansion order:

$$\left(\frac{dW}{dX}\right)_i = W'(X_i) \approx \frac{W_{i+1} - W_{i-1}}{2H} \tag{2.89}$$

$$\left(\frac{d^2W}{dX^2}\right)_i = W''(X_i) \approx \frac{W_{i+1} - 2W_i + W_{i-1}}{H^2} \tag{2.90}$$

$$\left(\frac{d^3W}{dX^3}\right)_i = W'''(X_i) \approx \frac{W_{i+2} - 2W_{i+1} + 2W_{i-1} - W_{i-2}}{2H^3} \tag{2.91}$$

The third difference at node i is the arithmetic mean of the two third differences calculated from the difference of the second differences for two points $H/2$ on each side of node i. A similar process gives the fourth difference.

$$\left(\frac{d^4W}{dX^4}\right)_i = W^{iv}(X_i) \approx \frac{W_{i+2} - 4W_{i+1} + 6W_i - 4W_{i-1} + W_{i-2}}{H^4} \tag{2.92}$$

The method can best be described by working an illustrative example. We will take the problem of a beam of finite length L resting on a Winkler foundation and loaded at one end, $X = 0$, by a point load P. The boundary conditions for this problem are:

$$\text{at} \quad X = 0, \quad W'' = 0, \quad W''' = \frac{P}{EI\alpha^2} = \frac{P}{\sqrt{kEI}} \tag{2.93}$$

$$\text{and at} \quad X = L, \quad W'' = 0, \quad W''' = 0$$

The parameter $P/\sqrt{kEI}$ is seen to represent a dimensionless load. Although it would be possible to apply the finite difference form of the fourth derivative as in equation (2.92) to successive points at intervals of $H(\alpha h)$ along the beam, it is more convenient, from a calculational point of view, to make the substitution

$$V = W'' \tag{2.94}$$

and to operate with this equation and that produced by substituting it into equation (2.88)

$$V'' + W = 0 \tag{2.95}$$

Only second differences are then required, and the solution will have direct physical meaning since V is proportional to the bending moment in the beam. The boundary conditions become

$$\text{at} \quad X = 0, \quad V = 0, \quad V' = \frac{P}{\sqrt{kEI}},$$

$$\text{and at} \quad X = L, \quad V = 0, \quad V' = 0 \tag{2.96}$$

If a computer were to be used, the beam might be divided into 10 intervals, but to simplify the calculations so that they can be done by hand, we will take two intervals only. In dimensionless terms, the length of each interval is then $L/2$ ($\alpha l/2$), and there are three node points 0, 1, and 2 as shown on Figure 2-8. The object of the calculations is to determine the values of W and V at each of these node points.

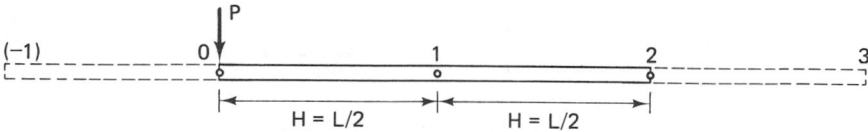

Figure 2-8. Node numbering in finite difference solution.

There are therefore initially six unknowns: W_0 W_1, W_2, V_0, V_1, and V_2; however, the first boundary condition at each end requires $V_0 = V_2 = 0$, so that four unknowns are left to be determined by the successive applications of equations (2.90) and (2.95) in finite difference form to the node points. First, we must translate the remaining two boundary conditions into finite differences. At $X = 0$

$$V' = \frac{V_1 - V_{-1}}{L} = \frac{P}{\sqrt{kEI}}$$

Thus

$$\tag{2.97}$$

$$V_{-1} = V_1 - \frac{PL}{\sqrt{kEI}}$$

Here we have introduced a fictitious nodal point (-1) one interval to the left of the end of the beam (node 0), in order to construct the finite difference at point 0.
At $X = L$

$$V' = \frac{V_3 - V_1}{L} = 0$$

Thus

$$\tag{2.98}$$

$$V_3 = V_1$$

Another fictitious node, 3, has been brought in to satisfy the conditions at the right-hand end of the beam.

The next step is to apply equations (2.95) and (2.94) to the nodal points of the beam. We will take equation (2.95) first.

At point 0 we have

$$\frac{4(V_{-1} - 2V_0 + V_1)}{L^2} + W_0 = 0 \tag{2.99}$$

but $V_0 = 0$, and equation (2.97) expressed V_{-1} in terms of V_1 and P, so that we can substitute V_{-1} in equation (2.99) to get

$$8V_1 + L^2 W_0 - \frac{4PL}{\sqrt{kEI}} = 0 \tag{2.100}$$

At point 1 equation (2.95) gives

$$\frac{4(V_2 - 2V_1 + V_0)}{L^2} + W_1 = 0 \tag{2.101}$$

Here $V_2 = V_0 = 0$, so that we have the equation

$$-8V_1 + L^2 W_1 = 0 \tag{2.102}$$

We repeat this procedure for node 2 and substitute $V_3 = V_1$ and $V_2 = 0$ to get finally

$$8V_1 + L^2 W_2 = 0 \tag{2.103}$$

The other relation, equation (2.94), serves to define the conditions at the ends, nodes 0 and 2, and only contributes information at the intemediate node 1. At this point it gives rise to the equation

$$4W_2 - 8W_1 + 4W_0 - L^2 V_1 = 0 \tag{2.104}$$

The four equations, (2.100), (2.102), (2.103), and (2.104), can be solved for the four unknowns to give

$$W_0 = \frac{4P}{L\sqrt{kEI}}\left[1 - \frac{32}{(128 + L^4)}\right] \tag{2.105}$$

$$W_1 = \frac{128P}{L\sqrt{kEI}(128 + L^4)} \tag{2.106}$$

$$W_2 = -W_1 \tag{2.107}$$

$$V_1 = \frac{16PL}{\sqrt{kEI}(128 + L^4)} \tag{2.108}$$

From the relation, moment, $M = EIw''$, the moment at point 1, can be expressed in the following terms:

$$\left(\frac{M}{Pl}\right)_1 = \frac{\alpha EI}{Pl} V_1 = \frac{16}{128 + L^4} \tag{2.109}$$

and the deflections can also be written

$$\left(\frac{wkl}{P}\right)_0 = 4\left(1 - \frac{32}{128 + L^4}\right) \tag{2.110}$$

$$\left(\frac{wkl}{P}\right)_1 = \frac{128}{128 + L^4} \tag{2.111}$$

$$\left(\frac{wkl}{P}\right)_2 = -\left(\frac{wkl}{P}\right)_1 \tag{2.112}$$

Since such a coarse interval, one-half the beam length, would not be expected adequately to describe deflections changing rapidly over short distances, it follows that these results would correspond better to an exact solution for stiff rather than very flexible beams. Stiff beams are represented by small values of L (αl); we will tabulate some solutions from the exact theory for $\lambda l = 2$ and $\lambda l = 4$, respectively, for comparison purposes, in Table 2.2. It is apparent that the bending moments are obtained more precisely than are the deflections and that the numerical solutions get progressively worse as L increases. It should also be pointed out that it is known from the exact solution that the *maximum* bending moment does not occur at the midpoint of the beam but at a point generally nearer the load. To obtain an indication of the maximum moment by this numerical method, a finer subdivision of the beam would be required. However, in view of the extreme coarseness of the interval adopted, and the small computational effort involved, the agreement of both the calculated moment at the midpoint and the end

TABLE 2.2

Finite Difference Values (Two Intervals) Versus Exact Solution

Values of M/Pl at Midpoint of Beam		
λl [see equation (5.12)]	2	4
$L = \alpha l$	2.828	5.657
From equation (2.109)	0.083	0.014
Exact equation (8.131)	0.109	0.031
Values of wkl/P at Load		
λl	2	4
$L = \alpha l$	2.828	5.657
From equation (2.110)	3.333	3.889
Exact equation (8.130)	4.550	8.006

deflection for the stiffer beam with the "exact" values is quite reasonable. The finite difference technique is useful in the solution of problems involving mats or rafts on a Winkler foundation (1, 10).

2.6 DISCUSSION

It is apparent that in the problems studied none of the approximate techniques gives particularly good results. Errors of from 30 percent to 40 percent occur in many of the solutions when they are compared to exact analyses. It should be pointed out, however, that particularly simple functional representations were chosen to illustrate the methods. A few more terms, with two or three more unknown coefficients to be determined, still lie within the limits of desk calculation. More suitable functions than polynomials might also be selected. These improvements will generally greatly reduce the errors. It is difficult to decide, in the collocation methods, which collocation points will give the best final answer in a given situation. Another, but substantial, difficulty lies in choosing the approximate function to begin with, so that it fits all the boundary requirements in the residual methods and some of them in the energy approach. Consider, for example, the case of the beam on the elastic foundation subjected to a point load at the center (or some other point). The ends are free, with the boundary conditions that moment (ψ'') and shear (ψ''') are zero at each end. For collocation, a function ψ has to be selected whose derivatives meet these requirements. In this problem, because of symmerty, usually one-half the beam will be taken. In this case, ψ'' and ψ''' are zero at the free end, but at the midpoint different conditions prevail. There, the slope, ψ', is zero, and the shear, ψ''', is equal to one-half the applied load. Obviously, the function will have to be symmetric in order to meet the slope condition, but the other properties are hard to match. If the Ritz method is to be used, the requirements are eased somewhat, but the essential conditions in the displacement formulation consist of the deflections at the beam ends. Here, however, the displacements are unspecified. What conditions, therefore, must be imposed on the approximate displacement function for the entire beam or one-half of it?

If the complementary approach is adopted in this problem, whereby the equation is couched in terms of the dimensionless moments, μ, in the beam (Section 5.1.1), then an approximate moment function must be assumed. In collocation, using one-half the beam, the condition of moment (μ) and shear (μ') being zero at the free end is not perhaps so hard to meet, but at the center only the shear is known, as well as the condition of zero slope, which here translates as $\mu''' = 0$. Again, choice of a suitable function is apparently not easy. With the complementary energy method, things are simplified, since we need only make moments zero at the beam ends. However, knowing the exact solution to the distribution of moments in the center-loaded beams (see

Chapter 5) indicates that realistic moment functions may be complicated to select.

These difficulties are related to the practical foundation requirements that must be imposed on beam and pile, where the loads (moments) are nearly point loads (moments) or are distributed over short lengths of the member, and the deflections occur freely under these conditions. It is not difficult to pick configurations for which residual and Ritz methods all work out well, without the difficulties encountered here, and with greater accuracy for less effort. These easier problems involve (a) uniform or continuously distributed symmetric loadings and (b) constraints on the deflection of the member. For example, in Crandall (3), the beam on the elastic foundation is finite, uniformly loaded, and supported on pins at the ends so that the end deflections are zero. Sinusoidal approximate deflection functions therefore suggest themselves, since for this problem deflections, slopes, and moments all vary smoothly without discontinuities throughout the beam. If the beam is built in at each end so that displacement and slope there are both zero, cosine functions for displacement can be employed. It is also evident in such problems that the beam section can vary symmetrically (so that EI changes) without introducing much more complexity into the analysis.

The advantage of these approximate analysis methods is that they enable one to make calculations when analytical exact solutions are very difficult or impossible. Frequently the design variables of beam dimensions, say, appear explicitly in the approximate, instead of implicitly as in the exact solutions (for example, in the λ-parameter of Chapter 5). Examples of exact solutions which are difficult to obtain in beam or pile problems are when the soil property varies in some prescribed manner along the pile or beam length, in which case the problem remains linear, and when the soil behavior is plastic, so that the problem becomes nonlinear. In such cases, a solution that is within 50 percent of the correct but unattainable answer may still be very valuable. More terms and greater exactness may always be obtained by using computers. A number of these problems is discussed in later chapters. The other prime example of when formal analysis breaks down is when the problem must be described in two or three dimensions. In practice, in the majority of such cases, analysis now proceeds by computer with either finite element or, less commonly, finite difference methods.

PROBLEMS

1. Solve the pile problem of Section 2.3.1 using the various collocation techniques, and a polynomial equation [equation (2.4)] with four terms.
2. Suppose the second term in the pile equation (2.1) has the form $s\psi$ because the soil resistance increases linearly with depth. When the tip of the pile is

fixed obtain the solution by a collocation method, using a polynomial with three terms. For the case when the dimensionless length of the pile is 2, compare your answer with the exact solution (see Section 8.2.6).

3. Take the form of the exact solution to the pile in uniform soil, equation (8.35) and solve the problem 2 above, using the Ritz method. Compare your result with the exact solution (see Section 8.2.6).

4. Consider a uniformly-loaded beam on an elastic foundation. Its ends are pinned. Assume a suitable form of solution matching the essential boundary conditions, and obtain a solution by Ritz's method.

5. Would a better choice of a collocation point or points improve the accuracy of the pile problem? How would you select them, if this were the case?

6. How would you use a finite element elastic analysis of an earth dam to give a lower bound to a stability computation? Illustrate by an example.

7. An embankment is constructed of soil identical in properties to those of the foundation material in two ways: (a) by placing the soil in layers on the foundation soil, or (b) by *removing* the foundation soil in layers. In both cases the resulting embankment slope and layer thickness is the same. What differences would you make in preparing these two problems for a finite element solution? Would you expect the stresses in the embankment to turn out differently in the two cases? If you have a finite element program available, choose some reasonable properties, and solve the two problems.

8. Are there any difficulties in using the finite element method to solve problems involving incompressible ($v = \frac{1}{2}$) linearly elastic media? Show how this material property enters into the finite element steps.

9. Obtain the exact solution for the rigid beam on the Winkler foundation from Section 4.3 when the beam is loaded at its end point (Figure 4-5). Compare the end deflections and the central moment with the solutions obtained by the finite difference technique, equations (2.109) and (2.110) when the latter are specialized to the rigid beam case.

10. Solve the problem of the centrally-loaded beam on an elastic foundation by the finite difference method. Note symmetry, and divide the beam into four equal parts.

11. Compare the exact solution for the central moment in the end-loaded finite beam with $\lambda l = 1.0$ with the finite difference solution, Table 2.2, and the rigid beam solution of Problem 9.

12. A rigd retaining wall rotates about its base and is subjected to a horizontal force P at its top. In order to calculate the relation between P and displacement at the top of the wall, make a finite element solution using the single triangle 1–2–3 shown in Figure P2-12. Outside the triangle the soil is rigid. Assume the wall is frictionless and the displacements are linear in the coordinates. For added realism take the elastic modulus of the soil to vary linearly with depth, z, from the surface. Ignore weight effects.

13. In the finite element problem of Section 2.4.2, the base of the dam was considered to be bonded to the foundation. As a more realistic condition, consider the base boundary condition to be one where uniform shearing stress τ_0 is applied between the points 3 and 4 as shown in Figure P2-13. Modify the solution taking the shearing stress into account. As a further practical limita-

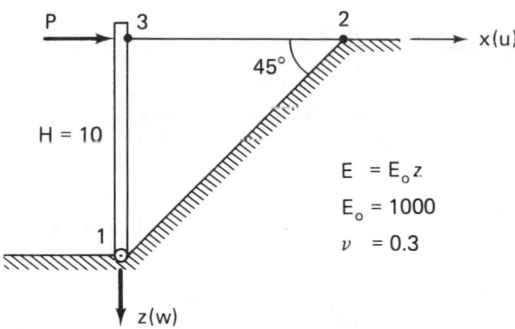

Figure P2-12.

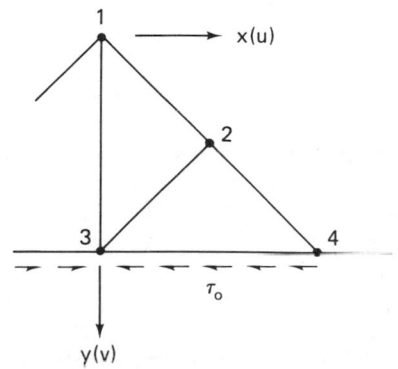

Figure P2-13.

tion, τ_0 cannot be greater than μ times the normal stress σ_{yo} on the base, where μ is the friction coefficient between dam and foundation. How would you obtain a solution for the limiting case where $\tau_0 = \mu\sigma_{yo}$ exactly?

14. Consider the finite element problem on the triangular dam in Section 2.4.2. Assume that half the dam is represented by one element and that horizontal displacements are negligible (zero) with respect to vertical displacement. This time take a more sophisticated model for vertical displacements, namely that $v = a(1 - y^2/H^2)$. (a) Carry through the same analysis as before to determine the actual vertical displacements in the structure under gravity loading. In particular, determine the vertical displacement at the crest, and compare it with the value in the text. The dam properties remain as before. (b) Comment on the assumed displacement functions and how they fit both stress and diaplacement boundary conditions. How do these considerations match with the solution you get?

15. Seepage is occurring in the nonhomogeneous medium of unit thickness shown in Figure P2-15 under the boundary conditions shown. It is desired to find the total head at point 2 by a finite element method using the triangles indicated. The total head is assumed to vary *linearly* in each triangle according

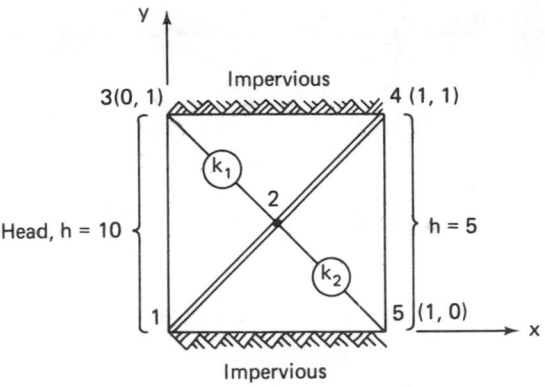

Figure P2-15.

to the relation

$$h = ax + by + c$$

The functional to be minimized is given by equation (2.67). Using the axes and coordinates shown, solve for the total head at point 2. Show the suggested paths of 1 flow line and 1 equipotential. The material in region 134 is isotropic of permeability k_1, that in 145 is of permeability k_2 and $k_2 = 0.5k_1$.

16. Consider the simple triangular dam of the finite element example of Section 2.4.2. *Instead* of the gravity loading of that problem, we are going to displace the dam with a water load (from the reservoir) on one side of the dam as indicated in Figure P2-16. Two cases are to be examined: (a) The upstream surface of the dam is lined with an impervious membrane, Figure P2-16(a). (b) The dam is homogeneously permeable and has a rock drain at the toe of the same mechanical properties as the dam. In this case the water seeps through the dam. Take the flow net as shown (incorrectly, but adequately) in Figure P2-16(b). Assume that the mechanical properties of the dam material are the same in cases (a) and (b). Show whether or not the displacements which would result from two finite element analyses (a) and (b) would be the same. The actual analyses need not be followed through except to the extent required to answer the above question.

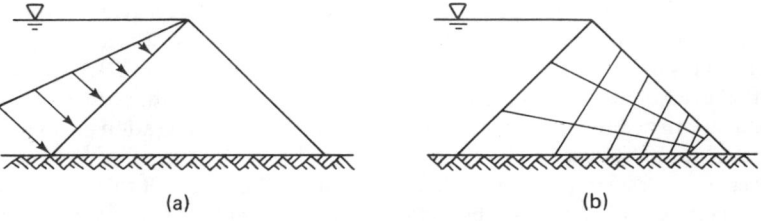

       (a)   (b)

Figure P2-16.

17. How would you employ a three-dimensional finite element model in studies of railroad track-tie-ballast behavior? Include a discussion of the size of the model, the loads, the effect of repetitive loads, and the particular features you would look for in the results. Would a linear analysis be of value?

REFERENCES

[1] ALLEN, D. N. DEG., AND R. T. SEVERN. "The Stress in Foundation Rafts," Part I, *Proc. I.C.E.*, *15*, 35–48 (Jan. 1960); Part II, *Proc. I.C.E.*, *20*, 293–304 (Oct. 1961).

[2] COLLATZ, L. *The Numerical Treatment of Differential Equations*, 3rd ed. Berlin: Springer-Verlag, 1960.

[3] CRANDALL, S. H. *Engineering Analysis*. New York: McGraw-Hill, 1956.

[4] DESAI, C. S., AND J. F. ABEL. *Introduction to the Finite Element Method*. New York: Van Nostrand Reinhold., 1972.

[5] KETTER, R. L., AND S. PRAWEL, *Modern Methods of Engineering Computations*, New York: McGraw-Hill, 1969.

[6] MATLOCK, H., AND L. C. REESE. "Generalized Solutions for Laterally Loaded Piles," *Proc. ASCE, 86*, SM5, 63–91 (Oct. 1960).

[7] MILNE, W. E. *Numerical Solution of Differential Equations*. New York: John Wiley, 1953.

[8] MUSKHELISHVILI, N. I. *Some Basic Problems of the Mathematical Theory of Elasticity*, 2nd English ed. Groningen: Noordhoff, 1963.

[9] SCOTT, R. F. *Principles of Soil Mechanics*. Reading, Mass.: Addison-Wesley, 1963.

[10] TENG, W. C. *Foundation Design*. Englewood Cliffs, N. J.: Prentice-Hall, 1962.

[11] VERRUIJT, A. *Theory of Groundwater Flow*. London: Gordon and Breach Science Publishers, 1970.

[12] ZIENKIEWICZ, O. C. *The Finite Element Method*, 3rd ed. London: McGraw-Hill, 1977.

Additional Analysis Methods

3

3.1 INTRODUCTION

In the last chapter several methods of attacking soil engineering problems in an approximate way were described. Here we discuss two more approaches that use the same basic principles but we apply them in different ways. The first takes up the potential minimization technique used in the Ritz and finite element methods, but it obtains the deflection function in a more systematic fashion than before. In the second procedure the finite element scheme is combined with some results obtained from analytical solutions in solid mechanics. They each possess advantages with respect to particular problems.

3.2 HETENYI'S TECHNIQUE

An alternative method of solving a variety of soil-structure interaction problems such as beam deflection on soil foundations again involves a variational approach. Once more, we minimize the potential π, where

$$\pi = U_v - W \tag{3.1}$$

Here we will develop the technique in terms of the situation of a loaded beam on an elastic foundation [equation (2.86)]. Considering, respectively, the strain energy in beam bending and deflection of the subgrade (soil), we can represent the total strain energy U_v along the length of the beam and foundation in dimensional form

$$U_v = \frac{1}{2} \int_0^l EI\left(\frac{d^2w}{dx^2}\right)^2 dx + \frac{1}{2} \int_0^l kw^2 \, dx \qquad (3.2)$$

The parameters EI and k are the beam and foundation constants, respectively, w is the vertical displacement, x is distance, and l is the length of the beam.

If a point load P acts on the beam at a particular coordinate, $x = s$ say, then the work done by that load as the beam deflects a distance $w(s)$ under the load is $Pw(s)$. The work done by a moment M acting at $x = s$ is the product of the moment and the rotation at $x = s$, $\theta(s)$, or $M\theta(s)$. If a distributed load $q(x)$ is acting on the beam between the locations $x = a$ and $x = b$, the work done by the distributed load is given by the integral of the product $q(x)$ and the deflection over that section of the beam or $\int_a^b q(x)w(x) \, dx$. Rather than write down the work done by all these forms of loading, we will consider them to be represented in the expression $P_s w(s)$ for the work done by a generalized load P_s acting at a point or over a portion of the beam indicated by the coordinate, s. Thus, the expression for the potential energy of the entire beam will be formed from the difference between the stored strain energy and the work done:

$$\pi = \int_0^l \left[\frac{EI}{2}\left(\frac{d^2w}{dx^2}\right)^2 + \frac{1}{2}kw^2\right] dx - P_s w(s) \qquad (3.3)$$

This expression is equivalent in dimensional form to equation (2.38). The solution to the problem then consists, as before, in finding an equation for the deflected shape of the beam $w = w(x)$ which minimizes the potential, π, in equation (3.3).

In Section 2.4.1 we examined trial solutions to the deflected shape in the form of power series

$$w = a_0 + a_1 x + a_2 x^2 + \ldots \qquad (3.4)$$

or an exponential series

$$w = a_1 e^{-x} + a_2 e^{-2x} + \ldots \qquad (3.5)$$

For a small number of terms, they yielded generally fairly inaccurate solutions. A larger number of terms would lead to a number of equations large enough to require computer evaluation.

For preliminary design calculations, however, it is useful to be able to perform at least *some* calculations by hand. The assumption of a rigid beam may be possible, and this is dealt with in the next chapter. If the beam is so flexible with respect to the soil that this is not considered appropriate, a method suggested by Hetenyi provides solutions obtainable by a relatively modest expenditure of effort. Hetenyi (6) proposes to represent the deflected shape of the beam by the series

$$w(x) = \sum_{n=1}^{\infty} X_n \tag{3.6}$$

Where the coordinate x has its origin at the left end of the beam and the X_n are the shapes of the normal modes of vibration of the beam; the reasons for this are not important here.[1] We can simply consider the X_n as convenient forms for the deflected shape which are superimposed to obtain the final deflection. In equation (3.6) it is convenient to represent the X_n as the following functions:

$$X_n = C_{1n}\left(\cos r_n\frac{x}{l} + \cosh r_n\frac{x}{l}\right) + C_{2n}\left(\cos r_n\frac{x}{l} - \cosh r_n\frac{x}{l}\right)$$
$$+ C_{3n}\left(\sin r_n\frac{x}{l} + \sinh r_n\frac{x}{l}\right) + C_{4n}\left(\sin r_n\frac{x}{l} - \sinh r_n\frac{x}{l}\right) \tag{3.7}$$

The coefficients C_{1n}, C_{2n}, ..., and r_n are constants which are determined by the boundary conditions of the problem at hand.

It will be seen that this formulation has an advantage over the analytical solutions given in a later chapter in that the parameters describing the beam and foundation properties will appear explicitly in the resulting equations rather than in the arguments of trigonometric and hyperbolic functions.

The successive derivatives of equation (3.7) are required for the ready solution of typical problems; they are

$$\frac{l}{r_n}X_n' = C_{1n}\left(-\sin r_n\frac{x}{l} + \sinh r_n\frac{x}{l}\right) + C_{2n}\left(-\sin r_n\frac{x}{l} - \sinh r_n\frac{x}{l}\right)$$
$$+ C_{3n}\left(\cos r_n\frac{x}{l} + \cosh r_n\frac{x}{l}\right) + C_{4n}\left(\cos r_n\frac{x}{l} - \cosh r_n\frac{x}{l}\right) \tag{3.8}$$

$$\frac{l^2}{r_n^2}X_n'' = C_{1n}\left(-\cos r_n\frac{x}{l} + \cosh r_n\frac{x}{l}\right) + C_{2n}\left(-\cos r_n\frac{x}{l} - \cosh r_n\frac{x}{l}\right)$$
$$+ C_{3n}\left(-\sin r_n\frac{x}{l} + \sinh r_n\frac{x}{l}\right) + C_{4n}\left(-\sin r_n\frac{x}{l} - \sinh r_n\frac{x}{l}\right) \tag{3.9}$$

---

[1]For the complete analysis, the reader is referred to the original paper; however, it contains a number of typographical errors.

$$\frac{l^3}{r_n^3} X_n''' = C_{1n}\left(\sin r_n \frac{x}{l} + \sinh r_n \frac{x}{l}\right) + C_{2n}\left(\sin r_n \frac{x}{l} - \sinh r_n \frac{x}{l}\right)$$

$$+ C_{3n}\left(-\cos r_n \frac{x}{l} + \cosh r_n \frac{x}{l}\right) + C_{4n}\left(-\cos r_n \frac{x}{l} - \cosh r_n \frac{x}{l}\right)$$

(3.10)

The next differentiation gives

$$\frac{l^4}{r_n^4} X_n^{iv}$$

which is found to be equal to

$$\frac{l^4}{r_n^4} X_n$$

so that the original Winkler foundation differential equation, equation (2.86), is satisfied by the functions X_n.

The application of the method to a number of cases is discussed in detail by Hetenyi, but we will illustrate the approach by considering here only the case of the finite beam of length l supported on an elastic foundation and unrestrained (free) at both the ends $x = 0$ and $x = l$. In this case, the end conditions are that the end moments and shears are zero, i.e.,

$$w''(0) = 0 \qquad w'''(0) = 0$$
$$w''(l) = 0 \qquad w'''(l) = 0$$

(3.11)

Thus, from equation (3.6) we have

$$X_n''(0) = 0 \qquad X_n'''(0) = 0$$
$$X_n''(l) = 0 \qquad X_n'''(l) = 0$$

(3.12)

Looking back at the derivatives obtained above, we see that the first two of these conditions require that $C_{2n} = C_{4n} = 0$. The second two lead to the result that the ratio of C_{3n} to C_{1n} must be equal to a constant, say $-A_n$:

$$(-A_n) = -\frac{\cosh r_n - \cos r_n}{\sinh r_n - \sin r_n} = -\frac{\sinh r_n + \sin r_n}{\cosh r_n - \cos r_n}$$

(3.13)

This equality is satisfied if

$$\cosh r_n \cos r_n = 1$$

(3.14)

which determines the successive values of r_n, and thus, through equation (3.13), the values of A_n. In this case, Hetenyi finds that the r_n and A_n are given very nearly by the relations:

$$r_n = \frac{(2n + 1)\pi}{2}; \qquad A_n = 1.000$$

(3.15)

The deflection of a flexible beam supported only by the subgrade should include terms allowing for a vertical rigid-body translation and also rotation. With this in mind and considering the surviving constants C_{1n} and $C_{3n} = -A_n C_{1n}$ the deflection of the beam is given in general by the expression

$$
w = A + Bx + \sum_{n=1}^{\infty} C_n \left[\cosh r_n \frac{x}{l} + \cos r_n \frac{x}{l} \right.
$$
$$
\left. - A_n \left(\sinh r_n \frac{x}{l} + \sin r_n \frac{x}{l} \right) \right]
\tag{3.16}
$$

This equation can now be substituted in the integral for the determination of the strain energy in the beam and foundation. When the loading system is included, the constants A, B, and C_n can be evaluated by the method of minimizing the potential energy, as described earlier. It can be seen that this method has the further advantage of giving rise to only the three linear algebraic equations required for the determination of the three constants; *all* the C_n are obtained from the solution of these three equations.

In the present case, if the beam is subjected to a vertical load P applied at a distance c from the left end, the values of the constants are found to be

$$
A = \frac{2P}{kl}\left(2 - 3\frac{c}{l} \right)
\tag{3.17}
$$

$$
B = -\frac{6P}{kl^2}\left(1 - 2\frac{c}{l} \right)
\tag{3.18}
$$

$$
C_n = \frac{Pl^3}{r_n^4 EI + kl^4}\left[\cosh r_n \frac{c}{l} + \cos r_n \frac{c}{l} - A_n \left(\sinh r_n \frac{c}{l} + \sin r_n \frac{c}{l} \right) \right]
\tag{3.19}
$$

It will be observed that the term in brackets in equation (3.19) is of the same form as the summation term in equation (3.16). Consequently, if we were to tabulate the value of X_n in equation (3.7) and its successive derivatives for various r_n and A_n as a function of x/l in the range 0 to 1 for the case where $C_{2n} = C_{4n} = 0$, we would have all the values necessary to solve for deflections, moments, etc., in the present problem and all other problems for which $C_{2n} = C_{4n} = 0$. This has been done by Hetenyi whose table of values is reproduced in Table 3.1.

Thus C_n in equation (3.19) can be determined, for example, in the case where P is applied at distance c from the left end, by getting the value of X from the table at the appropriate ratio of c/l for each n and multiplying it by the load and material properties terms. To determine the deflection at a point x, the same table is used to give the value of X at the appropriate x/l for each n. Each term in the summation can then be constructed according to equation (3.16) and added to the linear part of the deflection to give the final deflection at x. If the moment at x were required, it would be obtained by multiplying the second derivative of w from equation (3.16) by $-EI$.

Thereafter, the same procedure as used to calculate w would be followed, with the same C_n, but we would employ values of $X''_n(x/l)$ from the table.

In the examples given by Hetenyi, inclusion of only the first term in the series gives deflections accurate to within a few percent, but it is suggested that for other problems at least the second term should be evaluated so that its contribution to deflection and moment can be assessed. For moment and shear, more terms are required than for deflection, when point loads are applied to the beam.

Although the solutions for other loading conditions can be calculated *ab initio*, we give here the results for one other case, that of the moment M_0 applied to the free-ended beam (as above) at the point $x = c$. Here

$$A = -\frac{6M_0}{kl^2} \tag{3.20}$$

$$B = \frac{12M_0}{kl^3} \tag{3.21}$$

$$C_n = \frac{M_0 l^2 r_n X'(c/l)}{r_n^4 EI + kl^4} \tag{3.22}$$

In his paper, Hetenyi gives solutions for beams with other forms of support and supplies additional tables of values of X_n when the other constants C_{mn} are zero.

3.3 COMPOSITE FINITE ELEMENT METHODS

Instead of setting up a complete finite element grid for the solution of certain problems, substantial economies in computing time may be obtained by doing some mathematical analysis prior to establishing the array of equations to be solved. It will frequently be possible to reduce the problem to few enough unknowns that a small (desk-sized) electronic computer with some programming capability can be used. In this method, which we will call a *composite finite element technique*, since it combines the approaches of mathematical analysis and finite element computations, the analytical results from an appropriate situation representing a subdivision of the problem at hand are combined with equilibrium and limited compatibility or consistency of deformation requirements to establish a system of equations for solution. The method is best illustrated by example.

Let us take the case of a beam of finite length resting on a subgrade and loaded on its upper surface. We consider beam and foundation material separately. The beam will be bent by the combination of the loads applied to its upper surface and the pressure distribution developed at the beam-soil contact by the subgrade. The loads are known, but the contact pressure distribution is not. Considering the soil subgrade, we can see it will be deflected by a pressure distribution equal to that which it imposes on the beam but in the

TABLE 3.1

Values of X_n, X_n', X_n'' and X_n''' for Free-free and Clamped-clamped Beams

$\frac{x}{\ell}$	X_1	X_2	X_3	X_4	X_5	X_1'	X_2'	X_3'	X_4'	X_5'
0.00	2.00000	2.00000	2.00000	2.00000	2.00000	−1.96500	−2.00155	−1.99993	−2.00000	−2.00000
0.05	1.53552	1.21590	0.90725	0.60415	0.31054	−1.96085	−1.98336	−1.95215	−1.90305	−1.83135
0.10	1.07433	0.45486	−0.10393	−0.58802	−0.96646	−1.93383	−1.87176	−1.67795	−1.38736	−1.00891
0.15	0.62399	−0.23500	−0.88309	−1.25518	−1.30025	−1.86666	−1.61620	−1.11054	−0.45011	0.25526
0.20	0.19545	−0.79450	−1.28572	−1.20092	−0.61048	−1.74814	−1.21002	−0.33199	0.58286	1.22851
0.25	−0.19839	−1.16950	−1.24229	−0.51204	0.55450	−1.57282	−0.68437	0.47744	1.27736	1.29326
0.30	−0.54401	−1.32402	−0.79387	0.45136	1.35061	−1.34076	−0.09872	1.10762	1.33056	0.43141
0.35	−0.82847	−1.24904	−0.08886	1.21281	1.20819	−1.05679	0.47145	1.38930	0.73172	−0.74127
0.40	−1.04050	−0.96605	0.65569	1.40010	0.22226	−0.73007	0.94823	1.24912	−0.22494	−1.39777
0.45	−1.17139	−0.52547	1.21525	0.91976	−0.91796	−0.37293	1.26438	0.73418	−1.07752	−1.08185
0.50	−1.21565	0.00000	1.42238	0.00000	−1.41386	0.00000	1.37532	0.00000	−1.41592	0.00000
0.55	−1.17139	0.52547	1.21525	−0.91976	−0.91976	0.37293	1.26438	−0.73418	−1.07752	1.08185
0.60	−1.04050	0.96605	0.65569	−1.40010	0.22226	0.73007	0.94823	−1.24912	−0.22494	1.39777
0.65	−0.82847	1.24904	−0.08886	−1.21281	1.20819	1.05679	0.47145	−1.38930	0.73172	0.74127
0.70	−0.54401	1.32402	−0.79387	−0.45136	1.35061	1.34076	−0.09872	−1.10762	1.33056	−0.43141
0.75	−0.19839	1.16950	−1.24229	0.51204	0.55450	1.57282	−0.68437	−0.47744	1.27736	−1.29326
0.80	0.19545	0.79450	−1.28572	1.20092	−0.61048	1.74814	−1.21002	0.33199	0.58286	−1.22851
0.85	0.62399	0.23500	−0.88309	1.25518	−1.30025	1.86666	−1.61620	1.11054	−0.45011	−0.25526
0.90	1.07433	−0.45486	−0.10393	0.58802	−0.96646	1.93383	−1.87176	1.67795	−1.38736	1.00891
0.95	1.53552	−1.21590	0.90725	−0.60415	0.31054	1.96085	−1.98336	1.95215	−1.90305	1.83135
1.00	2.00000	−2.00000	2.00000	−2.00000	2.00000	1.96500	−2.00155	1.99993	−2.00000	2.00000

TABLE 3.1 (Continued)

$\frac{x}{\ell}$	X_1''	X_2''	X_3''	X_4''	X_5''	X_1''	X_2''	X_3''	X_4''	X_5''
0.00	0.00000	0.00000	0.00000	0.00000	0.00000	0.00000	0.00000	0.00000	0.00000	0.00000
0.05	0.05160	0.13400	0.24694	0.38223	0.53246	0.41806	0.63116	0.79807	0.91666	0.98836
0.10	0.18910	0.45573	0.77005	1.07449	1.32178	0.72655	0.95776	1.01202	0.90088	0.65359
0.15	0.38675	0.84852	1.26761	1.49310	1.45002	0.92707	0.99764	0.72637	0.21017	-0.40502
0.20	0.61939	1.20674	1.50782	1.31923	0.67360	1.02342	0.79030	0.11050	-0.70122	-1.29164
0.25	0.86313	1.44486	1.37080	0.57035	-0.52789	1.02225	0.39794	-0.60491	-1.33577	-1.31986
0.30	1.09600	1.50550	0.86864	-0.42268	-1.33938	0.93338	-0.09916	-1.18057	-1.35944	-0.44262
0.35	1.29873	1.36498	0.13306	-1.19682	-1.20344	0.76976	-0.61167	-1.43035	-0.74612	0.73657
0.40	1.45545	1.03457	-0.62837	-1.39351	-0.22021	0.54723	-1.05271	-1.27099	0.21753	1.39584
0.45	1.55436	0.55724	-1.19633	-0.91715	0.91895	0.28401	-1.34960	-0.74365	1.07323	1.07503
0.50	1.58815	0.00000	-1.40600	0.00000	1.41457	0.00000	-1.45420	0.00000	1.41251	0.00000
0.55	1.55436	-0.55724	-1.19633	0.91715	0.91895	-0.28401	-1.34960	0.74365	1.07323	-1.07503
0.60	1.45545	-1.03457	-0.62837	1.39351	-0.22021	-0.54723	-1.05271	1.27099	0.21753	-1.39584
0.65	1.29873	-1.36498	0.13306	1.19882	-1.20344	-0.76976	-0.61167	1.43035	-0.74612	-0.73657
0.70	1.09600	-1.50550	0.86864	0.42268	-1.33938	-0.93338	-0.09916	1.18057	-1.35944	0.44262
0.75	0.86313	-1.44486	1.37080	-0.57035	-0.52789	-1.02225	0.39794	0.60491	-1.33577	1.31986
0.80	0.61939	-1.20674	1.50782	-1.31923	0.67360	-1.02342	0.79030	-0.11050	-0.70122	1.29164
0.85	0.38675	-0.84852	1.26761	-1.49510	1.45002	-0.92707	0.99764	-0.72637	0.21017	0.40502
0.90	0.18910	-0.45573	0.77005	-1.07449	1.32178	-0.72655	0.95776	-1.01202	0.90088	-0.65359
0.95	0.05160	-0.13400	0.24694	-0.38223	0.53246	-0.41806	0.63116	-0.79807	0.91666	-0.93836
1.00	0.00000	0.00000	0.00000	0.00000	0.00000	0.00000	0.00000	0.00000	0.00000	0.00000

opposite sense. The solution to the complete problem involves finding the soil-beam pressure distribution and the displacements in the system. As such, this is a difficult problem. It can be made, of course, as difficult as we like by considering such questions as: Is the beam attached to the soil so that shear stresses act at the interface? Is the beam flexible or rigid in the transverse direction? Should soil at the beam-soil interface be permitted to separate from the beam if the interface pressures become negative? And so on. As a separate topic, the soil properties may also be developed to some degree of complexity. The soil may, of course, be linear, isotropic, and homogeneous for the usual first cut at the problem, or varying degrees of inhomogeneity, anisotropy, or finally nonlinearity may be introduced to make the problem simultaneously more realistic and harder to solve. In the technique under discussion, some of these aspects may be introduced in a limited way without losing hope.

First we look at the idea of limited compatibility. When the beam deflects under generally vertical loads there will be, in any precise analysis, both vertical and horizontal displacements to take into account. Similarly, these will also develop when the underlying soil is subjected to pressure by the beam. Since, for design purposes, we are going to be interested in the moment in and the vertical deflection of the beam, we should first inquire if the horizontal displacements that will occur will have an important effect on these quantities. Without attempting to prove the point, we will invoke intuition to decide that for "normal" beams it is unlikely that horizontal displacements are important. A similar assumption is adduced for the Vlasov-Leontiev work in the following chapter. Therefore, it will not be necessary to require the horizontal displacements of beam and soil to match. The beam is considered, in effect, to be smooth at the beam-soil interface, and neither horizontal displacements nor shearing stresses at the interface will be taken into account. For other geometrical arrangements, similar considerations will apply. In addition, we may choose to let the pressure between the beam and soil be either positive only or both positive and negative (tensile). However, it is fair to assume that negative pressures will not occur because of the stresses due to the weight of the beam. Thus, it is required that beam and soil vertical deflections be compatible and that only vertical deflections be taken into account.

If we use simple beam theory and consider the soil subgrade to be a linearly elastic half-space, it will be possible to write general expressions for (a) the vertical deflection of the beam under the loading conditions and (b) the vertical displacement of the soil. These will involve the unknown pressure distribution at the interface. By requiring that the two sets of vertical displacements be everywhere compatible, we will arrive at an equation in terms of the unknown pressure distribution. This will generally be of integro-differential form, and its solution will, in general, not be easy. Things can be simplified if, as we have done before, we divide the problem up into finite rather than infinitesimal length intervals and consider that a uniform pressure (or in some circumstances, displacement) acts over each interval.

For illustration, we will work through a problem in which the beam is divided into four segments of equal length; this represents perhaps the easiest problem we can tackle, but which still has results of some interest. The configuration is shown in Figure 3-1(a) and (b) in elevation and plan views, respectively. We will load the beam at the center with a point load P and

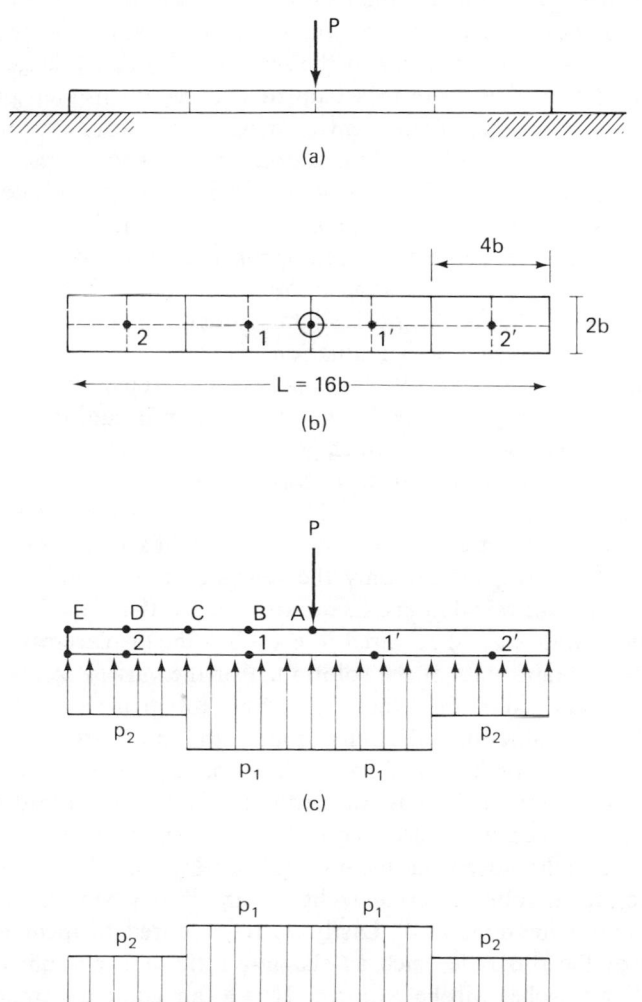

Figure 3-1. Beam subdivision for composite finite element analysis: (a) loaded beam, elevation; (b) plan view, showing dimensions for sample problem; (c) load distribution on beam; (d) pressure distribution on ground.

choose a beam whose length-to-width ratio is 8. If the beam is to be loaded at an arbitrary point other than the center, or by a moment rather than a vertical load, the method can still be employed. If the number of elements is to be kept fairly small for a hand or desk calculator computation, it may be necessary to make the elements of unequal size; this makes for only a little more arithmetic. As usual, the more elements assigned to the problem, the more accurate will be the solution, but the computational complexity increases. A few more elements than we have chosen here are still within the computational capability of a programmable desk-top computer.

Ten elements or more demand the services of a fairly large computer. It is obvious that if we were to set up this three-dimensional problem for solution by the complete finite element method, we would require perhaps several hundred or more three-dimensional elements to represent the beam and foundation soil. The final result would involve the solution of more than a thousand simultaneous algebraic equations. It is, of course, quite feasible to do this, but even with the appropriate program supplied, several hours of work are needed to establish the network of points and their coordinates. If a variety of problems of slightly different configurations or material properties (the usual design situation) is to be studied, some effort in establishing the initial network of points is usually repaid later. In finite element solutions, one always obtains not only the information of interest but also a vast heap of other data, which generally is of no value to the problem at hand and is always discarded. This includes displacements and stresses at all node points and elements in addition to the pertinent ones.

It is this last aspect that makes the composite discrete element approach attractive and useful; usually only the relevant stresses and displacements are employed in the calculations. Its disadvantage is that a distressing amount of analytical work may be required to establish the displacement-stress relations for the components of the solution. For the variety of beams that we are usually interested in, the difficulties are not too great; a number of books (4 and 11) gives tables of deflections, slopes, and moments in terms of standardized loading conditions. For the half-space, things are less formally available, but many of the relevant problems have been solved (5, 10, and 12). However, when we wish to consider problems in which the subgrade consists of a nonhomogeneous medium (elastic layer on rigid base or layered half-space), fewer solutions are in sight (3), and it may be necessary to resort to some approximations (13). Lastly, if it is desired to include nonlinear behavior of the subgrade, such as elastic-plastic deformation, intuitive or empirical approaches will be required. It is a fair guide to remark that if a solution to the desired beam on half-space load-deflection relation is not readily available, the problem should not be tackled by hand.

In each case, the function we are seeking is an *influence function*, which demonstrates the displacement effect, at a variety of points, of a load at one point. It may even be useful for some problems to generate the influence function by a full finite element solution for the performance of the subgrade if that material is to remain unchanged in subsequent analyses. The influence function thus obtained can be employed with a variety of beam configurations in subsequent composite solutions which consume less calculating time than would successive full finite element computations.

The beam of Figure 3-1 is considered to rest on an elastic half-space. For the symmetrical loading and subdivision selected for the problem, we will be concerned with two levels of uniform pressure p_1 and p_2 acting on the beam as shown in Figure 3-1(c). The pressures are assumed to be uniform across the width of the beam, which is, in consequence, taken to be flexible in this direction. This is not a very good postulate, since, in fact, the beam is more likely to be rigid in this direction, but it makes the analysis easier at this stage, and, moreover, it has been demonstrated that the difference between the assumptions of rigid and flexible cross sections has only a small effect on the moments in the longitudinal direction. The reader can choose the justification which best fits his or her philosophy.

Next, we must select points at which we will make the displacements compatible in the two separate systems. In addition to the notion of limited compatibility discussed earlier, we have now reduced each problem to discrete segments and we cannot expect that the form of the displacement functions will be the same in beam and subgrade; they can no longer be matched everywhere. Instead, we must limit our analysis still further to a limited discrete compatibility or consistency of displacements. Vertical displacements may be identified at any of the points A, B, C, D, or E in the two systems of Figure 3-1(c). Since we have two unknowns, we require two equations and thus two displacements. We might choose point pairs A and C, C and E, or B and D. Some logic may be associated with taking the vertical deflections at the midpoints of the loaded areas, so for our example we will employ the displacements at points B and D, which we will number 1 and 2, respectively; their mirror images are $1'$ and $2'$. Now let us look at the way the computation is going to proceed. Taking the subgrade first, we have a load p_1 extending from A to C and producing deflections w_{11} at point 1 (the first subscript indicates the point at which the deflection occurs; the second the load causing it), w_{21} at point 2, $w_{1'1}$ at point $1'$, and $w_{2'1}$ at point $2'$. These values are obtained from the influence line, or function, of vertical displacements at the surface of an elastic half-space arising from the imposition of a uniform loading over a rectangular area on the surface of the half-space. To the deflection produced at each point by p_1 we have to add the deflections caused by the other loaded areas p_2, $p_{1'}$ ($=p_1$), and $p_{2'}$ ($=p_2$) to get total deflections w_1 and w_2 at the two points 1 and 2. The deflections at points $1'$ and $2'$ will be the same as at points 1 and 2, respectively, by symmetry. These vertical displacements are absolute with respect to the undisplaced horizontal surface of the half-space far away from the loaded area.

Leaving the subgrade for the moment, we look at the beam. Again, because of the symmetry of the loading conditions, there will be zero slope to the beam at the midpoint A; therefore, the beam can be considered as if it were fixed or built-in (*encastré*) at A and free at E; only one-half of the beam need be analyzed. There are a number of books which give equations for the deflection of such a cantilever beam under a variety of loading conditions (4, 11). We may either employ such equations, being careful that they contain no errors (this is unfortunately frequently the case), or work out solutions analytically from elementary beam theory. The last procedure may be mandatory in the case of beams of variable cross section or subject to certain loading conditions. From either result, we obtain expressions for the total deflections at points 1 and 2 due to the loads p_1 and p_2 over the appropriate stretch of beam. We may notice that the beam deflections are not absolute in the same sense as are the subgrade displacements. The beam movements are referred to the zero deflection at point A, but point A can be anywhere; in our problem it does not correspond to the free surface level of the half-space but to the deflection at the center of the loaded area, a displacement which we have not evaluated. Thus, we can only say that the *difference* between the deflections at points 1 and 2 of the beam must match the corresponding difference between the half-space displacements at points 1 and 2. This gives one equation. So far we have not used the load P in the calculation. Obviously, it enters from the point of view of equilibrium. The total load due to the pressures p_1 and p_2 must equal P; this is the other equation here. From the two simultaneous equations the two pressures p_1 and p_2 can be calculated. Once they are known, the deflections, moments, and shears follow. We proceed with the calculations.

(a) Half-space displacement. The soil surface is displaced vertically by uniformly distributed loads acting over rectangular areas at the surface. Solutions for the vertical settlement at the corner of such a loaded area have been published by a number of authors (5, 10, and 12). It is convenient to write the displacement in the form

$$w = \frac{B(1 - v_s^2)p}{E_s} w_0\left(\frac{L}{B}\right) \tag{3.23}$$

where B is the width and L the length of the area loaded by p, v_s and E_s are the usual elastic properties of the subgrade, and $w_0\,(L/B)$ is a settlement function which depends only on the ratio of foundation length to width. It has been tabulated in the references cited above for cases both of the half-space and an elastic layer over a rigid surface. In the latter case, w_0 also depends on the ratio of foundation width to layer thickness. Looking at the subdivision of the problem in Figure 3-1(b), we see that the settlement at point 1 due to the load p_1 can be obtained from the summation of the displacements at the corners of four rectangular areas loaded with p_1 and of dimension $2b$ by b. The ratio L/B is thus 2 and we have

$$w_{11} = \frac{4b(1 - v_s^2)p_1}{E_s} w_0(2) \tag{3.24}$$

In this analysis it is a waste of time to try for too much precision, and we will therefore employ values of w_0 to two decimal places only. From tables in references (10, 12) we find that w_0 (2) is equal to 0.76, and so we have

$$w_{11} = \frac{2b(1 - v_s^2)}{E_s}(1.52p_1) \tag{3.25}$$

By adding and subtracting deflections below the corners of rectangularly loaded areas (in the fashion usually employed for calculating vertical stresses below foundation slabs in settlement analyses) we can arrive similarly at the other deflections:

$$w_{12} = \frac{2b(1 - v_s^2)}{E_s}(0.34p_2) \tag{3.26}$$

$$w_{11'} = \frac{2b(1 - v_s^2)}{E_s}(0.34p_1) \tag{3.27}$$

$$w_{12'} = \frac{2b(1 - v_s^2)}{E_s}(0.16p_2) \tag{3.28}$$

Summing equations (3.25) through (3.28) gives the total displacement at point 1, w_1:

$$w_1 = \frac{2b(1 - v_s^2)}{E_s}(1.86p_1 + 0.50p_2) \tag{3.29}$$

Going through the same process for point 2, we eventually get

$$w_2 = \frac{2b(1 - v_s^2)}{E_s}(0.50p_1 + 1.61p_2) \tag{3.30}$$

Equations (3.29) and (3.30) are the two half-space displacement-pressure relations we need for the solution of the problem. We can see that they are written here in the following form: displacement equals compliance times load. It would, of course, be possible to invert the equations to obtain the loads in terms of a stiffness times a displacement. Either form can eventually be used, but in the present case, operating on equations (3.29) and (3.30) preserves the physical interpretation.

(b) Beam displacements. The beam is displaced vertically by uniformly distributed loads acting over portions of its surface. As with the soil subgrade, we consider the problem in steps. First we consider the cantilever beam fixed at point A of Figure 3-1(c) and subjected to the load p_1 alone acting over one-half its span, the distance $4b$. We calculate the deflections at points 1 and 2 due to this load by using displacement equations (4, 11). The result, after substituting in the appropriate lengths, reduces to the form

$$w_{11} = -\frac{bl^4}{24EI}\left(\frac{17}{128}p_1\right) \tag{3.31}$$

where E and I are the modulus and moment of inertia, respectively, of the beam, and l is the *half* length of the beam [distance AE in Figure 3-1(c)]. The minus sign indicates the upward displacement. The deflection at point 2 due to p_1, w_{21}, is given by the equation

$$w_{21} = -\frac{bl^4}{24EI}\left(\frac{5}{8}p_1\right) \tag{3.32}$$

For the additional deflections at points 1 and 2 produced by the load p_2 acting on the outer half of the cantilever, we get

$$w_{12} = -\frac{bl^4}{24EI}\left(\frac{1}{2}p_2\right) \tag{3.33}$$

and

$$w_{22} = -\frac{bl^4}{24EI}\left(\frac{433}{128}p_2\right) \tag{3.34}$$

Thus, the total deflection at point 1, w_1, is

$$w_1 = -\frac{bl^4}{24EI}\left(\frac{17}{128}p_1 + \frac{1}{2}p_2\right)$$

or

$$w_1 = -\frac{bl^4}{24EI}(0.13p_1 + 0.50p_2) \tag{3.35}$$

and at point 2, w_2, is

$$w_2 = -\frac{bl^4}{24EI}(0.63p_1 + 3.38p_2) \tag{3.36}$$

(c) Equilibrium. The equilibrium condition is easily written down:

$$P = 8b \times 2b \times p_1 + 8b \times 2b \times p_2$$

or

$$\frac{P}{16b^2} = p_1 + p_2 \tag{3.37}$$

(d) Compatibility of beam and soil displacements. We pointed out earlier that the two equations describing the problem would be obtained from the equilibrium condition, equation (3.37), and the requirement that the *difference* between the displacements at points 1 and 2 would be the same for soil and beam. For the soil, the difference is $w_1 - w_2$, since w_1 has the greater deflection; thus,

$$(w_1 - w_2)_{\text{soil}} = \frac{2b(1 - v_s^2)}{E_s}(1.36p_1 - 1.11p_2) \tag{3.38}$$

With the beam, since w_2 is the larger displacement, we have to calculate $w_2 - w_1$ to get

$$(w_2 - w_1)_{\text{beam}} = \frac{bl^4}{24EI}(0.49p_1 + 2.88p_2) \tag{3.39}$$

Equating the right sides of equations (3.38) and (3.39), and rearranging, we have

$$0.49p_1 + 2.88p_2 = \frac{768EI(1 - v_s^2)}{E_s L^4}(1.36p_1 - 1.11p_2) \tag{3.40}$$

It is convenient to substitute F for the coefficient of the right-hand side of equation (3.40), which can then be manipulated to give

$$p_2 = \frac{1.36F - 0.49}{1.11F + 2.88}p_1 = Ap_1 \tag{3.41}$$

where

$$A = \frac{1.36F - 0.49}{1.11F + 2.88} \tag{3.42}$$

and both A and F are parameters indicating the relative stiffness of beam and soil. The larger the value of both parameters, the stiffer is the beam compared to the soil. For a completely rigid beam, A has its maximum value of $1.36/1.11 = 1.23$. For a very flexible beam relative to the soil, the value of F tends toward zero and A takes its minimum value of $-0.49/2.88 = -0.17$.

Substituting equation (3.41) in equation (3.37), we find that

$$\left. \begin{aligned} p_1 &= \frac{1}{(1 + A)}\frac{P}{16b^2} \\ p_2 &= \frac{A}{(1 + A)}\frac{P}{16b^2} \end{aligned} \right\} \tag{3.43}$$

or

$$p_1 = \frac{1}{(1+A)Lb}P = \frac{1}{(1+A)}\frac{16P}{L^2} \left.\vphantom{\frac{1}{1}}\right\}$$
$$p_2 = \frac{A}{(1+A)Lb}P = \frac{A}{(1+A)}\frac{16P}{L^2} \left.\vphantom{\frac{1}{1}}\right\}$$

$$(3.44)$$

Examining the maximum and minimum values of A, we can see that p_1 and p_2 have extreme values as follows:

Rigid: $\quad \left\{ p_1 = 0.45\frac{P}{16b^2} \right.$ Flexible: $\left\{ p_1 = 1.21\frac{P}{16b^2} \right.$
$A = 1.23 \quad \left\{ p_2 = 0.55\frac{P}{16b^2} \right.$ $A = -0.17 \quad \left\{ p_2 = -0.21\frac{P}{16b^2} \right\}$ (3.45)

Thus, for a rigid beam ($A = 1.23$), the pressure p_2 at the ends of the beam is higher than the pressure in the middle section. We recognize that this is the correct trend from analysis of the interaction of rigid punches with an elastic half-space. On the other hand, under the central load P, the very flexible beam tries to lift up at the ends and is restrained by a negative pressure p_2. In this case, the middle-section pressure p_1 is high.

Of interest in the design of the beam is the maximum moment, developed, in this case, at the loading point. Considering the pressures illustrated in Figure 3-1(c), we obtain the moment as follows:

$$M_{max} = 16b^3p_1 + 48b^3p_2 \qquad (3.46)$$

Substituting equation (3.43) in this equation gives

$$M_{max} = \left(\frac{A}{1+A} + \frac{3A}{1+A} \right)Pb$$
$$= \frac{(1+3A)}{(1+A)}Pb = \frac{(1+3A)}{(1+A)}\frac{PL}{16}$$

$$(3.47)$$

This equation shows how the maximum moment in the beam varies with the stiffness of the beam with respect to the soil. Substituting the extreme values of A discussed in equation (3.47) indicates the range of maximum moments possible. First, for the rigid beam, $A = 1.23$, we find $M_{max} = 0.131PL$. With the very flexible beam, $A = -0.17$ and $M_{max} = 0.037PL$.

It is of interest to see how the values of maximum moment calculated from such a crude model compare with those obtained by other, more exact analyses. There is a difficulty here in that the *exact* analysis of a finite beam on a half-space has not been performed. The nearest available solution is that of Biot (1) who studied the problem of an infinitely long beam on a half-space. Biot's solution would come closest to ours for the case of the very flexible beam. By a method identical to that described here, Brown (2) has studied the same problem of a beam loaded with a point load at a range of positions on the beam. In his case, the length-to-width ratio of the beam was 10. Included is the case of the centrally located load, with which we can compare our solution. Brown expressed his results in terms of the parameter K, where

$$K = \frac{16EI(1 - v_s^2)}{\pi E_s L^4}$$
(3.48)

This conveys the same interdependence of beam and soil properties as our coefficient F in equation (3.41). Solutions were obtained by Brown for values of $K = 0.001$ (flexible beam), 0.01, 0.1, and 100 (essentially rigid beam). We will use Brown's results for one comparison.

For our relatively poor model, the response of the foundation soil has been taken from that of an elastic half-space. The alternate, and most commonly employed method of representing a soil subgrade is the *Winkler* foundation, in which the soil is represented as a bed of springs, with the spring constant, or coefficient of subgrade reaction k_0. This is treated extensively in Chapter 5 where it is shown that identification of the maximum moment obtained from Biot's solution with that calculated by the analytical solution for an infinite beam on a Winkler subgrade requires that

$$k = k_0 B = \frac{0.95 E_s}{(1 - v_s^2)} \left[\frac{E_s B^4}{(1 - v_s^2)EI} \right]^{0.108}$$
(3.49)

The maximum moment in an infinite beam on a Winkler subgrade and loaded by a point load is given by equation (5.20). The maximum moment in a centrally loaded finite beam on a Winkler subgrade is given in Appendix C for various ratios of beam-to-soil stiffness. The extreme moment occurs in the case of a rigid beam. For it the pressure distribution is uniform and equal to P/BL, so that the maximum moment is $PL/8$.

The maximum moments from these different solutions can be compared with that from our elementary composite finite element analysis described above, as shown in Tables 3.2 and 3.3.

TABLE 3.2

Finite Beam

| | | | M_{max}/PL | | |
| | | λL | | Elementary | |
K (Brown)	A	(Winkler)	Brown	Analysis	Winkler
0.001 (flexible)	−0.17	5.79	0.046	0.037	0.0431
0.01	0.34	3.06	0.093	0.094	0.0890
0.1	1.14	1.62	0.13	0.129	0.1205
100 (stiff)	1.23	0.24	0.14	0.131	0.125

TABLE 3.3

Infinite Beam

K (Brown)	A	Brown	M_{max}/Pb Elementary Analysis	Biot	Winkler
0.001 (flexible)	−0.17	0.92	0.60	0.863	0.862

For the infinite beam, the moment has to be expressed in terms of Pb rather than PL. Since the ratio of beam length L to *half*-width b in Brown's calculations is 20, and those of this section is 16, the moments from these solutions can be expressed in the alternate form M_{max}/Pb by substituting in the values from the top line of Table 3.2 to give Table 3.3. Since in Table 3.3 the maximum moment is expressed in terms of the half-width of the beam, b, the results there have been obtained from the top row of Table 3.2 by multiplying by the appropriate ratio of L/b: 20 in the case of the Brown and Winkler moments and 16 for the moment from the elementary analysis. The Winkler moment for the finite beam was obtained by using the value of k from equation (3.49). Since equation (3.49) was obtained by making the Biot and Winkler infinite beam maximum moments equal, it is to be expected that the Biot and Winkler moments in Table 3.3 are closely similar.

What is interesting is the remarkable correspondence between the results of the elementary analysis and the other values. Except for the extremely flexible beam, the very approximate solution is within a few percent of the ten-element solution of Brown. It would appear that the equations given above for the problem of the centrally loaded beam of length of the order of 10 times its width are good enough to be used for design purposes. It can also be seen that the value of k required to equate the maximum moments in the two infinite-beam cases also gives acceptable values of the maximum moment in the case of the finite beam on the Winkler subgrade.

If additional accuracy is required in some circumstances in the approximate analysis of the composite finite element type, there are two ways to go. (This is also true in the finite element approach.) One is to increase the number of elements into which the problem is divided: from four in our simplified analysis to six, ten, or more. Obviously, this increases the computational time. The other is to refine the analytical model used as the basis for analysis. If a solution could be found or calculated for the surface settlement under one corner of a rectangular area subjected to a vertical pressure increasing linearly from one side to the other, a better representation of the soil-beam interaction pressure, as shown in Figure 3-2, would be obtained. In this case, three unknown pressures would be required to describe the interaction conditions, and thus three simultaneous equations would result.

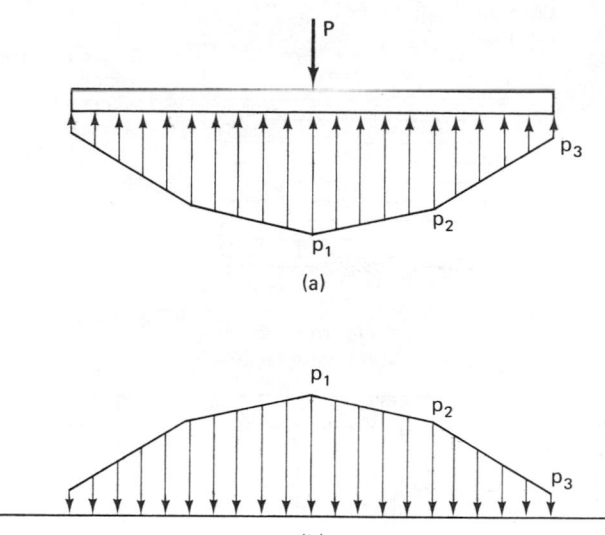

Figure 3-2. Alternative loading scheme for composite finite element analysis: (a) load distribution on beam; (b) pressure distribution on ground.

The technique described in this chapter has been used by Poulos (8 and 9) in his studies of pile-soil interactions. Poulos obtains his subsurface influence functions from integration of Mindlin's (7) equations. Selection of ten elements to represent the pile in such problems required the use of a computer, but, as indicated here, the solutions are much more economical than complete finite element analyses. The results of Poulos' investigations are examined in Chapter 8.

PROBLEMS

1. A method of reinforcing earth embankments has been suggested, as shown in Figure P3-1. If you were to employ this method, how would you calculate (a) the thickness and spacing of the reinforcing strips, (b) the stability of the slope? Is it preferable to have the strips rigid longitudinally with respect to the soil?

2. An anchored sheet pile wall in sand is installed by driving the interlocking sheet pile elements, laying out tie-rods to anchors, and finally excavating the soil. Considering this sequence of events, suggest a design/analysis method to arrive at the maximum moment in the wall. You may base your approach on any *one* of the techniques of finite elements, finite differences, or Winkler foundation models. Consider carefully how the loading applies to the wall. See Figure P3-2.

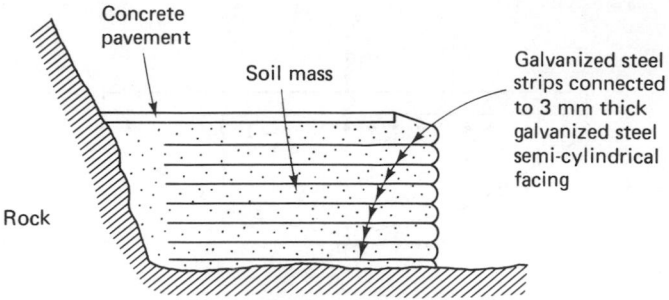

Figure P3-1.

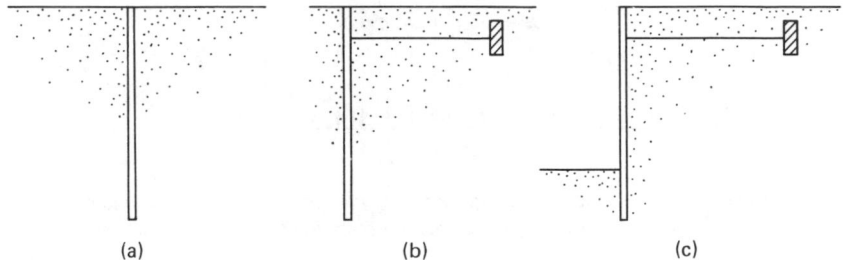

(a)   (b)   (c)

Figure P3-2. (a) Driving; (b) Anchors installed; (c) Excavation complete.

3. Perform a composite finite element analysis for the beam footing of Section 5.5.5, in which computations were carried out according to Baker's method, by estimating a value for the soil E_s from one of the equations of Section 5.2.4. Compare your maximum moment with those obtained in that section by both Baker's approach and using the Winkler model. Discuss the difference in terms of the equation you used to relate k and E_s. Take $v_s = 0.3$ and the beam properties as in Section 5.5.5.

4. Calculate the deflections w_1 and w_2 from your analysis of Problem 3, and estimate from them the maximum and differential settlement to compare with the solution of Section 5.5.5. What, in this case, was the effect of the equation relating k and E_s?

5. For a centrally-loaded rigid beam on a Winkler foundation, calculate the maximum (central) moment and deflection. The beam has the proportions of the composite finite element beam of Section 3.3. Derive an expression relating k and E_s by comparing your results with those of Section 3.3 in which the beam is taken to be rigid. How does your result compare with those of Vesic of Section 5.2.4?

6. Carry through a composite finite element analysis for the beam of Section 5.5.5, but apply the load to the *end* of the beam.

7. Analyze, by the composite finite element method, the second of the two examples of Baker's analysis in Section 5.5.5 for the case where the subgrade reaction coefficient k_0 is uniform and has the value 56 pci (15.2 MN/m³).

Use one of the equations of Section 5.2.4 (or some other method, such as relating k_0 to an estimated soil type and properties) to establish a value for E_s in your problem. Assume $v_s = 0.3$ and the beam properties as in Section 5.5.5 (also in Problem 3). This is a more difficult problem than before, and it will be necessary to choose three reaction pressures. Take advantage of the symmetry with a coordinate x ft ($0.3x$ m) from the center in Figure 5-19, and in it is suggested that p_1 run from $x = 0$ to $x = 5$ (1.5 m), p_2 from $x = 5$ (1.5 m) to $x = 15$ (4.6 m), and p_3 from 15 to 20 ft (4.6 to 6.1 m). Assume the beam terminates at the end column load. Three deflections must be calculated; they might conveniently be located at points with coordinates $x = 2.5$, 10, and 17.5 ft (0.76, 3.05, and 5.33 m).

8. Independently solve the problem of the beam on the elastic foundation, loaded centrally by P, using Hetenyi's approach, equation (3.16), and performing the minimization of the potential energy. Check the constants you obtain with the specialized forms of equations (3.17), (3.18), and (3.19).

9. Evaluate the approximate solution, equation (3.16), for (a) a centrally-loaded beam and (b) a beam loaded at the left end, by comparing deflection and moment at the center with those obtained from the exact Winkler solutions. First use one, and then two terms in your solution. Note the symmetry of the problem. How much improvement does a second term give?

10. Solve the general problem of a loaded beam on a Winkler foundation by Hetenyi's method when the ends of the beam are pinned.

11. If the moment of inertia of a foundation beam changes continually along its length, or abruptly because the beam width is greater over some portion of its length, standard solutions are no longer available in reference works. Assume the moment of inertia of the beam of Figure 3-1 is twice as great over the portion on which p_1 acts than elsewhere, and obtain the *beam* response by a finite difference method.

12. How would you modify Hetenyi's solution technique if the coefficient of subgrade reaction varied linearly along the length of the beam?

13. Compare the one-term maximum moment in the solution to Problem 9 with that obtained in the composite finite element solution. How does the computational effort compare in the two techniques?

14. How would you carry out a composite finite element solution for the case of a pile embedded in an elastic half-space and subjected to: (a) an axial load; (b) a lateral load at the top?

REFERENCES

[1] BIOT, M. A. "Bending of an Infinite Beam on an Elastic Foundation," Transactions ASME, *Journal of Applied Mechanics, 59,* A1–A7 (1937).

[2] BROWN, P. T. "Strip Footings with Concentrated Loads on Deep Elastic Foundations," University of Sydney, School of Civil Engineering Research Report No. R225, Sept. 1973.

[3] BURMISTER, D. M. "Stress and Displacement Characteristics of a Two-Layer

Rigid Base Soil System: Influence Diagrams and Practical Applications," *Proceedings Highway Research Board, 35,* 773 (1956).

[4] FLÜGGE, E., ed., *Handbook of Engineering Mechanics.* New York: McGraw-Hill, 1962 (Chapter 32).

[5] HAMPTON, D., SCHIMMING, B. B., AND E. L. SKOK, JR. "Stress and Displacement Bulletin," Report of Subcommittee SGF-B5C1, National Academy of Science Highway Research Board (Jan. 1966).

[6] HETENYI, M. "Series Solutions for Beams on Elastic Foundations," *Journal of Applied Mechanics,* 507–514 (June 1971).

[7] MINDLIN, R. D. "Force at a Point in the Interior of a Semi-Infinite Solid," *Physics 7,* 195–202 (1936).

[8] POULOS, H. G. "Torsional Response of Piles," *Proc. ASCE, 101,* GT10, 1019–1035 (Oct. 1975).

[9] ———, AND E. H. DAVIS. "The Settlement Behavior of Single Axially-Loaded Piles and Piers," *Geotechnique, 18,* 351–371 (1968).

[10] ———. *Elastic Solutions for Soil and Rock Mechanics.* New York: John Wiley, 1974.

[11] ROARK, R. J. *Formulas for Stress and Strain,* 3rd ed. New York: McGraw-Hill, 1954.

[12] SCOTT, R. F. *Principles of Soil Mechanics.* Reading, Mass.: Addison-Wesley, 1963.

[13] STEINBRENNER, W. "Tables for Settlement Calculations," *Die Strasse, 1,* 121–124 (1934) (in German).

Elastic Foundations: Beams

4

Gravitational and other loads are transferred through a steel or concrete structure into the soil foundation, which is required to support the super-structure without failing and with settlements inside certain acceptable limits. The structure in contact with the soil can be as apparently simple as an asphalt or concrete pavement, loaded by vehicular or aircraft wheels, or it can be the complicated basement of a multistory building. In any event, at the most basic level we would like to know the distribution of stresses and displacements resulting from the imposition of a load in an area at or near the surface of a half-space. For simple individual footings, this information may be sufficient; in more complex circumstances, it can be employed as a component of the solution, as in the composite finite element method dis-cussed previously. The loaded area, in practice, may have any shape, but it can usually be reduced to a circular or elliptic area (in the case of tires, for example) or combinations of rectangular regions where building foundations are concerned. The area may not be uniformly loaded nor may the load in practice be all normal to the surface. Although in many cases, including wheels, machine foundations, and seismic conditions, the dynamic behavior of the soil foundation is important, we will consider only static loads here.

The material forming the half-space will be nonhomogeneous, although we may ignore this in analysis, and it will, in practice, consist of layers of material of different constitutive relations and properties. Also, it usually is anisotropic. Since in real life the appropriate constitutive relations will be complex and unknown or even undeterminable, simplifications must be made. For mathematical analysis, therefore, all the soil subjected to stress is taken to be linearly elastic or rigid-plastic in its behavior. If numerical analysis by means of a digital computer is undertaken, the material response may be represented as elasto-plastic, bilinear, or multilinear. Such limited nonlinear behavior, however, is frequently based on the axial force-deformation relations observed in triaxial tests; the general three-dimensional response is unknown and ignored.

The simplest idealized class of structure-soil problem is that in which a loaded area is applied to a homogeneous half-space. The first relevant question is whether or not the soil will yield under the loaded area. In this case, the material property is taken as rigid-plastic, and all the results of soil mechanics bearing capacity theory can be brought to bear to achieve an answer. We will not discuss this further here, but it will be mentioned in a later section. Similarly, no attention will be given to the related question of how many repetitions of a load somewhat smaller than the failure load are required to produce unacceptably large permanent deformations of the soil, since this also falls into the realm of constitutive relations.

A pavement (highway or airfield) consists of, for example, a layer or layers of stiffer and stronger material, the "structure," overlying the natural soil subgrade, the "foundation." We must try to assess for a given configuration the stresses that the load will generate in the system and see if they exceed the yield value for the material. If they do, the layer thicknesses must be adjusted until yield does not occur under the design load (including a load factor). At this stage the resulting deflections are examined to ensure that they are within acceptable limits. This is the design process. One possible design would be that in which the design load just stressed each layer up to its yield stress simultaneously. Within the confines of our ability to characterize the materials present, this implies that we should perform a rigid-plastic or elasto-plastic analysis of the continuum for the postulated design. This would be very difficult and has not been done for this problem. Instead, for analysis, the material is generally taken to be linearly elastic. The stresses which result are then required to be lower than those which would cause yielding in each level of material (including the case of cracking in concrete if it constitutes the upper layer). A solution does not hold beyond the load which first causes yielding in one of the materials, so that a lower bound on the load is given. In practice, this load may be used in order to give adequate protection against the effects of repeated loads, for which an analysis cannot be carried out.

For a linearly elastic material, where all layers have the same elastic properties, this is the Boussinesq problem. The solution is extensively discussed in soil mechanics books. When two or three layers of differing but still linearly elastic properties are present, the case reasonably approximating to practice, the solution was obtained by Burmister (1). Even with tabulation of some of the computed values for a range of cases, the solution is complicated and it is not apparent that it has ever been employed in design. Currently, it would be faster and more convenient to obtain numerical (finite element) solutions to specific cases than it would be to use Burmister's results. However, finite element solutions are not cheap, particularly if trial-and-error processes are employed, and substantial costs might be involved before a satisfactory design evolved in a particular case. A guide to the selection of either preliminary or final designs is required. Within a theoretical framework, the only way this can be done is through a further simplification of the material model.

A calculational technique therefore is required that concentrates on the behavior of the upper stiff layer, which we consider here to be the *structure*, and relegates to the soil only a supporting role. It is not possible to do this by using the continuum representation of the half-space. A convenient model substitution was presented by Winkler in 1867[1] and has been widely used since then; the foundation material representation which he adopted is now called a *Winkler foundation*. It has been of value in many diverse circumstances and will therefore be discussed in some detail in this book.

In this model the deflection of the ground at a point, w, is assumed to depend only on the pressure, p, acting at that point through a proportionality constant, k_0, so that

$$p = k_0 w \qquad (4.1)$$

The constant, k_0, has been given a variety of names. It is the foundation stiffness, but it has been called the *coefficient of subgrade reaction* or *subgrade modulus*. Its dimensions are FL^{-3}. Winkler's assumption was first used to analyze the stresses and deflections of railroad tracks. Since in the Winkler representation only one parameter is used to describe the behavior of a half-space, which at the least requires two descriptive numbers (for example, E and v), there are difficulties in assuring the correctness of calculated quantities. It is usually employed in a semi-empirical way. Variations or improvements on the spring model have been suggested to enable us to make adjustments or corrections. Most of these have been postulated on an intuitive basis, but a theoretical justification, due to Vlasov and Leontiev (8), is available. Since it is a logical approach and permits us to solve a variety of inter-

---

[1]See footnote 1 in Chapter 5.

esting and relevant problems, the method of Vlasov and Leontiev will be
presented before further discussion on the subject of structures or the
Winkler foundation is continued.

4.2 ELASTIC FOUNDATION ANALYSIS ACCORDING TO VLASOV AND LEONTIEV

The Vlasov-Leontiev approach will be discussed with reference to
Figure 4-1, which shows a plane strain section of a foundation layer of depth
H and elastic properties E and v; it is subjected to an arbitrary load $p(x)$ at
its upper surface. The displacements u and w correspond to the coordinate
directions x and z as shown in the diagram. The section is taken to have unit
thickness perpendicular to its plane. One method of solving this problem in
terms of displacements, strains, and stresses within the scope of the theory of
linear elasticity is to represent the displacements by the finite series

$$u(x, z) = \sum_{i=1}^{m} U_i(x)f_i(z) \qquad (i = 1, 2, 3, \ldots, m)$$

$$w(x, z) = \sum_{k=1}^{n} V_k(x)g_k(z) \qquad (k = 1, 2, 3, \ldots, n)$$

(4.2)

in which $f_i(z)$ and $g_k(z)$ are dimensionless functions representing the variation
of the displacement in the z-direction, and $U_i(x)$ and $V_k(x)$ have the dimension of length and indicate the magnitude of the displacement in the x-direc-
tion. Vlasov and Leontiev assume a form for the functions f and g in general,

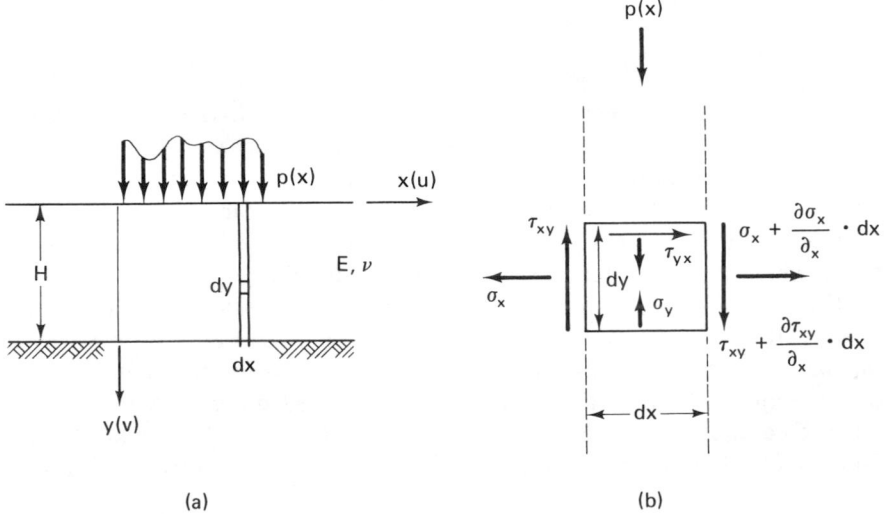

(a)   (b)

Figure 4-1. Approximate analysis after Vlasov and Leontiev: (a) load applied
to linearly elastic layer; (b) stresses on an element.

and then they proceed to calculate U and V. By truncating the series, equation (4.2), at a limited number of terms, an approximate solution of the elastic problem is obtained. The series can be made as simple as required to reduce the calculational effort involved. Depending on the loading and boundary conditions, the functions f and g can be taken to be constant, to vary linearly or exponentially with depth, to be sinusoidal functions, or to be in a variety of other forms, a number of which are examined by Vlasov and Leontiev.

The procedure is to assume forms for f and g and to obtain the functions U and V by a variational technique. The basic equations may be derived by considering the principle of virtual work [equation (A.59)]:

$$\delta U_v - \delta W = 0 \qquad (4.3)$$

where δU_v and δW are the increments in strain energy and applied body and surface forces developed by a small or virtual arbitrary displacement. We can express equation (4.3) in the usual way, with the strain energy term first:

$$\int \sigma_{ij}\, \delta\epsilon_{ij}\, dv - \int F_i\, \delta u_i\, dv - \int_S T_i\, \delta u_i\, dS = 0 \qquad (4.4)$$

in which F_i are the body forces and T_i are the surface forces acting on the boundary S. Since we are considering a plane strain problem here, we represent the volume element dv as $dx \cdot dz \cdot 1$ and can then proceed to write the individual terms of equation (4.4) as follows:

$$\int \sigma_{ij}\, \delta\epsilon_{ij}\, dv = \int_{-\infty}^{+\infty}\int_0^H \sigma_x\, \delta\epsilon_x\, dx\, dz + \int_{-\infty}^{+\infty}\int_0^H \sigma_z\, \delta\epsilon_z\, dx\, dz$$
$$+ \int_{-\infty}^{+\infty}\int_0^H \tau_{xz}\, \delta\gamma_{xz}\, dx\, dz \qquad (4.5)$$

However, we can express the stresses in terms of the strains by the usual elastic relations, here set out in terms of the plane strain condition:

$$\sigma_x = \frac{E}{(1+v)(1-2v)}[(1-v)\epsilon_x + v\epsilon_z]$$

$$\sigma_z = \frac{E}{(1+v)(1-2v)}[v\epsilon_x + (1-v)\epsilon_z] \qquad (4.6)$$

$$\tau_{xy} = \frac{E}{2(1+v)}\gamma_{xz}$$

In addition, we have

$$\epsilon_x = \frac{\partial u}{\partial x} = \sum_{i=1}^m \frac{\partial U_i}{\partial x} \cdot f_i$$

$$\epsilon_z = \frac{\partial w}{\partial z} = \sum_{k=1}^n V_k \frac{\partial g_k}{\partial z} \qquad (4.7)$$

$$\gamma_{xz} = \frac{\partial u}{\partial z} + \frac{\partial w}{\partial x} = \sum_{i=1}^m U_i \frac{\partial f_i}{\partial z} + \sum_{k=1}^n \frac{\partial V_k}{\partial x} \cdot g_k$$

Using the basic equation (4.2) for the displacements and the derivatory expressions, equation (4.7) for the strains, we obtain the virtual strains:

$$\delta\epsilon_x = \delta\sum_i \frac{dU_i}{dx} f_i = f_j \frac{d}{dx}(\delta U_j)$$

$$\delta\epsilon_z = \delta\sum_k V_k \frac{dg_k}{dz} = \frac{dg_h}{dz}\delta V_h$$

$$\delta\gamma_{xz} = \delta\left(\sum_i U_i \frac{df_i}{dz} + \sum_k \frac{dV_k}{dx}g_k\right) \tag{4.8}$$

$$= \frac{df_j}{dz}\delta U_j + g_h \frac{d}{dx}(\delta V_h)$$

which can be substituted, in turn, into each term on the right-hand side of equation (4.5). Thus,

$$\int_{-\infty}^{\infty}\int_0^H \sigma_x\,\delta\epsilon_x\,dz\,dx = \int_0^H\int_{-\infty}^{\infty}\sigma_x f_j \frac{d(\delta U_j)}{dx}\,dx\,dz \tag{4.9}$$

$$= \delta U_j \int_0^H \sigma_x f_j\,dz\Big|_{x=-\infty}^{x=+\infty} - \int_{-\infty}^{\infty}\int_0^H \frac{\partial\sigma_x}{\partial x} f_j\,\delta U_j\,dz\,dx$$

upon integration by parts. The boundary conditions of the problems we are interested in, with localized surface loads, require that $\delta U_j = 0 = \delta V_h$ at $x = \pm\infty$, so that only the double integral remains. By applying the same integration process to the other terms of equation (4.5), we obtain

$$\iint \sigma_z\,\delta\epsilon_z\,dx\,dz = \iint \sigma_z \frac{dg_h}{dz}\delta V_h\,dx\,dz \tag{4.10}$$

$$\iint \tau_{xz}\,\delta\gamma_{xz}\,dx\,dz = \iint \tau_{xz}\frac{df_j}{dz}\delta U_j\,dx\,dz - \iint g_h\frac{\partial\tau_{xz}}{\partial x}\delta V_h\,dx\,dz \tag{4.11}$$

In addition, the body and surface force terms can be expanded in the following form:

$$-\int F_i\,\delta u_i\,dv = -\int p(x,z)f_j\,\delta U_j\,dx\,dz - \int q(x,z)g_h\,\delta V_h\,dx\,dz \tag{4.12}$$

and

$$-\int_s T_i\,\delta u_i\,dS = -\int_{-\infty}^{+\infty} P(x)f_j(0)\,\delta U_j\,dx - \int_{-\infty}^{+\infty} Q(x)g_h\,\delta V_h\,dx \tag{4.13}$$

Since the surface force can be expressed as an equivalent body force $[P(x)\,\delta(z) = q_e(x,z)]$, equation (4.13) can be omitted and the surface traction effects included in equation (4.12). We can now rewrite equation (4.3) by using the results of the manipulations described by equations (4.9) through (4.12):

$$\int_{-\infty}^{+\infty}\left[\int_0^H \frac{\partial\sigma_x}{\partial x}f_j\,dz - \int_0^H \tau_{xz}\frac{df_j}{dz}\,dz + \int_0^H p(x,z)f_j\,dz\right]\delta U_j\,dx$$

$$+ \int_{-\infty}^{+\infty}\left[\int_0^H \frac{\partial\tau_{xz}}{\partial x}g_h\,dz - \int_0^H \sigma_z\frac{dg_h}{dz}\,dz + \int_0^H q(x,z)g_h\,dz\right]\delta V_h\,dx = 0$$

$$(4.14)$$

Now, since δU_j and δV_h are independent variations and may separately be set equal to zero, it follows that

$$\int_0^H \frac{\partial\sigma_x}{\partial x}f_j\,dz - \int_0^H \tau_{xz}\frac{df_j}{dz}\,dz + \int_0^H p(x,z)f_j\,dz = 0 \qquad (4.15)$$

and

$$\int_0^H \frac{\partial\tau_{xz}}{\partial x}g_h\,dz - \int_0^H \sigma_z\frac{dg_h}{dz}\,dz + \int_0^H q(x,z)g_h\,dz = 0 \qquad (4.16)$$

Equations (4.15) and (4.16) were derived by Vlasov and Leontiev in a different way.

Now, substituting equation (4.7) in equation (4.6) gives

$$\sigma_x = \frac{E}{(1+v)(1-2v)}\left[(1-v)\sum_{i=1}^m \frac{\partial U_i}{\partial x}f_i + v\sum_{k=1}^n V_k\frac{\partial g_h}{\partial z}\right]$$

$$\sigma_z = \frac{E}{(1+v)(1-2v)}\left[v\sum_{i=1}^m \frac{\partial U_i}{\partial x}f_i + (1-v)\sum_{k=1}^n V_k\frac{\partial g_k}{\partial z}\right] \qquad (4.17)$$

$$\tau_{xz} = \frac{E}{2(1+v)}\left[\sum_{i=1}^m U_i\frac{\partial f_i}{\partial z} + \sum_{k=1}^n \frac{\partial V_k}{\partial x}\cdot g_k\right]$$

Equation (4.17) can be inserted into equation (4.15) which, after rearranging, becomes

(1st member of σ_x)   (1st member of τ_{xz})

$$\frac{E(1-v)}{(1+v)(1-2v)}\int f_i f_j\,dz \sum_{i=1}^m \frac{\partial^2 U_i}{\partial x^2} - \frac{E}{2(1+v)}\int \frac{\partial f_i}{\partial z}\cdot\frac{\partial f_j}{\partial z}\,dz \sum_{i=1}^m U_i$$

(2nd member of σ_x)   (2nd member of τ_{xz}) (4.18)

$$+ \left[\frac{vE}{(1+v)(1-2v)}\int \frac{\partial g_k}{\partial z}f_j\,dz - \frac{E}{2(1+v)}\int g_k\frac{\partial f_j}{\partial z}\,dz\right]\sum_{k=1}^n \frac{\partial V_k}{\partial x}$$

$$+ \int p(x,z)f_j\,dz = 0$$

and, similarly, by substitution in equation (4.16), we obtain

(1st member of σ_z)   (1st member of τ_{xz})

$$-\left[\frac{Ev}{(1+v)(1-2v)}\int f_i\frac{\partial g_h}{\partial z}\,dz + \frac{E}{2(1+v)}\int \frac{\partial f_i}{\partial z}\cdot g_h\,dz\right]\sum_{i=1}^{m}\frac{\partial U_i}{\partial x}$$

(2nd member of τ_{xz})   (2nd member of σ_z)   (4.19)

$$+\frac{E}{2(1+v)}\int g_k g_h\,dz\sum_{k=1}^{n}\frac{\partial^2 V_k}{\partial x^2} - \frac{E(1-v)}{(1+v)(1-2v)}\int \frac{\partial g_k}{\partial z}\cdot\frac{\partial g_h}{\partial z}\,dz\sum_{k=1}^{n}V_k$$

$$+\int q(x,z)g_h\,dz = 0$$

In equations (4.18) and (4.19) the terms involving integrals are constants which can be evaluated once the functions f and g have been selected. Consequently, considering that u and w are represented by the series, equation (4.2), the two equations (4.18) and (4.19) will give rise, in general, to a number of simultaneous second-order ordinary differential equations in the unknown functions U_i and V_k. Solving these equations with the boundary conditions imposed at each end of the plate gives the solutions for U_i and V_k, and thus, since f and g are known, u and w are determined. The stresses may be calculated by substitution in equation (4.17).

The process can be illustrated, as done by Vlasov and Leontiev, by a simple example, which also has some intrinsic interest. When a symmetrical vertical load is applied to the surface of a half-space, or the surface of the plate of thickness H in Fig. 4-1, the resulting horizontal displacements are smaller, and they are also of less engineering interest, in general, than the vertical displacements. We can thus apply our intuition to simplify the problem. As a first approximation, we may assume that the horizontal displacements are zero, so that

$$u(x, z) = 0 \tag{4.20}$$

For a further simplification, we will take only one term in the series for w; thus,

$$w(x, z) = V_1(x)g_1(z) \tag{4.21}$$

The function $g_1(z)$ has to be selected, leaving $V_1(x)$ as the unknown variable which must be calculated. We must choose g_1 to fit the boundary conditions, which consist of a unit value at ground surface, and zero at depth H in a finite layer, or at infinity for the half-space. For the finite layer, Vlasov and Leontiev assume the function

$$g_1(z) = \frac{\sinh[\eta(H-z)]}{\sinh \eta H} \tag{4.22}$$

based, they indicate, on experimental evidence. However, Jones and Xeno-phontos (5), adopting a different variational approach which includes a loaded flexural plate in the calculations, show that equation (4.22) is the exact solution to the finite layer problem, although, strictly speaking, their equations for g_1 and V_1 are coupled. They also demonstrate that, in this equation, the parameter η, which determines how rapidly the vertical dis-placements diminish with depth, varies with the shape and flexibility of the plate and also with the form of the loading. Although Jones and Xenophon-tos do not discuss the question of the half-space, it is apparent from their reasoning that, in this case, the displacement function solution is

$$g_1(z) = e^{-\mu z} \tag{4.23}$$

in which μ again will be a function of the properties of load and plate. How-ever, for our example, the computations are simplified if a finite layer is chosen with an unrealistic linear variation

$$g_1(z) = 1 - \frac{z}{H} \tag{4.24}$$

We will carry through the calculations with this distribution.

Since u is zero, the first equilibrium equation (4.15) vanishes and the second equation (4.16) becomes, under the loaded area,

$$\frac{E}{2(1+v)} \int_0^H g_1^2 \, dz \, \frac{d^2 V_1}{dx^2} - \frac{E(1-v)}{(1+v)(1-2v)} \int_0^H \left(\frac{dg_1}{dz}\right)^2 dz \, V_1$$
$$+ q(x)g_1(0) = 0 \tag{4.25}$$

in which the last term represents the work done by a *horizontal* load $q(x)$ acting at the surface. When there is no surface load, the last term is omitted and the equation has its homogeneous form. Performing the integrations, using equation (4.24), we get

$$\int_0^H g_1^2 \, dz = \frac{H}{3}; \qquad \int_0^H \left(\frac{dg_1}{dz}\right)^2 dz = \frac{1}{H} \tag{4.26}$$

so that equation (4.25) becomes

$$\frac{EH}{6(1+v)} \frac{d^2 V_1}{dx^2} - \frac{E(1-v)}{H(1+v)(1-2v)} V_1 + q(x)g_1(0) = 0 \tag{4.27}$$

This is written by Vlasov and Leontiev in the form

$$2t V_1'' - k V_1 + q_1 = 0 \tag{4.28}$$

and is referred to by them as a foundation model with two characteristics, t and k. The symbols for the characteristic constants were chosen to illustrate the relationship of their model to the Winkler material described previously. If $t = 0$, then the Winkler representation is recovered (k is the Winkler foundation spring stiffness); for t present, the term including the second derivative of the vertical displacement takes into account some shearing interaction between foundation elements. Previously, other investigators had introduced such a second derivative, on an intuitive or physical basis, to describe the behavior of a string or membrane stretched over the surface of the bed of Winkler springs to give the model referred to as a *Pasternak* (7) *foundation*. Since it is a two-parameter model, it may be expected to give a better representation of the response of an elastic half-space. The present discussion gives an analytical basis for the presence of the second derivative and also gives an indication of how the constant coefficient of this derivative and the value of k might be related to the elastic properties of the subgrade material. For the improved distributions of equations (4.22) and (4.23), it has been shown (5) that the parameters η and μ depend on the properties of the plate and load. It follows that this is also true for k and $2t$.

Following Vlasov and Leontiev, we can obtain the vertical displacements in our two-dimensional example when the surface is subjected to a vertical line load P per unit of length, at $x = 0$. The solution of the homogeneous form of equation (4.28) is

$$V_1 = C_1 e^{-\alpha x} + C_2 e^{\alpha x} \tag{4.29}$$

where

$$\alpha = \sqrt{\frac{k}{2t}} = \frac{1}{H}\sqrt{\frac{6(1-v)}{(1-2v)}} \tag{4.30}$$

Since the vertical displacement for large x is zero, it follows that $C_2 = 0$, so that

$$V_1 = C_1 e^{-\alpha x} \tag{4.31}$$

In this model, where the displacement is assumed to vary linearly with depth, it follows that the vertical strain ϵ_z at any vertical cross section is constant and that, in consequence, the normal stress σ_z is also constant (since there are no lateral strains ϵ_x). This means that the calculation will give finite values of the stress σ_z at the surface of the layer even where there is no load. The existence of these stresses is a consequence of the variational approach of the method as discussed in Chapter 3; no attempt is made to satisfy the equilibrium conditions at all points in the medium. Instead, the variational method attempts to fit the requirements of the problem on the average, once the displacement distribution functions have been selected.

A further implication is that care must be taken in determining the constants of the equations by applying the boundary conditions. Since the approach was one of considering the work done and energy stored, it is necessary to take into account the boundary conditions in the same way. Thus, in the present problem, we can establish the value of the constant C_1 by considering shear along a vertical section of the foundation material just to the right (or left) of the origin. This statement should make the reader pause and reflect. The problem is symmetrical, so that, along the axis of symmetry, there should be no shearing stresses. Indeed, in an exact linearly elastic solution to the continuum problem, the slope of the displacement *is* zero on the axis (except at the point of load application) and there are no shearing stresses there. However, our solution is approximate, and there is a discontinuous slope in the displacements along the axis, so that shearing stresses do exist there. This situation arises again because of the variational approach and parallels the existence of constant vertical stress in this model. We can equate the work done by the shearing stresses acting along this section with the work done by one-half the load.
Thus,

$$\int_0^H \tau_{zx} g_1 \, dz = \frac{P}{2} g_1(0) = \frac{P}{2} \tag{4.32}$$

We obtain τ_{zx} from equation (4.17) and substitute in the left side of this equation to get

$$C_1 = \frac{P}{4\alpha t} \tag{4.33}$$

or, in terms of the material properties,

$$C_1 = \frac{3(1 + v)\sqrt{1 - 2v}}{\sqrt{6(1 - v)}} \frac{P}{E} \tag{4.34}$$

Consequently, the expression for displacement w becomes, from equations (4.34), (4.26), and (4.21),

$$w = \frac{3(1 + v)\sqrt{1 - 2v}}{\sqrt{6(1 - v)}} \frac{P}{E} \left(1 - \frac{z}{H}\right) e^{-\alpha x} \tag{4.35}$$

where α is given by equation (4.30).
We can obtain the surface displacements from this equation by making $z = 0$ and representing the equation in the form

$$w = \frac{VP}{E} \tag{4.36}$$

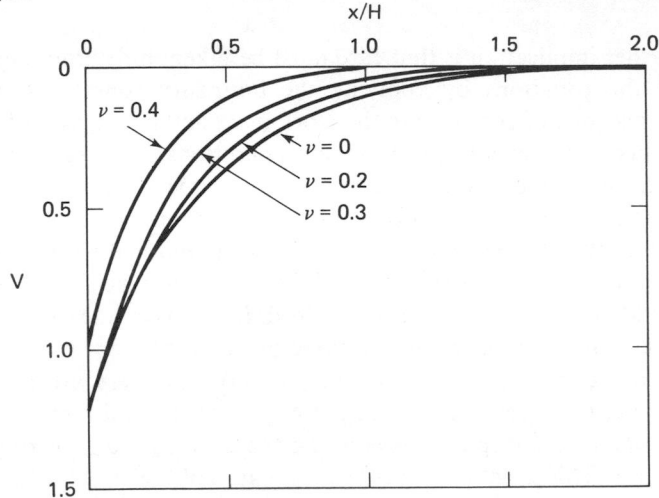

Figure 4-2. Deflections under line load at surface of elastic layer.

where V is a dimensionless coefficient. The variation of V with distance x is shown in Figure 4-2 for several values of Poisson's ratio. All the curves are, of course, exponential, and they differ from each other by a factor resulting from the Poisson's ratio expression in equation (4.35). In contrast to the Winkler model, in which displacement occurs only under the loaded area, it is seen here that the subgrade deflects to each side of the load point.

From this solution it is possible to derive the deflection for the case in which the load is uniformly distributed over a strip at the ground surface or has some other distribution. For example, when the load is uniformly distributed over a strip of width $2B$, as shown in Figure 4-3, the vertical displacement is obtained by integration of equation (4.35) to be

$$w = \frac{p}{2k}[2 - e^{-\alpha(B-x)} - e^{-\alpha(B+x)}]\left(1 - \frac{z}{H}\right) \qquad 0 < x < B \qquad (4.37)$$

$$w = \frac{p}{2k}[e^{-\alpha(x-B)} - e^{-\alpha(B+x)}]\left(1 - \frac{z}{H}\right) \qquad B < x < \infty \qquad (4.38)$$

where, as before,

$$k = \frac{E(1 - v)}{H(1 + v)(1 - 2v)}$$

The surface displacement is shown in Figure 4-4 for a range of ratios of B/H and $v = 0.25$. It is seen that the larger B/H becomes, the more uniform is the settlement over the width of the strip. This is, of course, similar to the behavior evidenced by a strip load on a linearly elastic layer.

An interesting result develops if we consider what will happen in the case of this subgrade model when it is loaded by a rigid beam instead of the

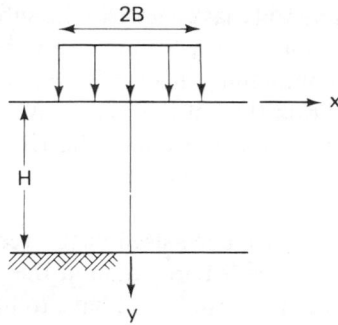

Figure 4-3. Arrangement of strip load on elastic layer.

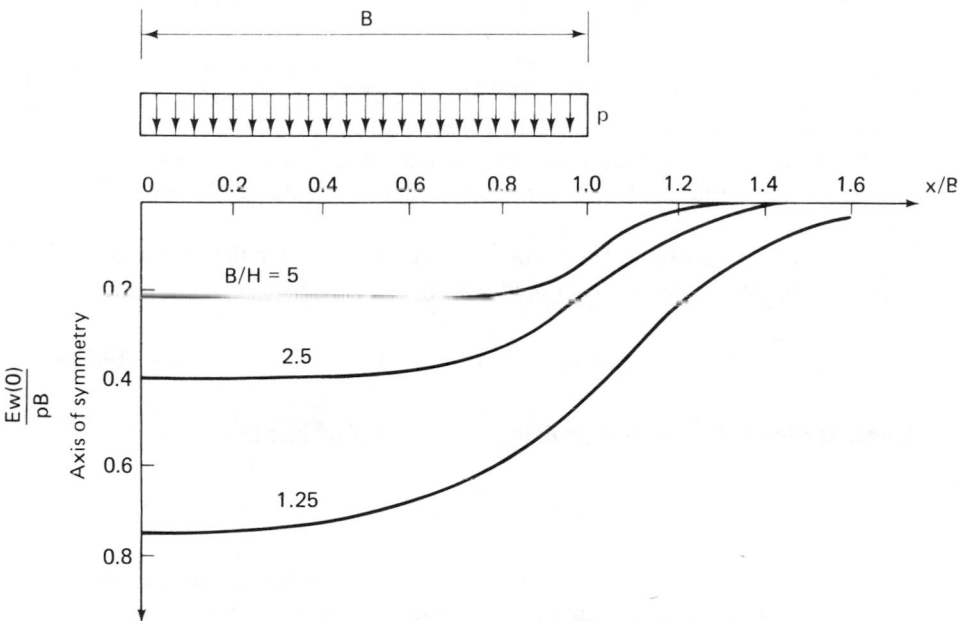

Figure 4-4. Deflections under strip load of width $2B$ at surface of elastic layer of depth H for Poisson's ratio = 0.25 and various ratios of H/B.

uniform pressure distribution over a strip considered above. We can see from Figure 4-4 that under the uniform pressure, a *realistic* (compared to the real elastic solid) displacement condition was obtained, where the settlements were greater under the center of the loaded area than at the edges. It is also a well-known result that when a rigid beam (or punch) is pressed into the surface of an ideal linearly elastic solid, the normal contact stresses go to infinity at the beam edges, for all values of the load applied to the beam. In real life, this means that any soil that comprises the subgrade will yield at the edges of the beam, but for soils such as stiff or overconsolidated clays, which approximate the elastic continuum, there will still be larger contact pressures at the edges of the beam than at the center. When we use the Winkler subgrade model the contact pressure under the rigid beam is, of course, uniform. Also, for the uniformly distributed load, the Winkler model gives a uniform settlement.

Since in the two-parameter models used in this section the displacement is less under the edges of the loaded area than at the center, it follows that, to obtain a uniform displacement, we would have to increase the pressure at the edges of the strip. Consequently, instead of trying directly to solve the problem of the two-parameter medium subjected to a uniform surface displacement over a strip, let us see what pressure distribution would be required to give a constant displacement over the strip width. Observing that the relationships in the brackets of equations (4.37) and (4.38) for the uniform loading case are exponentials, as are the relations in equation (4.35) for the line loading, we may try to find out what value of *line load* applied along the edges of the strip would make the displacement more uniform.

In the nature of a first approximation we will determine what load is needed to make the displacements equal under the center and edges of the loaded area.

From equation (4.37), the surface displacement under the center of the load, $x = 0$, due to the uniform load p alone is

$$w_{0p} = \frac{p}{k}(1 - e^{-\alpha B}) \tag{4.39}$$

Under the edge, $x = B$, due to p, the surface displacement is

$$w_{Bp} = \frac{p}{2k}(1 - e^{-2\alpha B}) \tag{4.40}$$

As a consequence of line loads P applied at the edges $\pm B$ (in Figure 4-3) of the loaded area, the vertical surface displacement under the center is, from equation (4.35),

$$w_{0P} = \frac{2P}{4\alpha t}e^{-\alpha B} \tag{4.41}$$

and at either edge the displacement is due to the load immediately above, plus the contribution of the other line load distant $2B$

$$w_{BP} - \frac{P}{4\alpha t}(1 + e^{-2\alpha B}) \tag{4.42}$$

Thus, equating the combined center and edge displacements, we have

$$\frac{p}{k}(1 - e^{-\alpha B}) + \frac{P}{2\alpha t}e^{-\alpha B} = \frac{p}{2k}(1 - e^{-2\alpha B}) + \frac{P}{4\alpha t}(1 + e^{-2\alpha B}) \quad (4.43)$$

from which we get

$$P = \frac{p}{\alpha} \quad (4.44)$$

Substituting this value back into the expressions for the displacements, it appears that the total (equal) edge and central displacements are

$$w = \frac{p}{k} \quad (4.45)$$

Although we began this as an approximate determination of these edge loads, it is not difficult to show that, by applying edge loads of the computed magnitudes, we have made the settlement uniform over the entire loaded area and equal to the amount given by equation (4.45). Outside the loaded area the surface settlement due to the combined uniform and edge loads is

$$w = \frac{p}{k}e^{-\alpha(x-B)} \quad (B < x < \infty) \quad (4.46)$$

This is an interesting and useful result, to which we will return later. It suggests that the theoretical, linearly elastic increase in contact stress at the edge of a rigid foundation can be approximated in this model by the application of a line load along the edge. For such a condition, the moments at the center of the loaded beam can be calculated from the combined uniform and line loads. Thus, the unconservative assumption of a uniform contact stress only acting below a rigid foundation resting on a stiff soil can be avoided without unduly complicating the calculations.

In the calculation we have included an assumed linear distribution of vertical displacement with depth, the result, equation (4.45), that two line loads of magnitude $P = p/\alpha$ make the surface settlement uniform over the loaded strip is independent of the nature of the vertical distribution of displacement.

In the general three-dimensional problem, when the horizontal displacements are taken to be zero and when the vertical displacement $w(x, y, z)$ includes only a single term of the series

$$w(x, y, z) = W(x, y)g(z) \quad (4.47)$$

the application of the virtual work principle gives the general equation

$$2t\nabla^2 W - kW + p = 0 \qquad (4.48)$$

where $2t$ and k have the same meaning and values as before, equations (4.27) and (4.28), p is the applied load and ∇^2 is the Laplacian operator. Useful problems to examine are those of axial symmetry; in these circumstances, W becomes the function $W(r)$ and the Laplacian is

$$\nabla^2 W(r) = \frac{d^2 W}{dr^2} + \frac{1}{r}\frac{dW}{dr} \qquad (4.49)$$

Thus, equation (4.48) becomes

$$\frac{d^2 W}{dr^2} + \frac{1}{r}\frac{dW}{dr} - \alpha^2 W + \frac{p}{2t} = 0 \qquad (4.50)$$

This is a Bessel equation whose homogeneous form has the general solution

$$W = C_1 I_0(\alpha r) + C_2 K_0(\alpha r) \qquad (4.51)$$

where I_0 and K_0 are modified Bessel functions of first and second kind and zero order. The functions I_0 and K_0 behave in a manner similar to e^{+x} and e^{-x}, respectively. Vlasov and Leontiev have used this equation to obtain a solution for the vertical displacement $w(r, z)$ of a half-space (their two-parameter model) subjected to a point load P

$$w(r, z) = \frac{P}{4\pi t}K_0(\alpha r)g(z) \qquad (4.52)$$

In this case, an assumption of a linear variation of g with depth is not realistic, and Vlasov and Leontiev present a solution in which

$$g = e^{-\mu z} \qquad (4.53)$$

where μ, as before, is a convenient parameter representing the rate of variation of vertical displacement with depth. The displacement under the point load is

$$w(r, z) = \frac{2P(1 + v)\mu}{\pi E}K_0(\alpha r)e^{-\mu z} \qquad (4.54)$$

A value of μ of about 1.0 gives a surface displacement distribution similar to but everywhere smaller than (partly because of the $u = v = 0$ restriction) that of the solution to the Boussinesq problem of a point load normal to the surface of a linearly elastic half-space.

We can arrive at a result analogous to that obtained for the plane strain case with a uniform displacement over a strip width by considering two more axisymmetric situations. The first is the displacement due to a uniformly loaded circular area at ground surface. Again, assuming zero lateral displacement, we see that the solution to equation (4.50) for W gives, according to Vlasov and Leontiev, the result

$$w(r, 0) = \frac{p}{k}\left[1 - \frac{K_1(\alpha R)I_0(\alpha r)}{D}\right] \qquad 0 < r < R$$

$$w(r, 0) = \frac{p}{k}\frac{I_1(\alpha R)K_0(\alpha r)}{D} \qquad R < r < \infty$$

$$(4.55)$$

where $D = I_0(\alpha R)K_1(\alpha R) + I_1(\alpha R)K_0(\alpha R) = 1/\alpha R$. It is understood that the function g describing the variation of w with depth has the value unity at the surface. For the case of a line load, P per unit of length acting along the circumference of a circle of radius R, the solution to equation (4.50) for W can be shown to give the result

$$w(r, 0) = \frac{PR}{2t}K_0(\alpha R)I_0(\alpha r) \qquad 0 < r < R \qquad (4.56)$$

$$w(r, 0) = \frac{PR}{2t}I_0(\alpha R)K_0(\alpha r) \qquad R < r < \infty \qquad (4.57)$$

If we proceed as before by evaluating these four functions at $r = 0$ and $r = R$, summing the two results at $r = 0$, $r = R$, and equating them, we find that this comes about if

$$P = \frac{p}{\alpha}\frac{K_1(\alpha R)}{K_0(\alpha R)} \qquad (4.58)$$

This is different from the plane strain case in which the required P was found to be independent of the size of the loaded area. The ratio K_1/K_0 diminishes from ∞ at $\alpha R = 0$ to unity at $\alpha R = \infty$, being within 5 percent of unity at $\alpha R = 10$. Thus, for large αR, the required P is essentially the same as for the plane strain case. For the exponential variation of g of equation (4.53), α is given by the relation

$$\alpha = \mu\sqrt{\frac{2(1 - v)}{(1 - 2v)}} \qquad (4.59)$$

so that, with μ of the order of 1 m^{-1}, α is about 2 m^{-1} for the usual range of values of Poisson's ratio. Thus, the plane strain value of P in equation (4.58) is reached for values of R of about 5 m or greater.

Once again, a closer examination of the equations reveals that the value of P given by equation (4.58) makes the displacement uniform over the whole circular area and equal to the amount

$$w(r, 0) = \frac{p}{k} \qquad 0 < r < R \tag{4.60}$$

In both the plane strain and axially symmetrical problems the uniform displacement effected by the prescribed edge loading is the amount that would occur, (p/k), if the uniform load were to be applied over the entire surface. This means that, as far as the loaded area is concerned in each case, the edge loading exactly simulates a uniform loading of intensity p applied everywhere *outside* the loaded area. It follows that by appropriate superposition of loadings and displacements the solutions for cases in which uniform settlement is caused by uniform loading applied outside a particular area, together with an appropriate edge load, can be obtained.

We will return to the method of Vlasov and Leontiev in the case of a pile problem subsequently.

In the next section the simplest case of soil-structure interaction, when the foundation is of the Winkler type, will be taken up. There the beam is considered to be rigid, a situation that will hold in practice for many short, stiff concrete footings, or it may apply in the case of short piles or caissons.

4.3 RIGID BEAM ON ELASTIC FOUNDATION

For the first part of the discussion, the response of the soil is taken to be represented by equation (4.1), the case known as the *Winkler foundation*. The "beam" either may be a real member of finite width b or may be considered to be a strip of unit width forming a typical element of a *slab* impressed with line loads, moments, or displacements, so that the complete problem is one of plane strain. In either of these cases, the situation is two-dimensional, and the basic subgrade modulus k_0 must be multiplied by the beam (b) or strip (1) width, to give the appropriate value for use in analysis:

$$k = k_0 b \quad \text{or} \quad k_0(1) \tag{4.61}$$

We will assume subsequently in this section that this has been done so that k in the following discussion will have the dimensions FL^{-2}. In a later chapter the determination of k_0 for a soil foundation will be described, and typical values will be presented. We will proceed with the mathematical problem first.

The analysis of a rigid beam on a Winkler foundation is not difficult, as can be seen in Figure 4-5. Here a load is placed a distance a from the left end of a beam of length l, with the usual arrangement of coordinates. With a

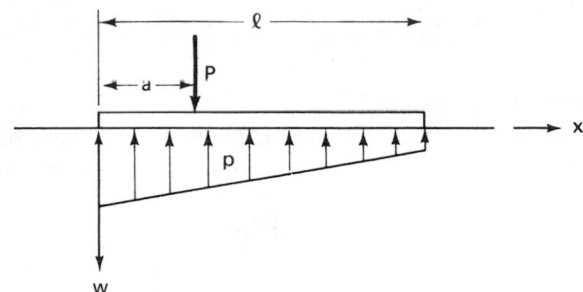

Figure 4-5. Rigid beam on Winkler foundation under load P arbitrarily located.

uniform coefficient k, the rigidity of the beam requires the deflection, and thus the pressure, to be linear functions of distance x. The conditions which apply to the beam and which are used to establish the solution are the equilibrium of vertical forces and moments. The downward vertical force is P and this is resisted by the pressure p which varies linearly from left to right. If the deflection at the two ends, w_0 and w_l are taken as the two unknowns in the problem, then p at the left end is equal to kw_0 and, at the right end, kw_l. The average pressure is $k(w_0 + w_l)/2$ and this, multiplied by the length l of the beam, gives the resisting force to balance P. This is the first equation. A similar operation with moments taken about say $x = 0$ gives the second equation.

The solution in terms of deflection, w, is

$$w = \frac{2P}{kl}\left[-3\left(1 - 2\frac{a}{l}\right)\frac{x}{l} + \left(2 - 3\frac{a}{l}\right)\right] \qquad (4.62)$$

When multiplied by k, equation (4.62) gives the pressure distribution on the subgrade and the moment and shear at a point x can be calculated from this and the load P.

If the load is due to a moment M_0 located at a distance a from the left end of the beam, vertical and moment equilibrium require that the deflection be given by the expression

$$w = \frac{6M_0}{kl^2}\left(2\frac{x}{l} - 1\right) \qquad (4.63)$$

which is independent of a, since for a rigid beam the deflection due to a moment must be independent of its point of application. This occurs because there is no external vertical force involved. The moment and shear can also be calculated as a function of x in this case; here the maximum moment, of course, is M_0. When the moment is applied at the left-hand end, the moment distribution is shown in Figure 4-3 and is given by equation (4.64):

$$\frac{M}{M_0} = 1 - 3\left(\frac{x}{l}\right)^2 + \left(\frac{x}{l}\right)^3 \tag{4.64}$$

By substituting x successively equal to zero and l in equation (4.63), it can be seen that at the left end of the beam (when the moment is positive when applied in a clockwise sense) the beam lifts up to a deflection

$$W_0 = -\frac{6M_0}{kl^2} \tag{4.65}$$

and the far end of the beam, at $x = l$, depresses by the same absolute amount. The pressure obtained by multiplying the deflections by k follows the same pattern and is consequently negative over the left-half of the beam, that is to say, the subgrade is pulling *down* or exerting a *tensile* stress on this side of the beam. In the case in which the beam represents a portion of pavement and is subjected only to the moment load considered, the tensile stress arrived at by this calculation could not exceed the stress due to the weight of the beam on the subgrade without the beam's lifting off the ground. If this possibility is to be realistically assessed, a different calculation must be performed, including the fact that tensile stresses cannot occur between the beam and the subgrade.

The analysis for a rigid beam proceeds as in the following example. In Figure 4-6, a rigid beam of weight γ per unit length and length l is subjected

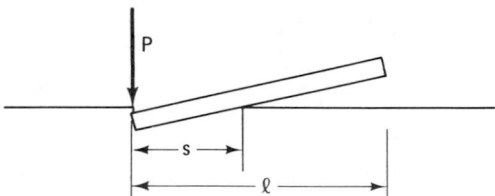

Figure 4-6. Deflection of rigid beam on Winkler foundation under end load P when no tensile stresses develop between beam and foundation.

to a load P at one end. It is assumed that no tensile stresses can exist between the beam and the subgrade and that the load P is consequently sufficient to lift the other end of the beam from the ground, so that only the length s remains in contact.[2] Over this length s the deflection and pressure distribution are linear with distance. By considering the vertical equilibrium of the beam, and taking moments about the left end, we find that s is given by the equation

$$s = \frac{3\gamma l^2}{2(P + \gamma l)} \tag{4.66}$$

---

[2]The case of the flexible beam for which tensile beam-subgrade stresses are not permitted to develop has been studied by Westmann and Tsai (9).

and that the equation of the beam deflection is

$$w = \frac{4(P + \gamma l)^2}{3k\gamma l^2}\left[-\frac{2(P + \gamma l)}{3\gamma l}\frac{x}{l} + 1\right]$$

(4.67)

In practice it is useful to obtain the moments in the beam, when lift-off does not occur. In the special case in which a load P is applied to the left-hand end of the rigid beam, the distance a in equation (4.62) is zero and the deflection becomes

$$w = \frac{2P}{kl}\left(-3\frac{x}{l} + 2\right)$$

(4.68)

and the moment in the beam at a section x is given by the equation

$$M = Pl\left[-\left(\frac{x}{l}\right)^3 + 2\left(\frac{x}{l}\right)^2 - \frac{x}{l}\right]$$

(4.69)

Here the maximum moment has a value

$$M_m = -\frac{4}{27}Pl$$

(4.70)

and occurs a distance $l/3$ from the left end of the beam.

It is of interest to compare the moment distribution obtained from the simple analysis involving a rigid beam with that given by the flexible beam theory described later for this case of a load P applied to the end of a finite beam of length l. The values are shown in Figure 4-7, from which it can be seen that by the time the relative stiffness of beam and soil reaches a value where $\lambda l = 2.00$, the maximum moment is within 10 percent of that given by the rigid beam theory. Another point of interest about equation (4.69) is that the moment is independent of the actual value of the subgrade reaction. This also holds true, of course, for the moment in the same beam loaded by P, which is arbitrarily located.

Some useful results applicable, for example, to short rigid piles or caissons can be established by considering the case in which the rigid beam is placed on a subgrade whose coefficient varies linearly from one end of the beam to the other, according to the equation

$$k = k_1 + k_2 x$$

(4.71)

in which the slope of the variation, k_2, may be positive or negative. If it is negative, it is unlikely that, in real circumstances, it would have such a value that k would become negative at some point in the beam. The variation can be due to a linear change in either the beam width or the subgrade reaction coefficient. For a pile, the latter is more likely to be the case.

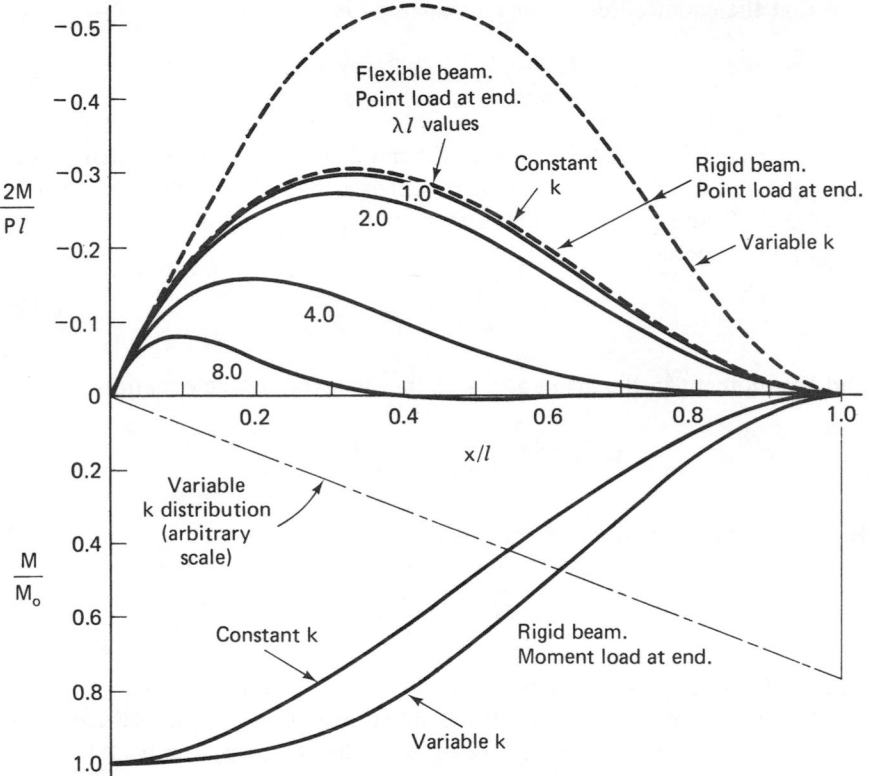

Figure 4-7. Moment distribution for rigid and elastic beams on Winkler foundation under various loading conditions. The elastic parameter λ is expressed by equation (5.12). Low values of λl represent stiff beams; high values indicate flexible beams.

Once again, the displacement of the beam under a load P arbitrarily located on the beam can be found from the requirement that vertical and moment equilibrium must be satisfied. The displacement is no longer a simple expression, but it will be given here since this general case can be simplified for application to a number of useful problems which will be considered later.

$$w = \left(-A\frac{x}{l} + B\right)\frac{6P}{l^2} \tag{4.72}$$

where

$$A = \frac{2(2k_2 l + 3k_1) - 6\frac{a}{l}(k_2 l + 2k_1)}{k_2^2 l^2 + 6k_1 k_2 l + 6k_1^2} \tag{4.73}$$

and

$$B = \frac{(3k_2l^2 + k_1l) - 2a(2k_2l + 3k_1)}{k_2^2l^2 + 6k_1k_2l + 6k_1^2} \qquad (4.74)$$

For this general case, the pressure distribution and moment can be evaluated from the deflection of the beam. However, for the present, we will proceed to a simpler case.

We will examine the problem of a beam loaded by P at the left-hand end, so that $a = 0$, and resting on a soil for which $k_1 = 0$ also. Putting these conditions into equations (4.72) through (4.74), we find

$$w = -\frac{24P}{k_2l^2}\frac{x}{l} + \frac{18P}{k_2l^2} \qquad (4.75)$$

so that the pressure distribution on the beam is

$$p = kw = -\frac{24P}{l}\left(\frac{x}{l}\right)^2 + \frac{18P}{l}\left(\frac{x}{l}\right) \qquad (4.76)$$

This equation shows that the maximum pressure has the value

$$P_m = \frac{27P}{8l} \qquad (4.77)$$

and occurs at a point located $3l/8$ along the beam from the left end. The moment acting in the beam is

$$M = Pl\left[-2\left(\frac{x}{l}\right)^4 + 3\left(\frac{x}{l}\right)^3 - \left(\frac{x}{l}\right)\right] \qquad (4.78)$$

Here the maximum moment is

$$M_m = -0.26Pl \qquad (4.79)$$

and occurs at the coordinate $x = 0.42l$ (approximately $27l/64$). The moment curve has been drawn on Figure 4-7 for comparison with the previous case of the beam supported on uniform soil. It is apparent once again that neither the pressure distribution nor the moment is dependent on the actual value of the coefficient of subgrade reaction or the parameter describing its variation with distance.

The same problem of the rigid beam on the soil with a linear variation of subgrade coefficient, in the special case in which $k_1 = 0$ but in which the load to be applied is a moment M_0 applied at the left end, can also be handled by the method described. The results for deflection and pressure are

$$w = \frac{12M_0}{k_2l^3}\left[3\left(\frac{x}{l}\right) - 2\right] \qquad (4.80)$$

and

$$p = \frac{12M_0}{l^2}\left[3\left(\frac{x}{l}\right)^2 - 2\left(\frac{x}{l}\right)\right] \tag{4.81}$$

In this case, as before, for the clockwise-acting moment, the left end of the beam lifts up and the right end is depressed. Because of the variation in the subgrade coefficient along the beam, the pressure between the beam and the soil is zero at the left-hand end, decreases to a tensile minimum of $-(4M_0/l^2)$ at a distance of $l/3$ from the left end, and then increases to a compressive maximum of $12M_0/l^2$ at the right-hand end of the beam. The moment on the beam is given by the expression

$$M = M_0\left[3\left(\frac{x}{l}\right)^4 - 4\left(\frac{x}{l}\right)^3 + 1\right] \tag{4.82}$$

and has, of course, the maximum value of M_0 at the left end of the beam. Moments for the moment loading case are also shown in Figure 4-7. Since the problem is still *linear*, all the values of deflection, slope, pressure, moment, and shear at a given beam section for a beam to which both moment and point loads are applied can be obtained by superposition of the values obtained from the point and moment loads considered separately.

A useful application of superposition for certain models of the foundation is discussed in the next section.

4.4   TWO-PARAMETER SOIL FOUNDATION MODEL

When a load is applied to the surface of a linearly elastic half-space or layer, the surface deflects, of course, under the load, but it also moves down in the unloaded areas adjacent to the load, with displacements diminishing with distance. This occurs because the material is a connected continuum. The Winkler model does not perform this way; only the immediately loaded region settles while the adjacent surface remains unchanged. A linearly elastic isotropic continuum is described by two material properties (for example, Young's modulus E and Poisson's ratio v; other pairs may be used, such as the Lamé constants), but the Winkler material is described by only one, the spring constant k. Thus, if a solution to a problem such as a loaded beam resting on a Winkler foundation is compared with that for the same beam as an elastic continuum, simultaneous correspondence among all derived quantities cannot be expected. Instead, for a particular E and v of the continuum, there will be a value of k which will render, say, maximum displacements equal in the two analyses. Another value of k will identify maximum moments, and so on. The analyst has to decide which condition is the important one for the design. We have seen that the Vlasov and

Leontiev approach has provided the addition of another property to the basic Winkler model. It enables deflections outside the loaded region to be effected and permits, for example, both deflections and moments to be matched. From usually physical points of view, various improvements along this line have been suggested by a number of authors, including Wieghardt (10), Filonenko-Borodich (2), Hetenyi (4), and Pasternak (7). Among the variety of expedients propounded, the most convenient mathematically is the one in which, in effect, the tops of the Winkler springs forming the ground surface are tied together by a stretched elastic string or membrane or shear beam. The tension in this string is the second foundation property; for the beam problems currently being investigated, we will designate it S (dimension F, equals $S_0 b$, as in the case of k and $k_0 b$). The effect of the stretched string on the mathematics is to modify the surface reaction of the foundation soil from the simple $p = kw$ of the Winkler spring to

$$p = kw - S\frac{d^2w}{dx^2} \tag{4.83}$$

It will be seen that the string with tension S is the physical representation of the result arrived at through the mathematical operation of Vlasov and Leontiev, equation (4.28), and that $S = 2t$ therein. Beyond the edges of the loaded area there is no normal surface stress and the left side of equation (4.83) becomes zero, so that the following equation holds

$$-S\frac{d^2w}{dx^2} + kw = 0 \tag{4.84}$$

It is apparent that this is the second step (the Winkler model is the first) in a general approach whereby the pressure exerted by the soil is given by the equation

$$p = aw^{(0)} + bw^{(2)} + cw^{(4)} + \cdots$$

where the superscripts denote the zeroth, second, fourth, etc., derivatives of deflection with respect to distance. We have recognized a physical significance to the first two of these, the bed of springs and the stretched string or membrane. The third would represent the behavior were the soil response to have a beam or slab component. The physical connection diminishes with higher derivatives.

The simplest situation in which the foundation model, equation (4.84), can be utilized is the case of the loaded rigid beam. Although we have developed some results of loading the surface of the two-parameter foundation in Section 4-2, it is instructive to repeat some of the calculations in a different form, emphasizing the presence of the structural member, as in Section 4.3.

4.4.1 Rigid beam on two-parameter foundation model

When the foundation properties are uniform, all loading conditions in the rigid beam can be resolved into the superposition of a symmetric and an antisymmetric deflection. We consider the simplest case, that of symmetry, first. The arrangement of beam, loading, and deflection is shown in Figure 4-8. The load P_0 shown represents any *symmetrical* load in the form of point or uniformly or even nonuniformly distributed applied forces. The gross downward load of any distribution is given by P_0. The deflection and soil pressure for any symmetric load P_0 are given in the following development, but the moment is worked out only for the point load P_0, applied at the beam center.

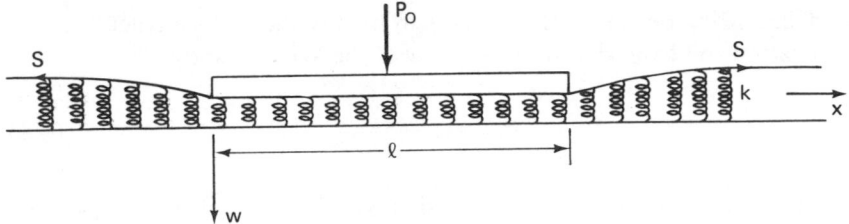

Figure 4-8. Symmetrically loaded rigid beam on two-parameter foundation.

To the left of the origin, $x < 0$, and to the right of the beam, $x > l$, there is no applied stress at the ground surface, so that equation (4.84) holds in these regions. The general solution to the equation is, as in equation (4.29),

$$w = Ae^{\alpha x} + Be^{-\alpha x} \tag{4.85}$$

where A and B are constants to be determined from the boundary conditions and α is a parameter given by the relation

$$\alpha^2 = \frac{k}{S} \tag{4.86}$$

and is thus the same parameter given by equation (4.30). In the region $x < 0$, x is, of course, negative, and we would expect the deflection (as well as all other quantities) to diminish with distance from the beam. It follows that $B = 0$ here. At the end of the beam the deflection, from equation (4.85), is therefore

$$w_0 = A$$

and the slope to the left is

$$\left(\frac{dw}{dx}\right)_0 = A\alpha \tag{4.87}$$

For conditions to the right, $x > l$, we must modify the solution, equation (4.85), to the form

$$w = Ce^{\alpha(x-l)} + De^{-\alpha(x-l)} \tag{4.88}$$

By the same argument as before, but with positive $(x - l)$ values, we conclude that here $C = 0$. Thus, from the right side, the deflection at $x = l$ is

$$w = D \tag{4.89}$$

However, the load is symmetric with respect to the beam, and so are the foundation properties; therefore, we would expect the vertical beam displacement to be constant along the beam. Thus, $D = A$. Consequently, at the right end of the beam the slope of the free surface (and attached string) is the same as at the left end, but, of course, of opposite sign:

$$\left(\frac{dw}{dx}\right)_l = -A\alpha \tag{4.90}$$

Since the membrane has a force S acting in it, the effect of its deflection by the beam is to develop a point or line load at each end of the beam with magnitude equal to the product of the string tension and the slope, assuming the slope to be small. This is the vertical component of the string force S acting at the beam ends and is given by the relation

$$R = SA\alpha \tag{4.91}$$

The force R is identical to that required at the edges of the uniformly loaded area to force uniform deflection in the Vlasov and Leontiev theory. Since the Winkler springs to the left and right of the beam ends are also compressed by the stretched strings, it follows that the summation of the Winkler spring forces in the region beyond the beam at either end is also equal to R. This can be demonstrated formally by the integration of the displacement curve, equation (4.85), when the spring constant k is included.

The stretched string contributes no other force to the beam since it is flat along the base and only turns through an angle at the beam ends.

The constant A can be obtained from the vertical equilibrium of the beam, by equating the force P_0 to the sum of the Winkler reaction along the beam length, kAl, and the two forces R:

$$P_0 = kAl + 2SA\alpha \tag{4.92}$$

from which

$$A = w_{\text{beam}} = \frac{P_0}{kl + 2\alpha S} = \frac{P_0}{kl\left(1 + \frac{2\alpha S}{kl}\right)} \tag{4.93}$$

The dimensionless term $\alpha S/kl$ in parentheses represents the relative contribution of string tension S to Winkler constant k in the behavior of the foundation; it is convenient to denote it as κ

$$\kappa = \frac{\alpha S}{kl} = \frac{\sqrt{S}}{l\sqrt{k}} \quad \text{or} \quad \kappa^2 = \frac{S}{kl^2} \tag{4.94}$$

in which the relation (4.86) has been employed. The pressure on the base of the beam is also constant and is given by the equation

$$p = kw_{\text{beam}} = \frac{P_0}{l(1 + 2\kappa)} \tag{4.95}$$

For the loading of Figure 4-8, the moment, M, in the region $0 < x < l/2$ is given by the equation

$$M = -\frac{P_0 l}{2(1 + 2\kappa)}\left[\left(\frac{x}{l}\right)^2 + 2\kappa\frac{x}{l}\right] \tag{4.96}$$

To the right of the load P_0 the moment is given by the same equation but in which $(1 - x/l)$ is substituted for x/l. Under the load P_0, the maximum moment is found by replacing x/l by the value $\frac{1}{2}$:

$$M_{\text{max}} = \frac{P_0 l}{8(1 + 2\kappa)}(1 + 4\kappa) \tag{4.97}$$

The corresponding Winkler foundation solutions for deflection, pressure, and moment can be found by taking S, and thus κ equal to zero in equations (4.93), (4.95), and (4.96).

We proceed to examine the antisymmetric case next. The situation is illustrated in Figure 4-9. In this case, obviously, there is no vertical resultant force on the beam, and thus the Winkler pressures and the string forces must balance themselves antisymmetrically as shown. The center of the

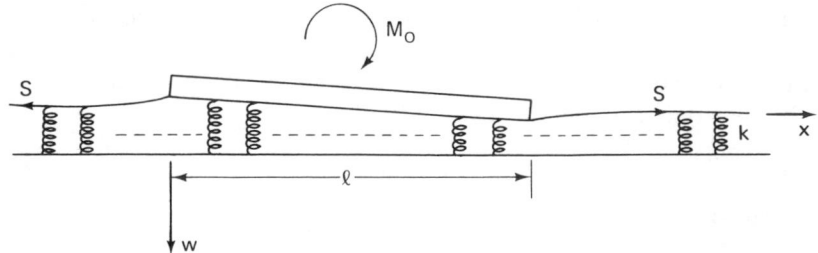

Figure 4-9. Moment loading of a rigid beam on a two-parameter foundation.

beam has zero deflection and is not subjected to any foundation pressure. A solution requires finding the deflection of the beam ends, or the rotation of the beam. With the same equations and conditions as before beyond the beam ends, except for symmetry, it follows that for $x < 0$,

$$w = -Ae^{\alpha x} \tag{4.98}$$

and at $x = 0$

$$w = -A \quad \text{and} \quad R_1 = -SA\alpha \tag{4.99}$$

At the other end, for $x > l$,

$$w = Ae^{-\alpha(x-l)} \tag{4.100}$$

and at $x = l$

$$w = A \quad \text{and} \quad R_1 = SA\alpha \tag{4.101}$$

In this case, the load R_1 is the vertical component of the string tension to the left and right of the ends of the beam. However, since the beam is tilted, the tension force S along the beam base is not directed horizontally and therefore also has a component in the vertical direction R_2, equal to the force times the angle or

$$R_2 = \frac{2SA}{l} \tag{4.102}$$

The total force R acting at each end of the beam, but in opposite directions, is then

$$R = R_1 + R_2 = SA\left(\alpha + \frac{2}{l}\right) \tag{4.103}$$

Taking moments about $x = 0$ again permits the calculation of the constant A:

$$M_0 = -\frac{kAl^2}{6} - SA\alpha l - 2SA$$

from which

$$A = -\frac{M_0}{\dfrac{kl^2}{6} + S(\alpha l + 2)} \tag{4.104}$$

The deflection of the beam is then given by the equation

$$w = \frac{6M_0}{kl^2\left(1 + 6\kappa + \dfrac{12\kappa}{\alpha l}\right)}\left(2\frac{x}{l} - 1\right) \tag{4.105}$$

and the pressure acting on it (not including the R forces at the ends) by the product of this quantity and k, as before. The moment in the beam can also be evaluated as a function of x.

As in the symmetrical case, the solution reduces to the one for the correspondingly loaded beam on the Winkler material only, if κ is put equal to zero.

The superposition of the two solutions for symmetrical and antisymmetrical loads can be illustrated by reference to one of the examples solved previously for the Winkler foundation, that of the beam loaded by a simple force P at its left end. The problem is represented diagrammatically in Figure 4-10. It is necessary to calculate the values of P_0 and M_0 required for

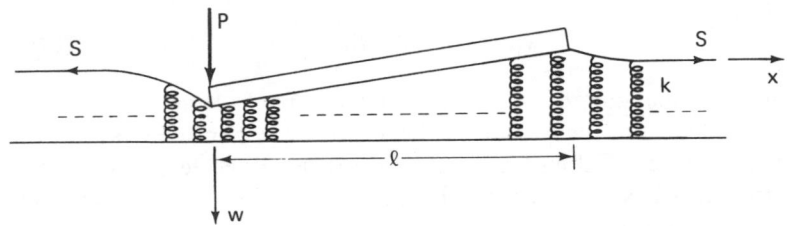

Figure 4-10. Point loading at the end of a rigid beam on a two-parameter foundation.

the solution. They are chosen so that the total vertical and moment equilibriums under the loads in the superposition are equal to those in the basic problem. From vertical equilibrium, it is apparent that $P_0 = P$, and taking moments about the left end of the beam, it follows that

$$0 = \frac{P_0 l}{2} + M_0$$

or

$$M_0 = -\frac{P_0 l}{2} = -\frac{Pl}{2} \tag{4.106}$$

Thus, the solution for the deflection of the beam loaded at the left end by the force P is obtained by summing the deflection for the two other cases, equations (4.93) and (4.105), respectively, but in which the substitutions P and $-Pl/2$ are made for P_0 and M_0, respectively. The pressure and moment can be arrived at by a similar summation.

When the "beam" studied here is taken to represent a pile in the ground, the physical identification of the second derivative with a stretched string disappears, particularly at the pile top. The "force," S, then becomes another fitting parameter to assist in the identification of the model behavior with either experimental results, or with those obtained from other models, such as, for example, the rod imbedded in an elastic half-space.

4.5 RIGID SLABS

In the case of the Winkler foundation model, the analysis of the rigid slab follows that for the rigid beam, since only the material directly below the slab responds to its presence. But when a two-parameter foundation is employed, the procedure is more complicated because edge reactions are present around the entire periphery.

Two cases are of interest: the circular slab and the general rectangular slab. We will discuss only the case of a symmetrically disposed load on each. In Section 4.2 the result was obtained that a uniform load p applied all over the surface of a circle of radius R on a two-parameter foundation together with a line load of magnitude P per unit length, where P is given by equation (4.58), acting along the circumference of the circle, produced a uniform displacement of the circle given by equation (4.60). The resulting situation is therefore identical to that of a rigid circular slab pushed into the surface with a load P_0 where

$$P_0 = \pi R^2 p + 2\pi R P \tag{4.107}$$

Substituting for P from equation (4.58) and then evaluating p and P for the given load P_0, we get

$$p = \frac{P_0}{\pi R^2 \left[1 + \dfrac{2K_1(\alpha R)}{\alpha R K_0(\alpha R)} \right]} \tag{4.108}$$

$$P = \frac{P_0}{\pi R^2 \left[1 + \dfrac{2K_1(\alpha R)}{\alpha R K_0(\alpha R)} \right]} \cdot \frac{K_1(\alpha R)}{\alpha K_0(\alpha R)} \tag{4.109}$$

where α is given by equation (4.86) and $K_{0,1}$ are the functions discussed in Section 4.2.

The deflection of the plate is given by equation (4.60). For the rigid slab, these relations are unaffected by the distribution of the load P_0 so long as it is symmetrical with respect to the axis of the slab.

When the load is applied symmetrically to a rectangular rigid slab, we distinguish two cases: in the first case, the slab has a length l relatively great (say, several times) in comparison with its width b; in the second case, the lengths l and b are nearly the same. In both cases, the deflected shape of the membrane that contributes the second parameter of the model through the membrane tension S per unit of length is quite complicated, particularly around the corners. In general, this deflected surface will provide a varying line load along the edges of the rigid footing and a point load at each corner. The point load will be small in comparison to the other forces involved in the problem. A simplification will be made in the case of the beam-like slab, in that the line force at each short end (across the width b) will be abbreviated to a concentrated force assumed to act at the middle of each end. The line force per unit length acting along each long edge in this case, and along all four sides in the case of the equidimensional slab, will be taken to be constant along each edge.

The solution for the deflection w of both slabs under a symmetrical load P_0 is (8)

$$w = \frac{P_0}{[kl + 2\alpha S(b + l) + 3S]} \tag{4.110}$$

where α is again given by equation (4.86) and $k = k_0 b$ as before. The term $3S$ in the denominator comes from the corner forces. The concentrated foundation reaction force T acting at each end of the *beam* is given by the equation

$$T = wS\left(\alpha b + \frac{3}{2}\right) \tag{4.111}$$

which includes an allowance for the corner forces, and the live reaction force per unit length, R, acting along each edge, is, as in the two-dimensional beam problem of Section 4.4,

$$R = \alpha wS \tag{4.112}$$

Each corner force Q is

$$Q = \frac{3}{4}wS \tag{4.113}$$

The moment at any cross section of the beam or slab can be calculated in the usual way by including the Winkler reaction stress $p = kw$, the effect of the end force T, and the long edge line load R. The analysis of the slab when a load is located antisymmetrically proceeds as in Section 4.4 by using reaction forces T of the same form as in equation (4.111) but acting in opposite directions at the ends of the beam.

This completes the discussion of rigid beams and slabs. We will return to a consideration of one- and two-parameter foundation models in connection with flexible beams in the next chapter.

PROBLEMS

1. What do you think would need to be done to make the two-parameter subgrade model of Vlasov and Leontiev a practical alternative to the use of the one-parameter Winkler model? Describe a systematic approach, including considerations of parameter determination for soils, analyses, and calculations.

2. How would you apply the model and method of V-L to the problem of determining displacements, stresses, etc., caused by a sheet pile wall embedded in the soil and subjected to a lateral load at ground surface (see Figure P3-2)?

3. Solve the problem of a line load p on a two-parameter foundation using equation (4.22), to represent the behavior of a half-space under a line load. Find the value of η which makes the result best match the Boussinesq solution.

4. Apply the Ritz method to obtain the solution to the problem of a rigid beam resting on a Winkler foundation whose coefficient varies from zero linearly with distance from one end at which a load is applied. Compare your result with equation (4.75).

5. Consider an infinite beam on the two-parameter foundation soil. Set up the beam equation, properly accounting for the edge loads due to the second parameter.

6. Solve the problem of a line load on an elastic layer of thickness H using the V-L model with two terms in the series representing the vertical displacement as a function of depth. Take the first term of the linear form

$$0 < z < \frac{H}{2} \qquad g_1(z) = 1 - 2\frac{z}{H}$$

and the second in two parts

$$0 < z < \frac{H}{2} \qquad g_2(z) = 2\frac{z}{H}$$

$$\frac{H}{2} < z < H \qquad g_2(z) = 2\left(1 - \frac{z}{H}\right)$$

7. Establish the equation for deflections V_1 in an elastic layer whose thickness decreases from left to right (i.e., H is linear in x) under the load. Assume a linear variation of the displacement function g_1 with depth. Solve the equation if you can.

8. How would you tackle the problem of a finite two-layer elastic foundation by the V-L method? (Hint: see Problem 6.)

9. A line load is applied at the surface of an elastic layer. Assume a variation in the displacement function g_1 with depth according to equation (4.22) and obtain the surface deflection profile.

10. Compare the solution, equation (4.54), to the deflection of the elastic half-space with a point surface load with both the Boussinesq and Westergaard exact solutions and find the values of μ in equation (4.23) which gives the best match of approximate and exact solutions.

11. For the finite rigid beam resting on a two-parameter foundation and loaded by a point load at one end, calculate the moments in the beam at four equally-spaced points along the beam for a range of values of α^2 ($k/2t$), and compare with the values for the beam on the one-parameter (k) foundation (Figure 4-7).

REFERENCES

[1] BURMISTER, D. M. "Stress and Displacement Characteristics of a Two-Layer Rigid Base Soil System: Influence Diagrams and Practical Applications," *Proceedings Highway Research Board*, *35*, 773 (1956).

[2] FILONENKO-BORODICH, M. M. "Some Approximate Theories of Elastic Foundation," *Uch. Zap. Mosk. Gos., Univ. Mekh.* *46*, 3–18, 1940 (in Russian).

[3] HARR, M. E., DAVIDSON, J. L., HO, D. M., POMBO, L. E., RAMASWAMY, S. V., AND J. C. ROSNER. "Euler Beams on a Two-Parameter Foundation Model," *Proc. ASCE*, *95*, SM4, 933–948 (July 1969).

[4] HETENYI, M. *Beams on Elastic Foundation*. Ann Arbor: University of Michigan Press, 1946.

[5] JONES, R., AND J. XENOPHONTOS. "The Vlasov Foundation Model," *Int. Journal of Mech. Sci.*, *19*, 317–323 (1977).

[6] KAMESWARA, RAO, N. S. P., DAS, Y. C., AND M. ANANDAKRISHNAN. "Variational Approach to Beams on Elastic Foundations," *Proc. ASCE*, *97*, EM, 271–294 (April 1971).

[7] PASTERNAK, P. L. *On a New Method of Analysis of an Elastic Foundation by Means of Two Foundation Constants*. Moscow: Gos. Izd. Lit. po Stroit i Arkh., 1954 (in Russian).

[8] VLASOV, V. Z., AND N. N. LEONTIEV. "Beams, Plates, and Shells on Elastic Foundations," translated from Russian by Israel Program for Scientific Translations, NTIS No. N67-14238 (1966).

[9] WESTMANN, R. A., AND N. C. TSAI. "Beam on a Tensionless Foundation," *Proc. ASCE*, *93*, EM5, 1–12 (Oct. 1967).

[10] WIEGHARDT, K. "Beams on Deformable Foundation," *ZAMM*, *2*, 165–184 (June 1922) (in German).

Flexible Beams and Slabs on Winkler Foundation

5

5.1 INTRODUCTION

Among the variety of models proposed to represent the half-space response under various structures, that attributed to Winkler[1] (28) has received the most attention and is widely used in the analysis of foundations. This is the case because the mathematics involved in a study of beam and slab behavior on a Winkler material is not too complicated, and many of the functions involved have been conveniently tabulated by a number of authors. The complexity of the analysis increases fairly rapidly for foundation models that possess more features than the simple Winkler representation.

Many investigators have devoted their efforts to the study of beams and slabs on an elastic foundation, and a number of books deal extensively with the problem, notably those by Hayashi (10), Hetenyi (11), Vlasov and Leontiev, Wölfer (30), and Sherif (23), the latter two of which contain exten-

---

[1]The spring model was first suggested by Euler and employed by his amanuensis, Füss, in 1801, so that Russian investigators refer to the representation by the name "Winkler-Füss." Since the model was used extensively by Zimmerman (32) a few years after Winkler treated it in his book, sometimes it is described by the name "Winkler-Zimmerman."

sive tables of numerical solutions for a variety of boundary conditions. Early studies of the deflections and stresses in railroad tracks and ties by Zimmerman (32) and Schwedler (22) used the Winkler model.

We take up the topic of Winkler and two-parameter models in this chapter, including some discussion of the question of their identification with more exact representations. The impact and practical aspect of identifying a particular value of the foundation constant with actual soil properties are left to a subsequent chapter.

5.1.1 Beam equations; general solution

With Winkler behavior, the relation between the vertical pressure, p, acting on the solid surface, and the vertical deflection, w, is, as in equation (4.1),

$$p = k_0 w \qquad (5.1)$$

It is convenient to begin a discussion with reference to the two-dimensional problem of a beam, as demonstrated by Hayashi (10) and Hetenyi (11). As in section 4.3, the subgrade modulus k_0 must be multiplied by the beam or strip width, to give the appropriate value for use in the analysis; thus, for the beam, $k = k_0 b$ and has dimensions FL^{-2}.

A beam or strip deflected under a variety of loads is illustrated in Figure 5-1. With the usual assumptions relating to a thin beam, the differential equation for the deflected shape of the beam away from the loaded area is

$$\frac{d^2 M}{dx^2} = pb \qquad (5.2)$$

When a load such as q is acting on the beam, the right-hand side is $(p - q)b$. Replacing p from equation (5.1), and including the beam width, we see that the equation becomes

$$\frac{d^2 M}{dx^2} = kw \qquad (5.3)$$

There are two ways to proceed from this equilibrium equation. The conventional approach is to formulate the equation defining the state of the beam

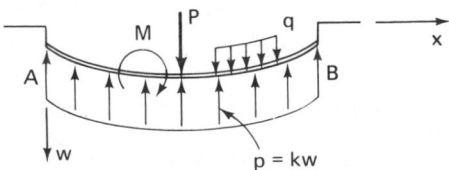

Figure 5-1.   Finite flexible beam on Winkler foundation; generalized loading.

in terms of displacement by using the bending moment-curvature relation

$$M = -EI\frac{d^2M}{dx^2} \tag{5.4}$$

differentiating this twice with respect to x, assuming EI to be constant, and substituting the result in equation (5.3) to get

$$EI\frac{d^4w}{dx^4} + kw = 0 \tag{5.5}$$

Most solutions for beam, plate, and pile problems are developed on this basis and are those used largely in this book.

   However, as a second description, sometimes it is convenient to use an equation based on moment, the complementary formulation of the state of the beam. This is obtained by differentiating equation (5.3) twice with respect to x to get

$$\frac{d^4M}{dx^4} - k\frac{d^2w}{dx^2} = 0 \tag{5.6}$$

assuming k to be independent of x. Now equation (5.4) can be used to substitute for the second term in equation (5.6), so that the result becomes

$$\frac{d^4M}{dx^4} + \frac{k}{EI}M = 0 \tag{5.7}$$

an equation of the same form as (5.5) but expressed in terms of the moment in the beam.

   The alternative expression is useful in a variety of problems. If the moment of inertia of the beam varies with x, $I(x)$, the procedure leading to equation (5.5) gives a more general fourth-order differential equation:

$$EI(x)\frac{d^4w}{dx^4} + 2EI(x)\frac{dI(x)}{dx}\frac{d^3w}{dx^3} + E\frac{d^2I(x)}{dx^2}\frac{d^2w}{dx^2} + kw = 0 \tag{5.8}$$

which must be solved to give deflections, moments, etc., in the loaded beam. On the other hand, the derivation of equation (5.7) did not include differentiation of $I(x)$, which therefore appears explicitly in the equation as it stands. A similar consideration applies in the case in which the soil-beam property k varies with x. No derivative of $k(x)$ was required in the generation of equation (5.5), and the k there represents $k(x)$ in the general case. Differentiation of $k(x)$ is, however, required in obtaining equation (5.7), which becomes more complicated in this case.

   If, therefore, we face a problem of a beam on an elastic foundation whose reaction coefficient k varies along the length of the beam, it is better to work with equation (5.5) since the required k-function appears directly there. Conversely, when the situation includes a beam whose moment of inertia varies with distance, the more convenient equation to use is (5.7). For example, k may vary linearly with x, say,

$$k = k_1 x$$

and equation (5.5) takes the form

$$EI\frac{d^4 w}{dx^4} + k_1 xw = 0 \qquad (5.9)$$

whose solution must be obtained. As discussed in Chapter 8, this has been done and numerical tables of the solution are available in Appendix C. With them, the usual quantities of displacement, moment, and so on, can be obtained for specific loading and boundary conditions. The same solution would also apply to the appropriate form of equation (5.7), but what would it mean in that case? A linear dependence of the coefficient of the second term on the left side, with k and E held constant, would imply that

$$I = \frac{I_h}{x} \qquad (5.10)$$

or that the moment of inertia varied hyperbolically along the beam. Since, for the assumption of constant subgrade coefficient, the beam width would have to remain constant, this might come about in an unreinforced beam by a variation of the beam depth with $x^{-1/3}$ or by some distribution of reinforcing steel. Such a beam would have a higher moment of inertia at the center and diminishing toward the ends. Although unusual, such an arrangement is possible, and it would have the advantage of placing the strongest section where, for centrally loaded beams, the greatest moment occurs.

Other variations of the different functions k and I with distance can be inserted in the appropriate equation, (5.5) or (5.7), depending on the problem.

There is one other aspect of utility in the complementary formulation when approximate solutions are being sought by the methods of undetermined coefficients. The essential boundary conditions for equation (5.5) are the displacements and slopes, whereas for equation (5.7) they are moments and shears. Consequently, if, for example, a beam with free ends resting on an elastic foundation and subjected to load is being studied by equation (5.5), none of the end conditions is essential (moment and shear equal zero at the ends). If, alternatively, the same problem expressed in terms of equation (5.7) is being examined, both end conditions are essential, since shears are the gradients of moment.

Returning to equation (5.5), we see that solutions for particular cases of loading can be obtained from the general solution to this equation. The solution is

$$w = (C_1 \cos \lambda x + C_2 \sin \lambda x)e^{\lambda x} + (C_3 \cos \lambda x + C_4 \sin \lambda x)e^{-\lambda x} \qquad (5.11)$$

where C_1 through C_4 are constants to be determined by the loading and end conditions, and

$$\lambda = \sqrt[4]{\frac{k}{4EI}} \qquad (5.12)$$

The parameter λ has dimensions L^{-1} and can be seen to include both the properties of the soil subgrade and the beam. The *reciprocal* of λ is a characteristic length of the soil-beam system. When the beam is very stiff in comparison with the soil, the characteristic length is large, implying that a load applied to the beam will cause deflections of the beam and the soil to a considerable distance from the point of action of the load. Alternatively, a soft beam and stiff soil combination will, in this Winkler model of soil behavior, result in a small characteristic length. In this case, the load will have only a local effect on deflections and stresses. The characteristic length $1/\lambda$ or λ itself is thus the measure of the interaction between the beam and the foundation. It can also be seen that all components of the solution are oscillatory and that two of them decay exponentially with distance, whereas the other two grow exponentially. Equation (5.11) with w replaced by M, and with the same value of λ as in equation (5.12), is also, of course, the general solution for equation (5.7).

5.2 SPECIFIC SOLUTIONS

Equation (5.11) describes the deflected shape of the beam between loading points or support constraints. The coefficients are determined from the boundary conditions, equilibrium, and continuity between adjacent sections of the beam. Specific examples will be used to illustrate the derivation of individual solutions.

5.2.1 Infinite beam

These considerations can be readily illustrated by considering the case of the infinitely long beam, loaded, in the first instance, by a vertical point load (line load in slab strip problem) P applied at the origin of the coordinates as shown in Figure 5-2. Since the vertical deflection will be zero at infinity in the positive x-direction, it follows that C_1 and C_2 must be equal to zero in equation (5.11) for $x > 0$. In addition, immediately under the load the slope of the deflected beam must be zero, for reasons of symmetry.

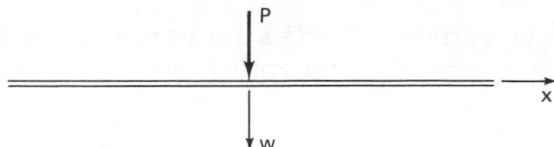

Figure 5-2. Infinite beam under point load.

Differentiating the remaining parts of equation (5.11) and setting the result equal to zero at $x = 0$ has the result that $C_3 = C_4 = C$, say. Thus, in this case, equation (5.11) is reduced to the form

$$w = C(\cos \lambda x + \sin \lambda x)e^{-\lambda x} \tag{5.13}$$

Vertical equilibrium requires that the active force P be resisted by the total subgrade reaction force. Thus, considering one-half of the beam, we have

$$\frac{P}{2} = \int_0^\infty kw \, dx = \int_0^\infty kC(\cos \lambda x + \sin \lambda x)e^{-\lambda x} \, dx$$

so that

$$P = \frac{2kC}{\lambda} \tag{5.14}$$

and, eventually, replacing C in equation (5.13) by its value in equation (5.14), we get

$$w = \frac{P\lambda}{2k}(\cos \lambda x + \sin \lambda x)e^{-\lambda x} \qquad x > 0 \tag{5.15}$$

The deflection is, of course, symmetrical about $x = 0$. The maximum deflection, w_m, occurs under the load P and is given by

$$w_m = \frac{P\lambda}{2k} \tag{5.16}$$

Zero deflections occur at points given by the zeros of equation (5.19), that is, at points where

$$\lambda x = \frac{3}{4}\pi, \qquad \frac{7}{4}\pi, \qquad \frac{11}{4}\pi \ldots$$

By differentiating equation (5.15) successively, equations for the slope, moment, and shear in the beam can be obtained:

$$\theta = \frac{dw}{dx} = -\frac{P\lambda^2}{k}e^{-\lambda x}(\sin \lambda x) \tag{5.17}$$

$$M = -EI\frac{d^2w}{dx^2} = \frac{P}{4\lambda}e^{-\lambda x}(\cos \lambda x - \sin \lambda x) \tag{5.18}$$

$$S = -EI\frac{d^3w}{dx^3} = -\frac{P}{2}e^{-\lambda x}\cos \lambda x \tag{5.19}$$

These results are shown in Figure 5-3(a). It is worth noting that displacement of the beam depends on $k^{-3/4}$, and it thus varies almost directly with the reciprocal of k, whereas the moment is much less sensitive to the reaction coefficient, being related to $k^{-1/4}$. The choice of subgrade reaction coefficient to represent a particular soil is thus more important for the calculation of settlements than of moments.

If instead of a force P applied at the origin in this problem, we were to apply a moment M_0, the variable parts of equations (5.17), (5.18), (5.19), and (5.15) *in that order* now represent the solutions for deflection, slope, moment, and shear, respectively, with the constants, in the same order $-P\lambda^2/k$, $P/4\lambda$, etc., replaced by the constants $M_0\lambda^2/k$, $M_0\lambda^3/k$, $M_0/2$, and $-\lambda M_0/2$, respectively. The solutions in Figure 5-3 can therefore also be used for the moment case.

The maximum moment, M_m, and shear, S_m, occur under the point of load application and are, for the case of loading by P:

$$M_m = \frac{P}{4\lambda} \tag{5.20}$$

$$S_m = -\frac{P}{2} \tag{5.21}$$

In analyses of subgrade and pavement behavior we will be interested in the maximum compressive stress in the subgrade and the maximum tensile stress in the beam, as well as in the maximum deflection as given by equation (5.16). Some practical results are given in Fig. 5-3(b).

The maximum subgrade stress, p_m, from equation (5.16), is

$$p_m = kw_m = \frac{P\lambda}{2} \tag{5.22}$$

and the maximum tensile stress in the beam or strip is

$$\sigma_m = \frac{M_m h}{2I} = \frac{Ph}{8\lambda I} \tag{5.23}$$

where h is the depth of the beam or slab. In the case of unit width of strip, a substitution can be made for I and this expression can be rewritten as follows:

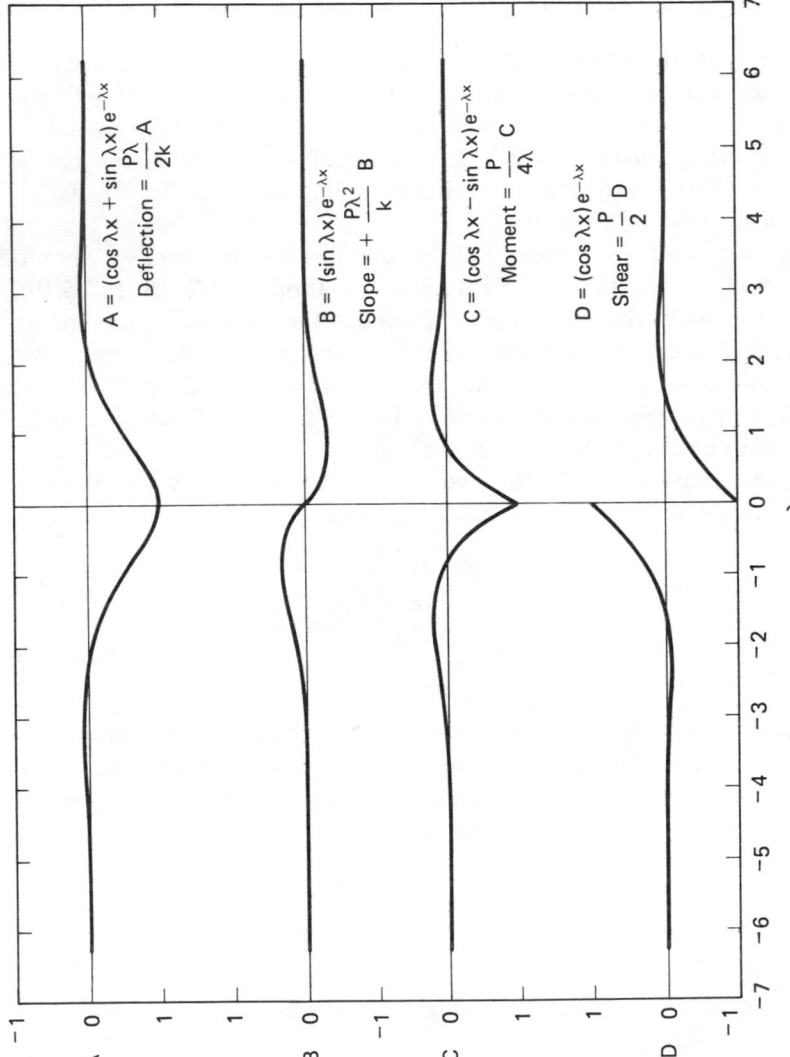

$A = (\cos \lambda x + \sin \lambda x) e^{-\lambda x}$

Deflection $= \dfrac{P\lambda}{2k} A$

$B = (\sin \lambda x) e^{-\lambda x}$

Slope $= + \dfrac{P\lambda^2}{k} B$

$C = (\cos \lambda x - \sin \lambda x) e^{-\lambda x}$

Moment $= \dfrac{P}{4\lambda} C$

$D = (\cos \lambda x) e^{-\lambda x}$

Shear $= \dfrac{P}{2} D$

λx

Figure 5-3. Deflection, slope, moment, and shear in infinite beam subjected to point load: (a) form of functions.

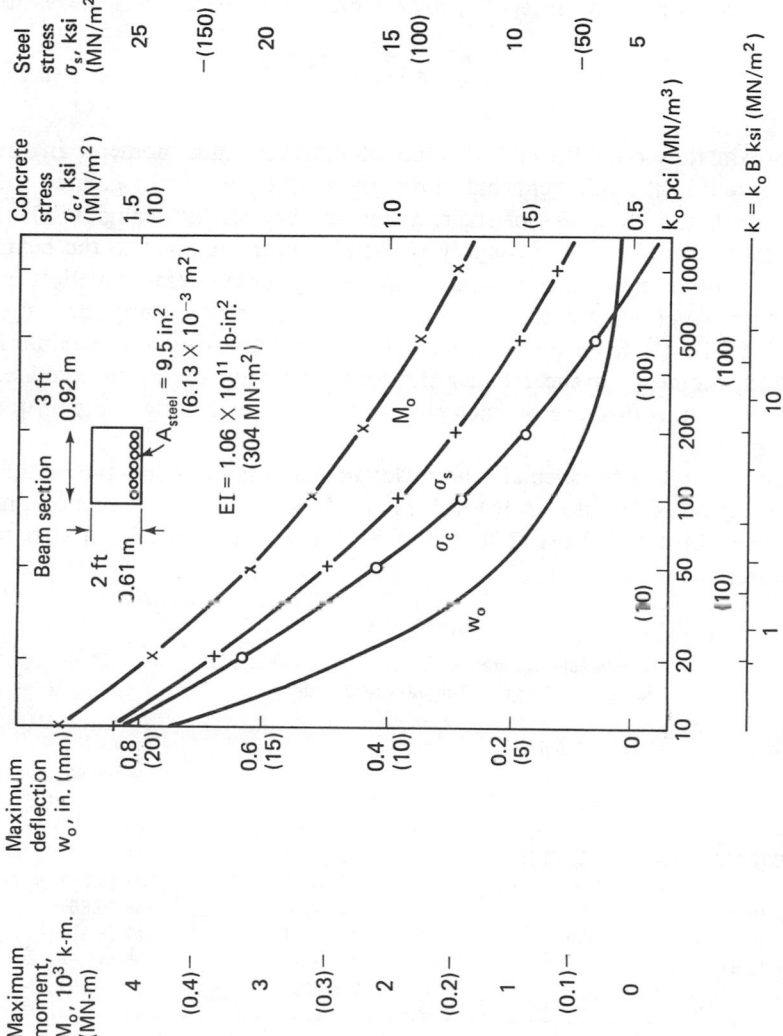

Figure 5-3. (b) Variation in maximum moment, deflection, concrete stress and steel stress with k_0 in infinite beam shown under 500 kip load.

$$\sigma_m = \frac{3P}{2\lambda h^2} \qquad (5.24)$$

It can be seen from Figure 5-3(a) that the maximum *negative* bending moment occurs at a distance of $\pm \pi/2\lambda$ from the load and has the value

$$M_n = -\frac{P}{4\lambda} e^{-\pi/2} \approx -\frac{0.052P}{\lambda} \qquad (5.25)$$

or approximately one-fifth of the value of the maximum moment. In the same figure it is also apparent that at distances of λx of 1.5π or greater, the values of all the quantities of interest become essentially negligible. This means that vertical loads of roughly equal magnitude applied to the beam at spacings of $1.5\pi/\lambda$ or greater would have a negligible interaction effect on each other. Alternatively, the nature in which the beam is supported at a distance of $1.5\pi/\lambda$ from the load makes little difference to the maximum deflection, moment, or shear under the load. In effect, the beam could be terminated at this distance on each side of the load with little effect on the loaded region.

It is of interest to calculate the order of magnitude of this interaction distance for practical cases. A few values are shown in Table 5.1. A modulus of elasticity of 4×10^6 psi (2.76×10^4 MN/m²) is assumed for a slab of *unreinforced* concrete.

TABLE 5.1

Interaction Distances for Various Unreinforced
Slab and Subgrade Properties

Beam Thickness, in. (m)	Subgrade Coefficient, pci (kN/m³)	1/λ in. (m)	Interaction Distance 1.5π/λ in. (m)
6 (0.15)	50 (135.8)	49 (1.24)	231 (5.87)
	500 (1358)	28 (0.71)	132 (3.35)
12 (0.31)	50	82 (2.08)	388 (9.86)
	500	46 (1.17)	217 (5.51)
24 (0.61)	50	139 (3.53)	655 (16.64)
	500	78 (1.98)	367 (9.32)

In the table the two selected subgrade moduli represent roughly soft and stiff soil behavior, respectively. The thin slab may be considered to be a pavement and the largest thickness that of a foundation slab. As is obvious from the fourth root appearing in equation (5.12), the effect of a wide range in subgrade property is relatively small. In the case of thin pavement, it is seen from this two-dimensional model that the individual wheels of typical vehicles cannot be taken to load the pavement separately; interaction effects are present.

Usually, columns or bearing walls in buildings are spaced at intervals of about 20 ft(6 m). The table shows that the slab provides sufficient continuity in the case of most materials and particularly softer soils (which are, of course, generally the reason for employing a continuous foundation) so that interaction effects *will* occur at the typical column spacing. We will examine this question again in the three-dimensional case.

In practice, the loads applied to beams, slabs, or pavements are distributed, rather than concentrated, and the calculations should be carried out for distributed loads. Since the system we have established is linear, the values of deflection, moment, etc., can be calculated by superposition for any form of distributed load. For our purposes here, it is sufficient to obtain the solution for a uniformly distributed load.

We represent the problem configuration in Figure 5-4. The distributed load is q per unit length of the beam and we want to know the values of the

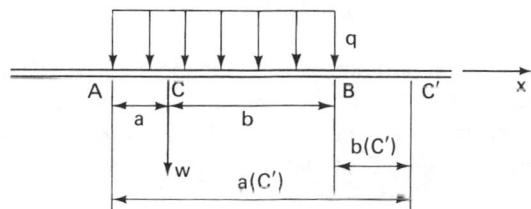

Figure 5-4. Uniform load on a portion of an infinite beam.

various quantities at point C which may be either within or outwith the loaded stretch of beam. It is taken inside the loaded area first. The point C is distant a from the left end of the load and b from the right end. The displacement solution is derived from the previous result, equation (5.15), through replacing P by $q\,dx$ and integrating within appropriate limits. We consider the deflection at C to be due to the deflection caused by the loaded area AC plus that caused by the loaded area CB. If we take the origin of the coordinates at C, the first integration should be performed between the limits $-a$ and zero. However, it can be observed that the resulting deflection would be the same if this portion of the load actually extended from the origin to a distance a to the right. Consequently, we have

$$w_C = \frac{q\lambda}{2k}\left[\int_0^a (\cos \lambda x + \sin \lambda x)e^{-\lambda x}\,dx + \int_0^b (\cos \lambda x + \sin \lambda x)e^{-\lambda x}\,dx\right]$$

or

$$w_C = \frac{q}{2k}[(1 - e^{-\lambda a}\cos \lambda a) + (1 - e^{-\lambda b}\cos \lambda b)] \qquad (5.26)$$

As before, differentiation gives the expressions for the remaining quantities at C:

$$\theta_C = \frac{q\lambda}{2k}[(\cos \lambda a + \sin \lambda a)e^{-\lambda a} - (\cos \lambda b + \sin \lambda b)e^{-\lambda b}] \qquad (5.27)$$

$$M_C = \frac{q}{4\lambda^2}(e^{-\lambda a}\sin \lambda a + e^{-\lambda b}\sin \lambda b) \qquad (5.28)$$

$$S_C = \frac{q}{4\lambda}[(\cos \lambda a - \sin \lambda a)e^{-\lambda a} - (\cos \lambda b - \sin \lambda b)e^{-\lambda b}] \qquad (5.29)$$

The maximum vertical displacement and moment occur when the point C is located below the midpoint of the loaded area, so that $b = a$ and a is one-half the loaded length:

$$w_m = \frac{q}{k}(1 - e^{-\lambda a}\cos \lambda a) \qquad (5.30)$$

$$M_m = \frac{q}{2\lambda^2}(\sin \lambda a)e^{-\lambda a} \qquad (5.31)$$

If the dimension a is comparable to the thickness of the beam, which by reference to Table 5-1 we will take for convenience equal to $\pi/12\lambda$ (this will approximately be the case for tire or column loads), the maximum deflection and moment at the midpoint C of the loaded area can be calculated explicitly:

$$w_m = \frac{0.26q}{k} = \frac{0.49Q\lambda}{k} \qquad (5.32)$$

$$M_m = \frac{0.10q}{\lambda^2} = \frac{0.19Q}{\lambda} \qquad (5.33)$$

where $Q = 2aq$ is the total load.

In comparing these with the maximum deflection and moment produced by the concentrated load, from equations (5.16) and (5.20), we see that the maximum deflection is essentially unchanged (as is the maximum pressure on the subgrade) but that the maximum moment has been reduced by about 25 percent.

When the point C is located *outside* the loaded area as shown at C' in Figure 5-4, different expressions result from the appropriate integration and subsequent differentiations. Here

$$w_{C'} = -\frac{q}{2k}(e^{-\lambda a}\cos \lambda a - e^{-\lambda b}\cos \lambda b) \qquad (5.34)$$

$$\theta_{C'} = \frac{q\lambda}{2k}[(\cos \lambda a + \sin \lambda a)e^{-\lambda a} - (\cos \lambda b + \sin \lambda b)e^{-\lambda b}] \qquad (5.35)$$

$$M_{c'} = \frac{q}{4\lambda^2}(e^{-\lambda a}\sin \lambda a - e^{-\lambda b}\sin \lambda b) \tag{5.36}$$

$$S_{c'} = \frac{q}{4\lambda}[(\cos \lambda a - \sin \lambda a)e^{-\lambda a} - (\cos \lambda b - \sin \lambda b)e^{-\lambda b}] \tag{5.37}$$

In practice, beams are generally finite in length and loads may be close enough to the ends that the assumption of an infinite beam is not valid. The conditions at the ends of the beam—specified deflections, slopes, moments, or shears—then control the constants in the solution. A simple illustration of the application of specific boundary conditions follows.

5.2.2 Semi-infinite beam

One of the problems we will be interested in later is that of the semi-infinite beam loaded by a vertical force P_0 and moment M_0 at the free end. Let us consider the two cases separately. The problem is illustrated by Figure 5-5. Once again, the conditions at infinity require C_1 and C_2 to be zero, but

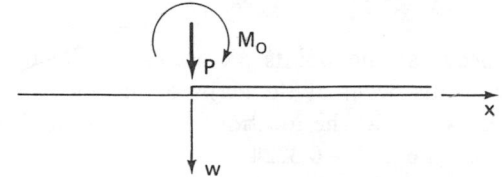

Figure 5-5. Force and moment loads on a semi-infinite beam.

C_3 and C_4 remain to be determined. Differentiating the displacement w in the remaining terms of equation (5.11) once, twice, and three times, respectively, with respect to distance x gives equations for the slope, moment, and shear in the beam as follows:

$$\theta = \frac{dw}{dx} = -\lambda e^{-\lambda x}[C_3(\cos \lambda x + \sin \lambda x) - C_4(\cos \lambda x - \sin \lambda x)] \tag{5.38}$$

$$M = -EI\frac{d^2w}{dx^2} = -2EI\lambda^3 e^{-\lambda x}[C_3 \sin \lambda x - C_4 \cos \lambda x] \tag{5.39}$$

$$S = -EI\frac{d^3w}{dx^3} = -2EI\lambda^3 e^{-\lambda x}[C_3(\cos \lambda x - \sin \lambda x) \\ + C_4(\cos \lambda x + \sin \lambda x)] \tag{5.40}$$

For the case of a load P_0 only acting at the origin, $M(0) = 0$, and $S(0) = -P_0$, so that, from equations (5.39) and (5.40), we find $C_4 = 0$ and

$$C_3 = \frac{P_0}{2\lambda^3 EI} = \frac{2\lambda P_0}{k} \tag{5.41}$$

Substituting in equation (5.11) gives the deflection

$$w = \frac{2\lambda P_0}{k} e^{-\lambda x} \cos \lambda x \qquad (5.42)$$

and the other quantities follow from the equations above:

$$\theta = -\frac{2\lambda^2 P_0}{k} e^{-\lambda x}(\cos \lambda x + \sin \lambda x) \qquad (5.43)$$

$$M = -\frac{P_0}{\lambda} e^{-\lambda x} \sin \lambda x \qquad (5.44)$$

$$S = -P_0 e^{-\lambda x}(\cos \lambda x - \sin \lambda x) \qquad (5.45)$$

The maximum deflection, by equation (5.42) occurs at the load and is given by the equation

$$w_m = \frac{2\lambda P_0}{k} \qquad (5.46)$$

Zero deflections occur at the points $\lambda x = \pi/2$, $3\pi/2$, etc. As usual, the extrema of the moments occur at the zeros of the shear, S, and equation (5.45) shows that these are at the locations $\lambda x = \pi/4$, $3\pi/4$, etc. The maximum moment, $M_m \lambda/P_0$, equals -0.3224.

Following the same reasoning, we can obtain the analogous quantities when the loading is due to a moment M_0 applied at the end of the beam, in a clockwise direction:

$$w = -\frac{2\lambda^2 M_0}{k} e^{-\lambda x}(\cos \lambda x - \sin \lambda x) \qquad (5.47)$$

$$\theta = \frac{4\lambda^3 M_0}{k} e^{-\lambda x} \cos \lambda x \qquad (5.48)$$

$$M = M_0 e^{-\lambda x}(\cos \lambda x + \sin \lambda x) \qquad (5.49)$$

$$S = -2\lambda M_0 e^{-\lambda x} \sin \lambda x \qquad (5.50)$$

It will be noted that the functions of λx occurring in these expressions are all the same as those appearing in equations (5.15) and (5.17) through (5.19) for the infinite beam case, although in different order. Accordingly, only the variable parts are plotted in Figure 5-6, which can then be used for a variety of cases.

For loading by the moment M_0 at the end of the beam, the maximum deflection again occurs at the load

$$w_m = -\frac{2\lambda^2 M_0}{k} \qquad (5.51)$$

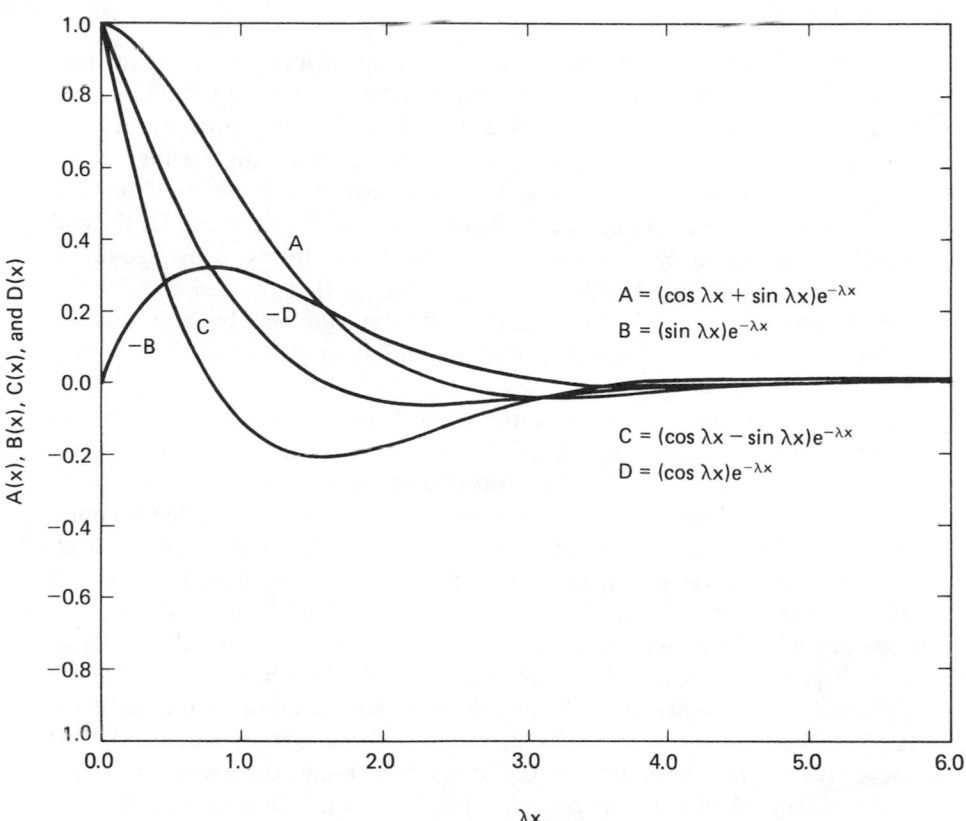

Figure 5-6. Variable parts of equations occurring in infinite and semi-infinite beam loading.

with zero deflections at $\lambda x = \pi/4$, $3\pi/4$, etc. The maximum moment in this case is, of course, caused by the load M_0 at the origin. Other maxima occur at the points $\lambda x = \pi$, 2π, etc.

Plots of deflection, slope, moment, and shear in dimensionless form for the semi-infinite beam are given in Figure 5-6.

We will return later to the semi-infinite beam solutions, which will prove of value in the analysis of the behavior of long piles under lateral and moment loads.

Footings and foundation beams are generally limited in length, and analysis must account for conditions at both ends of the beam, as discussed in the next section.

5.2.3 Finite beam

When the beam is subjected to a single, central load, there are, as previously, four constants to be determined in the equation for deflection. The conditions that are used to determine them in the direct approach are: (a) the equilibrium of the load system, where the applied load is equal to the foundation reaction plus any shear loads induced by support conditions at the beam ends; (b) zero slope of the beam at the load point; and (c) the end conditions of the beam, which can take a variety of forms. Most usually in foundation engineering the beam ends are free, so that the end condition is one of zero moment and shear. Occasionally the ends may be built in, presenting a requirement of zero displacement and slope. Other conditions may be encountered less commonly.

In the first case of zero end moment and shear, the three requirements listed above give rise to four equations in the four unknowns. The simultaneous solution of these is tedious, but it presents no particular difficulties. However, when a single load is located asymmetrically, the situation becomes more complicated if the conventional approach is adopted. Here, a deflection solution involving four constants applies to the displaced shape of the beam between the left-hand end and the load. The same form of solution, but with four different constants, is required to describe the displacement between the load and the right-hand end of the beam. Thus, there are eight unknown coefficients to be determined. They are obtained as follows: (a) Two boundary conditions exist, in general, at each end of the beam; these give rise to four equations. (b) At the load point, the deflection, slope, and moment calculated from the left side of the beam equal those calculated from the right; these give three additional equations. (c) The eighth equation is obtained either from the overall vertical equilibrium condition or from the requirement that at the load the shear discontinuity calculated from the difference between the shears calculated at this point from the left- and right-hand solutions, respectively, is equal to the applied load.

When the beam and soil properties are supplied in numerical terms, the result is a system of eight numerical equations in the eight unknowns. These may be solved by computer. But when a general solution is sought, the algebra is very tedious, although it has been carried through for a number of cases of end conditions by Hayashi (10). There is an alternative technique, however, involving the principles of superposition.

For finite beams Hetenyi (11) has presented a systematic approach based on the results determined for an infinite beam. This approach is called the *method of end conditioning*. The displacements, slopes, moments, and shears at the locations corresponding to the ends of the real finite beam are, in effect, calculated by using the expressions developed, as above, for the infinite beam. Then, in Hetenyi's method, fictitious moments and forces are applied just outside the locations in the infinite beam where the ends of the finite beam occur, in such a way as to achieve the desired end conditions of the real beam. The location infinitesimally beyond the ends avoids the ambiguities involved in those quantities which are discontinuous. If, for example, an end is free, a fictitious moment and shear must be applied just beyond this point in the infinite beam so as to cancel out the moment and shear produced there by the applied loading. The quantities of deflection, moment, etc., along the beam are obtained by the superposition of the real and fictitious quantities. If the end is hinged so that the vertical deflection and moment there are zero, this is the condition to be produced by the fictitious loading.

As an example, we set out the problem of the beam of length l loaded centrally, as in Figure 5-7. There the fictitious shears and moments S_{0A},

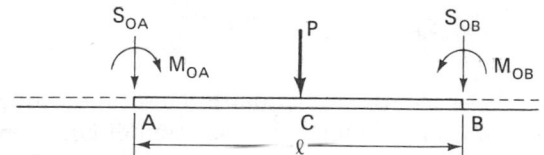

Figure 5-7.   Centrally loaded finite beam.

S_{0B}, M_{0A}, and M_{0B}, positive as shown, are placed just beyond points A and B. In general, for an asymmetric loading, the fictitious values S_{0A} and so on, must be calculated to produce the desired boundary conditions at A and B. We would thus obtain four equations in these four unknowns. In the present case, from symmetry $S_{0B} = S_{0A}$, $M_{0B} = M_{0A}$, and there are only two unknowns. We need only consider the end conditions at point A, therefore, to give the two equations required. The most common circumstance in foundation engineering is that ends A and B are free, with zero moment and zero shear at these points. All the quantities required are obtained from equations (5.18) and (5.19) for the point load on the infinite beam and from their counterparts for the moment load. Thus, considering first moment at point A, we have

(1)   (2)   (3)   (4)   (5)

$$M_A + \frac{S_{0A}}{4\lambda} + \frac{S_{0B}}{4\lambda}e^{-\lambda l}(\cos \lambda l - \sin \lambda l) + \frac{M_{0A}}{2} + \frac{M_{0B}}{2}e^{-\lambda l}\cos \lambda l = 0 \quad (5.52)$$

in which all quantities are the moments that would be produced at A by the individual loads were the beam infinite in length. Term (1) is the moment generated by the real load, in this case P; term (2) is the moment given by the fictitious load S_{0A} applied just to the left of A; term (3) is developed by fictitious S_{0B} just to the right of B; term (4) is the moment produced at A by the fictitious moment M_{0A} applied just to the left of A; and term (5) is that given by fictitious M_{0B} to the right of B. For the given applied load P, here, from equation (5.18)

$$M_A = \frac{P}{4\lambda} e^{-(\lambda l/2)} \left(\cos \frac{\lambda l}{2} - \sin \frac{\lambda l}{2} \right) \tag{5.53}$$

In the case of shear at A, the relevant equations are assembled to give the expression

$$S_A - \frac{S_{0A}}{2} + \frac{S_{0B}}{2} e^{-\lambda l} \cos \lambda l - \frac{\lambda M_{0A}}{2} + \frac{\lambda M_{0B}}{2} (\cos \lambda l + \sin \lambda l) e^{-\lambda l} = 0 \tag{5.54}$$

where, from equation (5.19), the shear S_A due to P is

$$S_A = + \frac{P}{2} e^{-(\lambda l/2)} \cos \frac{\lambda l}{2} \tag{5.55}$$

(positive because it is to the left of the load). Substitution for S_{0B} and M_{0B} in terms of S_{0A} and M_{0A} and simultaneous solution of equations (5.52) and (5.54) give values for the fictitious quantities S_{0A} and M_{0A}. Then the deflection, slope, moment, and shear in the beam are obtained by summing the relevant quantities obtained from all the loads, both fictitious and real.

Inspection of equations (5.52) and (5.54) reveals that a general algebraic solution for even this relatively simple case is laborious; it will not be given here. Instead, it is generally less tedious to solve each numerical problem individually. The plotted solutions of Appendix C can assist in choosing beam sections for a given loading initially. We present only the quantities of interest at the middle (C in Figure 5-7) and end points (A or B).

$$w_C = \frac{P}{2kl}(1 + a) \tag{5.56}$$

$$M_C = \frac{Pl}{4}(1 - b) \tag{5.57}$$

$$S_C = -\frac{P}{2} \tag{5.58}$$

$$w_{A,B} = \frac{2P}{kl} c \tag{5.59}$$

where

$$a = \frac{2 + \cos \lambda l - \sin \lambda l + e^{-\lambda l}}{\sinh \lambda l + \sin \lambda l} \tag{5.60}$$

$$b = \frac{\cos \lambda l + \sin \lambda l - e^{-\lambda l}}{\sinh \lambda l + \sin \lambda l} \qquad (5.61)$$

and

$$c = \frac{\cosh \dfrac{\lambda l}{2} \cos \dfrac{\lambda l}{2}}{\sinh \lambda l + \sin \lambda l} \qquad (5.62)$$

The slope at the load point is, of course, zero.

A numerical solution is given to illustrate the method. A plain concrete beam is selected, approximately 24 in. (0.6 m) deep, such that a 12-in. (0.3 m) width of the beam has a moment of inertia, I, equal to 1.56×10^4 in.^{4} (6.5×10^{-3} m^4). The load P is applied to the 12-in. (0.3 m) width (for greater widths, the load increases proportionately). The beam, of length $l = 240$ in. (6.1 m), rests on a soil of soft to medium stiffness with k_0 equal to 100 pci (2.7 kN/m^3). What are the maximum deflection of and the moment in the beam when the load is applied centrally and when the concrete has an elastic modulus, E, of 4×10^6 psi (2.76×10^4 MN/m^2)?

From equation (5.12) the value of $1/\lambda = 120$ in. (3.05 m) and thus $\lambda l = 2.00$. Substituting the appropriate values in equations (5.52) and (5.53) gives the moment equation

$$\frac{120}{4} S_{0A}(0.8207) + \frac{M_{0A}}{2}(0.9437) = -\frac{120P}{4}(-0.1109) \qquad (5.63)$$

and in equations (5.54) and (5.55), the shear force equation

$$-\frac{S_{0A}}{2}(1.0563) - \frac{M_{0A}}{240}(0.9333) = -\frac{P}{2}(0.1987) \qquad (5.64)$$

Simultaneous solution of (5.62) and (5.63) yields

$$S_{0A} = 0.2213P; \qquad M_{0A} = -4.5040P \qquad (5.65)$$

The positive sign on S_{0A} indicates that the shears act down as shown in Figure 5-7, whereas the negative value of M_{0A} shows that M_{0A} is actually counterclockwise, and M_{0B} clockwise, opposite the directions illustrated in the figure.

With these values for the fictitious forces, the deflections, slopes, etc., in general can be calculated. In particular, the deflection at the midpoint can be obtained from equation (5.15) and the corresponding equation for an applied moment as follows:

$$w_C = \frac{P\lambda}{2k} + \frac{2S_{0A}\lambda}{2k} e^{-(\lambda l/2)} \left(\cos \frac{\lambda l}{2} + \sin \frac{\lambda l}{2} \right) + \frac{2M_{0A}\lambda^2}{k} e^{-(\lambda l/2)} \sin \frac{\lambda l}{2} \qquad (5.66)$$

Substituting the appropriate values gives

$$w_C = (3.4722 \times 10^{-6} + 0.7812 \times 10^{-6} - 0.1614 \times 10^{-6})P \qquad (5.67)$$

illustrating the different contributions of the components, or

$$w_C = 4.0920 \times 10^{-6}P \qquad (5.68)$$

Application of the formula, equation (5.56), gives the same value. The first term of equations (5.66) and (5.67) is the deflection at the load point calculated as if the beam were infinite. It is seen that it is modified only slightly, even in this fairly stiff beam, by the freeing of the ends, whose contribution appears in the second and third terms. The parameter k appearing in these equations is, as usual, equal to k_0B, or here 1200 psi (8.3 MN/m^2). If the beam were, say, 36 in. (0.91 m) wide, k would increase proportionately, as would I, so that λ would remain the same. The load P as employed here is the load per 12 in. (0.31 m) of width. A load of, say, 30 kips (134 kN), on the 12-in. (0.31 m) width would give, according to equation (5.68), a deflection of 0.12 in. (3 mm).

Again, from equation (5.18) and the corresponding expression when the load is a moment, an equation for the maximum moment can be constructed

$$M_C = \frac{P}{4\lambda} + \frac{2S_{0A}}{4\lambda}e^{-(\lambda l/2)}\left(\cos\frac{\lambda l}{2} - \sin\frac{\lambda l}{2}\right) + \frac{2M_{0A}}{2}e^{-(\lambda l/2)}\cos\frac{\lambda l}{2} \qquad (5.69)$$

When the numerical values are substituted, we get

$$M_C = (30 - 1.4725 - 0.8949)P = 27.63P \qquad (5.70)$$

as is confirmed by the use of the relation (5.57). An even larger proportion of the moment arises from the infinite beam contribution at the load point. For a 30-kip (134 kN) load the maximum moment would be 8.29×10^5 lb-in. (94 kN-m).

For simplicity in this example, the concrete beam was taken to be unreinforced. The above load would induce a tensile stress in the concrete of about 638 psi (4.40 MN/m^2), which is approximately the real (no safety factor) tensile strength of concrete. The beam would, in all likelihood, crack under this load.

It is seen from equations (5.59) and (5.62) that the term $\cos(\lambda l/2)$ appears in the numerator of the expression for the deflection of the ends of the centrally loaded beam. In consequence, the deflection there is zero when the beam length is π/λ. It follows that the ends of a weightless, centrally loaded, longer beam will lift off the ground and that only the length π/λ will remain in contact. This is called the *effective* length of a longer beam. In practice, for foundation beams, this rarely or never occurs because of the weight of the real beam and the distribution of loads. However, it can occur with railroad ties, for example, and it may develop in other circumstances of the application of the Winkler model.

Numerical solutions have been worked out for a number of other cases such as a point load applied to various stations along the length of a finite beam. The relevant equations and diagrams are given in Appendix C. Tables of numerical values are also given by Wölfer (30) and Sherif (23). The reader is referred to Hetenyi (11) for more details on the end-conditioning method.

Another approach using the boundary conditions at the end of a beam has been employed, mostly by Russian investigators. It proceeds as follows. Equation (5.11) with its coefficients C_n represents the displaced shape of the beam from its left-hand end up to the point at which the first load (force, moment, pressure) is applied, beyond which point the form of the equation still holds, but the constants are altered by the load. The new form of the equation stands for the next portion of the beam from this point to the next load section, and so on. The left-end conditions can be described by equation (5.11) and its derivatives representing slope, moment, and shear when the coordinate distance x is made zero, so that

$$w_0 = C_1 + C_3$$

$$\theta_0 = \left(\frac{dw}{dx}\right) = \lambda(C_1 + C_2 - C_3 + C_4)$$

$$M_0 = 2\lambda^2 EI(-C_3 + C_4) \tag{5.71}$$

$$S_0 = 2\lambda^3 EI(C_1 - C_2 - C_3 - C_4)$$

Since at least some of the end conditions are known (for example, M_0 and S_0 equal zero for a free end), we may solve these equations to obtain the C_n in terms of the end parameters, y_0, θ_0, etc. Once the expressions for C_n have been written, they can be substituted back in the original equation (5.11) to give an expression, after some rearranging, in the form

$$w = w_0 f_1(\lambda x) + \frac{1}{\lambda}\theta_0 f_2(\lambda x) - \frac{1}{\lambda^2 EI}M_0 f_3(\lambda x) - \frac{1}{\lambda^3 EI}S_0 f_4(\lambda x) \tag{5.72}$$

where the f_n are functions of λx. Other equations for θ, M, and S in general can be derived from this expression. We can consider that each term in equation (5.72) indicates the effect of the appropriate end condition on the elastic curve of the beam to the right. Thus, the end deflection w_0 has the influence $w_0 f_1(\lambda x)$ on the deflection to the right; end moment M_0, the effect

$$-\frac{f_3(\lambda x)M_0}{\lambda^2 EI}$$

and so on. If any end conditions are known, for example, $M_0 = S_0 = 0$, the related terms disappear from this portion of the deflection function. It follows that other requirements placed at other locations on the beam would modify the elastic shape similarly. The terms in equation (5.72) are therefore influence functions. A load P_1, for example, located at the point $x = x_1$ will modify the deflection to the right of x_1 by the amount (as in the case of shear)

$$+\frac{f_4[\lambda(x-x_1)]P_1}{\lambda^3 EI}$$

positive since the shear induced in $x > x_1$ by P_1 acting downward in the usual sense is negative. Thus, equation (5.72) becomes, for $x > x_1$ in this case of a load P_1 at x_1,

$$w = w_0 f_1(\lambda x) + \frac{1}{\lambda}\theta_0 f_2(\lambda x) - \frac{1}{\lambda^2 EI}M_0 f_3(\lambda x)$$
$$-\frac{1}{\lambda^3 EI}S_0 f_4(\lambda x) + \frac{1}{\lambda^3 EI}P_1 f_4[\lambda(x-x_1)] \qquad (5.73)$$

The consequence of a moment M_1 applied at x_2 can be developed similarly by adding the appropriate term to equation (5.73) which then holds in its modified form for $x > x_2$. Companion equations for slope, moment, and shear follow for each interval of the beam between loads.

Eventually, at the right-hand end of the beam, $x = l$, all four equations, modified by the applied loads, can be written to give w_l, θ_l, M_l, and S_l. Two of these will normally be known, for example again, for a free end $M_l = S_l = 0$. Consequently, in the equations for M_l and S_l, the left sides will be zero, and, if the left end is also free, we will have M_0 and S_0, which appear in these equations, also equal to zero. The only unknowns remaining in all equations are w_0 and θ_0 which are obtained by solving the two (M_l and S_l equal to zero) equations. Substitution of the determined w_0 and θ_0 in the other general equations for w, θ, M, and S in all the beam intervals supplies the complete solution to the problem.

This method of initial conditions or parameters has been extended by Vlasov and Leontiev who have included the two-parameter foundation model in its development.

Generally, structural members resting on or in the ground have uniform properties along their lengths. However, on occasion, the section may change. Methods of handling the problem when this happens are detailed by Hetenyi and by Vlasov and Leontiev, but they will not be discussed here except to say that a number of the numerical techniques described earlier can be applied in such situations. On the other hand, whereas the *soil* properties giving rise to the spring constant k_0 for beams or slabs resting at or near ground surface are usually fairly constant laterally, they may vary with depth in the case of piles. It would be expected, for instance, that a beam in the form of a vertical pile imbedded in sand would experience a subgrade reaction coefficient that would increase in some fashion with depth. This behavior is too important to ignore in analysis and solutions are therefore required for the case of a loaded beam supported by a soil whose spring constant varies with distance along the beam. Such cases will be treated in Chapter 8.

5.2.4 Considerations of the subgrade reaction coefficient

In reality, the soil subgrade is a continuum, generally neither homogeneous nor isotropic. Even if the same soil in a fairly uniform state is present to a considerable depth, the increasing overburden pressure with depth causes the stiffness of the material to increase with depth. It would seem that a representation of the soil by some kind of continuum would be more suitable than the use of the Winkler foundation model. However, even if a linearly elastic semi-infinite continuum is employed, the mathematical solution of the problem of the deflection of a loaded beam or plate resting on it is not easy. Because of the convenience of the simple Winkler solution, two related questions have therefore been asked by investigators: (a) How different are the results obtained by using a continuum representation and a Winkler model for the subgrade under the same beam and the same load? (b) In comparison with tests of beams on real soils, which model better represents the behavior of a real subgrade soil in its most uniform state?

The second question will be answered first. In spite of the importance of this problem, few attempts have been made to treat it experimentally. The most pertinent studies have been made by Vesic (24) who finds that Winkler models represent the behavior of beams on soil fairly well. Investigators on slabs [studies described by Vesic (25)] find that the actual maximum moment in large slabs appears to be somewhat less than that calculated by the Winkler model, although this depends on the selection of the subgrade reaction coefficient. Correlation of the two-parameter model with beam or slab tests does not appear to have been attempted.

The first question develops into a problem of identifying the two models. In the case of the beam on the homogeneous Winkler representation, there is one property k; the linearly elastic isotropic continuum requires two, a modulus and Poisson's ratio. A comparison between results in the two cases will depend on the values selected for these properties. The problem of an infinite beam loaded by a point load and resting on a linearly elastic three-dimensional subgrade was solved by Biot (6) who evaluated only the maximum bending moment in the beam. Biot's solution was extended by Vesic (24) who gave the distribution of deflection, slope, moment, shear, and pressure along the beam. When the supporting pressure at any point along the beam was divided by the deflection, Vesic found the ratio to be essentially constant along the beam, indicating a fairly close correspondence between the continuum solution and a Winkler solution in which an appropriate value of k was employed. The evaluation enabled Vesic to choose a relation between k to be used in the simpler problem and the material properties in the continuum as follows:

$$k = \frac{0.65E_s}{1 - v_s^2} \sqrt[12]{\frac{E_s B^4}{EI}} \qquad (5.74)$$

where E_s and v_s are the Young's modulus and Poisson's ratio for the subgrade, respectively, and B is the beam width. This value for k causes the corresponding values of all the variables in the two problems to be close. The actual differences, described as an "error" of the Winkler results using the continuum solution as a reference, are functions of the ratio of the characteristic length $(1/\lambda)$ of the beam to its width and are shown by Vesic (24) to be generally less than 10 percent.

If, of course, an exact correspondence between the two values of a selected variable such as maximum deflection or maximum moment were desired, a different value of k would be required for the Winkler solution. For example, Biot made the maximum *moment* in the beam identical in the two models by choosing

$$k = \frac{0.95E_s}{(1 - v_s^2)} \left[\frac{E_s B^4}{(1 - v_s^2)EI} \right]^{0.108} \qquad (5.75)$$

For typical material and beam properties,[2] the term in brackets in equation (5.75) is close to unity. It follows that the maximum *moments* obtained from the Winkler and continuum solutions will be nearly identical if k is taken equal to E_s. However, the selection of k by equation (5.75) increases the differences in the other variables.

In the method of Vlasov and Leontiev, the value of k, the coefficient of subgrade reaction, is arrived at rationally. It is given by the expression

$$k = \frac{E_s(1 - v_s)B}{(1 + v_s)(1 - 2v_s)} \int_0^\infty (g')^2 \, dz \qquad (5.76)$$

in which g is the assumed function describing the variation of vertical displacement with depth in the supporting layer, which may be either finite or infinite in depth. The material properties E_s and v_s represent the Young's

---

[2]This does not apply to members whose EI is very great, such as large offshore piles.

modulus or Poisson's ratio of the underlying medium in a plane strain problem. In a plane stress condition they are replaced, by E_0 and v_0, where

$$E_0 = \frac{E_s}{1 - v_s^2}; \qquad v_0 = \frac{v_s}{1 - v_s} \tag{5.77}$$

As described in Chapter 4, a variety of functions may be selected for g: One linear with depth is appropriate for a shallow soil layer; another varying exponentially with depth would suit a deep layer or the semi-infinite medium. Considering the semi-infinite medium, we may choose

$$g = e^{-\mu z} \tag{5.78}$$

as in equation (4.23) in which z is depth and μ is a constant of dimension L^{-1} expressing the rate at which vertical displacement decays with z. It can be employed as a matching constant to enable a particular exact solution to be fitted, or it can be employed for other reasons.

By using equation (5.78), the integral in equation (5.76) may be evaluated; substituting in that equation gives

$$k = \frac{E_s(1 - v_s)}{(1 + v_s)(1 - 2v_s)} \frac{\mu B}{2} \tag{5.79}$$

or, in terms of the plane stress solution, with soil properties E_0 and v_0,

$$k = \frac{E_0}{(1 - v_0^2)} \frac{\mu B}{2} \tag{5.80}$$

The only beam property involved is the width. If we compare equation (5.79) with equation (5.74) for a general correspondence between the Winkler solution and Biot's solution or equation (5.75) for the moment identification, we will see that the particular value of μ required depends on the Poisson's ratio of the soil but that, in general, values of μ satisfying the relation

$$1 < \mu B < 2$$

are appropriate. In particular, for example, in a soil with a Poisson's ratio of $v_s = 0.3$, and ignoring the twelfth root as being close to unity, in general, we get

$$\frac{(1 - v_s)^2}{(1 - 2v_s)} \frac{\mu B}{2} = 0.65$$

or

$$\mu B = \frac{1.3}{1.225} = 1.06$$

These results apply to infinitely long beams, and, as demonstrated earlier, can be expected to apply to finite beams with length (or spacing between loads) greater than $1.5\pi/\lambda$. For shorter beams down to lengths of $\pi/4\lambda$, Vesic (24) recommends using the analysis presented by Hetenyi and the value of k given by equation (5.74). Vesic suggests that very short beams, defined to be less than $\pi/4\lambda$ in length, be considered as rigid. It is almost as precise and easier to remember the criterion that rigid beams are less than $1/\lambda$ in length. Some of these considerations will appear again in the case of loads applied to *slabs* resting on a Winkler foundation, which will be treated in the next section of this chapter.

One method of overcoming the problem of matching deflection, pressures, and moments in the Winkler model with those of the presumably more correct half-space foundation is to construct a more complex representation by adding a stretched string or shear beam to the surface of the Winkler springs to form the two-parameter model discussed in Section 4.3 in relation to rigid beams. If the string is stretched by a total tensile load S, applied across the width of the beam, the value of S can be taken as an additional subgrade property to be varied, along with the spring constant k, to match either a particular requirement of the half-space foundation model or other requirements. The two-parameter and other representations have been summarized by Fletcher and Herrmann (9) who have also compared their behavior with that of the beam imbedded in or adhering to the surface of a linearly elastic continuum. In the Winkler model one property or parameter describes the subgrade material; when the string is included, two are required. Fletcher and Herrmann treat a further three-parameter case.

For footings or foundations, frequently the beam or slab may be assumed rigid and the subgrade response may be taken to be that of the Winkler model. For longer beams or slabs, flexible beam theory will be required, but generally the accuracy of knowledge of the subgrade properties in either case precludes the justifiable assumption of any model more complicated than the Winkler one. In special cases, perhaps for machine foundations located at some depth below the surface, sufficient information may be available to justify the employment of a two-parameter model. It is very unlikely that a case can ever be made for using a three-parameter foundation representation. A brief exposition of the flexible beam on the two-parameter foundation model follows.

The modified general beam equation away from loaded areas in this case is obtained by adding the appropriate relationship describing the membrane behavior as in equation (4.83) to the previous expression, equation (5.5), to get

$$EI\frac{d^4w}{dx^4} - S\frac{d^2w}{dx^2} + kw = 0 \qquad (5.81)$$

For many cases of interest in practice, $S < \sqrt{4kEI}$, and the general solution to this equation is

$$w = (C_1 \cos \beta x + C_2 \sin \beta x)e^{\alpha x}$$
$$+ (C_3 \cos \beta x + C_4 \sin \beta x)e^{-\alpha x} \tag{5.82}$$

where

$$\alpha = \left(\lambda^2 + \frac{S}{4EI}\right)^{1/2} \tag{5.83}$$

$$\beta = \left(\lambda^2 - \frac{S}{4EI}\right)^{1/2} \tag{5.84}$$

and

$$\lambda = \left(\frac{k}{4EI}\right)^{1/4} \tag{5.85}$$

as before [(equation (5.12)]. Applying this solution to the case of the infinitely long beam loaded by the vertical load P at $x = 0$, as we have done previously, we have again the result that $C_1 = C_2 = 0$.

From the zero slope condition at $x = 0$ and vertical equilibrium, we find eventually that

$$w = \frac{P\lambda^2}{2k\alpha\beta}(\beta \cos \beta x + \alpha \sin \beta x)e^{-\alpha x} \tag{5.86}$$

The maximum deflection, w_m, occurs at $x = 0$ and is given by

$$w_m = \frac{P\lambda^2}{2k\alpha} \tag{5.87}$$

Since α is larger than λ, by equation (5.83), we can see, by comparison with equation (5.16), that the deflection, in this case, is smaller, as it should be, than when S, representing the membrane, is absent from the problem.

The pressure between the beam and the subgrade, at the interface represented now by the upper surface of the membrane, is no longer given by the subgrade coefficient times the deflection, but it must also take into account the pressure required to deflect the membrane. Thus, the pressure is given by the expression

$$p = kw - S\frac{d^2w}{dx^2} \tag{4.83}$$

The other quantities, slope, moment, and shear, can be obtained by successive differentiation of the deflection in equation (5.86). For example, the maximum moment under the load in this case is

$$M_m = \frac{P}{4\alpha} \tag{5.88}$$

In their comparison of various foundation models, Fletcher and Herrmann give curves for the selection of the moduli k and S, to make deflection in the one- and two-parameter cases fit the behavior of the semi-infinite continuum. Their results were calculated for the problem of a vertical point load on an infinitely long beam. The curves are shown in Figure 5-8, where it can be seen that the subgrade reaction coefficients are given as ratios k/E_s and S/E_sB^2 versus the Poisson's ratio of the subgrade. There E_s is the Young's modulus of the soil and B is the beam width.

It is apparent, therefore, that in this comparison the properties of the beam do not play a part. Consequently, there must be some restriction on the results. Fletcher and Herrmann say that their results apply to cases in which the ratio of subgrade Young's modulus E_s to the beam modulus E is less than 0.01. With a modulus of concrete in the range of 3 to 4 × 10⁶ psi (2 × 10⁴ to 3 × 10⁴ MN/m²), this requires the average modulus of the sup-

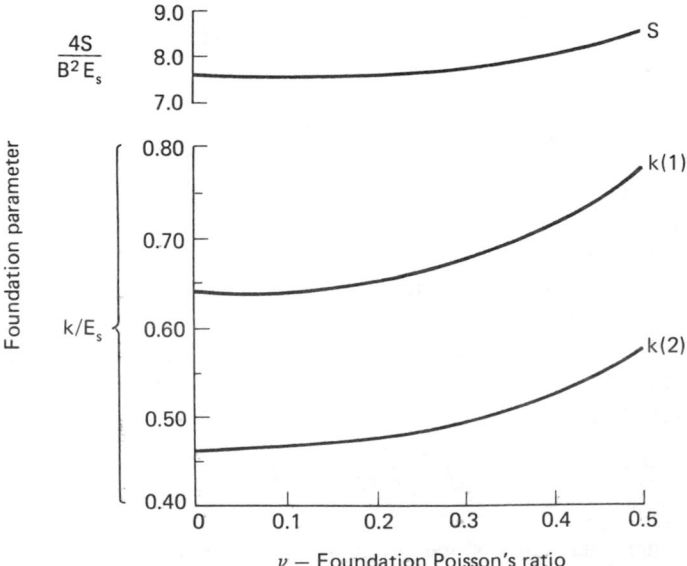

k(1)   Winkler k (= k₀ B) — one-parameter model

k(2), S   Pasternak — two-parameter model

Figure 5-8. Values of constants in two-parameter foundation model (9).

porting soil to be less than 3 to 4×10^4 psi (2×10^2 to 3×10^2 MN/m²).
Such a value is relatively high and would be representative of a medium-
dense to dense sand at typical foundation pressures, or a somewhat overconsolidated or desiccated clay. Figure 5-8 therefore would be applicable to
a wide range of soil foundation materials in practice. In Figure 5-8, at a
Poisson's ratio in the usual range of 0.25 to 0.3, it can be seen that k/E_s is
about 0.67 for the one-parameter (Winkler) model. In comparison with
equation (5.74), this implies that

$$\frac{1}{(1 - v_s^2)} \sqrt[12]{\frac{E_s B^4}{EI}} \approx 1 \tag{5.89}$$

Substitution of values of $E_s = 0.01E$ or less, and typical dimensions of the
order of a foot for the breadth and depth of the beam, into this equation
indicates that the assumption is approximately correct, mostly because the
twelfth root of any reasonable number inside the radical sign is close to
unity.

For the same range of Poisson's ratio, Fig. 5-8 also gives values of k/E_s
and $4S/E_s B^2$ of about 0.49 and 7.6, respectively, for the two-parameter
model. The $4/B^2$ comes about because of the different definition of the
second parameter in Fletcher and Herrmann's work. Their results indicate
that such values give a very satisfactory representation of infinite beam deflection as compared with the continuum case, but they do not evaluate and
compare moments. An analysis of the maximum moment under the point
load for both one-parameter and two-parameter representations with the
constants given in Figure 5-8 shows that the moment is about 6 percent too
high for the first and about 6 percent too small for the second compared to
the beam on the elastic half-space solution of Biot. Using k as given by
equation (5.74) with the Winkler model gives a maximum moment about
9 percent too high.

Again, a comparison with the values obtained from Vlasov and Leontiev's method is interesting. The constant S in their approach is given by
the equation

$$S = \frac{E_0 B}{2(1 + v_0)} \int_0^\infty g^2 \, dz \tag{5.90}$$

With the same assumption for g from equation (5.78), as before, the integral
has the value $1/2\mu$, and thus,

$$S = \frac{E_0}{4(1 + v_0)} \frac{B}{\mu} \tag{5.91}$$

However, the conversion between plane strain and plane stress does not
change the form of the equation this time, so that it can be rewritten as

$$S = \frac{E_s}{4(1 + \nu_s)} \frac{B}{\mu} \qquad (5.92)$$

For their two-parameter fit to the infinite beam case when $\nu_s = 0.3$, Fletcher and Herrmann obtained $k/E_s = 0.49$. Matching this to equation (5.79), we find that μB has the value 0.728. When this is substituted in equation (5.92) to derive the value of S, it appears that

$$\frac{4S}{E_s B^2} = 1.057$$

which is considerably different from the value of 7.6 obtained by Fletcher and Herrmann for a best fit of deflection in the two problems. It may therefore be more desirable to leave S as an independently variable fitting parameter rather than to tie it to k through the depth relation, equation (5.78), which is almost as arbitrary and involves another constant, μ, to be determined.

If both parameters k and S are to be used in an infinite beam solution, then the membrane tension must be applied both along and transverse to the beam, to represent in the latter circumstance the deflection of the medium to each side. This means that along each edge of the beam there will be a line load R_0 per unit length where R_0 is given by the product of the membrane tension force per unit of length, S_0, and the slope of the adjacent surface as in equation (4.91). However, from equation (4.90), the slope is equal to the local deflection times the exponential constant α ($= \sqrt{k/S}$), so that the total line load from both sides of the beam, $2R_0$, is equal to a constant times the beam deflection. The result is equivalent to another term additional to the subgrade reaction coefficient k in equation (5.81), which then becomes $k' = (k + 2\alpha S_0)$, or with substitution for α, equals $(k + 2\sqrt{k_0 S_0})$. For the infinite beam case, the solution to equation (5.81) remains therefore in the same form as before, but with the replacement of k by k'. For the Vlasov and Leontiev values of k and S given by equations (5.79) and (5.92) (using the plane strain results along the beam edge as an approximation) and the above modification to equations (5.81), it is found that the maximum moment, equation (5.88), is smaller than the Biot infinite beam moment for all values of μB, with the best results obtaining for μB equal to 0.5. The error depends on the relative stiffness $EI/E_s B^4$ and ranges from 10 percent to 20 percent; it is smaller for the stiffer beams. Deflections have not been examined.

Although few analytical solutions are available to the problem of a loaded member on an elastic continuum, a number of numerical solutions have been described in the literature. One of these, due to Brown (7), has been discussed previously. Another was presented by Barden (3 and 4) who derived an approximate method of obtaining the contact pressures and deflections by means of influence coefficients which he presented in tabular form. For this approach, the soil foundation material may be homogeneous and either isotropic or anisotropic. In Barden's method the beam is divided into ten equal increments of length, and the average pressure is given for each increment for unit loads placed at various positions on the beam. The beam-soil system is characterized by the parameter ϕ, where

$$\phi = \frac{E_s l^4}{4EI} \cdot \frac{1}{J} \tag{5.93}$$

in which E_s is the Young's modulus of the soil, EI is the beam property, J is a dimensionless factor related to the behavior of the subgrade, and l is the length of each beam *increment*. Since the analysis assumes ten increments, equation (5.93) is rewritten in terms of the beam length L:

$$\phi = \frac{E_s L^4 10^{-4}}{4EIJ} \tag{5.94}$$

For an isotropic soil, $J = (1 - v_s^2)/\pi$, where v_s is Poisson's ratio. In the case of a cross-anisotropic soil, J must be evaluated in terms of the five elastic coefficients. Barden presents the relevant equations.[3] For a particular beam and soil, ϕ is first evaluated from equation (5.94) and the pressure distribution is then obtained by interpolating among Barden's tables of pressure for different values of ϕ. The tables apply strictly to beams of length-to-width ratios of 10, but they can be employed for values in the range 5 to 20. When the soil modulus increases with depth, a modified relation for ϕ can be obtained, but a new set of influence tables is required. Barden also studied the case of the finite beam resting on a *Winkler* subgrade and found by comparison with the continuum solution that the subgrade reaction coefficient could be described in general by the relation

$$k = \frac{0.45 E_s}{J} \sqrt[3]{\frac{B}{L}} \tag{5.95}$$

in which B is the width of the beam. This may be compared with the previous relation, equation (5.74), given by Vesic. The two relations are in close agreement because the factor

$$\left(\frac{E_s B^4}{EI}\right)^{1/12}$$

cannot in practice have values much different from unity.

---

[3]There is an error in Barden's work, as pointed out and corrected by Dooley (8). Dooley's comments apply generally to materials with cross-anisotropic properties, which have frequently been derived erroneously.

For application of this or similar approximate methods in which use has to be made of the deflected shape of the ground surface under rectangular or circular loaded areas and the soil properties are cross-anisotropic or inhomogeneous, it is convenient to employ the following result obtained by Barden. He found that the surface deflections in, for example, a particular cross-anisotropic case can be obtained by multiplying the deflections arising in the corresponding isotropic case by the factor $J\pi/(1 - v_s^2)$. Various tables for settlements on isotropic half-spaces are given in the literature (19).

5.3 SLABS

When the structural member supporting a load or loads extends a considerable distance horizontally in both directions and individual loads of small extent are applied to it, the problem of the deflections, moments, and stresses induced in the member becomes a three-dimensional one. Examples of such circumstances are, of course, highway and airfield pavements and the mat or raft foundations of buildings. In this case, the equation describing the deflection of the structural plate or slab resting on a uniform Winkler subgrade whose reaction coefficient is again k_0 is

$$D\left(\frac{d^4w}{dx^4} + 2\frac{d^4w}{dx^2\,dy^2} + \frac{d^4w}{dy^4}\right) + k_0 w = 0 \tag{5.96}$$

in regions away from loads. In this equation the parameter D is the flexural stiffness of the slab, analogous to the property EI for beams considered previously, and is given by the expression

$$D = \frac{Eh^3}{12(1 - v^2)} \tag{5.97}$$

where E, v, and h are the slab modulus, Poisson's ratio, and thickness, respectively.

Here we note that k_0 no longer includes a beam width but is expressed in units of FL^{-3}. When the loading and slab are axially symmetrical, equation (5.96) takes the radial form:

$$D\left(\frac{d^4w}{dr^4} + \frac{2}{r}\frac{d^3w}{dr^3} - \frac{1}{r^2}\frac{d^2w}{dr^2} + \frac{1}{r^3}\frac{dw}{dr}\right) + k_0 w = 0 \tag{5.98}$$

Again analogously to the beam problem, we can establish a characteristic parameter β with dimensions of L^{-1} expressing the relative stiffness of the beam and foundation by the equation

$$\beta = \sqrt[4]{\frac{k_0}{D}} \tag{5.99}$$

The solution to equation (5.98) can be written in the form

$$w = C_1 Z_1(\beta r) + C_2 Z_2(\beta r) + C_3 Z_3(\beta r) + C_4 Z_4(\beta r) \qquad (5.100)$$

in the same way as for the solution to the beam equation. Here, as before, the C_n are constants, and the functions Z_n are related to Bessel functions. The functions Z_1 and Z_2 are the real and imaginary parts, respectively, of Bessel functions of the first kind and zeroth order $J_0(x\sqrt{i}\,)$; Z_3 and Z_4 are the real and imaginary parts of Bessel functions of the third kind (Hankel functions) and zeroth order, $H_0(x\sqrt{i}\,)$, where $i = \sqrt{-1}$. The solution is sometimes given in the form

$$w = A_1 \, \text{ber} \, (\beta r) + A_2 \, \text{bei} \, (\beta r) + A_3 \, \text{kei} \, (\beta r) + A_4 \, \text{ker} \, (\beta r) \qquad (5.101)$$

where the Kelvin functions ber, bei, ker, and kei are also related to the real and imaginary parts of Bessel functions. In particular ber $(x) = Z_1(x)$, bei $(x) = -Z_2(x)$, kei $(x) = -(\pi/2)Z_3(x)$, and ker $(x) = -(\pi/2)Z_4(x)$. The solution, equation (5.100) or (5.101), is set out in the same way as the original solution, equation (5.11) to the beam equation, in that all the terms behave as wave functions, with the first two terms oscillating but increasing in exponential fashion and the second two terms also oscillating but decreasing approximately exponentially. The Z or Kelvin functions have been tabulated in a number of textbooks (11, 18, and 21) and tables (1).

Paralleling our previous calculations, let us consider the case of an infinite slab loaded by a point load P acting at $r = 0$. We follow Hetenyi's analysis (11). As before, the fact that both the deflection and slope must be zero at infinity requires that $C_1 = C_2 = 0$ in equation (5.100). Since at the origin the slope must also be zero, we find that $C_4 = 0$; the solution then becomes

$$w = C_3 Z_3(\beta r) \qquad (5.102)$$

We can obtain C_3 by equating the total downward load P with the total subgrade reaction to deflection. Considering an annulus below the plate of radius r and width dr, the soil force p acting on it is

$$p = 2\pi r \, dr \, k_0 w$$

and thus the total reaction force is obtained by integration and must be equal to P

$$P = 2\pi k_0 \int_0^\infty r w \, dr$$
$$= 2\pi k_0 C_3 \int_0^\infty r Z_3(\beta r) \, dr \qquad (5.103)$$

Since we have (1),

$$\int_0^\infty x Z_3(x)\, dx = \frac{2}{\pi}$$

we get

$$C_3 = \frac{\beta^2 P}{4 k_0} = \frac{P}{4 \beta^2 D} \tag{5.104}$$

and therefore

$$w = \frac{P}{4 \beta^2 D} Z_3(\beta r) \tag{5.105}$$

The maximum deflection, w_m, occurs under the load, and since $Z_3(0) = \frac{1}{2}$, is given by

$$w_m = \frac{P}{8 \beta^2 D} \tag{5.106}$$

The slope, moments, and shear in this problem are described by the following equations:

$$\theta = \frac{dw}{d} = \frac{P}{4 \beta D} Z_3'(\beta r) \tag{5.107}$$

$$M_r = -D\left(\frac{d^2 w}{dr^2} + \frac{v}{r}\frac{dw}{dr}\right) = -\frac{P}{4}\left[Z_3''(\beta r) + \frac{v}{\beta r} Z_3'(\beta r)\right] \tag{5.108}$$

$$M_\theta = -D\left(v \frac{d^2 w}{dr^2} + \frac{1}{r}\frac{dw}{dr}\right) = -\frac{P}{4}\left[v Z_3''(\beta r) + \frac{1}{\beta r} Z_3'(\beta r)\right] \tag{5.109}$$

$$S = -D\left(\frac{d^3 w}{dr^3} + \frac{1}{r}\frac{d^2 w}{dr^2} - \frac{1}{r^2}\frac{dw}{dr}\right)$$

$$= -\frac{P\beta}{4}\left[Z_3'''(\beta r) + \frac{1}{\beta r} Z_3''(\beta r) - \frac{1}{(\beta r)^2} Z_3'(\beta r)\right] \tag{5.110}$$

In these expressions, Z_3', Z_3'', and Z_3''' are the first, second, and third derivatives of Z_3, respectively. For this geometrical arrangement, there are two moments of interest, one radial, M_r, and the other tangential, M_θ. In a slab the moment is expressed as a couple per unit length, or in dimensions FLL^{-1}, and thus has the dimensions of force. Similarly, the shear in the slab is expressed as a shear force per unit length, or in dimensions FL^{-1}. For the infinite flexible slab under a point load, plots of the quantities in equations (5.105), (5.107), (5.108), (5.109), and (5.110) are given in Figure 5-9 as functions of the dimensionless distance variable βr.

From the figures it can be seen that most of the quantities are again quite small at distances of about 5 for βr. This, then, is the interaction dis-

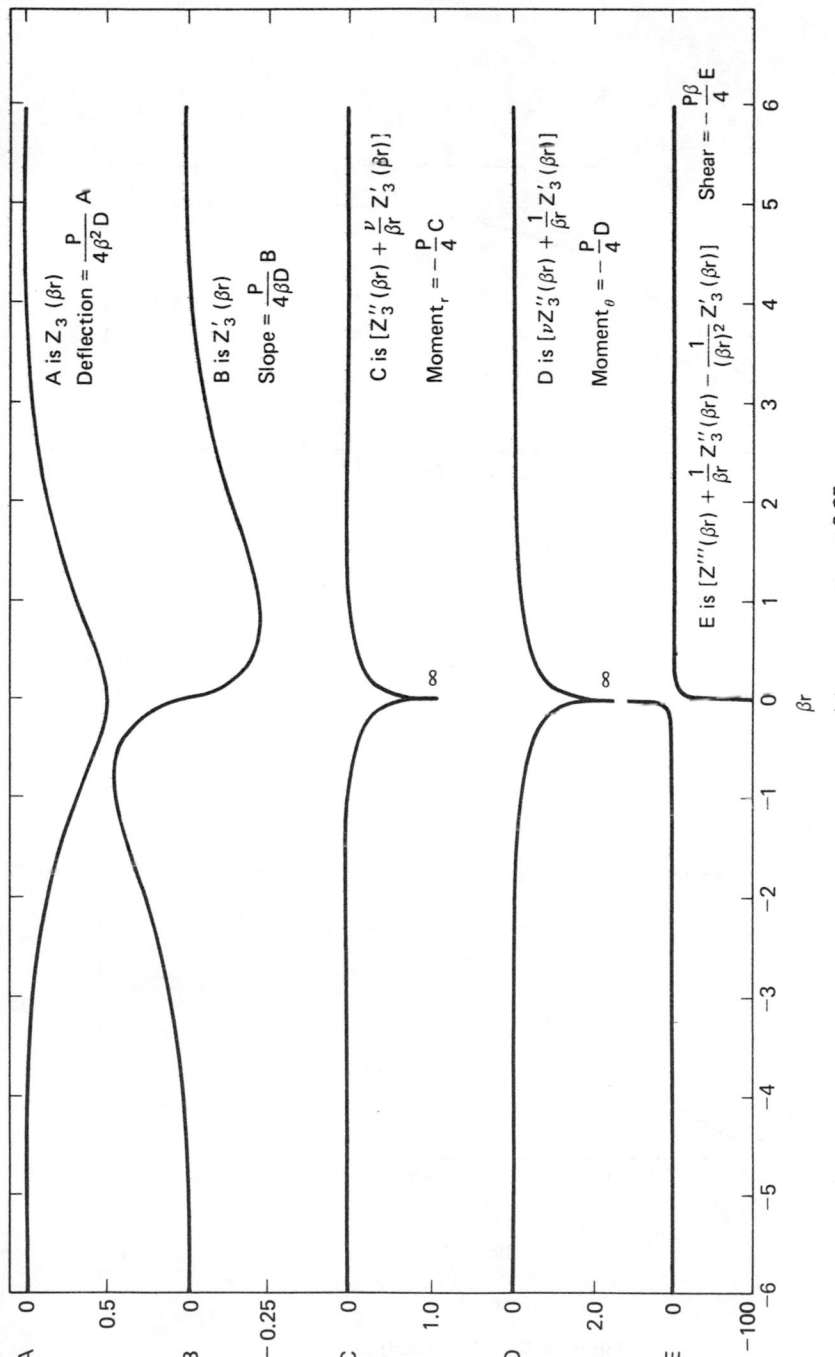

Figure 5-9. Deflection, slope, radial and tangential moments, and shear in a slab under point load: (a) form of functions; (b)—(f) enlarged scales.

$\nu_{slab} = 0.25$

(a)

A is $Z_3(\beta r)$

Deflection $= \dfrac{P}{4\beta^2 D} A$

B is $Z_3'(\beta r)$

Slope $= \dfrac{P}{4\beta D} B$

C is $[Z_3''(\beta r) + \dfrac{\nu}{\beta r} Z_3'(\beta r)]$;

Moment$_r = -\dfrac{P}{4} C$

D is $[\nu Z_3''(\beta r) + \dfrac{1}{\beta r} Z_3'(\beta r)]$

Moment$_\theta = -\dfrac{P}{4} D$

E is $[Z'''(\beta r) + \dfrac{1}{\beta r} Z_3''(\beta r) - \dfrac{1}{(\beta r)^2} Z_3'(\beta r)]$

Shear $= -\dfrac{P\beta}{4} E$

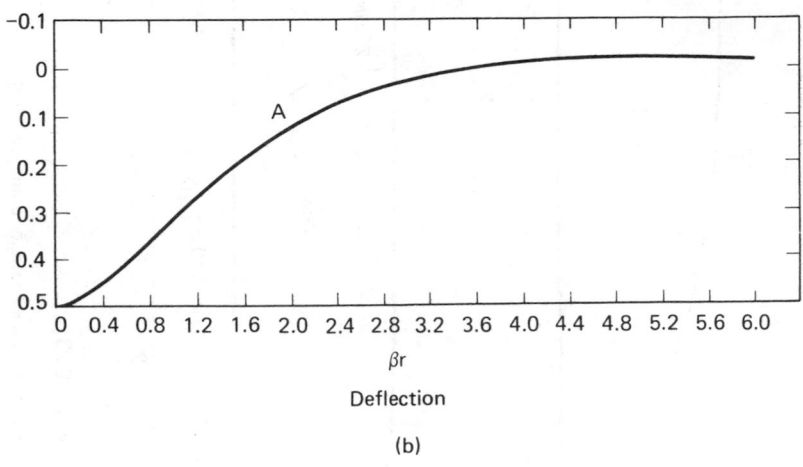

Deflection

(b)

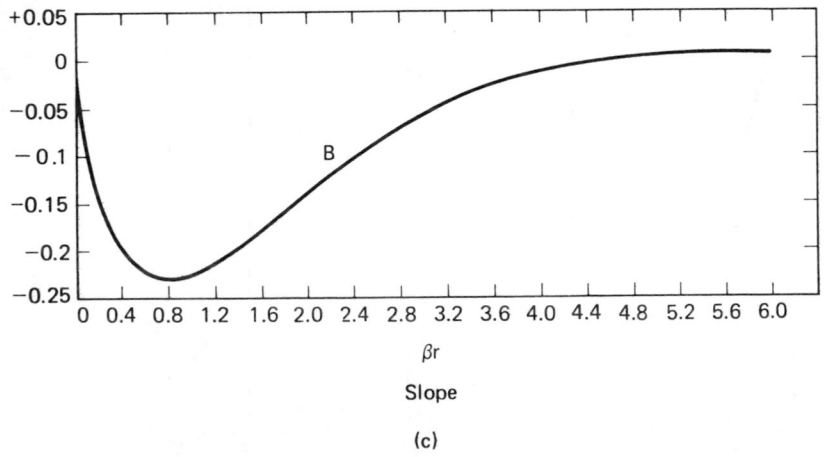

Slope

(c)

Figure 5-9. (Continued)

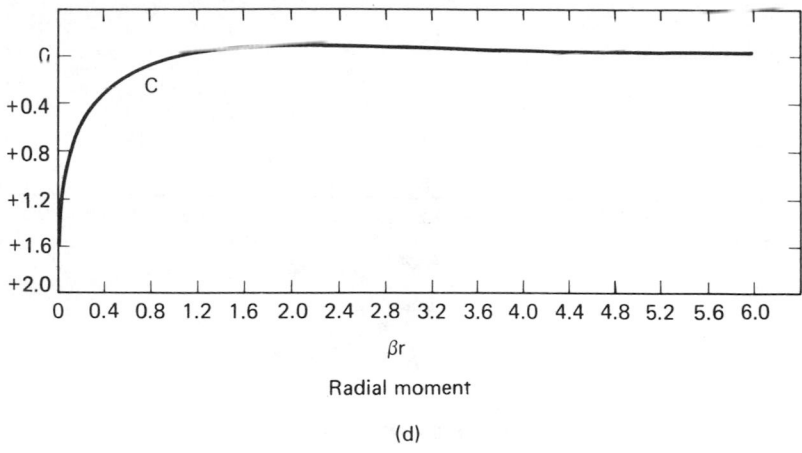

Radial moment

(d)

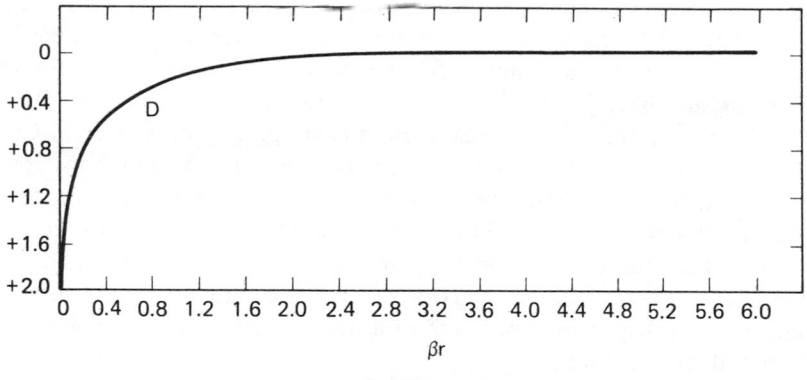

Tangential moment

(e)

Figure 5-9. (Continued)

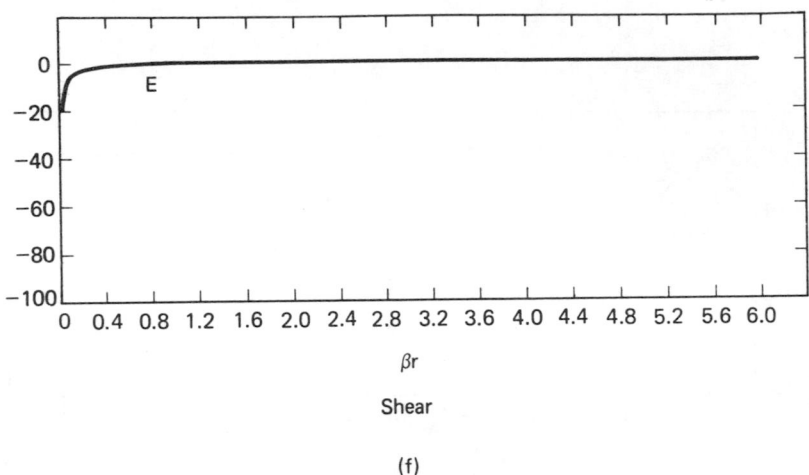

βr

Shear

(f)

Figure 5-9. (Continued)

tance for point loads on a slab analogous to the 1.5 π observed for the beam. Loads at spacings of this distance or greater have little or no effect on each other's local deflections or moments. With the same modulus of elasticity for concrete as before, a Poisson's ratio of 0.15 and the same thickness and subgrade coefficient range as in Table 5.1, it is found from equations (5.97) and (5.99) that the characteristic distances $1/\beta$ for a slab are almost identical to the values in Table 5.1 for a beam. Since the value of 5 estimated for the interaction distance in terms of βr is close to 1.5π for a beam, it follows that the interaction distances for slabs correspond to those for beams in Table 5.1.

There is, however, a major difference between slab and beam for the loading case we have just worked out. The second derivative of Z_3, Z_3'', in equations (5.108) and (5.109) goes to infinity as βr approaches zero. Consequently, both moments (and the shear) become infinite under the load. The moments therefore cannot be employed in a calculation of the maximum stresses in the slab. A point load is, of course, unrealistic in terms of the load distributions encountered in practice; the latter are usually distributed over a finite area of the slab's surface. To get the stresses in a practical situation, therefore, we must solve the problem of a plate loaded over some area by a uniformly distributed load.

For a number of practical problems, a uniform normal load applied over a circular area will be a convenient idealization. For a column base or array of columns, loading over a series of rectangular areas will be more appropriate. In both cases, the stress of interest will be the maximum tensile stress that is developed in the base of the slab under the center of the loaded area. The solution is obtained from the integration of the point load solution. For a circular area, subject to a total load P, uniformly distributed, the maximum moment in the plate at the center of the circle is approximately (26):

$$M_m = \frac{(1 + v)P}{4\pi}\left(\log_e \frac{1}{\beta c} + 0.616\right)$$ (5.111)

Here the term 0.616 comes from numerical constants appearing in the solution.[4] In the calculations leading to this result it was required that the radius of the loaded area, c, be small in comparison with the characteristic length $1/\beta$. This requirement comes about because in the derivation of equation (5.111) a term of approximate value $0.1\beta^2 c^2$ was omitted, in comparison with a term of unit value. The effect of the omission will become important at the 10 percent level when c approaches a dimension of about 0.7 times the characteristic length.

If the uniformly loaded area is a square of side length s, an approximate expression for the maximum moment can be obtained by substituting the value 0.575s for c in equation (5.111). The areas of the loaded square and circle are thereby equated. From equation (5.111) the maximum tensile (or compressive) stress can be obtained in the usual way by multiplying the moment by $h/2I$ where h is the thickness of the slab and I is its moment of inertia. Assuming a rectangular section of unit width, this product has the value $6/h^2$. The maximum tensile stress is thus

$$\sigma_r = \frac{6M_m}{h^2} = \frac{0.477P(1 + v)}{h^2}\left(\log_e \frac{1}{\beta c} + 0.616\right)$$ (5.112)

When the value of the radius c is made small, so that the loaded area tends toward a point, the bending moment and the tensile stress in the plate at the load increase toward infinity, as indicated by equation (5.108). This is a result of using elementary plate theory in the derivation of the part of the equation contributed by the slab. Consequently, the use of the more complicated two-parameter foundation model in analysis would give rise to the same result. The use of the incorrect plate model means that the calculated stresses in the slab are indicated to be higher than those the slab would experience or than those which a better analysis would provide. In actuality,

---

[4]The full approximate solution is

$$M_m = \frac{(1 + v)P}{4\pi}\left(\log_e \frac{2}{\beta c} + \frac{1}{2} - \gamma\right)$$

in which γ is Euler's constant, 0.5772157.... The summation ($\log_e 2 + 1/2 - \gamma$) gives the resulting numerical term 0.616.

the tensile stress at the base of a real plate does not, of course, become infinite when a point load is applied. Its value ought to be found by considering the plate an elastic continuum instead of a simple bending slab and by analyzing the continuum under appropriate loading and boundary conditions.

When the slab rests on a Winkler foundation and is subjected to a load over a small circular area, the region of the slab at some distance from the load can be satisfactorily represented by the elementary theory; only the area close to the load requires more detailed study. This suggests a continuum analysis of a portion of the slab in the vicinity of the load impressed by a uniform pressure over a small area of radius c and by the subgrade reaction pressure. For simplicity, we can take the boundary conditions to be provided by the simple plate requirements outside the area in question. This region is shown in Figure 5-10 as a circular disc of radius n times the thickness h; the boundary conditions consist of the usual normal beam stresses varying linearly from compression to tension over the thickness and a simple support reaction around the circular boundary. For the region below the load, the simple support will have approximately the same effect as a boundary shear. For relatively small values (of the order of 5 or less) of n, the subgrade reaction is small over this area in comparison to the total load acting on the upper surface and can presumably be neglected, so that

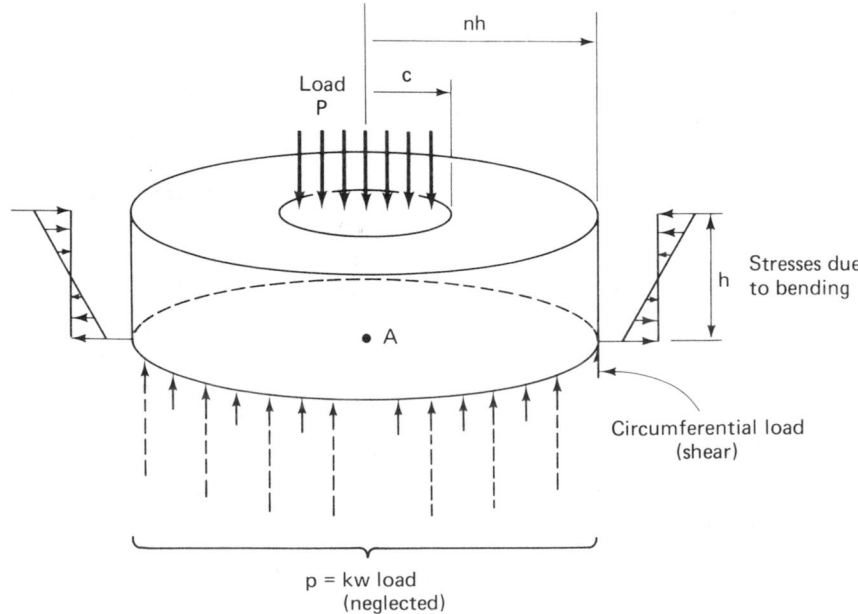

Figure 5-10. Uniform load over circular portion of circular disk.

the simple boundary reaction and moment support the load. In the resulting form the problem was solved by Woinowsky-Krieger.

The form of the exact solution is a complicated expression involving Bessel functions. At the time it was obtained, it was popular to obtain simpler equations in such cases, so that the result would be readily available for engineering calculations without having to use tables not normally accessible in engineering design offices. Woinowsky-Krieger (29) found that the following equation represented the more complicated result for tensile stress at the bottom of the plate directly below the load, with an accuracy better than 1 percent:

$$\sigma_m = \frac{P}{h^2}[(1 + \nu)(0.485 \log_e n + 0.52) + 0.48] \qquad (5.113)$$

In this form the equation is purely empirical (that's what Woinowsky-Krieger said himself). The radius of the loaded area c does not appear in this equation because Woinowsky-Krieger found that the maximum tensile stress was independent of the ratio c/h when c was sufficiently small. It will also be observed that the stress is independent of the Young's modulus of the plate. Since the supporting foundation did not appear in the problem configuration, it follows, therefore, that the subgrade reaction coefficient k_0 is also absent from equation (5.113). With the disc supported in such a way that the sum of the radial and tangential moments at the periphery is zero, a problem with the same geometry was solved somewhat earlier by Nádai (17).

When we were examining the problem of a loaded beam on an elastic foundation, we devoted some space to the identification of the deflection and moments with those of the same beam but resting on a homogeneous linearly elastic continuum. This subject also arises in connection with the loading of slabs. We still leave, for the present, the question of which model best represents the behavior of a real soil subgrade.

For the comparison we need to know the deflection of an infinite slab of thickness h resting on a homogeneous isotropic linearly elastic solid with properties Young's modulus E_s and Poisson's ratio ν_s and loaded by a point load P. This problem has been analyzed by Hogg (12 and 13) and the results have been discussed by Vesic (25) as follows:

The slab deflection w is given by the equation

$$w = \frac{P}{2\pi D \bar{\beta}^2} \int_0^\infty \frac{J_0(\alpha r \bar{\beta})}{1 + \alpha^3} \, d\alpha = \frac{P}{D \bar{\beta}^2} I_w \qquad (5.114)$$

in which D is the flexural stiffness of the slab [equation (5.97)] and the parameter $\bar{\beta}$ is the reciprocal of a characteristic length as before, except that it is now defined as

$$\bar{\beta} = \sqrt[3]{\frac{E_s}{2D(1 - v_s^2)}} \quad \text{or} \quad \frac{1}{\bar{\beta}} = 0.55h\sqrt[3]{\frac{E(1 - v_s^2)}{E_s(1 - v^2)}} \qquad (5.115)$$

In equation (5.114) J_0 is the Bessel function of first kind and zeroth order and α is an integration variable. Including the term $1/2\pi$, evaluation of the integral gives the deflection influence function I_w as a function of $\bar{\beta}r$, as shown in Figure 5-11. The subgrade reaction pressure is obtained from the solution in the form

$$p = \frac{P}{\bar{\beta}^2}I_p \qquad (5.116)$$

in which the pressure influence function I_p, also a function of $\bar{\beta}r$, is shown in Figure 5-11. It can be seen in the figure that the ratio p/w is not a constant

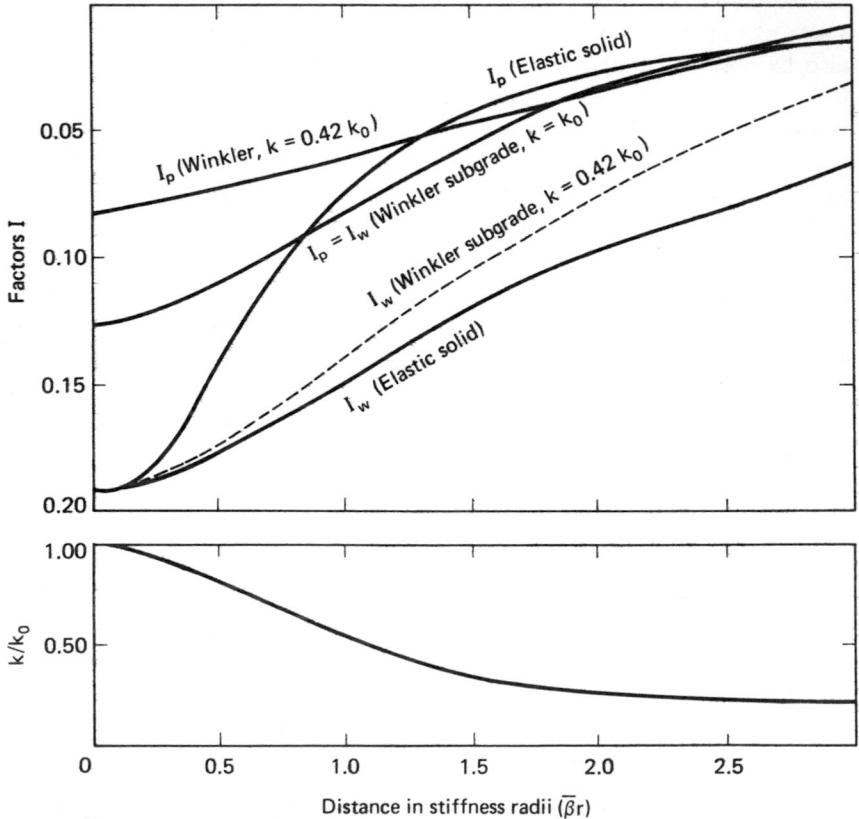

Figure 5-11. Comparison of deflection and pressure influence coefficients for point load on a plate over an elastic solid and over a Winkler foundation.

but a function of radius. Consequently, an attempt to represent this model by a Winkler foundation with constant k_0 is bound to lead, as with the beam, to poorly matched deflections and pressures. If we define

$$k^0 = \frac{p(0)}{w(0)} \qquad (5.117)$$

(which in this case is also equal to $D\bar{\beta}^4$, since $I_p(0) = I_w(0)$ from the solution) and calculate the pressures and deflections for the same slab resting on a Winkler subgrade characterized by the reaction coefficient $k_0 = k^0$, by substituting $k^0 = D\bar{\beta}^4$ in equations (5.116) and (5.114), we find that neither variable w nor p agrees with those obtained for the slab resting on a continuum. The results are shown on Figure 5-11 for $k_0 = k^0$ in the form of a curve of $I_p = I_w$ for this case. If the Winkler solution is obtained for a value of $k_0 = 0.42k^0$, which makes maximum *deflections* equal in the two models, then it will be found that the deflections match quite well (Figure 5-11) but that the pressures are poorly described by the Winkler model. A value of $k_0 = 2.37k^0$, which makes the maximum pressures under the load equal, matches pressures fairly well on the average, but deflections disagree.

Vesic concluded that the *moments* were best matched in the two cases if the value of k^0 was used in the Winkler model. Since we have found previously that the maximum moment produced in an infinite plate on any foundation by a point load is infinite if simple plate theory is used, the basis for the comparison is not clear. It is apparently obtained by matching the moments as a function of radius at distances greater than $\bar{\beta}r = 0.1$. If we accept it, then it requires a value of k_0 given by the equation

$$k_0 = k^0 = D\bar{\beta}^4 = \frac{0.91E_s}{h(1 - v_s^2)}\sqrt[3]{\frac{E_s(1 - v^2)}{E(1 - v_s^2)}} \qquad (5.118)$$

This indicates that one cannot arrive at the value to use for k_0 in a given case by considering the subgrade alone; instead, k_0 is seen to depend on the relative properties of slab and subgrade, and, in particular, to vary inversely with the thickness of the slab. Because of the matching problem, when the above value is used, the calculated *deflections* will be only about 65 percent of those that would be observed if the subgrade material actually behaved in accordance with the continuum model.

The effect of a finite depth of linearly elastic continuum underlying the slab has also been examined by Vesic. The material below the elastic layer is considered to be rigid. He found that, as the depth of the elastic layer decreases, there is little effect on the moments in the slab which, therefore, continue to be well-represented by the Winkler model with the value of $k_0 = k^0$ given above. However, the continuum model deflections decrease until, when the elastic layer thickness is about 2.5 characteristic lengths $(2.5/\bar{\beta})$, the deflections in the continuum model closely match those in the

Winkler representation. For smaller thicknesses, Vesic calculated that good matching of all values of deflection, pressure, and moment could be obtained if a modified value of k_0 was used in the Winkler analysis as given by the relation

$$k_0 = \frac{1.38E_s}{(1 - v_s^2)H} \tag{5.119}$$

where H is the layer thickness. This equation holds, therefore, for $\bar{\beta}H <$ 2.5. Kokusho (15) has made a finite element analysis of infinite plates on both Winkler and continuum elastic foundations. His conclusions as to the value of k to use for finite and semi-infinite layers agree with those ascribed to Vesic, above. Vesic further concluded that if the subgrade consists of soil whose modulus E_s varies, but increases with depth, the value of k_0 to be used in a Winkler model should be calculated from the average E_s down to a depth of $\bar{\beta}H = 2.5$.

Here we have concentrated on analytical solutions to the slab or plate problem, but many of the numerical techniques described in early chapters, such as finite difference and finite element methods, are also applicable. Lam (16) has adapted an approach, the edge function method devised by Quinlan (20) in his studies of plates and shells, to the problem of loaded slabs on an elastic foundation.

5.4 SLAB ON TWO-PARAMETER FOUNDATION

When it is proposed to model the foundation material with both the Winkler springs and the stretched membrane, equation (5.96) is modified by an additional term representing the contribution of the membrane or shear slab to the foundation reaction

$$D\nabla^4 w - S\nabla^2 w + k_0 w = 0 \tag{5.120}$$

where D is the flexural stiffness of the plate as given before in equation (5.96) and S is the tension in the membrane or shear beam modulus in dimensions of force per unit length. In this equation ∇^4 and ∇^2 are the biharmonic and Laplacian operators, respectively. The first of these has been given previously in expanded form for Cartesian and polar coordinates in equations (5.96) and (5.98), respectively; the second, in Cartesian form, is

$$\nabla^2 w(x, y) = \frac{\partial^2 w(x, y)}{\partial x^2} + \frac{\partial^2 w(x, y)}{\partial y^2} \tag{5.121}$$

and, for polar geometry,

$$\nabla^2 w(r) = \frac{d^2 w(r)}{dr^2} + \frac{1}{r}\frac{dw(r)}{dr} \tag{5.122}$$

As before, in the treatment of such problems, only normal stresses are assumed to act between the plate and the foundation material. Frictional stresses presumably act at the interface in real life, but they are taken to make a negligible contribution to the deflection of and moments in the plate under the vertical loading conditions considered here.

When w has been obtained by the solution of equation (5.120) and the inclusion of the boundary conditions for a particular problem, the pressure distribution between the plate or slab and its foundation can be obtained from the following equation:

$$p = -S\nabla^2 w + k_0 w \tag{5.123}$$

Similarly, the moments are calculated from the expressions on the left side of equations (5.108) and (5.109). We will consider only the axially symmetric plate and loading conditions here.

Dividing equation (5.120) throughout by D and making the substitution, as before, of $\beta = (k_0/D)^{1/4}$, we get

$$\nabla^4 w - \frac{S}{D}\nabla^2 w + \beta^4 w = 0 \tag{5.124}$$

which may be rewritten conveniently in dimensionless form by making the substitution $R = \beta r$ and

$$2T^2 = \frac{S}{\beta^2 D} \tag{5.125}$$

$$\nabla^4 w_R - 2T^2\nabla^2 w_R + w_R = 0 \tag{5.126}$$

in which it is understood that the derivatives are all taken now with respect to the dimensionless distance R. The parameter T is dimensionless and describes the contribution of the membrane relative to the Winkler springs in forming the foundation stiffness. The factor 2 is added for convenience in the mathematical manipulations. For the usual range of foundation soil behavior, T will lie between 0 and 1, being zero for the Winkler case. Limiting the upper value of T to unity and substituting in equation (5.125) leads to the condition

$$S \le 2\beta^2 D$$

which, upon substitution from equation (5.99), becomes

$$S \le \sqrt{4k_0 D} \tag{5.127}$$

analogous to the solution requirement following the two-dimensional equation (5.81). A general solution to equation (5.124) can be written in the form

$$w = C_1 \mathrm{Re} J_0(e^{i\varphi}R) + C_2 \mathrm{Im} J_0(e^{-i\varphi}R)$$
$$+ C_3 \mathrm{Re} H_0^{(1)}(e^{i\varphi}R) + C_4 \mathrm{Im} H_0^{(2)}(e^{-i\varphi}R) \tag{5.128}$$

where, as before, C_1 through C_4 are constants which will be evaluated for particular boundary conditions, and the variables are real and imaginary parts of complex zero-order Bessel and Hankel functions of the first and second kind. The exponent φ represents the addition of the membrane to the foundation system in the following way:

$$\varphi = \frac{1}{2} \tan^{-1}\left(-\frac{\sqrt{1 - T^4}}{T^2}\right) \tag{5.129}$$

Table 5.2 relates φ to T and $2T^2$, for a range of practical values of φ and T, and Figure 5-12 shows graphically the relation between the two variables. It can be seen from equation (5.129) that when $T = 0$ (no membrane force

TABLE 5.2

Effect of Membrane Property

$\varphi°$	45	46	47	48	49	50	55	60	65	70
T	0	0.1868	0.2641	0.3233	0.3731	0.4167	0.5848	0.7071	0.8017	0.8752
$2T^2$	0	0.0698	0.1395	0.2090	0.2784	0.3473	0.6840	1.0000	1.2856	1.5321

present, Winkler foundation case), $\varphi = 45°$, and for the relative maximum membrane contribution, $T = 1$ and $\varphi = 90°$. The functions in equation (5.128) are arranged, as in previous equations, so that the first two oscillate and increase without limit as R increases, whereas the second two oscillate and decrease toward zero with increasing R. They are therefore similar in their behaviors to those of the corresponding functions in the simpler beam on a Winkler foundation case. When $T = 0$, $\varphi = 45°$, the solution becomes identical to the Kelvin or Z-function solution for the slab supported by the Winkler foundation alone.

As in the case of the slab on the Winkler foundation, there are a variety of ways of representing the solution to equation (5.126). Yu (31) writes the solution in terms of the real and imaginary parts of the modified Bessel functions I_0 and K_0 of complex argument; these are related to the Hankel functions given above. Because of the identification of the functions with the Kelvin (Thomson) functions ber, bei, ker, and kei in the Winkler case when $S = 0$, Yu refers to these solutions as *generalized* Kelvin functions.

The solutions for finite circular slabs with various loading conditions are complicated, involving, as they do, the membrane-spring model outside the slab boundaries, and will not be described further here. We will treat only the infinite slab subjected to a point load, P.

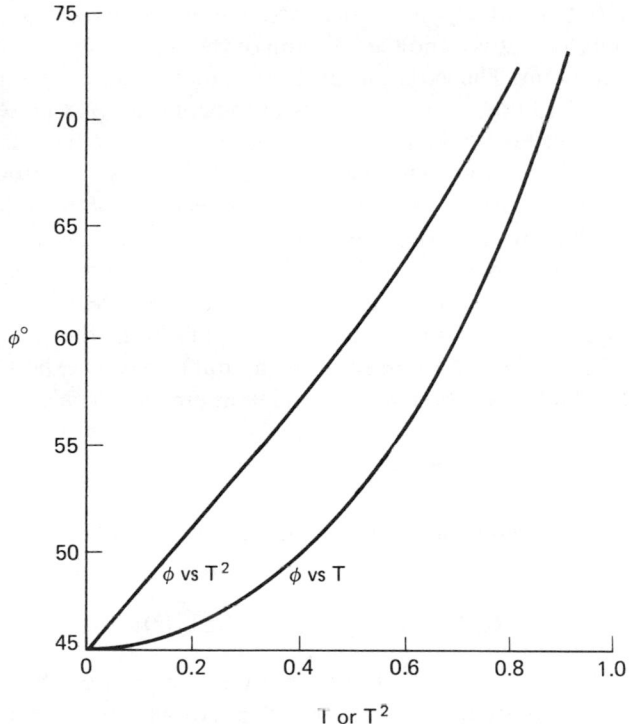

Figure 5-12. Relationship of constants T and φ in two-parameter theory for a plate.

As in previous cases, the nature of the first two terms in the solution, equation (5.128), precludes their matching the boundary condition at infinity, while the last term, although tending to zero at infinite radius, is infinite at zero radius. Thus, only the third term remains and the solution is

$$w = C_3 \mathrm{Re} H_0^{(1)}(e^{i\varphi}R) \tag{5.130}$$

From consideration of the vertical equilibrium of the plate, the constant C_3 is found to be

$$C_3 = \frac{P\beta^2}{4k_0 \sin 2\varphi} = \frac{P}{4D\beta^2 \sin 2\varphi} \tag{5.131}$$

Thus, the deflection w, in dimensional terms, is given by the equation

$$w = \frac{P\beta^2}{4k_0 \sin 2\varphi} \mathrm{Re} H_0^{(1)}(e^{i\varphi}R) \tag{5.132}$$

where $R = \beta r$. The slope, moments, and shear can be expressed in terms of the usual derivatives; the expressions are identical to equations (5.107) through (5.110) but with Z_3 and its derivatives replaced by the above Hankel function and its derivatives and the addition of the sin 2φ term in the denominator of the constant. The variation of the different quantities with radius is given in Figure 5-13 for a range of values of the constant φ, representing the ratio of membrane to Winkler spring properties in the model as given in Table 5.2 and Figure 5-12. The maximum deflection, w_m as usual, occurs at r or $R = 0$, the point of load application as seen in Figure 5-13(a).

Figure 5-13 shows that the effect of the force S becomes progressively diminished with the succession of derivatives, until it is almost negligible in its influence on the shear in the slab. Once again, because of the thin slab model employed, the maximum moments are infinite at the point of load application. The tension S diminishes all the quantities as expected.

Now, the Hankel function with complex argument z is also expressible by the relation

$$H_0^{(1)}(z) = J_0(z) + iY_0(z) \tag{5.133}$$

where J and Y are Bessel functions of first and second kind. Each of them can be written as a sum of real and imaginary parts so that

$$\mathrm{Re}H_0^{(1)}(z) = \mathrm{Re}J_0(z) - \mathrm{Im}Y_0(z) \tag{5.134}$$

When z is replaced by $e^{i\varphi}R$, it is found that the first term on the right-hand side has the value unity for zero R and all φ. An asymptotic expansion of the second term for small R indicates that it is equal to $2\varphi/\pi$, so that for zero R, equation (5.134) becomes

$$\mathrm{Re}H_0^{(1)}(z)_{R=0} = 1 - \frac{2\varphi}{\pi} \tag{5.135}$$

Thus, from equation (5.132), the maximum deflection at $R = 0$, w_m is given by the equation

$$w_m = \frac{P\beta^2}{4k_0 \sin 2\varphi}\left(1 - \frac{2\varphi}{\pi}\right) = \frac{P}{4\beta^2 D \sin 2\varphi}\left(1 - \frac{2\varphi}{\pi}\right) \tag{5.136}$$

When $\varphi = 45°$, this equation becomes identical to the previous equation (5.104), which holds in the absence of the tension S.

Finally, with respect to the two-parameter model, the question arises as to how well it can be made to match the plate on the elastic half-space solution of Hogg. It was found earlier that if equation (5.118) was used to obtain a value for k_0, the Winkler plate moments matched the Hogg solution moments, but the deflection differed considerably. It is seen how the presence of the membrane force S decreases all the quantities of deflection, slope, moment, etc., in the slab, so that the value of the subgrade coefficient must

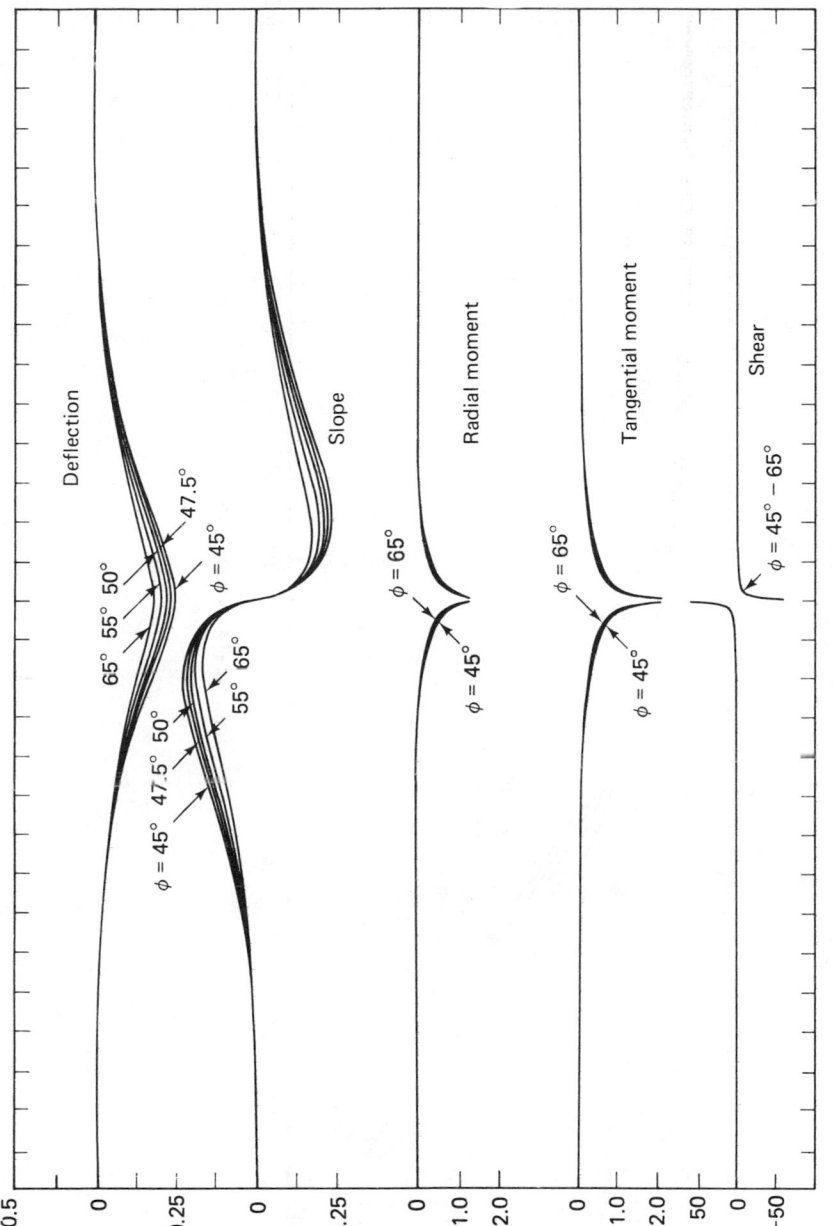

Figure 5-13. Deflection, slope, radia and tangential moments, and shear in a slab under point load; two-parameter model.

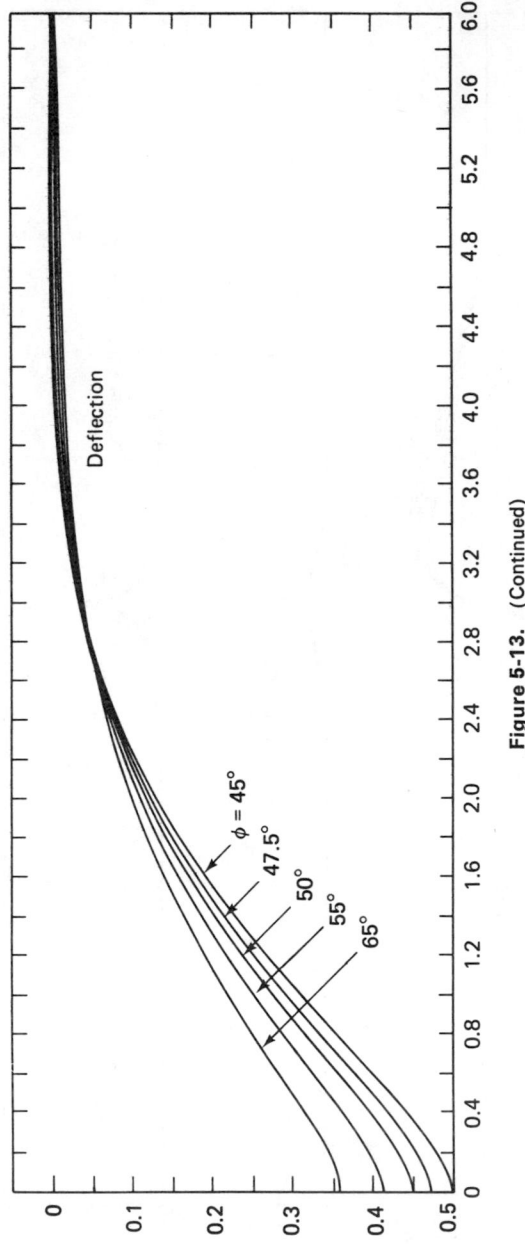

Figure 5-13. (Continued)

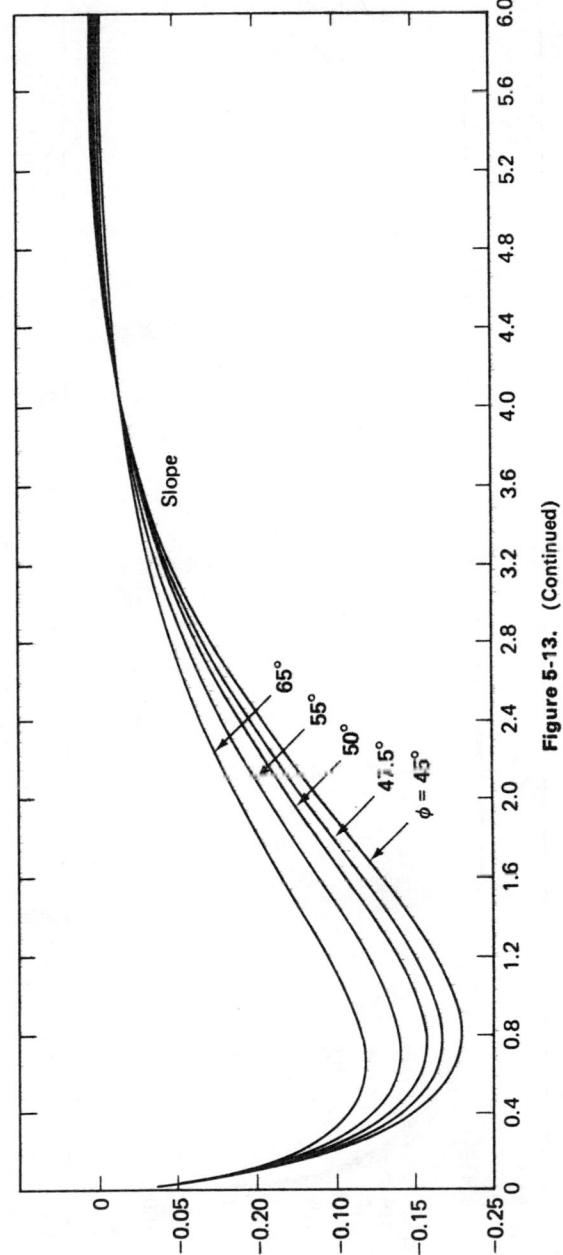

Figure 5-13. (Continued)

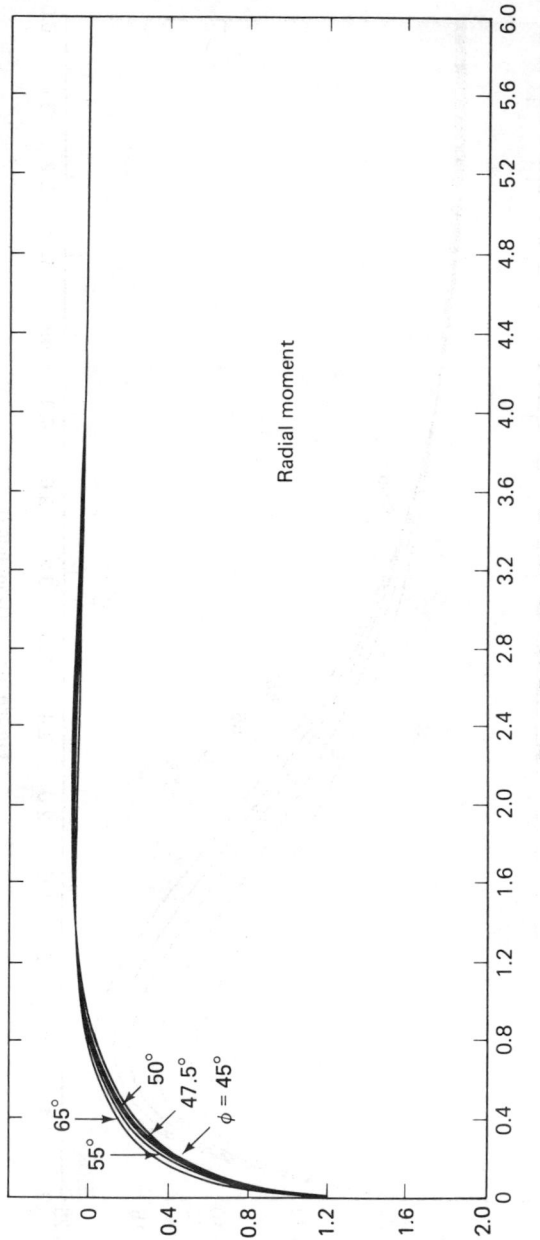

Figure 5-13. (Continued)

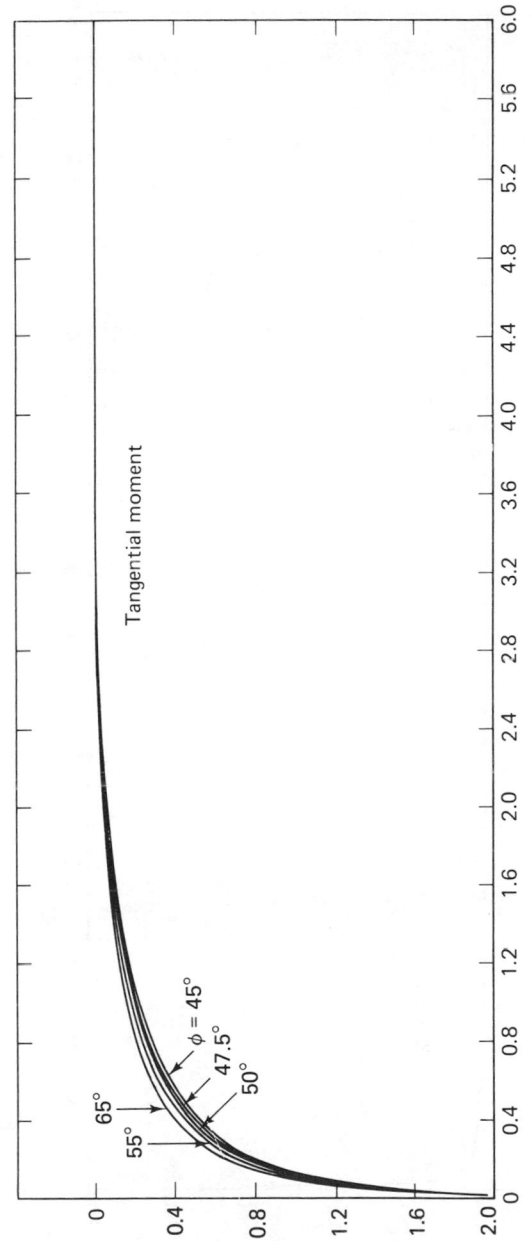

Figure 5-13. (Continued)

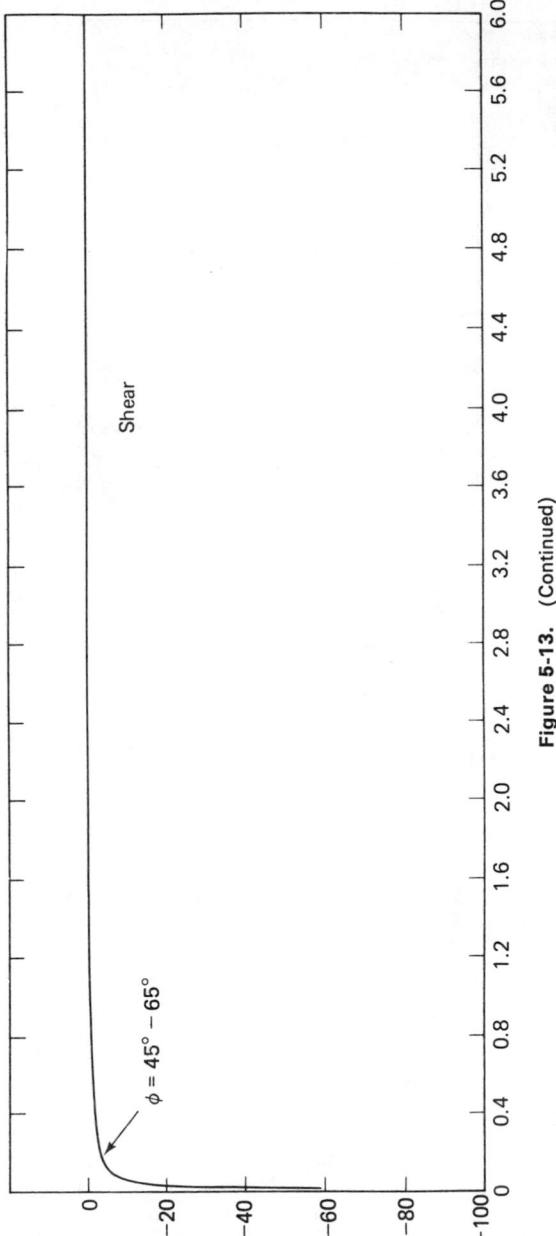

Shear

$\phi = 45° - 65°$

Figure 5-13. (Continued)

be decreased relative to that in equation (5.118) to enable a match to be made. With inspection, it appears that a suitable value is one-quarter of the former value of equation (5.118) or

$$k_u = k^0 = \frac{0.227E_s}{h(1 - v_s^2)} \sqrt[3]{\frac{E_s(1 - v^2)}{E(1 - v_s^2)}} \tag{5.137}$$

If this value is used in conjunction with $\varphi = 60°$, it is found that both deflections and moments in the Hogg solution are closely matched. From Table 5-2, this value of φ corresponds to a T of 0.707.

5.5 MAT OR RAFT FOUNDATIONS; BAKER'S METHOD

In Section 3.3 an approximate method of analysis for a loaded beam on a continuum was presented. In this analysis the effect of the overall deformation of the foundation material was taken into account, in contrast with the uniform Winkler model, which includes only local deflections. Although the subgrade reaction coefficient can be varied along a beam or across a slab to account for general deformations, analysis is considerably complicated thereby. The moments generated as a consequence of large-scale relative deflection or settlement of a beam are substantial and may dominate for beam design. If the overall relative settlement of the ends of a loaded beam with respect to its center is say 1 in. (25 mm) due to softer soil at either the center or ends (a result either of really softer soil or of continuum effects), the bending moment generated at the beam center can be of magnitude equal to that caused by the primary column loads.

An approximate method of analysis which can take into account the variation of soil properties across a building site without unduly complicating the calculational effort was proposed by Baker (2). It is simple and straightforward in application, giving results which compare well with the exact Winkler analysis in comparable situations, and will be described in this section. The essence of the method is to assume a specific shape of pressure distribution, of unknown magnitude, between soil and beam. The pressure distribution is applied to the beam and the maximum relative displacement of the beam ends expressed as a function of the pressure magnitude parameter. Since the beam is elastic, the relation, called the *beam equation*, is linear. The same pressure distribution is considered to be applied to the foundation soil, and the maximum differential settlement is related to the pressure magnitude. This relation, referred to as the *soil equation*, may or may not be taken to be linear. Simultaneous solution of the two equations gives the magnitude of the pressure and the relative settlement or displacement which satisfies both expressions. From these values the moments generated in the beam can be calculated for addition to the moments developed by the column loads.

It is seen that the technique is similar to that described at the end of Chapter 3; the difference is in the assumption of a fixed shape of pressure distribution and in the method in which the soil response is arrived at. Obviously, in this method, the detailed shape of the deflected beam does not conform to the undulations of the displaced soil; only the maximum differential movement is taken as the matching criterion. More complex models could be obtained which would give a better matching of shapes, but they will result in a larger number of equations to be solved. It will be seen that the simple assumptions of Baker's method give remarkably good correspondence with exact Winkler's solutions in cases in which the soil behavior can be identified with the Winkler representation. Another advantage of the method is that it can make use of subgrade reaction coefficients (adjusted for the beam dimensions) if they are available at the site, or it can make use of other data, obtained, say from the measured performance of structures previously built on the same soil. We proceed to a detailed description of the method followed by some example calculations.

5.5.1 Systems of forces

The discussion begins by considering a continuous beam foundation, which may be an element of a slab and beam raft foundation, and which is loaded by a number of column loads as shown in Figure 5-14(a) in which the column rows are identified by letters $a, b, \ldots,$ and numbers $1, 2, \ldots,$ respectively. In the upper diagram the entire raft is shown and is subjected to the real dead and live loads of the structure T_{ij}. In the case of a low, relatively flexible building, such as, say, a one- to three-story storage or office structure, the loads T_{ij} are essentially constant and independent of the differential settlements of the foundations. If the building is relatively stiff, the load reaching each column base will depend on the settlement of that base with respect to those of other bases. This aspect of the problem lies more within the domain of structural mechanics and will not be dealt with here, although it can be handled through the development of a frame equation analogous to the beam equation mentioned above and treated later.

In the lower part of Figure 5-14(a) is shown a single foundation beam, 2, isolated from the foundation raft. Normally, such a beam will not have a uniform width in contact with the soil along its entire length, but it will be underlain by a slab whose width may vary from place to place. The actual width must be accounted for in practical analysis through the assignment of a subgrade reaction coefficient, or in resultant earth pressure. For convenience in discussion here, we will assume a uniform contact width. The beam 2 is loaded by forces P_{i2} acting at the column bases. These forces are the dead and live loads of the upper diagram, modified by the reactions of the

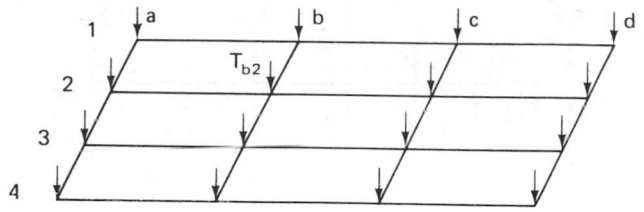

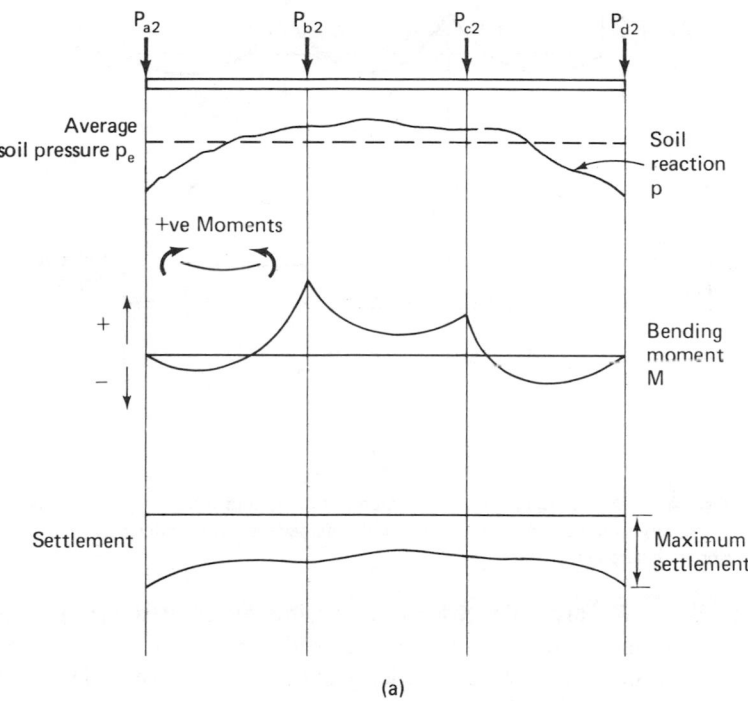

(a)

Figure 5-14. Mat foundation analysis: (a) total mat and beam loading moments and settlements.

transverse beams a, b, etc., at each of the intersections. The reactions arise from the differential settlements and soil reactions on the bases and foundation beams in the transverse direction. For example, the force P_{b2} arises from the dead and live load (if the building is taken to be flexible) T_{b2} adjusted by the reactions of transverse beam segments b_1–b_2 and b_3–b_2 on the beam 2 at column location b. These reactions will be discussed later after the calculation has been broken down into separate steps for solution. If the beam is a single member acting on its own, the reactions are, of course, absent. The

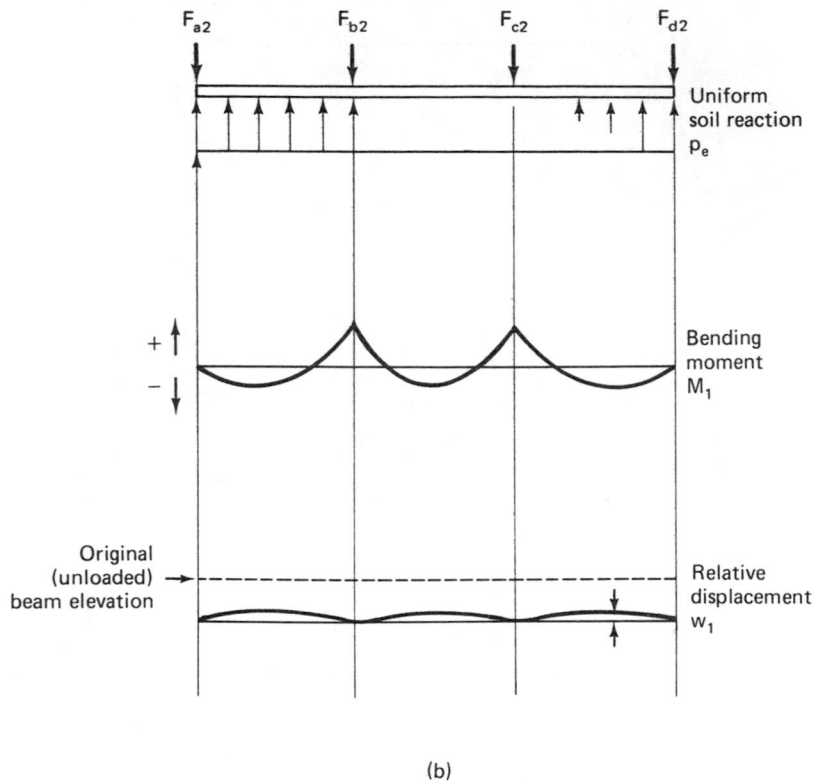

(b)

Figure 5-14. (b) Beam loads F_{ij}, moments M_1, and differential displacements w_1, when supports are kept at the same level and uniform soil reaction p_e is applied.

column forces P_{i2} are balanced by a varying distributed soil pressure p acting on the underside of the beam. The soil pressure is a consequence of the possibly varying soil properties along the beam, the overall layered half-space reaction to the building load, and any eccentricity in the loading that would cause generally greater pressures toward some edge of the building than elsewhere. We will assume here that the load resultant is close to the center of the structure. As a consequence of the forces and reactions, the beam deflects in some fashion as shown schematically in Figure 5-14(a) and is subject to the moments illustrated. For design of the beam, we need to calculate these moments. In order to protect against excessive distortions, in addition, which may give rise to panel cracking or other difficulties, it is also desirable to calculate the differential settlements. Gross overall settlement is arrived at by the usual soil mechanics calculations.

To solve the problem, Baker divided it up into several subsidiary problems, each consisting of a force system in equilibrium, and all of which, in

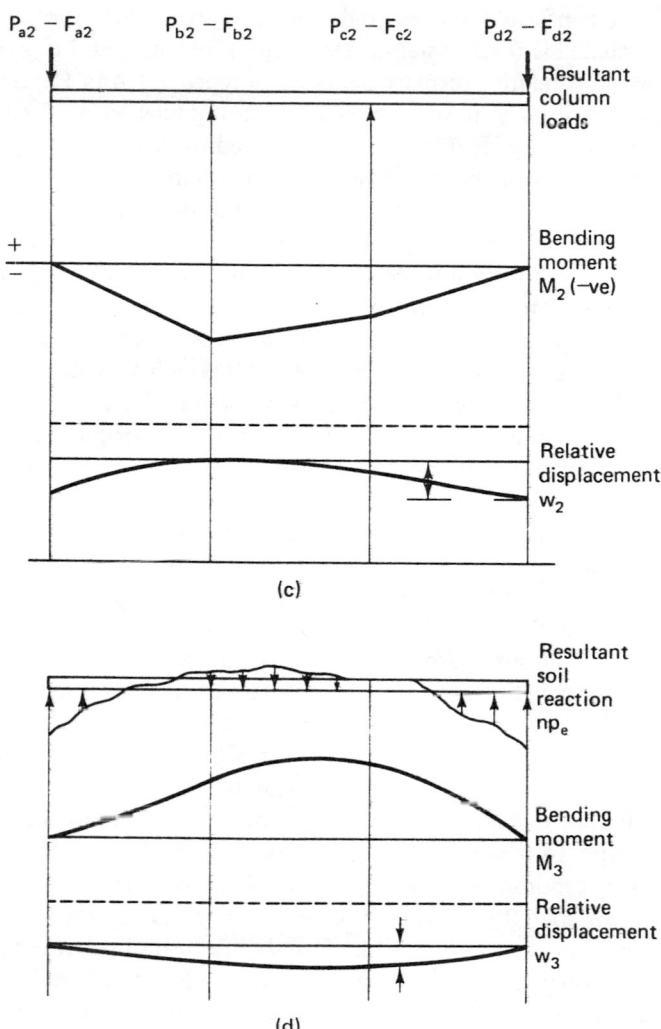

Figure 5-14. (c) Difference loads (*P–F*), moments M_2, and differential displacements w_2; (d) moments M_3 and differential displacements w_3, when beam is loaded by resultant soil reaction np_e.

sum, give the total result of Figure 5-14(a). The first of these is shown in Figure 5-14(b) which shows the beam 2 with its column bases held at the same level and subjected to a uniform upward soil pressure p_e. Here, p_e is the average soil pressure obtained by dividing the building load by the bearing area; it is shown in Figure 5-14(a). It is uniform across the building because the building load acts through the center of the foundation structure. The loads F_{ij} are calculated as the reactions to p_e required to maintain the supports at a constant level less the reaction forces from the intersecting beams a, b, etc. This force system is therefore in balance, and it gives rise to moments M_1 and relative displacements w_1 as represented in Figure 5-14(b). The usual convention of positive moments causing tension on the underside of the beam is employed; this is also considered to give rise to positive differential settlements in the beam (ends higher than the center). It will be recognized that the relative displacements of the beam in this first loading configuration are very small, typically of the order of a hundredth of an inch.

The second equilibrium force system is illustrated in Figure 5-14(c) and consists of the differences in forces $P_{ij}-F_{ij}$ acting at each column location on the beam. These are, in effect, the differences between the actual column loads and the loads in system 1, Figure 5-14(b). This loading causes the largest relative displacements in the beam; these may be as large as an inch or two in typical foundation beams. Beside them, the displacements of the first system are negligible in most cases. The bending moments from loading 2 are also substantial.

Finally, the remaining force system, 3, to be considered is the nonuniform part of the soil reaction, as shown in Figure 5-14(d). It consists of the difference between the total soil reaction pressure p per unit length of the beam of Figure 5-14(a) and the uniform component of the pressure p_e of Figure 5-14(b). The nonuniform pressure is thus also in force equilibrium and causes relative displacements, w_3, and moments, M_3, as sketched in Figure 5-14(d). These displacements and moments are also relatively large. The total forces and moments acting on the beam are given by the sum of the separate forces and moments in the systems 1, 2, and 3.

Baker divides the soil response in system 3 into three further balanced force subgroups. In the context of this book they can be identified with: (a) the half-space response of the soil; (b) a uniform Winkler behavior component; and (c) a spatially varying Winkler response. They are analogous to the three force systems applied to the foundation beam in that the soil reaction (only the varying part) is considered to develop as a result of (a) the pressure acting on a rigid foundation slab, (b) the pressure due to the foundation deflection, when the response is uniform along the beam or over the slab, and (c) the pressure component caused by soil properties varying across the site. The sum of the three reactions gives the resultant pressure system 3 in Figure 5-14(d). The contribution from subsystem (a) will be different for cohesive and cohesionless soils, but the others are dependent only on the measured or estimated soil behavior. In the case of a rigid slab resting on and near the surface of a cohesive soil, we would expect the pressure acting at the slab edges to be higher than in the interior, since the material behaves approximately as an elastic or elastic-plastic continuum. In sand, however, we would find the edge pressures lower than those at the center because of the relatively lower confinement at the edges.

5.5.2 Beam equation

The forces, moments, and deflections of systems 1 and 2 can all be obtained by the normal methods of structural mechanics; it is the corresponding quantities of the third group, the soil interaction area, which are harder to determine. The simplifying assumption, which resolves this problem, was introduced by Baker. He assumed that the soil reaction in Figure 5-14(d) could be represented by a simple triangular distribution, as shown in Figure 5-15. The upper diagrams in Figure 5-15(a) and (b) show the total soil reaction pressure distribution of Figure 5-14(a); the lower diagram gives the varying component, as shown in Figure 5-14(d), and this is represented by the triangular shape, whose magnitude is characterized by n, which may be positive or negative. The reason that the soil reaction is of the general shape shown in Figure 5-15 is that the overall soil response is dictated by the deflected shape of the beam. This is principally controlled by the force system 2, whereas the force system 1 results in only small local deflection variations along the beam. As far as the beam is concerned, we consider the soil pressure distribution of Figure 5-15(a) to be positive, since it induces positive beam moments and deflections. The pressure in Figure 5-15(b) is, by the same criterion, negative.

Now the behavior of the entire beam 2 can be expressed as the sum of the relative deflections of systems 1, 2, and 3. However, as pointed out earlier, the system 1 deflections may be neglected, and so the maximum relative deflection of the beam, w, can be written in terms of the maximum relative displacements, w_2, due to the loads, and w_3, due to the soil pressure variation. The positive triangular pressure distribution of Figure 5-15(a) gives a maximum displacement w_3 of the beam

$$w_3 = \frac{7}{1920} \frac{npl^4}{EI} \qquad (5.138)$$

where p is the uniform load per running foot of the beam, equal to Bp_e. The maximum moment (at the center) is

$$M_3 = \frac{1}{24} npl^2 \qquad (5.139)$$

where l and B are the length and width of the beam, respectively, and EI is the usual product of beam Young's modulus and moment of inertia. The numerical coefficient in equation (5.138) is equal to 0.00365, but in considering different realistic forms of pressure distribution that could act on the beam under various conditions, Baker concludes that a conservative value to use would be 0.0035. The actual numerical value varies to a surprisingly small extent for widely different shapes of the distributed pressure. In the

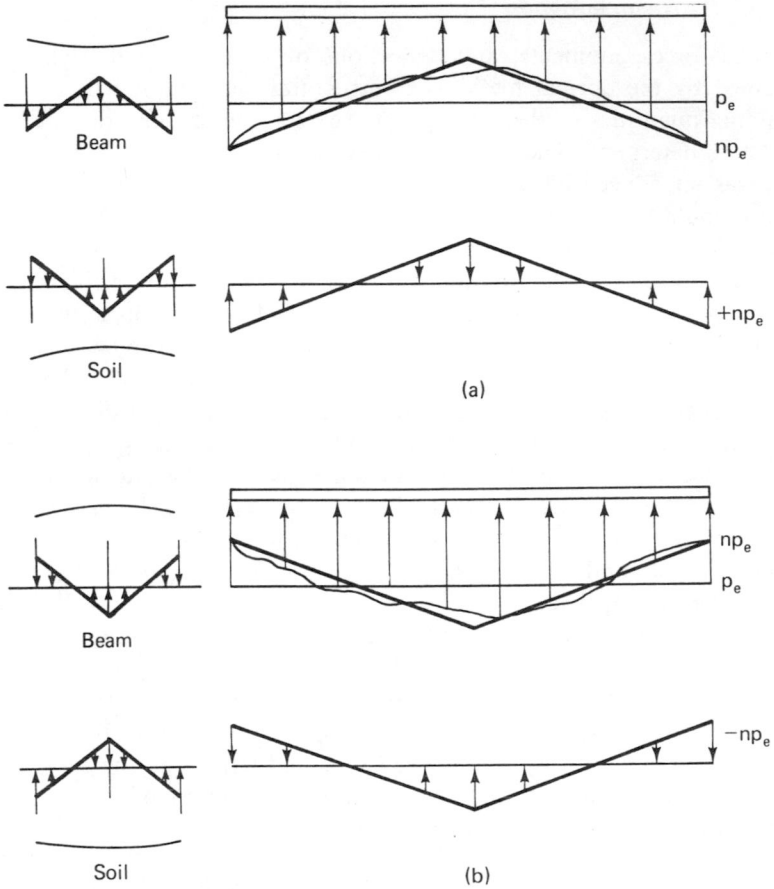

Figure 5-15. Triangular pressure distributions: (a) positive distribution, cohesive soil case; (b) negative distribution, cohesionless soil case.

moment equation (5.139) the coefficient equals 0.042, and this value is retained in future calculations.

Thus, under the loading conditions considered, the maximum relative displacement of the beam is approximately

$$w = w_2 + 0.0035\frac{npl^4}{EI} \tag{5.140}$$

since w_1 is negligible. This is the *beam equation*, which relates beam deflection to applied loads and idealized soil reaction, n. An equivalent expression is required to represent the soil's response.

5.5.3 Soil equation

For the positive pressure distribution of Figure 5-15(a), the reaction pressure on the soil will be downward, or compressive at the position of the beam ends, and upward or tensile at the beam center. This will induce a convex upward curvature of the soil surface that corresponds in our convention to a *negative* relative deflection. If, as a simple first case, we consider the soil behavior to correspond exactly to that of a uniform Winkler material with subgrade reaction coefficient $k_0(= k/B)$, then the downward deflection at the ends will be equal to np_e/k_0 and the upward deflection at the center will be equal to $-np_e/k_0$. The maximum relative soil deflection will therefore be

$$w = -\frac{2np_e}{k_0} = -\frac{2np}{k} \tag{5.141}$$

where $k = Bk_0$, as usual. This is a *soil equation*. However, we wish to include either a real variation of k_0 across the site, or an estimate of the variation we might expect in k_0 along the length of the beam due to the continuum response of the soil to the loaded raft. In this case, we return to the total idealized pressure distribution in the top diagram in Figure 5-15(a) and apply it to soil that has a subgrade reaction coefficient k_0 at the beam ends and sk_0 at the beam center. (Alternatively, if we wish to restrict the value of s to be greater than one, k_0 is the value at the beam center and sk_0 is the value at the ends.) The analysis is equally valid. In the displacement diagram in Figure 5-14(c) in which the minimum displacement is w_s and the maximum relative displacement is w, at the beam center, approximately, the total displacement is caused by the pressure $p_e(1 - n)$ applied to soil whose reaction coefficient is sk_0, so that

$$w_s = \frac{p_e(1 - n)}{sk_0} \tag{5.142}$$

At the ends a pressure $p_e(1 + n)$ is applied to the soil subgrade coefficient k_0 and we have

$$w_s + w = \frac{p_e(1 + n)}{k_0} \tag{5.143}$$

Then the maximum relative settlement w is obtained by subtraction:

$$-w = \frac{p_e(1 + n)}{k_0} - \frac{p_e(1 - n)}{sk_0} \tag{5.144}$$

Since the resulting settlement shape is negative by our definition, a negative sign has been added to the w. Adaptations of this analysis can be made to include the possibility of hard or soft spots existing at particular places along the beam. Equation (5.144) is a generalized soil equation for relative displacement w as a function of n. The values of k_0 and s in it, when it is used to express the soil response of a single beam in a raft foundation, will be different depending on whether the beam is internal, as beam 2 in Figure 5-14, or lies along an edge, as beam 1 does. For a raft on clay, higher values of k_0 would be required to describe the behavior of an edge compared to a center beam. In sand, the reverse would be true. In each case, the depth of embedment would modify the values selected.

It is recommended that soil values selected for incorporation in equation (5.144) should lie within a range of numbers describing the site properties. They should be employed in such a way that two soil equations can be developed to represent upper- and lower-bound soil conditions. Ultimately, the equations can be used to develop a range of expectable beam moments and deflections. In general, the soil behavior is not linear, so that the bounding equation could also include the effect of expected nonlinearities. Where measured or estimated subgrade reaction coefficients are used, they should be modified to reflect the width and length of the beam according to the discussion in Chapter 7.

At many sites the values of subgrade reaction coefficient are not measured, and the soil equation must be arrived at by other means. Normally, a site investigation includes boreholes and testing of soil samples. The properties can be used in the following way to establish soil equations and settlement bounds. The material properties are used to calculate the settlements that a uniformly loaded flexible mat would undergo at the site when loaded with the average pressure that the structure exerts over the building area. This flexible structure could be an oil tank, for example. Both edge and center settlements are required; the difference between them is the differential displacement of the site under uniform loading, for which the value of n is zero. Thus, one point in the soil equation, at the value $n = 0$, is obtained. For another point, the soil properties are used in a calculation of the pressure distribution that a rigid mat of the size of the proposed mat foundation would encounter on the site soil. The value of n for this pressure distribution then corresponds to zero relative settlement. With some limits on the values assigned to the two points, equations for the soil line are obtained. Depending on the history of construction at the site, previous measurements of building settlements may also be employed in constructing the soil equation.

Now that two equations have been assigned to beam and soil, the equations can be solved simultaneously to give values of the pressure variation parameter n and the maximum relative settlement w at which the loads and deflections of beam and soil are compatible in the limited sense of this analysis. It is now necessary to assess the bending moments that will act on the beam in consequence.

5.5.4 Beam bending moments

The principal influence on the distribution of soil pressure in system 3 is the relative displacements of the beam caused by the balanced load system 2. Thus, with the approximations adopted, Baker shows that the bending moment diagram developed by the soil pressure 3 is similar in shape to M_2, but is, of course, opposite in sign and of smaller magnitude. In other words, the effect of the distribution of soil reaction is to reduce the bending moments in system 2. Because of the very small deflections induced by the first set of loads, their bending moments remain substantially unchanged by the soil reaction.

If we assume that the two bending moment diagrams M_2 and M_3 are exactly similar and of opposite sign, and that, moreover, the bending moments are proportional to the maximum relative deflection of the beam, then we can express this dependence by the equation

$$\frac{M_2}{M_2 + M_3} = \frac{w_2}{w_2 + w_3} = r, \text{ say} \qquad (5.145)$$

where M_2 and M_3 are the values at a point on the beam and w_2 and w_3 are the maximum relative values of deflection. In this equation, w_2 is calculated for the load system 2 directly, and $w_2 + w_3$ is obtained from the simultaneous solution of the soil and beam equations. Thus, the reduced final moment sum $M_2 + M_3$ is derived by taking the ratio $r = w_2/(w_2 + w_3)$ from the solution and dividing the moments or moment diagram M_2 by r. The final moment distribution on the beam for design is then $M_1 + M_2/r = M_1 + M_2 + M_3$. Compared to performing a Winkler analysis of the structure, this is a very abbreviated and rapid technique.

In his exposition, Baker draws the beam and soil equations, which he refers to as the beam and soil *lines*, on a diagram of w versus n, such as shown in Figure 5-16. The effect of changes in either equation can be readily visualized. Suppose, for example, the raft of which the beam is a component will rest on a medium-stiff clay deposit and that a large oil tank adjacent to the site of dimensions and loading comparable to the proposed raft exhibited a differential settlement of 2 in. (50 mm) greater at the center than at the edges. Since the oil tank is uniformly loaded, $n = 0$, and the differential settlement is positive, we can plot the point A in Figure 5-16 from this data. From other measurements or from soil data we estimate that a loaded rigid mat at the site would undergo soil reaction pressures at its edges 30 percent greater than the uniform pressure and at the center 30 percent less; thus, for this case, $n = 0.3$ for $w = 0$, since there are no differential settlements. This gives the point B, and the soil equation is represented by the line AB, marked $S1$. From the loading conditions and the tentatively designed beam

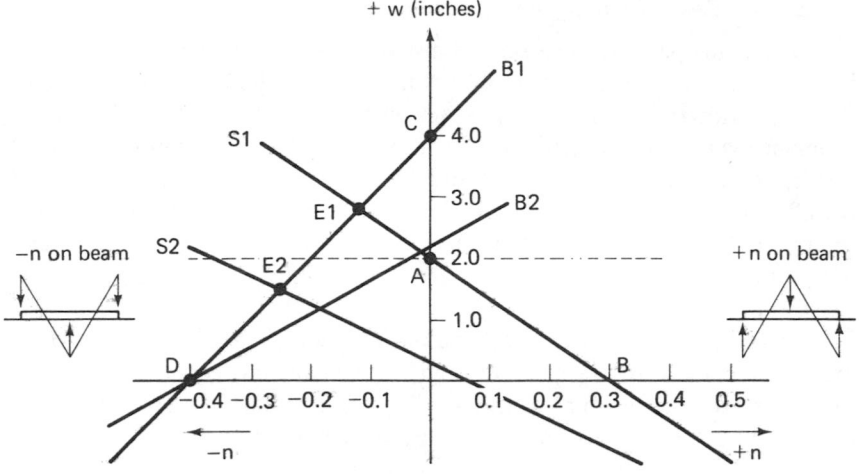

Figure 5-16. Beam and soil equations.

properties, the beam equation is determined and plotted as line B1, CD in Figure 5-16.

The two equations as shown are:

$$\text{Soil:} \quad w = -6.67n + 2$$
$$\text{Beam:} \quad w = 10n + 4$$

(5.146)

The soil and beam equations have a common point E1 in the figure, at which $w = +2.8$, $n = -0.12$. From equation (5.140), the intersection of the beam line with the w-axis, point C, occurs at the value of w_2, which is 4 in. in this example. The ratio r is therefore obtained to be

$$r = \frac{4}{2.8} = 1.43$$

so that the sum of the moments $M_2 + M_3$ is equal to the moment M_2 alone divided by 1.43.

If there were several oil tanks at the site exhibiting different amounts of differential settlement, and if the measurement of soil properties indicated a range of values of n for the pressure distribution on the hypothetical rigid foundation, the extreme range of soil behavior could be represented in Figure 5-16 by upper and lower bounding lines, say S1 and S2. Very stiff soil is indicated by a line tending toward the horizontal, very soft material by a line tending toward the vertical. Similarly, a stiff beam has only a small slope, a flexible beam a steeper slope. For the range in soil properties indicated and the single beam equation B1, the two intersections E1 and E2 indicate the range in moments and relative deflections that it could experience. The stiffer soil gives, as we would expect, a greater moment reduction and smaller relative deflection. Selection of a stiffer beam, indicated by line $B2$, causes smaller moment reductions and a smaller differential settlement. For the same set of column loads, lines $B1$ and $B2$ and beam lines for any other stiffness of beam pass through the point D on the n-axis. Baker demonstrates that if the beam represented by line $B2$ is q times as stiff as beam $B1$, it will be subject to m times the moments of $B1$ for the same column loads. The factor m depends on the softness of the soil, approaching unity for very soft soils, and q for very stiff soils.

A raft is usually considered for a site, because bearing capacity considerations dictate that individual footings will occupy more than 50 percent of the building area. Provision of the raft will then normally ensure that individual beams or inverted T-beams cause stresses in the soil well below failure values. The next consideration is the gross settlements that will develop; if these are not a concern, design of the raft may proceed.

In the design of the beam or beams for the soil conditions illustrated in Figure 5-16, a first check is made that the moments do not exceed the safe moments for the particular stiffness of beam examined, say $B1$ in the figure. Let us say that this beam meets the moment requirement. Another criterion that must be met concerns the relative deflections of the structure. This is discussed in more detail in Chapter 7, but, in general, to avoid architectural damage, the ratio of differential settlement between two adjacent columns to the distance between the columns should not exceed about 1:300 for typical buildings. If the beam of Figure 5-16 is part of the foundation structure of Figure 5-14, we might conclude that the maximum differential settlement of the entire beam should not exceed 2 in. (50 mm) if local distressful differential movements are to be avoided. In this case, reference to Figure 5-16 shows that the beam should be at least as stiff as $B2$ even though it would then be subject to higher moments than $B1$ would see.

Other considerations will develop in the course of the examples.

5.5.5 Example calculations

Baker includes two illustrative problems that are interesting to compare with Winkler solutions for the same beams, loading conditions, and soil. The first is a single footing carrying a column load of 80,000 lb (356 kN); it is illustrated in Figure 5-17. The footing is concrete, 20 ft by 4 ft by 2.5 ft ($6.1 \times 1.2 \times 0.8$ m) deep. The moment of inertia is 108,000 in.^{4} (4.5×10^{-2} m^4) and Young's modulus E for concrete is taken to be 3×10^6 psi (2×10^4 MN/m^2). A brief study of the problem shows that in this case there can be no separation into load systems 1 and 2, since there is only one column. The deflections under loading system 1 are not negligible here, and they take the place of the w_2 which control in more complicated structures.

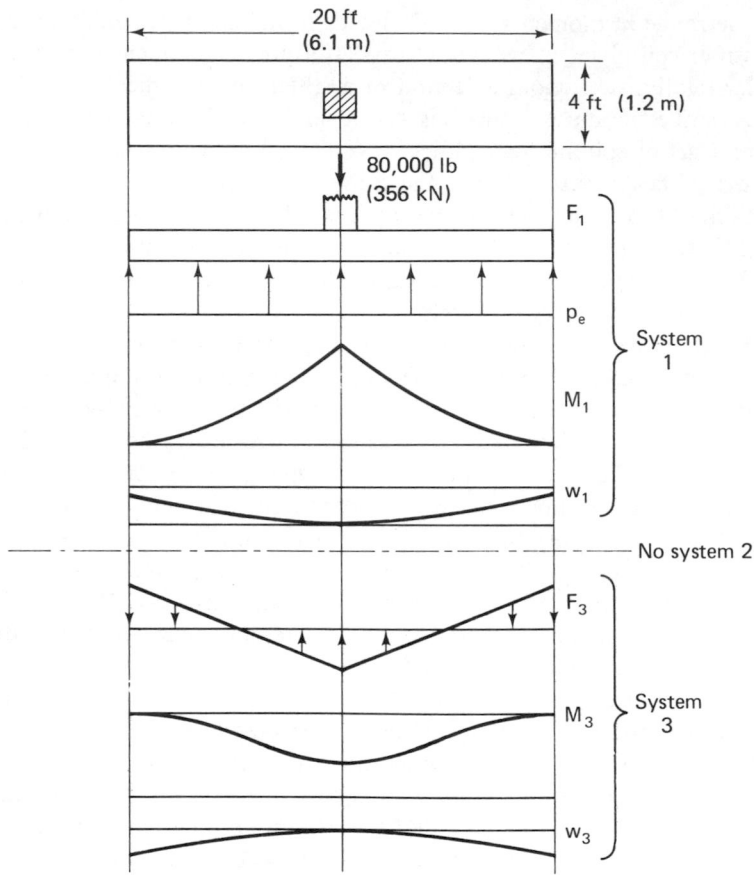

Figure 5-17. Single footing example.

For system 1, the uniform soil reaction loading p is $80,000/240 = 333.3$ lb per in. (58 kN/m) run of the beam, and, from the deflection of a uniformly loaded cantilever beam of length l

$$w_1 = \frac{1}{128}\frac{pl^4}{EI} = 0.0267 \text{ in. } (0.68 \text{ mm})\tag{5.147}$$

From equation (5.140), the beam equation is

$$w = 0.0267 + 0.0035\frac{pl^4}{EI}n$$

or

$$w = 0.0267 + 0.0119n\tag{5.148}$$

when l is taken as the *half*-length of the footing. In Baker's example the subgrade reaction coefficient, k_0, is given as uniformly 2000 pci (543 MN/m³), an unrealistic value. This, for the 48-in. (1.22 m) width of the beam, corresponds to a k_{01} of about 4400 to 7000 tons per cubic ft (1.38×10^3 to 2.2×10^3 MN/m³) depending on whether the soil is sand [equation (7.44)] or clay [equation (7.47)]. According to Tables 7.2 or 7.3, along with equations (7.44) or (7.47), this value describes the response of an extremely dense sand (soft sandstone) or very hard clay (shale).

Since $k = k_0 B = 96,000$ psi (660 MN/m²) here, equation (5.75) indicates this to be approximately the Young's modulus of the foundation material. Typically, the *strength* of soil materials is about a factor of several hundred smaller than the modulus, so that the strength of this soil or rock is a few hundred psi (2 to 3 MN/m²). Since the bearing pressure of the footing is about 7 psi (48 kN/m²), it would be extremely unlikely for a designer to choose such a large base for this column. It could be borne quite safely, as far as the soil is concerned, on a footing a few feet (1 m) square. However, we continue with the analysis, for illustration of some aspects of the method.

A uniform value of k_0 indicates that the soil behaves as a Winkler foundation, in which case the soil equation is given by equation (5.141)

$$w = -\frac{2p}{k}n = -0.0069n \tag{5.149}$$

with no constant term.

Simultaneous solution of equations (5.148) and (5.149) gives a relative settlement of $w = w_1 + w_3 = 0.0098$ in. (0.25 mm) at a value of n equal to -1.42. In this case, the ratio

$$r = \frac{w_1}{w_1 + w_3} = \frac{0.0267}{0.0098} = 2.72 \tag{5.150}$$

Now the moments M_1 generated by the uniform load p are parabolic:

$$M_1 = \frac{px^2}{2} \tag{5.151}$$

where x is the distance measured from either end of the beam toward the center. In particular, the maximum moment at the center of the footing is

$$M_{1\max} = \frac{pl^2}{8} = 2.4 \times 10^6 \text{ lb-in. (271 kN-m)} \tag{5.152}$$

These moments, for the approximate solution, are reduced by division by r, so that the maximum moment becomes

$$M_{1\,max} = \frac{2.4 \times 10^6}{2.72} = 0.88 \times 10^6 \text{ lb-in. (99 kN-m)} \qquad (1.153)$$

The derived value of -1.42 for n is very large; it indicates that the pressures under the column are 2.42 times the average pressure, and that at the footing edges the pressure is $-0.42p_e$ or tensile. The latter pressure is -2.90 psi (-20 kN/m²), and if this exceeds the pressure due to the weight of the concrete, the footing in practice will lift off the ground. The concrete pressure can easily be calculated to be about 2.60 psi (18 kN/m²), so that the ends of the footing are in a state of barely touching the soil surface. This has not been accounted for in the analysis.

We proceed to the exact Winkler evaluation of the same problem. Here we find for the relevant parameters

$$\lambda^4 = \frac{2000 \times 48}{4 \times 3 \times 10^6 \times 1.08 \times 10^5}; \qquad \lambda = 1.6497 \times 10^{-2} \text{ in.}^{-1}$$

$$\therefore \lambda l = 3.9594 \qquad (5.154)$$

For a centrally loaded finite weightless beam, of length such that $\lambda l > \pi$, the ends lift off the surface, which confirms the previous conclusion. The maximum central bending moment is given by the expression

$$M_{max} = \frac{P}{4\lambda}\left(\frac{\cosh \lambda l - \cos \lambda l}{\sinh \lambda l + \sin \lambda l}\right)$$

$$= 1.212 \times 10^6 \times (1.0562) \qquad (5.155)$$

$$= 1.28 \times 10^6 \text{ lb-in. (145 kN-m)}$$

which compares with the approximate value from equation (5.153) of 0.88×10^6 lb-in. (99 kN-m), which is therefore 30 percent low by comparison.

The maximum central deflection of the Winkler beam is

$$w_{max} = \frac{P\lambda}{2k}\left(\frac{\cosh \lambda l + \cos \lambda l + 2}{\sinh \lambda l + \sin \lambda l}\right)$$

$$= 0.0069 \times (1.0811) \qquad (5.156)$$

$$= 0.0074 \text{ in. (0.19 mm)}$$

However, the ends come slightly off the ground; their deflection is

$$w_{end} = \frac{P\lambda}{2k}\left(\frac{\cosh \dfrac{\lambda l}{2} \cos \dfrac{\lambda l}{2}}{\sinh \lambda l + \sin \lambda l}\right)$$

$$= 0.0069 \times (-0.0576) \qquad (5.157)$$

$$= -0.0004 \text{ in. (-0.01 mm)}$$

so that the relative settlement is 0.0078 in. (0.2 mm) for comparison with the estimated amount of 0.0098 (0.25 mm) which is therefore about 25 percent too high.

The beam and soil lines are shown in Figure 5-18. It is obvious that the only reason the uniform pressure bending moment is reduced for this beam is the extremely flat slope of the soil line. This occurs because of the very high k_0 assigned to the soil. Were a more realistic value of k_0 selected, appropriate to the footing size and load, say in the order of 50 pci (13.6 MN/m³), the soil line would be so steep that it would intersect the beam line very close to the w-axis. Thus, r would be close to unity and n to zero. A little reflection shows that this is to be expected—a soft soil would exert a quite uniform pressure under the relatively stiff beam, and this would give the moment M_1.

A more realistic combined footing example is used as a second illustration. The configuration is shown in Figure 5-19. Three columns are attached to a beam of the same width and depth as in the previous example. Loads and lengths are shown in the figure. Here all components of the problem are present. Baker presents two cases for analysis: (a) a uniform value of $k_0 = 56$ pci (15.2 MN/m³) throughout; (b) k_0 has the value 112 pci (30.4 MN/m³) at the beam center diminishing linearly to 28 (7.6 MN/m³) at the ends. The first of these two cases can be analyzed by the exact Winkler theory; the second would be much more difficult to solve, but it might represent a case in which the beam in question was part of a larger mat foundation, as in Figure 5-14, resting on sand which exhibits less resistance to deformation near the edge of the mat.

In this problem, once again the uniform pressure, p, is 333.3 lb per in. (58 kN/m) run, so that the system of forces F_1 (see Figure 5-19) consists of the beam carrying a uniformly distributed pressure p, with the supports maintained at constant elevation. The solution for reactions or moments can be obtained *ab initio* or from tables from the behavior of a cantilever (one-half of the beam) supported at its free end. The reaction loads and the peak moment values are shown in Figure 5-19. The general moment is given by the quadratic

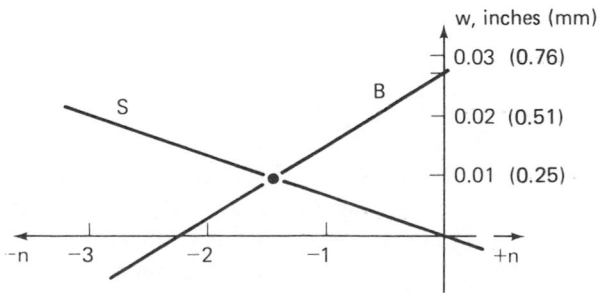

Figure 5-18. Beam and soil lines for first example.

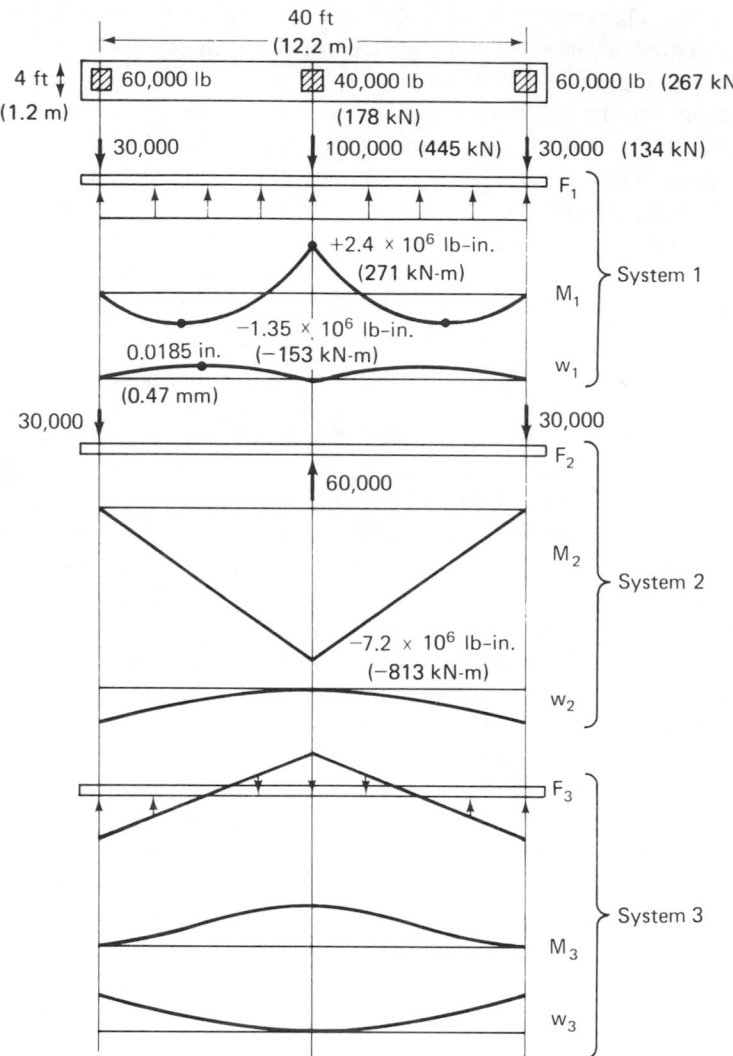

Figure 5-19. Combined footing example.

$$M_1 = \frac{1}{8} p x \left(\frac{3l}{2} - 4x \right) \tag{5.158}$$

where x is measured from either end of the beam toward the center. The peak negative moment occurs at $x = 3l/16$ from either end (the total beam length is $l = 40$ ft (12.2 m). The maximum deflection under this loading occurs a distance of 0.21 l from either end and is equal to 0.0185 in. (0.47 mm). Subtracting the loads F_1 from the original column loads gives the loading system F_2, which can be considered to be a 60,000-lb (267 kN) point load on a simply supported beam. The maximum moment M_2 is negative, as shown on Figure 5-19. Deflection w_2 is given by the expression

$$w_2 = -\frac{Pl^3}{48EI} = -0.43 \text{ in. } (-10.9 \text{ mm})$$ (5.159)

where P is the point load, 60,000 lb (267 kN). The relative deflection is negative because the loading induces tension on the *top* of the beam.[5] From equation (5.140), the beam line is then

$$w = -0.43 + 0.19n$$ (5.160)

For case (a), the uniform value of $k_0 = 56$ pci (15.2 MN/m³) gives again a soil equation as a line passing through the origin

$$w = -0.25n$$ (5.161)

from equation (5.141). The point common to the two equations is $n = +0.98$, $w = -0.24$ in. (−6.1 mm). Using equation (5.145) gives

$$r = \frac{-0.43}{-0.24} = 1.79$$ (5.162)

which reduces the negative moment M_2 from -7.2×10^6 lb-in. (−814 kN-m) to $-7.2 \times 10^6/1.79 = -4.02 \times 10^6$ lb-in. (455 kN-m). The design moment diagram for the beam is obtained from the algebraic sum of M_1 [unchanged, since the system 1 relative deflections with a maximum of 0.0185 in. (0.47 mm) are essentially negligible in comparison with the combined systems 2 and 3 deflection of 0.24 in. (6.1 mm)] and the reduced M_2 moments. This gives an approximate moment diagram as shown in Figure 5-20.

Before going on to consider the nonuniform k_0 distribution, it is interesting to compare the above calculations with the exact Winkler result. The exact analysis can be assembled from the superposition of two configurations with known solutions. The first of these is a beam subjected to a point load P_1, at both ends; for this problem, the bending moment is

$$M = -\frac{P_1}{\lambda}\left[\frac{\sinh\left(\frac{x}{l}\lambda l\right)\sin\left[\left(1 - \frac{x}{l}\right)\lambda l\right] + \sinh\left[\left(1 - \frac{x}{l}\right)\lambda l\right]\sin\left(\frac{x}{l}\lambda l\right)}{\sinh \lambda l + \sin \lambda l}\right]$$ (5.163)

where x is measured from a loaded end.

For the example given, λ and λl are different from before

$$\lambda^4 = \frac{k}{4EI} = \frac{56 \times 48}{4 \times 3 \times 10^6 \times 108 \times 10^5}; \quad \lambda = 6.748 \times 10^{-3} \text{ in. } (0.171\text{mm})$$ (5.164)

---

[5] The deflection w_2 is incorrectly given as positive by Baker. The resulting beam equation and its intersection with the soil line are also incorrect. The numerical results for the uniform k_0 case are correct in magnitude but of the wrong sign.

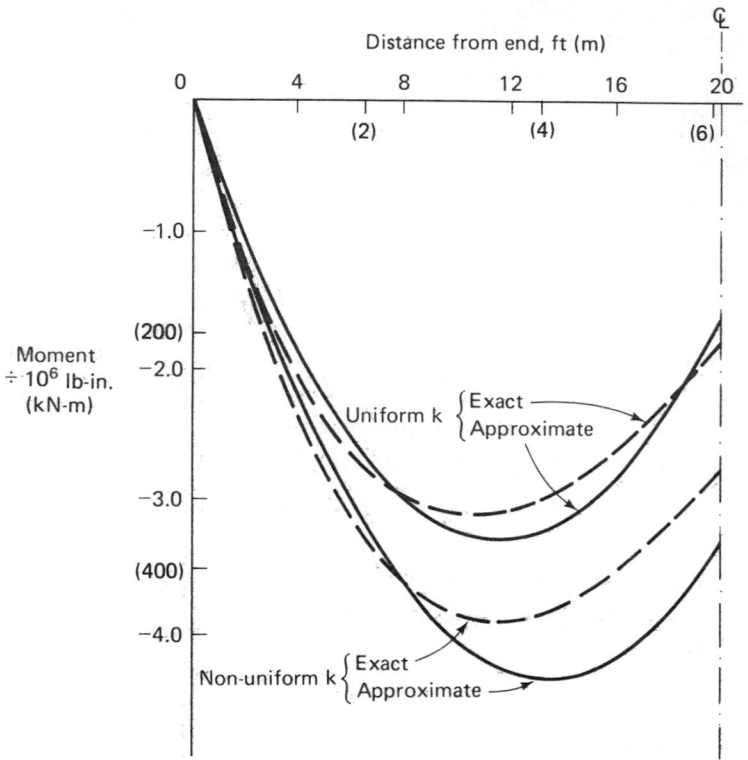

Figure 5-20. Comparison of approximate and exact solutions for second example.

so that, with a beam length of 480 in. (12.2 m)

$$\lambda l = 3.2393$$

The same constants hold for the second superposition configuration, that of the beam loaded with P_2 centrally, as for the single-column loaded footing. The general bending moment for this case is given by the expression

$$
\begin{aligned}
M = \frac{P_2}{4\lambda} \Bigg\{ & \frac{1}{(\sinh \lambda l + \sin \lambda l)} \Bigg[\sinh \frac{x}{l}\lambda l \left(\sin \frac{x}{l}\lambda l - \sin \left(1 - \frac{x}{l}\right)\lambda l \right) \\
& - \cosh \frac{x}{l}\lambda l \left(\cos \frac{x}{l}\lambda l + \cos \left(1 - \frac{x}{l}\right)\lambda l \right) \\
& + \sin \frac{x}{l}\lambda l \left(\sinh \frac{x}{l}\lambda l - \sinh \left(1 - \frac{x}{l}\right)\lambda l \right) \\
& + \cos \frac{x}{l}\lambda l \left(\cosh \frac{x}{l}\lambda l + \cosh \left(1 - \frac{x}{l}\right)\lambda l \right) \Bigg] \Bigg\}
\end{aligned}
$$

(5.165)

The attractiveness of the approximate solution is becoming apparent! Summing moments from equations (5.163) and (5.165) with the appropriate substitutions $P_1 = 60{,}000$ lb (267 kN) and $P_2 = 40{,}000$ lb (178 kN) gives the exact solution at various distances along the beam, as displayed in Table 5.3 and graphed in Figure 5-20.

TABLE 5.3

Comparison of Exact and Approximate Solutions for Loaded Beam

Distance $\dfrac{x}{l}$		0	0.1	0.2	0.3	0.4	0.5
	Exact	0	−2.10	−2.98	−3.07	−2.60	−1.80
$\dfrac{M}{10^6}$, lb-in.			(−0.24)	(−0.34)	(−0.35)	(−0.29)	(−0.20)
(N-m)	Approximate	0	−1.86	−2.95	−3.28	−2.83	−1.62
			(−0.21)	(−0.33)	(−0.37)	(−0.32)	(−0.18)

It can be seen that the two solutions correspond remarkably well, with the maximum moment being only about 5 percent high. Without presenting all the equations for end and central deflections for the Winkler solution, the deflection result is presented. The maximum deflection of 0.2788 in. (7.1 mm) occurs at the beam ends and the minimum at the center, 0.0486 in. (1.2 mm), so that the maximum differential settlement in this case comes to −0.2302 in. (−5.8 mm) in favorable comparison with the result of the approximate analysis −0.24 in. (−6.1 mm).

Turning to part (b) of the example, where the subgrade reaction coefficient is higher, 112 pci (30.4 MN/m³), at the center than at the ends, 28 pci (7.6 MN/m³), we obtain the soil equation from equation (5.144)

$$-w = \frac{333.33(1 + n)}{28 \times 48} - \frac{333.33(1 - n)}{112 \times 48} \tag{5.166}$$

or

$$w = -0.31n - 0.186$$

The beam equation remains the same as before, (5.161), and simultaneous solution gives the intersection this time as $w = -0.34$, $n = 0.49$. Now the M_2 moment is only decreased by the factor $r = -0.43/-0.34 = 1.26$, and the maximum value goes from -7.2×10^6 to $-7.2 \times 10^6/1.26 = -5.71 \times 10^6$ lb-in. (−0.65 MN-m). The sum of the M_1 and modified M_2 moments is shown on Figure 5-20. At 0.34 in. (8.6 mm) the relative deflection of the beam, as expected, is greater than that for the beam on the uniform soil.

From these examples, it is apparent that a very long continuous beam, subjected to a number of column loads, has to be designed for the moments that a uniform soil reaction would impose on it. This is the case, because, as we have seen, a reduction in moments can only be imposed if the soil is very stiff. But if it *is* stiff, single footings are used to support the columns, and a continuous foundation is unnecessary.

The nonuniform case can be checked analytically in an interesting example of the solution to the Winkler equation with linearly varying subgrade coefficients. Tables of this solution are given in Appendix C. For the problem, we take one-half of the beam of Figure 5-19, say the left half. This configuration is then a beam loaded at the left with 60,000 lb (267 kN), and at the center with one-half of the former load, 20,000 lb (89 kN). These constitute two of the required boundary conditions. From symmetry, the slope of the half beam at the right side is zero, and, of course, the moment at the left end is also zero; these are the other two boundary requirements. The beam rests on a soil whose subgrade reaction coefficient, k_0, increases linearly from a value of 28 pci (7.6 MN/m³) at the left end to 112 (30.4 MN/m³) at the right. Thus,

$$k = 28 \times 48 + \frac{84 \times 48}{240} x$$

$$= 1344 + 16.8x \tag{5.167}$$

where x is the distance coordinate measured from the left end of the beam. We will refer to the two numerical values on the right side of the equation as k_1 and k_2, respectively.

In Appendix C the solution is supplied for the case of equation (8.153) in which n is equal to unity, so that the solution applies for a problem in which k varies linearly from zero at the coordinate origin. We can convert the present situation to that case by coordinate transformation. Let

$$k_2 y = k_1 + k_2 x \tag{5.168}$$

where y is a new dimensional length variable. Then the controlling equation becomes

$$\frac{d^4 w}{dy^4} - \frac{k_2 y}{EI} w = 0 \tag{5.169}$$

which is now of the correct form. Making the dimensionless substitution

$$s = \lambda y \tag{5.170}$$

gives

$$\frac{d^4 w}{ds^4} + sw = 0 \tag{5.171}$$

when

$$\lambda^5 = \frac{k_2}{EI} \tag{5.172}$$

and equation (5.171) has the solution of Figure C-6 and Table C.3 in Appendix C. We must be careful with the coordinates in this transformation. When $x = 0$, we have

$$s = \lambda y = \frac{\lambda k_1}{k_2} \tag{5.173}$$

and, at the right end, when $x = l = 240$ in.,

$$s = \lambda y = \frac{\lambda k_1}{k_2} + \lambda l \tag{5.174}$$

For simplicity, we will use the notation $w(0)$, $w'(0) \ldots$ and $w(l)$, $w'(l) \ldots$, to mean the displacement, slope, etc., when $x = 0$, l, respectively, but from the tables we must look up the appropriate values of w, w', etc., at the values of s given by equations (5.173) and (5.174). For intermediate sections of the beam, the dimensionless coordinate is given by

$$s = \lambda y - \frac{\lambda k_1}{k_2} + \lambda x \tag{5.175}$$

As a first boundary condition, the moment at the left end is zero, so that, if the solution to equation (5.171) has the form

$$w = C_1 w_1 + C_2 w_2 + C_3 w_3 + C_4 w_4 \tag{5.176}$$

where the C_n are constants to be determined, the requirement becomes

$$C_1 w_1''(0) + C_2 w_2''(0) + C_3 w_3''(0) + C_4 w_4''(0) = 0 \tag{5.177}$$

At the left end the second boundary requirement is that the shear, S, equal the applied load P_1 of opposite sign:

$$S(0) = -EI\lambda^3 w'''(0) = -P_1$$

or

$$C_1 w_1'''(0) + C_2 w_2'''(0) + C_3 w_3'''(0) + C_4 w_4'''(0) = \frac{P_1}{EI\lambda^3} \tag{5.178}$$

For zero slope at the right end, we have

$$C_1 w_1'(l) + C_2 w_2'(l) + C_3 w_3'(l) + C_4 w_4'(l) = 0 \tag{5.179}$$

and, finally, the shear at the right end equals the applied load, P_2, one-half of the full beam load there:

$$S(l) = -EI\lambda^3 w'''(l) = P_2$$

or

$$C_1 w_1'''(l) + C_2 w_2'''(l) + C_3 w_3'''(l) + C_4 w_4'''(l) = -\frac{P_2}{EI\lambda^3} \qquad (5.180)$$

Inserting the various numerical constants for the problem, we get, in addition to k_1 and k_2,

$$\lambda = 8.7691 \times 10^{-3} \text{ in.}^{-1} (3.45 \times 10^{-4} \text{ mm}^{-1}); \qquad \frac{\lambda k_1}{k^2} = 0.7015;$$

$$\lambda l = 2.1046; \qquad \frac{\lambda k_1}{k_2} + \lambda l = 2.8061; \qquad \frac{P_1}{EI\lambda^3} = 0.2746; \quad (5.181)$$

$$\frac{P_2}{EI\lambda^3} = 0.0915$$

from which $s(0)$ is then 0.7015, and $s(l) = 2.8061$. Interpolating from the Appendix C table for the appropriate functions and derivatives for these values at the left and right ends of the beam, we have the following equations, which represent equations (5.177), (5.178), (5.179), and (5.180) in the same order.

$$
\begin{aligned}
0.9810C_1 + 0.0452C_2 - 1.6327C_3 + 0.9841C_4 &= 0 \\
-0.6621C_1 + 0.5726C_2 - 1.3687C_3 - 0.3777C_4 &= 0.2746 \\
0.1106C_1 - 0.0910C_2 - 8.5170C_3 - 1.1033C_4 &= 0 \\
-0.1502C_1 - 0.1890C_2 + 1.3920C_3 - 13.8837C_4 &= -0.0915
\end{aligned}
\qquad (5.182)
$$

Simultaneous solution of these four equations gives the constants

$$C_1 = -0.0290; \qquad C_2 = 0.4343; \qquad C_3 = -0.0051; \qquad C_4 = 0.0005$$

Combination of these constants with values of $w_1, w_2, \ldots$, or $w_1'', w_2'', \ldots$, at various s from the table, times the appropriate multiplying factors (1 for displacement, $-EI\lambda^2$ for moment), gives the distribution of displacement, moment, or any other desired variable along the beam. Moment is shown in Figure 5-20 for comparison with the approximate solution, which is not so accurate in this case.

In particular at the center of the beam, the exact solution moment is

$$M(l) = -EI\lambda^2(0.1106) = -2.76 \times 10^6 \text{ in.-lb} (-0.31 \text{ MN-m})$$

versus -3.31×10^6 in.-lb (-0.37 MN-m) for the approximate solution. At the end and center of the beam the displacements by the exact solution are 0.330 in. (8.4 mm) and 0.029 in. (0.74 mm), respectively. The relative displacement is then -0.301 in. (-7.6 mm) in comparison with the value of -0.34 in. (-8.6 mm) obtained by Baker's method.

In sum, the approach devised by Baker gives reasonable (± 30 percent) values of moments and displacements in comparison with more exact and lengthier analyses. The method is useful for preliminary estimates of the quantities in early stages of design when its advantage of indicating directly the effect on moment, displacement, and pressure distribution of changes in the beam dimensions or in soil properties is particularly apparent. It would be interesting to examine the consequences for the analysis of retaining the triangular load distribution but representing the soil response by two parameters, rather than the one-coefficient Winkler model. The shape of the ground deflection would thereby be rendered more realistically.

PROBLEMS

1. Could you make any improvements on Baker's method as a result of experiments (laboratory, centrifuge, or field)?
2. What, to you, is the most unsatisfactory aspect of the design of mat foundations, as discussed in this chapter?
3. If, over a long time, a clay soil creeps, or exhibits a viscous component in its behavior, how would you expect the pressure distribution on the base of a rigid slab founded on the clay to vary with time? Distinguish between viscous and consolidation time effects. What would be the effect on the bending moments in the slab as a function of time?
4. You are given the task of designing the *floor* of a rectangular grain silo to support the weight of the grain. The floor itself rests on beams and footings, so that it can be considered an inverted mat or raft. Would you be able to use the soil-line method to solve this design problem? Describe briefly the steps you would take, especially where they would differ from those involved in designing a conventional mat foundation.
5. Could you devise a variation of Baker's method for the design of a tied-back sheet pile wall? How would you obtain the soil properties for this case?
6. A flat steel sheet is placed on the surface of a dry uniform sand deposit in which the Standard Penetration Test gives an average of 10 blows per foot to a depth of 50 ft (15.2 m) below the surface. The sheet is loaded as shown in Figure P5-6. Design the sheet (find the thickness) so that *neither* a tensile stress of 20,000 psi (138 MN/m²) in the steel *nor* a maximum differential settlement of 3 in. (76 mm) is exceeded. For your design give the deflections and indicate the maximum bending moment in the sheet. Draw the pressure distribution between sheet and soil.
7. Carry through the analysis of Problem 6 for the case where the sheet rests on uniform clay with a cohesion of 1000 psf (48 kN/m²), and the moments, displacements, etc., are to be calculated soon after loading. What would happen to them as a function of time?

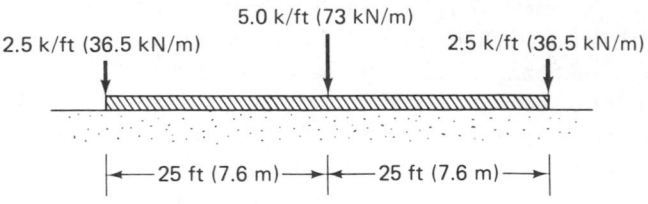

Figure P5-6.

8. In the mat foundation of Problems 6 and 7 the loads were independent of the mat and soil deflections. In a real structure supported on a mat this would not be the case. Figure P5-8 shows a more realistic version of this problem. Modify your calculations for this case, for both sand and clay soil.

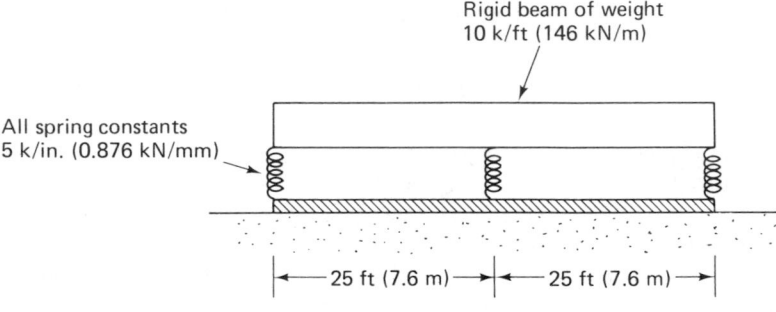

Figure P5-8.

9. Hayashi (10) presents a convenient equation for the deflection at the load in a centrally-loaded finite beam:

$$w_m = \frac{P}{kl} \frac{5[(\lambda l)^4 + 48]}{2[(\lambda l)^4 + 120]}$$

This result was obtained from a series solution of equation (5.5). It is also possible to obtain a similar result for the maximum moment under the load

$$M_m = \frac{Pl}{24} \frac{[(\lambda l)^5 + 360]}{[(\lambda l)^4 + 120]}$$

The deflection equation gives displacements accurate to within a few percent up to $\lambda l = 4$; the maximum moment expression, however, is only good to a λl value of about 2. Perform a series solution to verify these equations, and check their range of accuracy by comparing them with the exact solutions.

10. For a correct comparison of the plane strain two-parameter solution for an infinite beam loaded by a point load, what continuum solution is required? Could you obtain it?

11. Using Baker's method and the beam of Section 5.5.5 (same length, section modulus) calculate the moments and deflections if the beam is loaded by 2 60 kip (267 kN) loads, located 10 ft (3.05 m) in from each end.

12. In the design of railroad ties, two possibilities are apparent: equalizing the maximum positive and negative moments, or making the settlement uniform by equalizing the pressures at the ends and center of the tie. U.S. railroad tracks have a center-to-center distance of 5 ft (1.52 m) [the distance between the inside edges, the "gauge," is 4 ft $8\frac{1}{2}$ in. (1.43 m)]. If the ties are hardwood, 9 in. (229 mm) wide by 7 in. (178 mm) deep, with a modulus, E, of 1.5 $\times$ 10^6 psi (1.04 $\times$ 10^4 MN/m²) and the value of k_0 for the supporting ballast is 1000 pci (272 MN/m³), how long should the ties be to satisfy each of the above criteria? The ties are usually made 8 ft 6 in. (2.59 m) long. What do you conclude about (a) the moment distribution, (b) the pressure distribution in the real ties?

13. In cases where railroad track maintenance is poor, ties have been found to break along their centerlines (midway between the rails). Explain what happens to cause this.

14. In the railroad tie Problem 12, consider that the ties are spaced 0.5 m (1.64 ft) on centers, and support rails whose EI is 8 MN-m² (1.16 k-in.²). A typical wheel loading on the rail is 150 kN (34 kips). Calculate, for the conditions of Problem 12, the spring constant for a tie at the point of rail support, the deflection as the wheel passes over a tie location, and the maximum stress in the rail when the wheel is midway between two ties. If the distance between adjacent wheels is equal to four tie spacings, are the rail moments affected by interaction? Take the distance from the major axis to the extreme fiber as 80 mm (3.15 in.).

15. The floor plate of a 30 m (98.4 ft) diameter oil tank is made of steel, 15 mm (0.59 in.) thick and is subjected to a uniform oil pressure of 150 kN/m² (3130 psf). At the edge, where the plate is welded to the cylindrical side, take the boundary conditions to be clamped (zero slope) but unrestrained in displacement. Calculate the edge moment. What is the factor of safety against the initiation of yielding in the extreme fibers? Here k_0 is 50 MN/m³ (184 pci).

REFERENCES

[1] ABRAMOWITZ, N., AND I. A. STEGUN, *Handbook of Mathematical Functions.* Washington, D.C.: National Bureau of Standards, 1964.

[2] BAKER, A. L. L. *Raft Foundations*, 2nd ed. London: Concrete Publications, Ltd., 1948.

[3] BARDEN, L. "Stresses and Displacements in a Cross-Anisotropic Soil," *Geotechnique, 13*, 198–210 (Sep. 1963).

[4] ———. "Influence Coefficients for Beams Resting on Soil," *Civil Engineering*, (London), **58**, 601–605 (1963).

[5] BEAUFAIT, F. W. "Numerical Analysis of Beams on Elastic Foundations," *Proc. ASCE, 103*, EM1, 205–209 (Feb. 1977).

[6] BIOT, M. A. "Bending of an Infinite Beam on an Elastic Foundation," *Trans. ASME, Jour. Appl. Mech., 59*, A1–A7 (1937).

[7] BROWN, P. T. "Strip Footings with Concentrated Loads on Deep Elastic Foundations," University of Sydney, School of Civil Engineering Research Report No. R225 (Sept. 1973).

[8] DOOLEY, J. C. Correspondence on Barden's paper (3), *Geotechnique, 14*, 278–279 (1964).

[9] FLETCHER, D. Q., AND L. R. HERRMANN. "Elastic Foundation Representation of Continuum," *Proc. ASCE, 97*, EM1, 95–107 (Feb. 1971).

[10] HAYASHI, K. *Theory of Beams on Elastic Foundation.* Berlin: Springer-Verlag, 1921 (in German).

[11] HETENYI, M. *Beams on Elastic Foundation.* Ann Arbor: University of Michigan Press, 1946.

[12] HOGG, A. H. A. "Equilibrium of Thin Plate Symmetrically Loaded Resting on an Elastic Foundation of Infinite Depth," *Philosophical Magazine, 25*, 576–582 (1938).

[13] ———. "Equilibrium of a Thin Slab on an Elastic Foundation of Finite Depth," *Philosophical Magazine, 35*, 265–276 (1944).

[14] HOLL, D. L. "Thin Plates on Elastic Foundations," Proc. 5th International Congress on Applied Mechanics, Cambridge, Mass., 71–74 (1938).

[15] KOKUSHO, T. "Numerical Analysis of Plates Resting on a Soil Subgrade," Central Res. Inst. of Elec. Power Ind., Japan, Report E376002, (Dec. 1976).

[16] LAM, I. P. "Edge-Function Method Applied to Thin Plates Resting on Winkler Foundation," C. E. Thesis, Calif. Inst. Tech., Dec. 1975.

[17] NADAI, A. "The Bending Stress in Plates with a Single Point Load," *Schweizerische Bauzeitung 76*, 257 (1920) (in German).

[18] PANC, V. *Theories of Elastic Plates.* Leyden: Noordhoff International Publishing, 1975.

[19] POULOS, H. G., AND E. H. DAVIS. *Elastic Solutions for Soil and Rock Mechanics.* New York: John Wiley, 1974.

[20] QUINLAN, P. M. "The Edge-Function Method in Elastostatics," in *Studies in Numerical Analysis.* New York: Academic Press, 1974.

[21] SCHLEICHER, F. *Circular Plates on Elastic Foundation.* Berlin: Springer-Verlag, 1926 (in German).

[22] SCHWEDLER, J. W. "Correspondence on Iron Permanent Way," *Minutes of Proc. Institution of Civil Engineers, 67*, 95–118 (1882).

[23] SHERIF, G. *Elastically Fixed Structures.* Berlin: Ernst & Sohn Verlag, 1974.

[24] VESIC, A. S. "Bending of Beams Resting on Isotropic Elastic Solid," *Proc. ASCE, 87*, EM2, 35–51 (Apr. 1961).

[25] ———. "Slabs on Elastic Subgrade and Winkler's Hypothesis," 8th International Conference on Soil Mechanics and Foundation Engineering, Moscow (1973).

[26] WESTERGAARD, H. M. "Stresses on Concrete Pavements Computed by Theoretical Analysis," *Public Roads, 7*, 25 (1926).

[27] ———. "New Formulas for Stresses in Concrete Pavements of Airifields," *Trans. ASCE, 113*, 425–444 (1948).

[28] WINKLER, E. *Theory of Elasticity and Strength.* Prague: H. Dominicus, 1867 (in German).

[29] WOINOWSKY-KRIEGER, S. "The State of Stress in Thick Elastic Plates," *Ingenieur-Archiv, 4,* 305–331, (1931) (in German).

[30] WÖLFER, K.-H. *Elastically Supported Beams.* Berlin: Bauverlag GMBH, 1971.

[31] YU, Y. Y. "On the Generalized Ber, Bei, Ker, and Kei Functions with Applications to Plate Problems," *Quart. J. Mechanics and Appl. Mathematics, 10,* Part 2, 254–256 (May 1957).

[32] ZIMMERMANN, H. *Calculation of the Upper Surface Construction of Railway Tracks.* Berlin: Ernst & Korn Verlag, 1888 (in German).

Highway and Airfield Pavements

6

6.1 INTRODUCTION

For thousands of years loads have been borne in horse-drawn vehicles with wooden wheels. The difficulties of pulling carts and carriages along level ground or up hills consisting of bare soil or turf resulted eventually in the development of roads to avoid the rutting associated with an unmade surface. It was early recognized that large-diameter wheels were advantageous in these conditions. Where they existed, the seventeenth- and early eighteenth-century roads consisted of a gravel coating or layer, commonly about 15 in. thick, laid on top of the natural soil if the soil was deemed or found sufficiently strong or on a sand bed placed on the natural surface. On such a surface, a single horse could pull a wagon and load of a ton (10 kN), with a tractive force of from 100 to 150 lb (0.5 kN) on a level surface, at a mile or two per hour (2 to 3 km/hr). Relatively rigid, smooth stone surfaces were employed occasionally as narrow strips or tramways for the wheels of vehicles. However, ruts were rapidly worn in them by the iron-shod wheels. Concrete was used in road construction both in Roman times and in the nineteenth century, but it was used as a support for a broken stone or gravel layer instead of as the road surface itself.

Good drainage of the gravel roads was soon found to be desirable to prevent saturation and consequent softening of the underlying foundation soil. However, in pursuit of drainage, roads were frequently cambered to excess. As a consequence, drivers of carts and carriages would keep to the center or crown of the road because of the discomfort or even danger of toppling involved in passage along the edges. The ruts formed by one vehicle were thus deepened and enlarged by the subsequent movement of other wagons (8).

Frequent maintenance was necessary for the preservation of good roads, but the social structure did not permit it until the early nineteenth century. The Industrial Revolution, with the concomitant increase in traffic volume, led to the design improvements of McAdam and Telford, and attention to road problems reached a peak in Europe about the 1830's. By this time barges on an extensive network of canals were carrying the heavier loads, and the growth of railroads for mass human transportation inhibited further highway development until the first automobiles appeared at the beginning of the twentieth century. Once again problems of rutting and road maintenance appeared. This time the social consequence was that simple gravel-surfaced roads were no longer acceptable for intercity travel. In the period 1920 to 1930 the design principles of hard-surfaced roads were formulated, largely in the United States, on the basis of the profiles used by Telford and McAdam. The cross section of such a pavement is shown in Figure 6-1; current terminology is used in the figure. The natural soil is called the *subgrade;* on it is placed a *base course* in one or two layers of different materials (*subbase* and *base*). The hard traveling surface lies on top of the base course and may consist of several inches (0.1 m) of asphalt-impregnated gravel or a similar thickness of generally unreinforced or lightly reinforced concrete. The asphaltic or concrete surface is called the *wearing surface.* For reasons related to the development of the two industries, asphaltic pavements are frequently called *flexible,* and concrete pavements are termed *rigid.* The

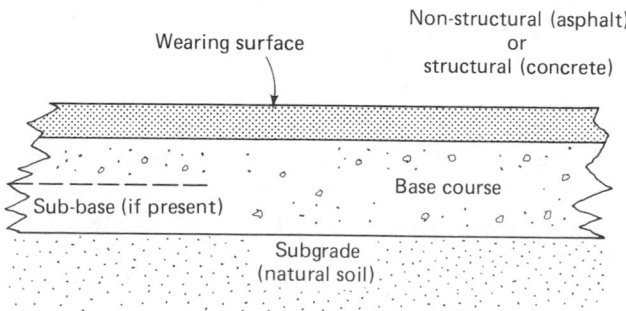

Figure 6-1. Cross section of typical highway or airfield pavement.

reader will recognize the inappropriateness of these terms. Both asphaltic and concrete pavements have a degree of flexibility related to the pavement material itself and the nature of the supporting soil, as discussed in Chapter 5. We will not treat the problem of asphaltic pavement design here, for it makes little use of the principles discussed.

While highway pavement design methods were being formulated, airplanes had become practical, and by 1930 passenger aircraft were becoming large enough to force consideration of the design of airfield pavements to take their wheel loads.

Airplanes built in the early years of the twentieth century were very light in weight, had low-pressure tires, and landed on grassy fields. The shearing strength of the top soil layers, bound together with the roots of the grass, was great enough to take the wheel loads without distress except in very wet weather. Operations were informal and the fields were large enough that a particular section of the landing area was not subjected to repetitive loading. Pavements were constructed only in airplane parking areas, because of convenience in loading, unloading, and handling operations and because loads were frequently applied to the same stretch of ground. As larger airplanes were built in the 1920's and early 1930's, the tire pressures remained essentially at the same relatively low values, and the wheels therefore had to be larger to transmit greater loads to the ground without breaking into the surface. Many airfield runways were still grass. However, with increasing rate of engine development and emphasis on speed, streamlining of aircraft became important, and retractable undercarriages were introduced. For the wheels to fit into the wings, or occasionally in the fuselage, the wheels had to be kept relatively small. However, the weight of aircraft continued to go up and thus higher tire pressures resulted.

Aircraft weights exceeded those of highway vehicles. The outbreak of World War II accelerated the development of large, heavy airplanes, and it became necessary to extend the highway pavement design procedures that had been developed. Accordingly, in the early 1940's extensive practical research was carried out, largely by the U.S. Army Corps of Engineers, to systematize airfield pavement design techniques.

Tire pressures have increased from the 15 or 20 psi (100 kN/m^2) of the early 1930's to 200 psi (1.4 MN/m^2) common in the 1980's; individual wheel loadings have gone from a few thousand pounds (30 kN) to 50,000 lb (250 kN) or higher. Thus, today very heavy wheel loads are being transmitted at high tire pressures through small imprint areas onto thick slabs. Because of the greatly increased aircraft loads with the introduction of "jumbo" jets that have to use existing airfields, aircraft designers have been forced to develop undercarriage assemblages instead of single wheels to transmit the loads to the pavement. Gear containing groups of two, four, or eight wheels are in operation, and undercarriages incorporating tracks and air cushions have been devised to permit aircraft to use unimproved landing fields. The necessity of operating military aircraft from temporary fields has also led to the development of reusable landing mats for lighter pursuit or fighter airplanes.

The principal problem in the area of landing field design of interest here is the interaction of stresses and displacements due to the presence of circular or oval loaded areas on the pavement surface. Consequently, we will examine the solutions that have been obtained for the problem of slab deformation under loads applied at first simply at various locations on a slab on a soil foundation. Occasionally it is desirable to calculate the limiting load that can be placed on an existing slab. The treatment of this aspect of slab loading will be left to a later chapter.

6.2 LOAD ON INFINITE SLAB; WESTERGAARD'S STUDIES

The point in analyzing the bending moments and stresses in a slab on a Winkler foundation is, in civil engineering, to give information of value to the design of, for the most part, loaded concrete slabs resting on the ground, either in its natural state or prepared by compaction to take the slab. Concrete is most sensitive to tensile stresses and thus a primary concern in design is the magnitude of these stresses. Under surface loads, the maximum tensile stress develops in the base of the slab under the point of load application and acts in the horizontal direction. This is the case for which equations (5.112) and (5.113) are of interest. The simplest practical application is a highway or airfield pavement loaded by wheels sufficiently far apart that they interact negligibly. Although the tire imprint is oval or elliptical, it may be reasonably approximated by a circle. The rapid development of highways in the 1910's and 1920's and then airfield pavements in the late 1930's and 1940's encouraged a search for suitable design methods. Attention was directed early to elastic solutions involving slabs on Winkler foundations. The application of this theory to the highway problem was examined by Westergaard.

In the first of his frequently-employed studies of the problem of stresses developed in concrete pavements by tires, Westergaard (16) used Nádai's problem and solution to arrive at values for the maximum tensile stress in the slab under a relatively concentrated load, although the relation of Nádai's edge boundary condition to the true condition around an isolated circular section of a simple plate is not clear.

What Westergaard apparently did, although his description of his procedure lacks detail, was to calcuate the maximum tensile stress on the underside of the slab below the load (point A in Figure 5-10) from Nádai's thick-plate solution for a given point load, radius c of loaded area, and slab thickness h. Then he determined, from the *elementary* theory of slabs on a Winkler foundation (Section 5.3) for the same slab thickness and total load, the fictitious radius b of a uniformly loaded area that would give the same maximum tensile stress. He appears to indicate that the tensile stress from his analysis of the Nádai problem was not very sensitive to the value of n in the vicinity of $n = 5$. Westergaard then plotted the actual dimensionless radius, c/h, versus the radius b/h required to give the same stress, and he fitted the observed relationship with a hyperbola, as shown in Figure 6-2, with the following equation:

$$b = 1.6c^2 + h^2 - 0.675h \qquad (6.1)$$

Finally, Westergaard recommended the determination of the actual maximum tensile stress by the use of equation (5.112) derived from elementary plate theory, but in which the actual radius c is replaced by the modified radius b according to the hyperbolic relation of equation (6.1). He said that

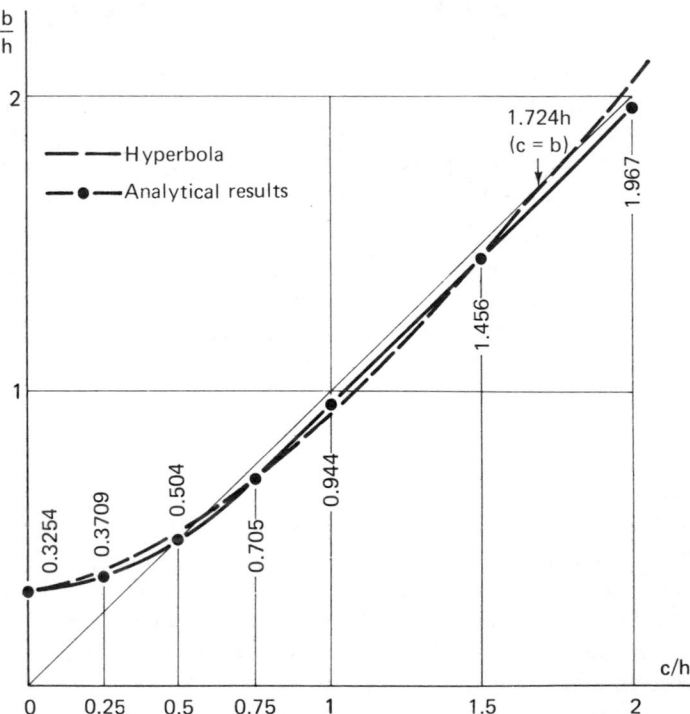

Figure 6-2. Relation among true radius c, equivalent radius b, and slab thickness h (16).

when c is less than $1.724h$, the corrected value b is to be employed, but for greater values of c, c itself may be used.

Since these relationships have been widely quoted and used in design calculations, some discussion of a number of interesting points is in order here.

In the "exact" analysis of the isolated portion of the plate, as shown in Figure 5-10 it is apparent that the value of the subgrade reaction coefficient will not appear, since it does not enter into the problem, and this is apparent in equation (5.113). However, in the parallel comparison by elementary theory to determine the value of radius b which will give the same maximum tensile stress, equation (5.112) is used and this, of course, includes the subgrade property k_0, as well as the plate properties. For the same plate, different values of k_0 will give different relations between b and c in Figure 6-2. The dependence on k_0 is not strong because of the way in which it enters into equation (5.112), but Westergaard does not mention what value he used in preparing Figure 6-2.

The above relationship between b and c was described empirically by Westergaard in his use of a hyperbola, equation (6.1), which matches the analytical points reasonably well in the range shown in Figure 6-2, but which diverges at *higher* values. The value of c equal to $1.724h$ chosen by Westergaard as the limit for the use of equation (5.112) is simply the value where $c = b$ in the hyperbolic equation. If one were to continue to employ the hyperbola for *higher* values of c than $1.724h$, the calculations would be in error because the hyperbola does not fit the calculated relationship of c and b above this value. The assignment of a number so precise as 1.724 for this point in the analysis is unnecessary, since it has no relation to the conditions under which the approximation of the elementary theory breaks down. In fact, it can be seen from Figure 6-2 that b and c are nearly equal all the way down to values of c less than $0.5h$. This means that in equation (5.112) there is no need to use b instead of c until c is less than about $0.5h$; equation (5.112) gives perfectly satisfactory values of stress. There does not seem to be a widespread understanding of these details of Westergaard's analysis.

The points may be clarified by referring to Woinowsky-Krieger's slightly different analysis, which is illustrated by Figure 5-10 and results in equation (5.113). It is of interest to compare values of the stress obtained from this equation with those calculated by Westergaard for different plate thicknesses. Westergaard presented a table of values of stress calculated for $P = 10,000$ lb (45 kN), $E = 3 \times 10^6$ psi (2.1×10^4 MN/m²), $\nu = 0.15$ and various slab thicknesses. Since he used equation (5.112) modified by substitution of the hyperbolic equation (6.1), his stresses depend on the Young's modulus of the slab and on the subgrade reaction coefficient. The values of stress are compared in Table 6.1. The value of n in equation (5.113) was taken to be 5 as in Westergaard's use of Nádai's results, although the two analyses, due to Nádai and Woinowsky-Krieger, are not strictly comparable because of

TABLE 6.1

Slab thickness in. (m)	4 (0.10)	6 (0.15)	8 (0.20)	10 (0.25)
Maximum Tensile Stress, psi (MN/m²)				
Woinowsky-Krieger Equation (5.113)	1235	549	309	198
	(8.52)	(3.79)	(2.13)	(1.37)
Westergaard Equations (5.112 and (6.1)				
k_0, pci (MN/m³) 50 (13.6)	1231	523	288	181
	(8.49)	(3.61)	(1.99)	(1.25)
200	1112	470	258	162
	(54.3)	(7.67)	(3.24)	(1.78)	(1.12)

the different boundary conditions involved.[1] In Westergaard's table and in equation (5.113) the loaded radius c was taken to be equal to zero.

It can be seen that there is not a great deal of difference between the values obtained by the two methods. As pointed out above, the variation of stress with k_0 in Westergaard's table is an artifact of the method of analysis in this particular case, although if the problem of the thick slab on a Winkler foundation were solved correctly, it would be expected that k_0 would enter into the result.

The question remains as to the conditions when the difference between the use of b according to equation (6.1) and c itself becomes important. According to the instructions accompanying the equations, c is used in place of b when c is greater than $1.724h$. However, the restriction in the development of the thin-plate equation (5.112) was that c should be small with respect to $1/\beta$. Can c meet this requirement in practical cases and still be greater than $1.724h$? Let us assume a slab to be made of concrete of Young's modulus 3×10^6 psi (2.1×10^4 MN/m²) and Poisson's ratio 0.15 (the latter value is relatively unimportant) and calculate the value of $1/\beta$ for various slab thickness and support conditions. The results are shown in Table 6.2. It appears that c can be greater than $1.724h$ and still be smaller than the characteristic length, although in the case of the stiffer subgrade the stresses would be affected to an extent greater than the 10 percent mentioned previously. The use of equation (5.112) for $c \geq 1.724h$ should therefore be handled with caution.

An automobile or aircraft tire imprint generally has an equivalent radius,

---

[1]It is not clear why Westergaard chose $n = 5$, since the value selected *does* affect the calculated stresses. For example, the stress from equation (5.113) for a 6-in. (0.15 m) thick slab and the other conditions of Table 6.1 is 577 psi (3.98 MN/m²) for $n = 6$ and 514 psi (3.55 MN/m²) for $n = 4$.

TABLE 6.2

Typical Range of Values of $1/\beta$ in. (m) for Pavements

Slab thickness, in. (m)	6 (0.15)		12 (0.30)	
Dimension 1.724h, in. (m)	10.3 (0.26)		20.7 (0.52)	
Subgrade reaction coefficient, pci (MN/m³)	50 (13.6)	500 (136)	50 (13.6)	500 (136)
Characteristic length $1/\beta$, in. (m)	32.42	18.23	54.53	30.66
	(0.82)	(0.46)	(1.39)	(0.78)

c, of a few inches (0.1 m) on a pavement slab from 6 to 12 in. (0.3 m) in thickness. In these circumstances, the use of Westergaard's modified equation to take into account very concentrated loads does not seem justified.

Instead, it is simpler to use Woinowsky-Krieger's equation (5.113) in any case when c is less than $0.25h$; here the calculated stress is independent of the size of the loaded area. For a larger radius of loaded area, equation (5.112) is applicable.

In summary, reexamination of Westergaard's original analysis indicates that the equation, (5.112), derived from elementary thin-plate theory will give the maximum tensile stress at the base of the slab with sufficient accuracy for a circular or square loaded area with radius as small as one-quarter of the slab thickness. For smaller loaded areas, Woinowsky-Krieger's result, equation (5.113) with an n of 5 can be employed. There seems to be no need to use equation (5.112) with the modified radius given by equation (6.1), as suggested by Westergaard.

One last point remains. The maximum tensile stress in the underside of the slab is frequently given in the form (14)

$$\sigma = \frac{0.275P(1+v)}{h^2} \log_{10} \frac{Eh^3}{k_0 c^4} \tag{6.2}$$

and attributed to Westergaard, although his original paper did not present his results this way. This equation can be rearranged to give

$$\sigma = \frac{0.477P(1+v)}{h^2} \left[\log_e \frac{1}{\beta c} + \frac{1}{4} \log_e 12(1-v^2) \right] \tag{6.3}$$

When the value of $v = 0.15$ is substituted in the last term in the square brackets, it is found to equal 0.616, so that, with this Poisson's ratio, the equation becomes identical to the correct equation (5.112). Thus, the value of $v = 0.15$ *must* be used in the logarithmic term of equation (6.2) to give the correct results, but it has been left open in the coefficient preceding the logarithm. If a value of v other than 0.15 is used throughout equation (6.2), the equation will give an incorrect value of the tensile stress. For exact calculations, or comparisons, therefore, or for analyses of slabs whose

Poisson's ratio is markedly different from 0.15, equation (6.2) should not be used.

The value of Poisson's ratio is not of great importance in most concrete construction and, as a consequence, there are few detailed studies of its value for concrete. In general, however, for static loading, the ratio falls in the range of from 0.15 to 0.20. Under dynamic conditions or repetitive loadings, higher values are obtained, with 0.25 being frequently quoted.

In the case of wide columns on slabs or large low-pressure aircraft wheels resting on an airfield pavement, it is possible for the wheel imprint radius to approach the characteristic length. As we have seen, equation (5.112) is not valid if this occurs because it gives stresses that are too low. For this condition, Westergaard (17) obtained a modified equation for the maximum stress:

$$\sigma_r^1 = \sigma_r + \frac{0.16Pc^2}{h^3}\left(\frac{k}{Eh}\right)^{1/2}(1+v) \tag{6.4}$$

which can be written in the form

$$\sigma_r^1 = \sigma_r + \frac{0.046P(\beta c)^2}{h^2}\left(\frac{1+v}{1-v}\right)^{1/2} \tag{6.5}$$

where σ_r is the stress calculated from equation (5.112).

Westergaard published this result in 1939, and it reflects the conditions in the immediately preceding period of aircraft development.

6.3 EDGE AND CORNER LOADS ON SLAB

Since pavements are built in sections, usually from slab elements from 15 to 20 ft in dimension, for both highways and airfields, it is likely that an edge or a corner of a pavement surface will be subjected to load. A concrete pavement and subgrade must be designed to withstand the stresses resulting from this loading. These cases of edge and corner loading of a slab on a Winkler foundation have also been studied by Westergaard (16) who presented the following equations for the maximum tensile stress developed in the slab. For the edge case, the pressure is considered to be applied uniformly over a semicircle of radius c whose center lies on the edge of the slab (Figure 6-3). The problem was tackled by considering the slab to be impressed by an infinite number of such loads spaced at equal intervals along the edge of the slab (13). A displacement solution was assumed in the form of a trigonometric series, and the final result was established for a single load by letting the distance between the loads go to infinity.

Westergaard recommended that the same conditions apply here to the calculation of the value of a fictitious b, related to c, as in the case of the infinite slab [equations (5.112) and (6.1) and related discussion]. Without

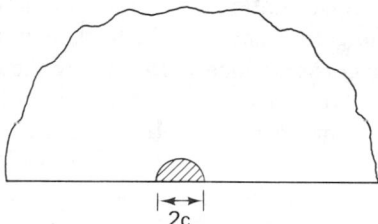

Figure 6-3. Semicircular loaded area at edge of semi-infinite slab.

further analysis, we will assume here, as before, that this modification is not required. Without adjustment, therefore, the equation for the maximum tensile stress σ_e for an edge-loaded slab was given by Westergaard in the form

$$\sigma_e = \frac{0.529P(1 + 0.548v)}{h^2}\left[4 \log_{10}\left(\frac{1}{\beta c}\right) + 0.359\right] \qquad (6.6)$$

or, if we are to be consistent in the use of $v = 0.15$,

$$\sigma_e = \frac{0.572P}{h^2}\left[4 \log_{10}\left(\frac{1}{\beta c}\right) + 0.359\right] \qquad (6.7)$$

In equations (6.6) and (6.7) the maximum tensile stress occurs in the base of the slab under the loaded area.

Kokusho (7) has used a discrete element scheme to study the distribution of pressures, displacements, and moments in a semi-infinite plate resting on both elastic solid and Winkler foundations and loaded by an edge point load. Unfortunately, of course, under the point load, the moments are infinite, and his results for elastic solid and Winkler cases can only be generally compared at some distance from the point of load application. No stress or moment can be derived at the load for comparison with the Westergaard value, equation (6.7). However, in general, he concludes that the Winkler results [due to Westergaard (16)] for deflections and bending moments can be modified to give a good comparison with those obtained by his numerical analysis for the same plate resting on a finite elastic layer of depth H, as long as H does not exceed 2.5 times the characteristic length $(1/\bar{\beta})$. The modification consists in multiplying the actual load by 0.6, and using the subgrade reaction coefficient k_0 of equation (5.119). When the soil (elastic) layer is greater than $2.5/\bar{\beta}$ in thickness, the same reduction in edge load is recommended in analysis, but the k_0 should be made equal to 0.54 of the value in equation (5.119).

The task of obtaining the deflections, moments, and stresses caused by a load at the corner of a sectorial segment on a Winkler foundation (usually a 90° corner) is more difficult and does not appear to have been solved analytically as yet. For a calibration problem, the plate is first taken to be unsupported by any subgrade material; instead, it is taken to be fixed at some distance from the corner, which is loaded by the concentrated load P, as in Figure 6-4(a). The moment at any section of the plate, distance x along the corner bisector from the corner, can be obtained by assuming the plate to be a simple cantilever and is equal to Px. For the right-angled corner, the width of the plate at this section is seen to be $2x$, and the moment of inertia is therefore $xh^3/6$, where h is the thickness of the plate. If follows in this case that the maximum tensile or compressive stress in the plate σ_m is given by the equation

$$\sigma_m = \frac{3P}{h^2} \tag{6.8}$$

which is independent of x, so that the maximum tensile stress occurs everywhere at the top surface of the plate.

When a subgrade supports the plate, the problem becomes a two-dimensional one in which displacements, say, vary both in the x-direction of Figure 6-4 and at right angles to it. The moments at right angles to the edges of the plate and the shears along the edges are both zero, and these constitute the boundary conditions for the problem. In the absence of a solution to this problem, Westergaard assumed that displacements would only be a function of the x-coordinate and would be the same at all points along every cross section at right angles to the x-axis. The deflection contours would then appear as the straight dashed lines in Figure 6-4(a).

Experiments with loaded wedges on simulated Winkler foundations by Gold et al. (4) have since shown that for wedges with angles up to 60°, this assumption is essentially correct; only minor variations were observed in the case of 90° wedges, the practical circumstance of interest here. The assumption converts the problem to a one-dimensional one of a beam of linearly varying moment of inertia resting on ground with a subgrade reac-

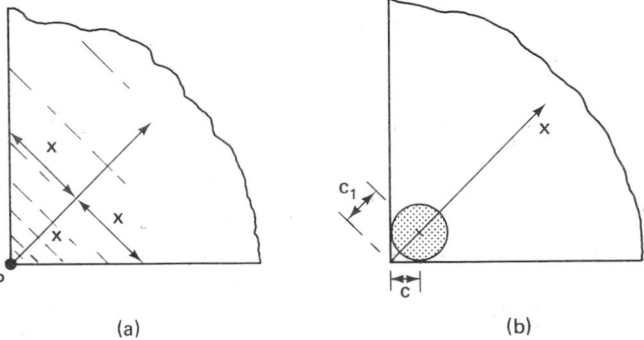

(a)   (b)

Figure 6-4.   Load at corner of quarter-infinite slab: (a) point load; (b) circular loaded area.

tion coefficient k_0: The subgrade coefficient, k, for the *beam* is therefore $2k_0x$, also varying linearly with distance.

We can obtain the equation for this problem from the general relation

$$-\frac{d^2M}{dx^2} + kw = 0 \tag{6.9}$$

in which

$$M = -EI\frac{d^2w}{dx^2} \tag{6.10}$$

$$I = I_0x = \frac{h^3x}{6} \tag{6.11}$$

and

$$k = 2xk_0 \tag{6.12}$$

Substituting equations (6.10) through (6.12) in equation (6.9), evaluating the derivatives, and dividing through by EI_0, we get the expression

$$\frac{d^4w}{dx^4} + \frac{2}{x}\frac{d^3w}{dx^3} + \frac{2k_0}{EI_0}w = 0 \tag{6.13}$$

With the substitution of an equation for the characteristic length of the beam $1/\lambda$ [not the same λ of equation (5.12)],

$$\lambda^4 = \frac{2k_0}{EI_0} \tag{6.14}$$

and replacement of w and x by their dimensionless forms $W = \lambda w$, $X = \lambda x$, the dimensionless equation

$$\frac{d^4W}{dX^4} + \frac{2}{X}\frac{d^3W}{dX^3} + W = 0 \tag{6.15}$$

is obtained. This equation, which has a singularity at the origin, has been solved by Dieudonné (2) in integral form, but the integrals have not apparently been evaluated (10 and 11).

Westergaard did not attempt the solution of equation (6.15); instead, he assumed an approximate expression for the displacement in the form of a truncated exponential series for the case in which the load is uniformly distributed over the circular area of Figure 6-4(b). He obtained the coefficients of the two terms in the series by minimizing the total potential energy of the system of slab, subgrade, and load, in the manner described in Section 2.4.1. The deflected shape was found to be

$$w = \frac{P\beta^2}{k_0}(1.1e^{-\beta x} - 0.88\beta c_1 e^{-2\beta x}) \tag{6.16}$$

where c_1 is equal to $c\sqrt{2}$ and is shown in Figure 6-4(b). Since this is an approximate solution, it does not match the boundary conditions at $x = 0$; in particular, the curvature is not zero there. In equation (6.16) Westergaard used the characteristic plate length $1/\beta$, as defined by equation (5.99) instead of the beam length $1/\lambda$ of equation (6.15). From equation (6.16) the moment per unit width can be calculated; Westergaard found it to have a maximum value

$$M = -\frac{P}{2}[1 - (\beta c_1)^{0.6}] \qquad (6.17)$$

at a distance x, approximately, where

$$x = 2\sqrt{\frac{c_1}{\beta}} \qquad (6.18)$$

When M is divided by the modulus of the section per unit width, the maximum tensile stress, σ_c, is given:

$$\sigma_c = \frac{3P}{h^2}[1 - (\beta c_1)^{0.6}] \qquad (6.19)$$

In the case of a point load at the corner, $c = c_1 = 0$, and the stress reduces to that given by the unsupported cantilever, equation (6.8). It is seen that in Westergaard's solution with the uniformly distributed load over a circular area, the stress is reduced from the cantilever value, as would be expected, by the $(\beta c_1)^{0.6}$ term.

For the problem of the sectorial beam on an elastic foundation as described by equation (6.15), a solution has been obtained by the numerical method of Franklin and Scott (3) when the beam is loaded by a point load P at the apex of the wedge.

From this solution, the tensile (compressive) stress in the beam was found to be a maximum at an approximate distance

$$x = \frac{0.25}{\lambda} \qquad (6.20)$$

in contrast to the unsupported cantilever beam solution. The maximum tensile stress, σ_c, was approximately

$$\sigma_c = 0.939\left(\frac{3P}{h^2}\right) \qquad (6.21)$$

The stress is independent of the elastic properties of the beam and subgrade, although the distance [equation (6.20)] at which it occurs varies. Apparently, the presence of the elastic foundation does not reduce the peak tensile stress more than about 6 percent below that of the unsupported slab.

The problems of a right-angled slab resting on both a linearly elastic solid and a Winkler foundation and subject to a point corner load have also been solved numerically by Kokusho (7). His results are presented in the form of dimensionless deflection, $W = Dw\bar{\beta}^2/P$, and moment M/P versus dimensionless distance along the bisector $X/\sqrt{2}$. The $\sqrt{2}$ is inserted in the original paper to compare moments and deflections with those occurring along an edge, since the solution is a two-dimensional one. In Kokusho's work $\bar{\beta}$ is defined as in equation (5.115). In order to compare elastic and Winkler subgrade solutions, he defines $k_0 = k^0$ as in equation (5.117). With the definition of λ as in equation (6.15), we have

$$\lambda^4 = \frac{2k^0}{EI_0} = \frac{2D\bar{\beta}^4}{EI_0} = \frac{1}{(1 - v^2)}\bar{\beta}^4 \tag{6.22}$$

so that, for v equaling about 0.15 for concrete, $\lambda = \bar{\beta}$ very nearly. This enables us to compare Kokusho's solutions for deflection and moment along the bisector axis with those obtained by the Franklin and Scott method. The comparison is shown in Figure 6-5, which also presents the results of Kokusho's analysis for the corner plate on elastic solid foundations of various thicknesses, $\bar{\beta}H$. It is seen that the two results for the plate on a Winkler foundation agree quite well but that they differ substantially from the elastic solid foundation values. Kokusho suggests retaining the subgrade reaction coefficient $k_0 = k^0$ for both deflection and moment in the Winkler analysis, but reducing the actual load P to $0.4P$ in the deflection case, and to $0.65P$ in the moment case, in order to get deflections and moments more closely agreeing with the elastic solid analysis.

The Westergaard deflections from equation (6.16) with $c_1 = 0$ and derived moment per unit width for the apical load case are also shown in Figure 6-5. In the case of deflections, agreement is quite good with the "exact" Winkler solutions, but the moment has a peak value at the corner and cannot correspond in detail to the other solutions.

Stresses occurring under elliptically shaped uniformly loaded areas were also calculated by Westergaard (18). Using Westergaard's results, Pickett and Ray (12) have presented influence diagrams for assessing the moments and deflections in a slab loaded internally and at an edge. For the loading far from an edge, separate charts for slabs supported by the Winkler model and an elastic half-space are given. In the case of edge loading, only the Winkler subgrade model is represented.

Since both slab and subgrade models in these representations are linearly elastic, the principle of superposition holds, so that the stresses caused by combinations of load on the slab or pavement surface may be obtained by adding together the stresses caused by the loads considered separately. This will be required when columns bearing on a beam or slab are more closely

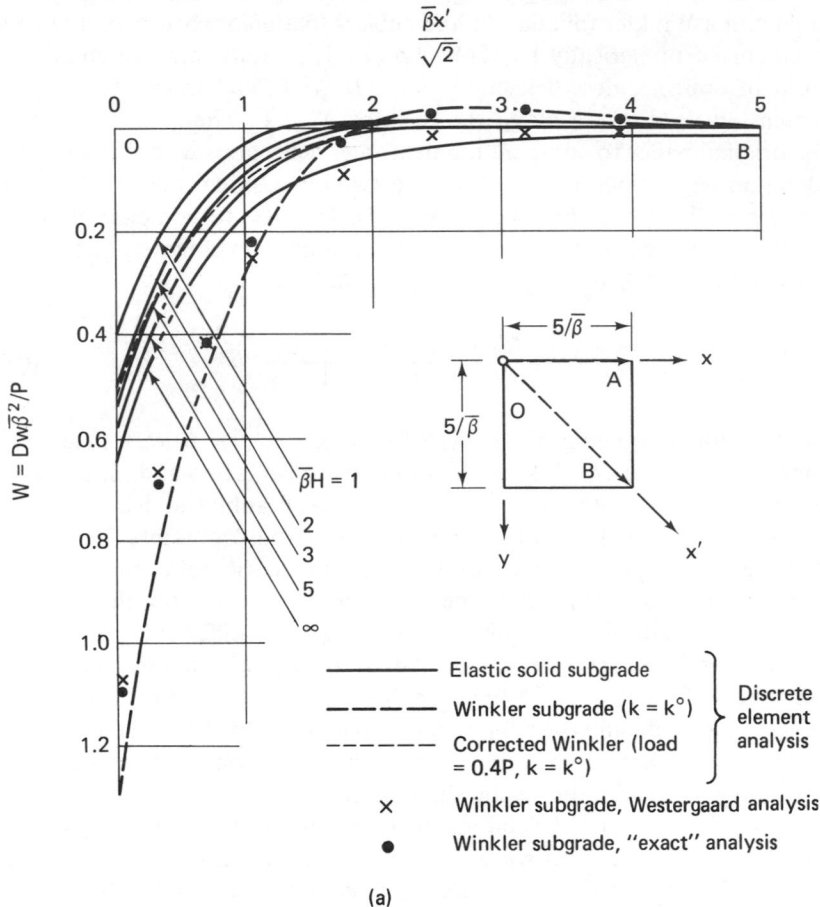

Figure 6-5. Comparison of corner loading results from the analyses of Kokusho (7), Franklin and Scott (3), and Westergaard (16): (a) dimensionless deflections.

spaced than the interaction distance of the system. In the case of airfield or highway pavements, load combinations are developed by closely spaced wheels that are designed to reduce the stress in the pavement. These may be truck wheels or the wheels on airplane landing gear. Of the two, the higher loads are generated by aircraft.

Other than single tires, aircraft wheel arrangements are varied: two wheels side by side are said to be *dual*; one behind the other *tandem*; a set of four in a square array on a single leg *twin-tandem* or *dual-tandem*. Dual- and twin-tandem arrangements are common, although other more complicated arrays have been devised. The arrangement of landing gear and wheels

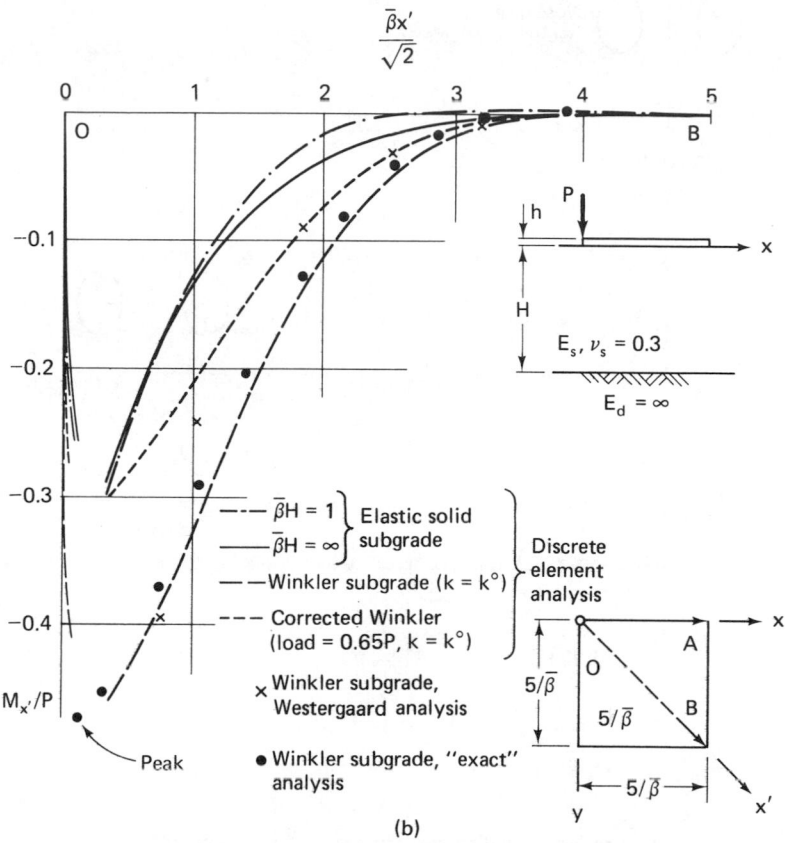

Figure 6-5. (b) Dimensionless moments per unit width.

on a single leg on the Boeing 747 is shown in Figure 6-6. In its heaviest configuration, the 747-SP, this aircraft places a load of 180 kips (800 kN) on a single leg, or 45,000 lb (200 kN) per tire. In some cases, the interaction distance may be such that the stresses from individual legs of this aircraft may require superposition. A way of dealing with the design of pavements to handle such aircraft load combinations is to calculate the load on a single fictitious wheel, the same size and tire pressure as the real wheels, which would give rise to the same stress in the pavement as the wheel group. This has been termed the *equivalent single-wheel load* (ESL). It is obtained from charts which have been calculated from the superposition of stresses caused by wheel groupings. The basic solution used for concrete pavements is Westergaard's for the load on the infinite slab. Charts have been prepared (1) for dual or tandem (the cases are the same) loads and for twin-tandem arrangements. A similar methodology has been applied to the design of

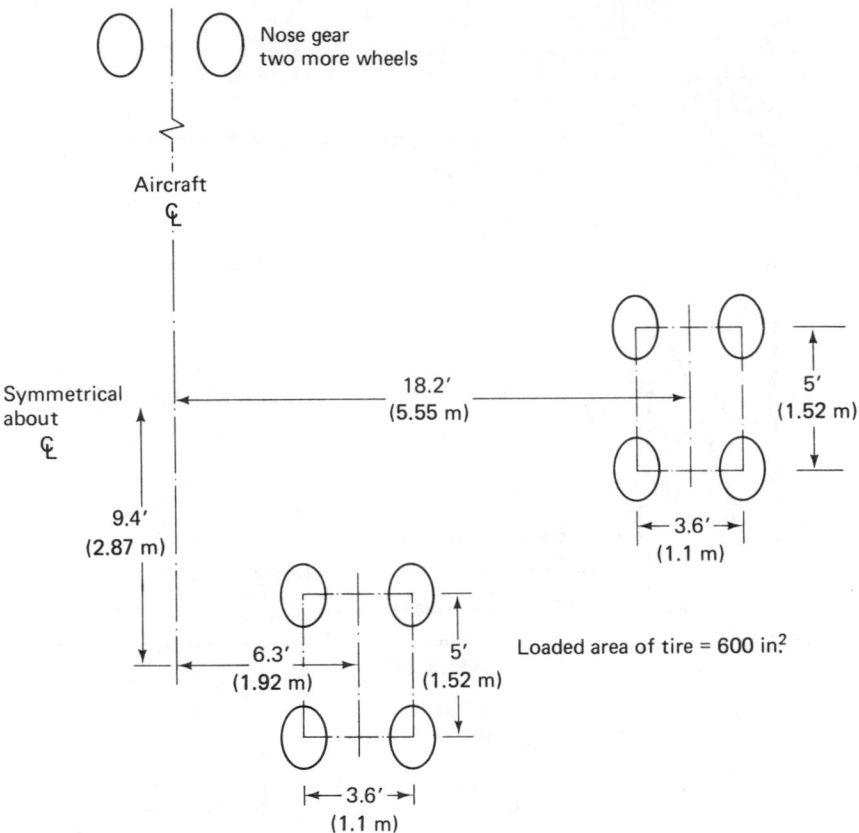

Figure 6-6. Landing gear arrangement on Boeing 747.

asphaltic pavements for multiple-wheel loads; in this case, the criterion of interest is generally the vertical stress transferred to the natural soil by the surface and base courses.

6.4 DISCUSSION; CURRENT CONSIDERATIONS

The three equations for slab stress developed by interior, edge, and corner loads [equations (5.112), (6.7) and (6.17) or (6.21), respectively] have been the subject of extensive discussions in the literature from points of view of both theory and experimental measurements [see, for example, (14)]. In practice, slabs exist in a complex environment of daily temperature and humidity changes. The subgrade properties depend on the water content of the soil in many cases, and this undergoes seasonal changes, particularly at slab edges and corners. The loads in the case of highways and aircraft parking areas in particular are repetitive, and the succession of stress changes in the pavement and soil in itself causes changes in the material properties. The overall temperature changes and temperature gradients that the slab experiences develop stresses, in some circumstances, of the same order as the wheel load stresses. Warping of the slab as a result of temperature gradients in it may cause loss of subgrade reaction under a load. It has been observed, for example, that the corner stresses may be as high as the simple cantilever stress $3P/h^2$, implying that no subgrade reaction is present. It was seen in connection with equation (6.21) that this stress is very nearly attained when the slab is loaded by a point load at the corner.

Experimental indications are that the edge and corner stresses experienced in practice are higher than the Westergaard equations indicate, but that the stresses generated by interior loads are lower than those given by the equations. Various empirical modifications have been suggested to account for these observations.

For many years the theoretical basis of concrete airfield and highway pavement design has been Westergaard's equations, as presented here, with adjustments to take into account some of the myriad variables that affect a pavement's performance, such as number of load repetitions, temperature, and other climatic conditions. The theoretical analysis is supplemented by experience of actual field performance of pavements under real loading conditions. As a consequence, the detailed methods that are used vary widely in different areas. Various equations have been presented for the determination of the quantity of reinforcing required to hold cracks tightly closed in either jointed or continuous reinforced concrete pavements. These equations include the effect of the variables: yield strength of the steel; tensile strength of the concrete; and the frictional resistance between the concrete and the base course, which tends to resist the contraction of the concrete. In practice, individual pavement slabs are connected with joints that are designed to transfer shear loads from slab to slab. These may be in the form of concrete keys, or they may be steel dowels accurately aligned at right angles to the slab edge.

The outward manifestation of developing pavement distress is an increasing roughness of the surface, which results in discomfort, or in extreme cases, danger to vehicular traffic. Formerly, pavements were considered either to be satisfactory or unsatisfactory. In the latter case, repairs were required. However, during the AASHO (American Association of State Highway Officials) Road Test at Ottawa, Illinois, in the early 1960's, it became apparent that a pavement whose design was adequate for initial loadings could deteriorate with an increasing number of passes of wheeled loads until the pavement eventually became unserviceable. The concept of a present pavement serviceability index (PSI) was therefore developed (6 and 15) to describe essentially the state of roughness of the pavement, as viewed by passengers in vehicles. In the United States a scale of from 0 to 5 is used for the PSI, with the value of 5 representing an extremely good surface. A rating is obtained by analyzing the reports of users. Although the question of surface roughness is a complex one, since discomfort can be experienced over the whole spectrum of roughness from bumps with a small wavelength to long wavelength waviness, the details are relatively unimportant to the rating system because the users assess the overall hazard caused by the surface to the passage of vehicles.

The results of the AASHO Road Test, surely the most costly soil mechanics experiment ever performed,[2] enable the declining serviceability of a pavement to be correlated with the other variables involved in its design. These correlations have led to the development of design equations including, among the other factors of pavement strength and subgrade reaction coefficient discussed earlier, the number of standard load repetitions required to reduce the PSI from some assumed initial value (usually 4.5) to a final value at which it is accepted that repairs will be required. Since traffic studies give some indication of the load repetition rate, the chosen number of road repetitions is related to an expected lifetime, usually 20 years, of the pavement.

PROBLEMS

1. Discuss some causes and mechanisms of "rigid" pavement failures (see Chapter 11 also).
2. A 50,000 lb (223 kN) load is applied to one leg of an aircraft undercarriage. During design, the landing gear is changed from the original conception of a single tire inflated to 75 psi (518 kN/m²), to smaller, dual tires, each inflated to 150 psi (1.04 MN/m²), with centers 4 ft (1.22 m) apart. Which system is preferable as far as tensile stresses in a concrete pavement 12 in. (0.305 m) thick are concerned? Illustrate your answer with calculations. Consider the cross-section both vertically beneath each dual tire, and halfway between them. Assume loaded areas are circular, and far from any edge of the slab.
3. If a Boeing 747 were to be parked on a continuous floating saltwater ice sheet, how thick would the ice have to be? Tensile strength of ice 100 psi (690 kN/m²); ice unit weight 55 lb per cubic ft (8.6 kN/m³).
4. You are made director of a research program to develop a lightweight, portable, reusable landing mat to take wheel loads of up to 20,000 lb (89 kN) at tire pressures of 200 psi (1.38 MN/m²) on relatively soft soils. List the steps you would take in setting up the appropriate tests and analyses.
5. Consider the quarter-plate on the Winkler foundation loaded at the corner. Analysis indicates the maximum tensile stress to occur at the corner, but a slab

---

[2]It may yet be matched by the investigation related to the Oosterschelde closure in the Netherlands.

on real soil, loaded at the corner until it fails, breaks some distance away from the corner. Give as many reasons as you can for this behavior in terms of the inadequacy of the model. Bear in mind that in two-dimensional plate problems, the moment is given by

$$M_x = D\left(\frac{\partial^2 w}{\partial x^2} + v\frac{\partial^2 w}{\partial y^2}\right)$$

6. What effect do the characteristics of user vehicles have on the design of a concrete pavement?

7. Solve the corner plate problem by the Ritz method assuming the displacement function to be of the (dimensionless) form

$$w = a_1 e^{-x} + a_2 e^{-2x}$$

Consider the load to be the resultant force acting through the center of a circle, radius c, located tangential to the sides at the corner, as shown in Figure P6-7. For plate thickness h calculate the maximum stress in the slab, and compare your result with Westergaard's, equation (6-19).

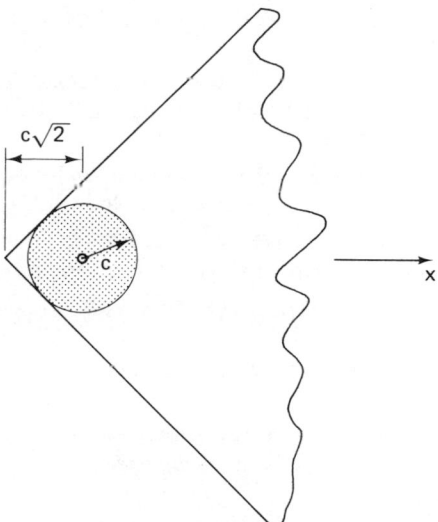

Figure P6-7.

8. Design an unreinforced concrete pavement on a medium-stiff clay for a fully-loaded Boeing 747. The maximum allowable tensile stress in the concrete is 250 psi (1.73 MN/m²). See Figure 6-6 for wheel configuration; the load on each wheel is 45 kips (200 kN).

9. How much thicker would the concrete of Problem 8 have to be if the wheel were placed (a) at a corner, (b) at an edge of the pavement?

REFERENCES

[1] COOPER, G. S. "The Influence of Multiple Wheel Undercarriages on the Design and Evaluation of Airfield Pavements," *Proc. Inst. Civil Engineers, 1,* Part 2, No. 2, 419–460 (June 1952).

[2] DIEUDONNE, J. "Study of the Equation $(d^4y/dx^4) + 2/x(d^3y/dx^3) + y = 0$, $x \geq 0$," unpublished manuscript, quoted in Ref. 10.

[3] FRANKLIN, J. N., AND R. F. SCOTT. "Beam Equation with Variable Foundation Coefficient," *Proc. ASCE, 105,* EM5, 811–827 (Oct. 1979).

[4] GOLD, L. W., L. D. BLACK, F. TROFIMENKOFF, AND D. MATZ. "Deflections of Plates on Elastic Foundation," *Trans. Engineering Inst. of Canada, 2,* No. 3, 123–131 (Sept. 1958).

[5] HEUKELOM, W., AND C. R. FOSTER. "Dynamic Testing of Pavements," *Trans. ASCE, 127,* Part 1, 425–457 (1962).

[6] Highway Research Board. "The AASHO Road Test," Report 61 (7 reports) (1961–1962).

[7] KOKUSHO, T. "Numerical Analysis of Plates Resting on a Soil Subgrade," Central Res. Inst. of Elec. Power Ind., Japan, Report E376002, (Dec. 1976).

[8] LAW, H. *Rudiments of the Art of Constructing and Repairing Common Roads.* First published 1855 by John Weale, London; reprinted 1970 by Kingsmead Reprints, Bath, England.

[9] LIVESLEY, R. K. "Some Notes on the Mathematical Theory of a Loaded Elastic Plate Resting on an Elastic Foundation," *Quart. J. Mechanics and Appl. Mathematics, 6,* Part 1, 32–44 (1953).

[10] NEVEL, D. E. "The Theory of a Narrow Infinite Wedge on an Elastic Foundation," *Trans. Engineering Inst. of Canada, 2,* No. 3, 132–140 (Sept. 1958).

[11] ——. "The Narrow Infinite Wedge on an Elastic Foundation," U.S. Army Snow, Ice, Permafrost Research Est., Tech. Rep. 56 (July 1958).

[12] PICKETT, G., AND G. K. RAY. "Influence Charts for Concrete Pavements," *Trans. ASCE, 116,* 49–73 (1951).

[13] TIMOSHENKO, S., AND S. WOINOWSKY-KRIEGER. *Theory of Plates and Shells,* 2nd ed. New York: McGraw-Hill, 1959.

[14] U. K. DEPARTMENT OF SCIENTIFIC AND INDUSTRIAL RESEARCH, ROAD RESEARCH LABORATORY. *Soil Mechanics for Road Engineers.* London: H. M. Stationery Office, 1957.

[15] WEISS, R. A. "Pavement Evaluation and Overlay Design: A Method That Combines Layered-Elastic Theory and Vibratory Nondestructive Testing," Transportation Research Record No. 700, Transportation Research Board, National Academy of Sciences, Washington, D.C. 20–34 (1979).

[16] WESTERGAARD, H. M. "Stresses in Concrete Pavements Computed by Theoretical Analysis," *Public Roads, 7,* No. 2, 23–35 (1926).

[17] ——. "Stresses in Concrete Runways of Airports," *Proceedings Highway Research Board, 19,* 197–205 (1939).

[18] ——. "New Formulas for Stresses in Concrete Pavements of Airfields," *Proc. ASCE, 73,* 687–701 (1947).

Settlements, Soil Tests, and Subgrade Coefficients 7

7.1 INTRODUCTION

The last few chapters have been devoted to deriving and discussing the mathematical relations that describe various models of structure and foundation. The ultimate aim, of course, is to use the equations in the process of designing structures, and it is not sufficient for this only to understand the relations and their applications. In fact, many engineers would claim that this analytical understanding is the least of all the difficulties a foundation designer faces; it is relatively easy to teach. The equations incorporate material properties and somehow these must be teased out of the mass of soil information available to the engineer. The soil properties are not easy to obtain in satisfactory numerical form, including all possible contingencies which may obtrude during the life of the proposed structure. Their evaluation involves experience, judgment, and intuition, all attributes highly valued in engineering; the older the engineer, the higher the valuation. The young engineer uses equations but has little knowledge of their limitations and less experience with the results of applying them in practice. The experienced engineer often virtually discards the formalism altogether and arrives at an estimation of the building settlement or the angle of the upstream slope of the earth dam by essentially an educated guess. The latter's mind works as an integrating engine, bringing together broad and deep experience in the past performance of similar structures on similar foundation soil; the equations are implicit in the functioning.

In reality, much of the experience claimed is spurious in a numerical sense. On how many structures are displacements, settlements, and stresses closely monitored, understood, and digested by designers? With the majority of structures, the post construction information is of a non-negative kind: Cracks were not observed; the slope did not fail; no distortions were apparent. In these cases, the margin of safety was unknown.

It is the point here that the mathematics of foundation analysis *and* the methods of determining and selecting the material properties merit equal attention. Both aspects contribute to satisfactory design. Wherever possible, numerical, calculated results should be compared against measured values in the finished structure. Far too few measurements on civil engineering structures are made; only the rare failure is studied exhaustively. The cost and level of effort required to instrument and study a large structure are such that this condition is likely to continue.

In this chapter we will attempt to summarize some of the experience pertinent to the problem of obtaining soil properties to be employed in conjunction with beam and slab analyses of previous chapters. First it is necessary to consider the situation with respect to the settlement of footings and foundations.

7.2 SETTLEMENT OF FOOTINGS

In soils, increasing overburden pressure causes stiffness and strength to become greater with depth. In a heavily overconsolidated clay, this effect can be slight or negligible, whereas in sands or normally consolidated clays, the two properties are virtually zero at ground surface, and the increase with depth is an important factor in assessing foundation behavior. When a footing is placed at or near the soil surface and loaded, the effective stresses are further increased by the stress field of the footing. In cohesionless soils the effective stress change is immediate; in clays the time process of consolidation is initiated, and the effective stresses develop more slowly. With highly overconsolidated clays, the effective stresses caused by the footing again may be relatively unimportant with respect to the behavior of the clay when the final effective stress is smaller than the past maximum consolidation stress.

Since the stiffness and strength are functions of the effective stress, the displacements generated in a particular soil by a loaded footing are not easy to assess. In particular, analytical half-space models of the soil taking into account local stress effects on the soil properties do not exist. The most frequently encountered solution is that for a uniformly loaded area on a homogeneous, isotropic, linearly elastic half-space, and this, from the discussion above, refers approximately only to the practical case in which the soil involved is heavily overconsolidated clay. Other analytical solutions have been developed for cases in which the linearly elastic half-space has an elastic modulus that increases linearly with depth from either zero or a finite value at the ground surface. This will be discussed in a later section of this chapter. Finite element computer solutions have been obtained for these and other cases and will also be examined later.

7.2.1 Overconsolidated clay

We will take up the simplest case first, that of the homogeneous half-space, corresponding to overconsolidated clay. The problems of interest are the settlement of foundations of various widths at the same average pressure, p and the pressure that can be allowed to act on a footing whose settlement is to be restricted to some specific amount. From the linearly elastic solution (26), we obtain the result for w, the deflection, when a uniform loading is applied to the surface of the half-space:

$$w = \frac{(1 - v^2)}{E} pBI \qquad (7.1)$$

where E and v are the Young's modulus and Poisson's ratio for the half-space (soil), respectively, B is the least dimension of the footing, and I is an influence factor that depends on the shape of the footing. If the loaded area is a circle, diameter D, I has the value 1, w is the central deflection, and $B = D$. If the loaded area is rectangular with length L, and w is the mean deflection, I varies with the ratio L/B from a value of 0.95 at $L/B = 1$ (square) through a value of about 2.25 at $L/B = 10$.[1]

When a load is symmetrically applied to the surface of a rigid plate resting on a half-space, and p is again the average pressure, equation (7.1) can still be used for the rigid plate deflection with the following values of I in the particular cases. If the plate is circular of diameter D, $B = D$ and $I = \pi/4$. If the plate is rectangular, as before, I varies with the ratio L/B from a value of 0.93 at $L/B = 1$ through a value of 2.24 at $L/B = 10$. It is seen that the differences between a uniformly loaded and a rigid plate area for the same pressure are not great and that in each case the deflection is linearly proportional to the width B of the footing.

If the properties of the soil are known, the displacement of a footing under a particular load can be estimated from equation (7.1) if the soil conditions match the required assumptions reasonably well. In the absence of property information, the deflection of a proposed building footing can be

---

[1]The deflection under a uniform load applied to an *infinite* strip is indeterminate.

related to the measured displacement of a standard area loaded in a field test. It is common in the United States to perform such a test on a 1-ft (0.3 m) diameter rigid circular plate or a 1-ft (0.3 m) wide square plate; in such a case, the relation of the settlement of the footing of interest, w, to the settlement, w_1, of the square footing at the same pressure is given by the relation

$$\frac{w}{w_1} = \frac{B}{B_1} \frac{I}{I_1} \qquad (7.2)$$

where I_1 is the influence factor (0.93), say, for the 1-ft square test plate, with $B_1 = 1$, and I and B are the influence factor and width of the footing, respectively. It should be borne in mind that such an equation can only be employed when the soil, such as heavily overconsolidated clay, corresponds to the assumptions involved, and, *most particularly*, that the same soil extends far enough (usually at least $2B$) below the *larger* footing to validate the assumptions.

When, as is usually the case, it is required that the settlement of the footing not exceed a specified value for structural reasons, then the w of equation (7.1) is fixed, and it follows that the relation between p and B is hyperbolic. When the level of pressure is specified in this way, it becomes an allowable pressure, p_a, or bearing capacity of the footing. However, another requirement must be remembered: The pressure on the footing must not reach a value at which extensive plastic flow or yielding of the soil develops below the footing. To safeguard against this, a factor of safety of at least 3 with respect to footing failure is required for most structures and soil conditions. It is of interest to see which criterion, excessive settlement or footing failure, controls the behavior of most footings in practice.

In a uniform soil possessing cohesion only, the bearing capacity of a near-surface footing in terms of stress p_u is found from plasticity theory to be approximately

$$p_u = 6c \qquad (7.3)$$

where c is the cohesion of the soil. The numerical factor preceding c in the equation depends somewhat on the shape of the footing; the value given is close enough for most footings for our purposes here. With a factor of safety of 3, the *allowable* bearing pressure based on failure becomes

$$p_a^f = 2c \qquad (7.4)$$

Now equation (7.1) can be rearranged in the form

$$p = \frac{E}{(1 - v^2)} \frac{w}{I} \cdot \frac{1}{B} \qquad (7.5)$$

We will select a value of the permissible settlement equal to 1 in. (25 mm), which results from studies of the effects of differential settlement on a wide variety of buildings (17). For many materials (steel, concrete), the Young's modulus E is approximately 1000 times the yield stress; in soils the ratio varies from 50 to 1000 depending on the plasticity index and overconsolidation ratio (10). For a heavily overconsolidated clay, we can use the value 1000, so that

$$E = 1000c \qquad (7.6)$$

in equation (7.5). For overconsolidated clays, a reasonable value of Poisson's ratio v is about 0.4, and, finally, for square or rectangular footings of common proportions, I is close to unity. Applying the above considerations to equation (7.5), we find that the allowable bearing stress with respect to settlement, p_a^s, is approximately given by the equation

$$p_a^s = \frac{100c}{B} \qquad (7.7)$$

In this form, it is clear that the smaller the breadth B of the footing, the higher the value of p_a^s. Above some breadth B, the allowable pressure based on settlement will fall below the pressure based on failure. This value of B is found by equating p_a^f and p_a^s from equations (7.4) and (7.7) to get $B = 50$ ft (15 m). For a smaller ratio of E/c in equation (7.6), the limiting B is smaller in proportion. Thus, for B less than, say, 20 to 50 ft (6 to 15 m), or in other words, for foundations smaller than rafts or mats, the controlling criterion for the allowable bearing pressure in the case of overconsolidated clays is the yield or failure condition. Settlements must be calculated in each case as a precautionary measure, but, in general, they will not be important for the usual sizes of individual or combined footings.

7.2.2 Normally consolidated clays

In normally consolidated clays, which are rarely found at ground surface in nature but which may be encountered on the ocean floor or on lake bottoms, the stiffness and strength of the soil increase more or less linearly from values close to zero at ground surface. When a footing applies load to the surface of such a deposit, instantaneous settlements are caused, and long-time displacements develop as the clay drains and consolidates. For the essentially instantaneous settlement, and the short-time bearing capacity, the clay properties are unchanged by the pressure of the load, and the material model is then one of linear increase of properties with depth from zero at the surface.

For an approximate analysis of the effect of footing width, we might consider that the settlement could be estimated on the basis of an average modulus of the material over some depth range proportional to the footing width. Alternatively, the displacement can be referred to a modulus at a particular depth, which is proportional to the footing width—greater for wider footings. In that case, the value of E to be employed in equation (7.1) can be written

$$E = dB \qquad (7.8)$$

where d is a constant of proportionality. Substituting this value for E in equation (7.1), we get the relation

$$w = \frac{(1 - v^2)}{d} pI \qquad (7.9)$$

which indicates that for constant footing proportions and pressure p, the immediate settlement is independent of the footing width. As will be seen in a subsequent section, this remarkable result is analytically exact. Its practical applicability is limited, but it expresses one of the extremes of settlement behavior that bound the performance of real soils; in this, it complements the results of the previous section, which represent the other extreme of behavior.

If a particular value of settlement is selected, then equation (7.9) sets a value for the allowable pressure, p_a^s, in this case, based on displacement.

In a consistent model of this kind, soil strength also increases with depth, and it would be expected by reasoning analogous to that given above that the allowable bearing capacity in terms of pressure p_a^f of a footing on such a soil would increase, perhaps linearly, with the footing width, having a value close to zero for footings of very narrow widths. Thus, here, as in the previous section, it would appear that when practical ranges of size and soil properties are employed the actual allowable bearing pressure must be based on the strength and failure condition instead of on a settlement requirement. The allowable bearing pressure, including the same factor of safety of 3, as before, is usually established from equation (7.4) but in which the value of c employed is the average from the surface to a depth B, which is equal to the width of the footing. Modifying factors are generally used (22) depending on the shape of the footing.

However, even at the modest surface pressures permitted by this evaluation, long-term consolidation settlements of footings may present difficulties for the structure and must always be calculated by the usual techniques (4).

7.2.3 Sands

Sands are perhaps the most difficult case of all to consider because, in them, the stress field of the footing causes immediate effective stress and, in consequence, stiffness and strength changes in the soil below the footing.

There have been many studies of the settlement behavior of footings placed on sand, some of which will be cited here. At the elementary level of the previous discussion, we can make the following argument. In the overconsolidated clay model the material properties are unchanged with depth. Those of the normally consolidated clay increase linearly with depth from zero at the surface. In sand, prior to footing placement, the latter condition may sometimes hold. Laboratory tests on the influence of confining pressure on the modulus of sands indicate that the modulus increases with some power of confining pressure; frequently 0.5 is used in analysis. However, when the footing is present and loaded, effective stresses are developed below it, and they increase the stiffness and strength of the soil. Thus, below the footing the material properties increase with depth but have finite values at the footing base. Such a situation could be considered to be intermediate between the previous two. It might be reasonable, therefore, to postulate that the settlement of a footing on sand varies with the square root of the width of the footing (22), or

$$w = CIp(B)^{1/2} \tag{7.10}$$

where C is a constant of proportionality, which depends on the relative state of density of the sand. In comparing footings of different sizes with the performance of the standard 1-ft (0.3 m) square test plate on the same soil at the same pressure at ground surface, equation (7.10) can be written in proportional form as before:

$$\frac{w}{w_1} = \left(\frac{B}{B_1}\right)^{1/2} \frac{I}{I_1} \tag{7.11}$$

In actual tests of footings, however, it is found that this relation is only approximately valid for footings larger than 1 ft (0.3 m) in diameter, and, to describe a compilation of such test results for the larger footings Terzaghi and Peck (34) suggested a relation different from equation (7.11), based on theoretical considerations expressed by Kögler (19):

$$\frac{w}{w_1} = \left(\frac{2B}{B+1}\right)^2 \tag{7.12}$$

This equation, in which B must be expressed in feet, is widely used in sand settlement analyses. It can be seen (22) that this can be obtained from a more general relation among the variables

$$w = Cp\left(\frac{B}{B+1}\right)^2 \tag{7.13}$$

in which C represents a soil compressibility. Again in this empirical relation, particular units of feet for B, tons per square foot for p, inches for w, and inches per ton per square foot for C must be employed. No influence functions appear in equations (7.12) and (7.13) because experiments showed that there was little difference in settlement between square and rectangular footings of the same width at the same pressure. Research in sands indicates that the stiffness and strength of sand in plane strain are greater than in the condition of axially symmetric compression. Consequently, the increased soil stress developed by a long footing as compared to a square footing is apparently compensated by the greater stiffness under the plane strain stress state induced by the long footing. The influence functions can also be eliminated from equation (7.11) for this reason.

It has been pointed out by Parry (23) that no single equation such as (7.12) or (7.13) can properly cover the relationship of settlement of a larger footing to that of a 1-ft (0.3 m) square test plate for all conditions of soil property variation with depth. Sands may exist in the field with the modulus constant, increasing, or changing arbitrarily with depth. The settlement behavior of a footing as a function of width, as we have seen, will be different for each profile.

For a prescribed value of settlement, equation (7.10) or a similar equation based on equations (7.12) and (7.13) indicates that the allowable bearing pressure based on settlement p_a^s must decrease approximately as the reciprocal of the square root of the footing width.

The bearing pressure required to cause failure of a footing at the surface of sand increases approximately linearly with the footing width. For typical soil properties, it is found that the allowable bearing pressure based on failure (including the same factor of safety of 3, as before) p_a^f is *smaller* than the allowable pressure based on settlement, p_a^s only for footings smaller than about 4 ft (1.2 m) in width. For larger footings, settlement is the controlling criterion. Thus, the situation for sands is quite different from that for clays.

In the field, in practice, it is time-consuming and expensive to obtain samples of sand representative of in-place unit weights and other properties. Consequently, it is difficult to obtain laboratory measurements of in-place sand properties that can be used in theoretical settlement or bearing capacity calculations. The size of the majority of footings employed in sands does not economically justify the drilling efforts that would be required to obtain undisturbed samples. For this reason, the properties of sands have been related to a number of field investigation techniques, which are readily and frequently employed. In turn, the observed settlements of footings have been correlated empirically with these properties in order to establish allowable bearing pressure criteria.

Laboratory investigations have shown that the stiffness and strength properties of a wide range of granular materials are similar at similar relative densities of the materials when relative density, D_r, is defined:

$$D_r = \frac{e_{max} - e}{e_{max} - e_{min}} \times 100\% \qquad (7.14)$$

or, in terms of dry density,

$$D_r = \frac{\gamma_d - \gamma_{dmin}}{\gamma_{dmax} - \gamma_{dmin}} \frac{\gamma_{dmax}}{\gamma_d} \times 100\% \qquad (7.15)$$

[The American Society for Testing and Materials (ASTM) has established a standard test D2049-64T for determining relative density.] In other words, two very different granular soils will have nearly the same modulus and friction angle if they are at the same relative density. This statement is only very approximately true in practice. The more the in-place properties of sand are investigated, the more complexity the material exhibits. In addition to its relative density, grain size, grain-size distribution, the in-place stress conditions, especially the horizontal stress, and degree of cementation all affect the soil's response to stress. In particular, subtle interparticle cementation has a considerable influence on soil behavior in the field, but it is generally totally destroyed in any sampling process. For this and other reasons, granular soils are frequently stiffer and stronger in place than is reflected by laboratory tests on retrieved samples.

There are two field tests which are widely used to obtain the properties of both granular and cohesive soils: the Standard Penetration Test (SPT) and the Cone Penetration Test (CPT). In the SPT, a standard sampling tool is driven into the soil by means of a standard weight. The weight of 140 lb (0.62 kN) is lifted and dropped a distance of 30 in. (0.76 m) to strike an impact collar at the upper end of the drive rods. The soil condition is assessed through the number of blows required (N) to drive the sampler a distance of 1 ft (0.3 m). The ASTM test D1586-67 specifies the procedure. Formerly, N was correlated directly with the relative density of the soil alone: less than 10 indicated a very loose to loose sand, from 10 to 30 indicated a medium-dense sand, and over 50 indicated a very dense sand. However, later research (13) showed that the overburden pressure must also be included in the relationship. It is now customary to correct the field N-values obtained at varying depths and overburden pressures to the value at a standard pressure [1 ton per square foot (96 kN/m²)] at which a correlation curve of N versus D_r has been established. The effect of overburden pressure on the assessment of blow counts is shown in Figure 7-1, from Gibbs and Holtz (13). In this respect, it should be pointed out that various investigators have reported a wide range of results. Schultze and Menzenbach (32) found the variation among N, D_r, and overburden pressure to be close to that of Holtz and Gibbs, but Meigh and Nixon (20) give a different result, with N being generally much higher at a given D_r. Still later, further tests by Schultze and Melzer (31) agreed with those of Meigh and Nixon. Schultze

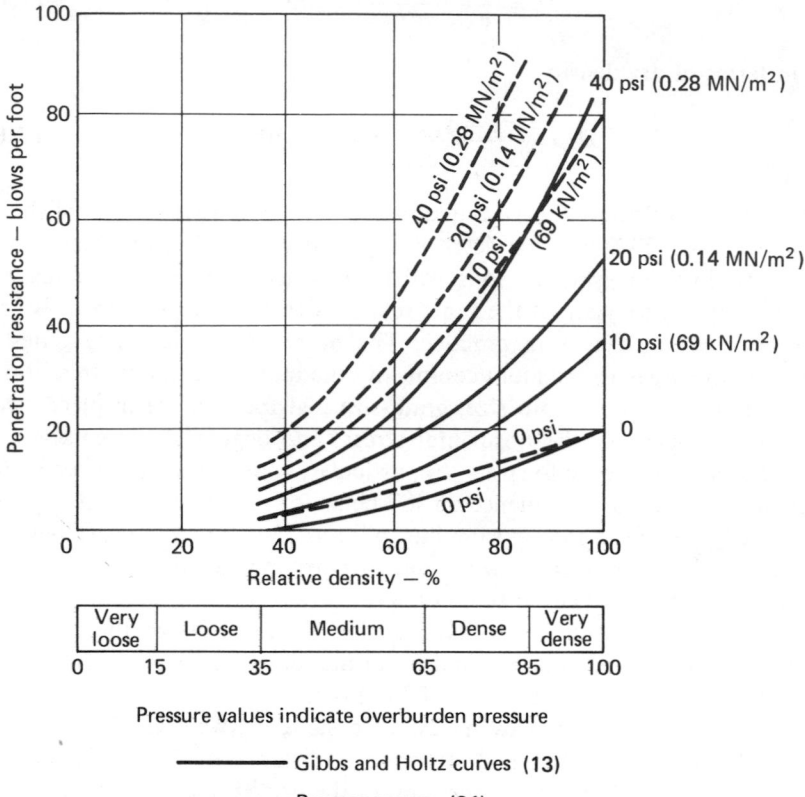

Figure 7-1. Relations among penetration resistance (blow count per foot, N), overburden pressure, and relative density for sand. The qualitative relative density scale is that used by the U.S. Bureau of Reclamation and employed by Gibbs and Holtz. Solid lines are from the experiments by Gibbs and Holtz (13); dashed lines are inferred by Holtz and Gibbs (18) from data by Bazaraa (3).

attributed this to better measurement of in-place soil density in the later tests. In a discussion of such tests, Peck and Bazaraa (24) presented other results, which are also fairly close to those of Schultze and Melzer and to Meigh and Nixon; these differed considerably from the relations suggested by Gibbs and Holtz. Some of these relations are shown later in Figure 7-3.

For the reasons given above relating the soil's properties to conditions other than relative density alone, it has become current practice to refer a sand's behavior to the SPT blow count alone without assessing a related relative density, which may be a misleading indicator of stiffness or strength (9). For example, in some desert alluvial granular soils, N-values of 200 or greater may be recorded in the field, whereas laboratory tests on returned, carefully obtained samples indicate that the relative density is from 85 to 90 percent.

The advantages of the SPT are that it is simple and is widely employed in soil investigations; the disadvantages are that information on the soil state is obtained only at intervals [2.5 or 5 ft (0.76 or 1.52 m)] and results may vary quite widely depending on the equipment and drilling crew engaged.

Using a factor of safety of 3 on failure, taking equation (7.12) for settlement, and incorporating a suggested maximum settlement of 1 in. (25 mm) to restrict differential settlement to values that would not distress common structures, Terzaghi and Peck (34) constructed a diagram of allowable bearing capacity p_a versus footing width B and SPT blow-count value N for footings on sand. The N-values were averaged to a depth of B below the footing and used as a measure of soil compressibility in entering the diagram to obtain p_a. Modifications to the values of p_a were suggested to account for an adjacent water table.

The relations employed to construct the figure can be construed (22, and 23) as follows:

$$\text{Footings} \begin{cases} B \leq 4\text{ ft (1.2 m)} & p_a^f = \dfrac{N}{8}(1) & (7.16) \\[2ex] 4 < B \leq 20 \text{ to } 30 \text{ ft (6 to 9m)} & p_a^s = \dfrac{N}{12}\left(\dfrac{B+1}{B}\right)^2 (1) & (7.17) \\ \qquad\qquad \text{approx.} \end{cases}$$

$$\text{Rafts or mats} \begin{cases} B > 20 \text{ to } 30 \text{ ft (6 to 9 m)} & p_a^s = \dfrac{N}{12}(1) & (7.18) \\ \qquad\qquad \text{approx.} \end{cases}$$

In these empirical equations, the units of p_a in tons per square foot and B in feet must be used, and the (1) in parentheses indicates that a settlement of 1 in. is implicitly assumed. It can be seen that equation (7.17) is a particular case of equation (7.13), written as follows:

$$p_a^s = \frac{Nw}{12}\left(\frac{B+1}{B}\right)^2 \tag{7.19}$$

in which the settlement w has been taken to be 1 in. (25 mm) and the compressibility C has been empirically related to the blow count N such that $C = 12/N$. For other criteria of settlement, the (1) term in the right-hand side of equations (7.16), (7.17), and (7.18) can be replaced with a general settlement w, as in equation (7.12). It is sometimes permitted, for example, to increase the allowable total settlement of a raft to 2 in. (50 mm), since the hazard of differential settlement is lessened with the stiff structure. If other systems of units, such as metric or SI, are to be used, the equations must be reformulated.

The diagram of Terzaghi and Peck has been widely employed for many years in footing design, but a number of studies has indicated that it may be too conservative. The conservatism results from an overestimate of footing settlements indicated by the use of equation (7.12).

It is considered on the basis of accumulating data (22) that settlements in sand may be only two-thirds of the values indicated by equation (7.13); alternatively, the allowable bearing pressure of equation (7.17) or equation (7.19) in general, may be increased by 50 percent. Thus, in this event, equation (7.19) becomes

$$p_a^s = \frac{Nw}{8}\left(\frac{B+1}{B}\right)^2 \tag{7.20}$$

With the same units as before. This relation for $w = 1$ in. (25 mm) is shown in Figure 7-2.

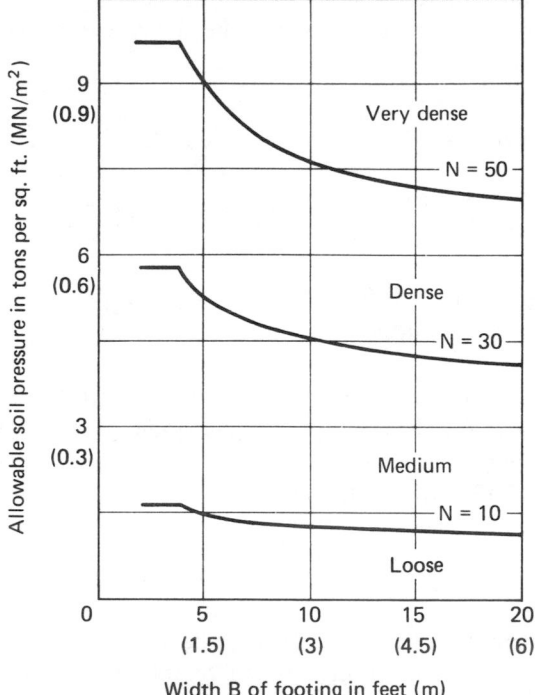

Width B of footing in feet (m)

Figure 7-2. Curves giving the soil pressure ("allowable") for a long footing of the given width, resting on sand, to produce a settlement of approximately 1 in. The sand is characterized by the average blow count, *N*, recorded to a depth of one footing width below the footing in a standard penetration test (SPT). The allowable pressures are 1.5 times those given in the original diagram of Terzaghi and Peck (33). The qualitative values of relative density are those assigned to the soil by Terzaghi and Peck.

A suggested procedure for determining allowable bearing pressure based on a SPT field investigation is as follows (24):

1. Modify, by using Figure 7-1, the average field blow counts N over the depth B below the proposed footing to the overburden pressure that will prevail below the planned footing level.
2. Enter Figure 7-2 with the modified blow-count value to obtain p_a for the proposed footing.
3. To account for the effect of the presence of a water table below the footing, multiply the value of p_a obtained from step 2 by a factor a, where a is the ratio of the effective overburden pressure at a depth $B/2$ below the footing in the presence of the water to the effective pressure in the absence of water.

Parry (23) takes a different approach to estimating relative settlement based on the SPT. He argues that the elastic modulus of the soil at any depth should simply be a linear function of the SPT N-value, since the latter reflects the pertinent sand properties of density and ambient stress at any level. Thus, equation (7.1) can be used in the modified form

$$w = \alpha \frac{pB}{N} \qquad (7.21)$$

when the elastic property $E/(1 - v^2)$ is replaced by N and when α is an empirical coefficient to be obtained from case studies. Although it also includes footing shape effects, these are not discussed by Parry. Field comparisons indicate that α is about 0.025 when p is in pounds per square inch. B is in feet, and w is in inches (α is 300 when p is in MN/m², B is in meters, and w in millimeters). The value of N is taken as a weighted average over a depth of $2B$ below the foundation, $N_m = (3N_1 + 2N_2 + N_3)/6$ where N_1 is the value from the foundation level to a depth of $2/3 \, B$ below it, N_2 from $2/3 \, B$ to $4/3B$, and N_3 from $4/3B$ to $2B$. When equation (7.21) is used to arrive at a settlement ratio w/w_1, α disappears, and the ratio of N_m for the full-size footing to N_m for the 1-ft square footing is employed. In this way, the actual N profile is taken into account in a fashion similar to the method developed by Schmertmann for the CPT, to be discussed later.

The measured values of N are used in the calculation of the settlement ratio; these are altered only if site conditions change due to excavation or if the ground water level changes. If the water level is at the base of both the footing and the test plate, the above method holds for saturated as well as dry soil. When the water level is at some depth below the foundation level, it will have more effect on the larger footing. Parry analyzes this effect, but it should also be reflected in the measured N-values.

It is now appropriate to consider the methods of settlement analysis and

allowable pressures based on the cone penetration test. As currently employed, the CPT gives a continuous record, in this case of force versus penetration
resistance of the soil. No sample is obtained. Instead, a conical point with a
60° tip angle and a base cross-sectional area of 10 cm² is pushed through the
soil at a rate of 2 cm/sec with a special arrangement of rods so that only the
tip resistance, and not the frictional resistance of the rods, is measured. In
its original form, the cone was devised in the Netherlands (although a variety
of cone penetration tests has flowered elsewhere; it is a natural test configuration to consider) and the force was measured mechanically. In its recent
form, the conical tip is followed by a short tubular section, called the *mantle*.
The force on the cone and the frictional force on the mantle are measured
separately and continuously. It has been found that the ratio of the two
forces at a particular depth indicates the type of soil through which the cone
is passing. The cone resistance, q_c, has usually been expressed in kilograms
per square centimeter, although units of MN/m² are being used more frequently.[2] The cone test and many applications are thoroughly described in
the book by Sanglerat (29).

With the widespread use of the CPT in Europe and its growing employment in the United States, some attention has been paid to the correlation
of the SPT and CPT. In Meyerhof's studies (21), which have been generally
confirmed by Schmertmann (30), it was found that for sands

$$q_c = 4N \, (\text{kg/cm}^2) \qquad\qquad (7.22)$$

However, Schmertmann found that the correlation was different depending
on the grain size of the soil, and he suggested that the relations in Table
7.1 are conservative.

TABLE 7.1

Relation Between Standard Penetration (N blows)
and Cone Penetration [q_c kg/cm²] Tests

Soil Type	q_c/N
Silts, sandy silts, slightly cohesive sand-silt mixtures	2.0
Clean fine-to-medium sands; slightly silty sands	3.5
Coarse sands; sands with a little gravel	5
Sandy gravels and gravel	6

The ratios in Table 7.1 are to be assumed to hold independent of the
depth at which the N-values were measured, the elevation of the water table,
and the relative density of the soil. In medium sands, therefore, the values of
q_c are associated with relative density approximately in the following way:
less than 40 kg/cm² (4 MN/m²) indicates a loose sand, 40 to 120 (4 to 12

---

[2]1 kg/cm² is approximately equal to 1 tsf; 10 kg/cm² is approximately 1 MN/m².

MN/m²) indicates a medium-dense sand, and over 200 (20 MN/m²) indicates a very dense soil.

Various relations between q_c and the state of a soil have been suggested (29). For sands, Meyerhof (21) gave the following table:

D_r	< 0.2	0.2–0.4	0.4–0.6	0.6–0.8	0.8–1.00
q_c, tsf	< 20	20–40	40–120	120–200	> 200
(MN/m²)	(< 2)	(2–4)	(4–12)	(12–20)	(> 20)

A diagram given here as Figure 7-3 expressing the results of statistical evaluation of a variety of tests was derived by Schultze and Melzer (31) to show

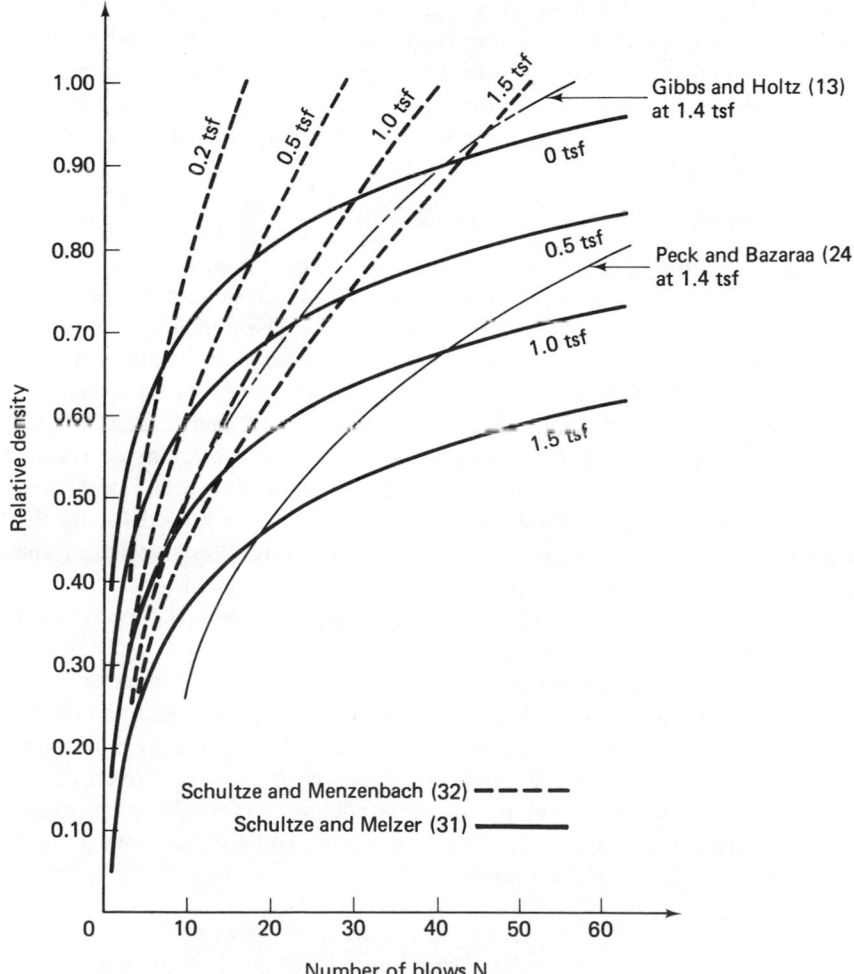

Figure 7-3. Relations among standard penetration resistance, overburden pressure, and relative density for sand according to various investigators (31).

the conflicting results of research on the effect of overburden pressure on the SPT values at depths less than the critical depth (described later). Comparing the table of Meyerhof with Figure 7-3, we see that Meyerhof's figures roughly refer to data at an overburden pressure of 1 tsf (0.1 MN/m²). This corresponds in dry soil to a depth of about 20 ft (6.1 m).

In clay, q_c is a function of the undrained strength or cohesion c. Theoretically, the tip bearing capacity in an infinite clay body is about $9c$ to $9.5c$, but, in practice, the cone is accompanied by a sleeve behind the tip. The soil must be sheared around the sleeve as well as the tip for the combination to move ahead. Thus, in practice, it is found that q_c is in the range $10c$ to $20c$.

Using the CPT as a basis, Schmertmann (20) has suggested a method for calculating the settlement of a rigid footing on sand; the method is simple, is rationally based, and overcomes many of the defects of other techniques. We describe it in the following paragraphs. Experiments and analytical and numerical calculations indicate that the vertical *strains* which give rise to the settlement of a footing are relatively small immediately below the footing, increase to a maximum value at a depth of about one-half the footing width, and diminish with depth below this. Although variations in detail occur in different models of soil behavior, this pattern holds. It was proposed by Schmertmann to approximate this behavior by a vertical strain *influence factor*, I_z, which is shown in Figure 7-4. As seen in the figure, the factor has zero value at the base of the footing, increases linearly to 0.6 at a depth of $B/2$, where B is the footing width, and decreases linearly to zero again at a depth of $2B$. For the reason given earlier in terms of the difference between axially symmetric and plane strain behavior of sand, this influence factor remains the same for footings of all shapes from circular or square through rectangular to very long. It is not altered by the presence of soil layers of differing density below the footing; they are accounted for by the modulus assigned to them. In particular, if rigid bedrock occurs below the footing within the $2B$ depth, the strain influence factor is simply truncated by the bedrock and only the portion extending through the soil zone is taken into account in the calculations.

Only in two circumstances has Schmertmann found it necessary to modify the influence factor I_z. The first of these allows for embedment of the footing. The shape of the factor with depth is retained, but its maximum value, 0.6, is reduced by a term depending on the ratio of the overburden pressure at the foundation level, p_0, to the net increase in foundation pressure $(p - p_0)$, where p is the pressure imposed on the soil by the footing. This correction C_1 is given by the equation

$$C_1 = 1 - 0.5\frac{p_0}{p - p_0} \tag{7.23}$$

but the value of C_1 is restricted to the range $1 > C_1 > 0.5$. Under long-term loads, footings on sand are found to settle with time. Various mecha-

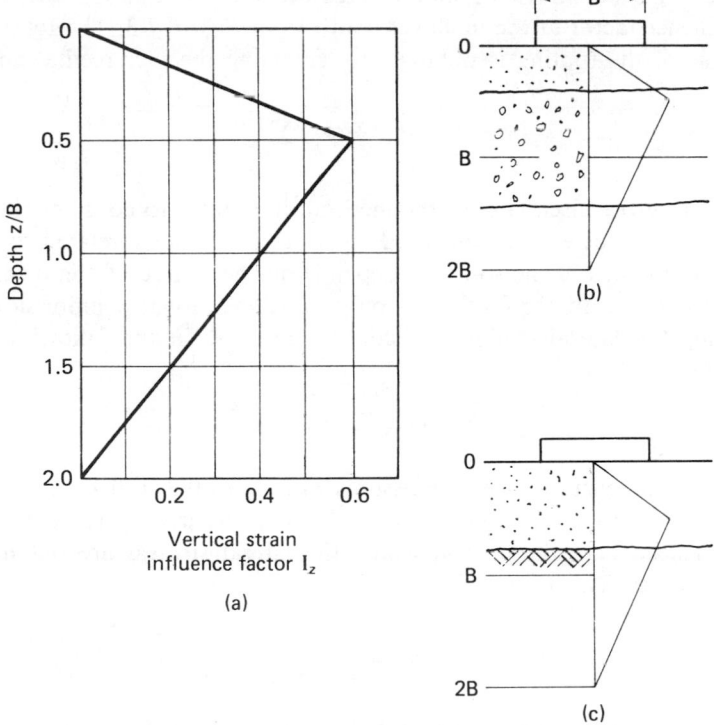

Figure 7-4. Vertical strain influence factor for use in settlement analysis of footings on sand (30): (a) general diagram; (b) application to layered profile; (c) use when bedrock is present.

nisms have been proposed to represent this behavior, considered as creep in the sand. Rather than a true viscous flow or creep behavior, however, it is more likely a consequence of the cumulation of individual sand grain movements arising from vibrations in the soil transmitted by the footing or otherwise. At any event, it has been observed to develop as a function of the logarithm of time, and it can be a significant contributor to the settlement of footings. Schmertmann added another modifier, C_2, to account for this influence of time:

$$C_2 = 1 + 0.2 \log_{10} (10t) \qquad (7.24)$$

where t is measured in years and the coefficient 10 comes from dividing t by a characteristic time, which Schmertmann has chosen to be 0.1 year.

With this method, the incremental settlement, Δw, caused by any homogeneous layer of soil of thickness Δz, at depth z within the depth $2B$ below the footing is given by the expression

$$\Delta w = C_1 C_2 (p - p_0) \frac{I_z}{E_s} \Delta z \qquad (7.25)$$

in which E_s is a compression modulus for the soil layer and I_z is the value of the influence factor at the midlayer depth from Figure 7-3. The total settlement, w, of the footing is obtained by summing the incremental amounts

$$w = C_1 C_2 (p - p_0) \sum_0^{2B} \frac{I_z}{E_s} \Delta z \qquad (7.26)$$

To complete the discussion of the method, it remains to consider the modulus E_s and its determination for a layer. It is obviously related to E, the Young's modulus of the soil, and depends on the degree of confinement of the soil underneath the footing. If only one-dimensional compression were occurring (no lateral strains), we could express E_s in the following form:

$$E_s = \frac{(1 - v)}{(1 + v)(1 - 2v)} E \qquad (7.27)$$

so that for the usual range of Poisson's ratio, v of 0.3 to 0.4 for soil, E_s is about 1.5 to 2 times E. However, lateral strains are usually present, and we might instead take the case in which the lateral *stresses* are one-half the vertical values. This would give

$$E_s = \frac{E}{1 - v} \qquad (7.28)$$

and E_s would range from 1.4 to 1.7 times E for the same range of v. Schmertmann chose to correlate E_s experimentally with cone tests, making use of the measured settlement of a loaded circular plate. He found that the relation could be expressed conveniently as follows:

$$E_s = 2q_c \qquad (7.29)$$

This expression agrees generally with results obtained by other investigators.
The settlement method can then be described in the following steps:

1. A CPT is performed, preferably at each proposed footing location, to a depth below the base of each footing considerably greater than $2B$. If the strength generally increases with depth, and no markedly softer soil is found below this depth, the soil between the footing base and a depth $2B$ greater than this is divided into homogeneous layers or zones, based on the fluctuations of the CPT results.
2. A q_c value is assigned to each layer, and equation (7.29) is used to obtain E_s for the layer.
3. For each layer, the average value of I_z at the mid-depth is obtained from Figure 7-4.

4. C_1 and C_2 are calculated from equations (7.23) and (7.24).
5. Equation (7.25) is used to calculate the settlement in each layer.
6. The settlements are summed to give the final value.

No account is taken of the location of the water table since the properties of the submerged soil are accounted for in the CPT values.

The method has been reported by Schmertmann to give improved results over previous techniques for footings or loaded areas ranging in size from just over a foot (0.3 m) in width to over a hundred feet (30 m). Other investigators (11 and 12) have confirmed these results. The improvement results from better correlation between the cone penetration tests and the property E_s used in equation (7.26) and in the ability of the method to include the soil response layer by layer, rather than as an overall average as is used in approaches that use the SPT. There is no modification to the basic theory of deformation incorporated in the Schmertmann model, since it uses results from linear elastic theory.

This can be seen, for example, by considering what would happen in uniform deposits of soil where we might assume that the CPT value q_c would vary continuously with depth. To account for such a variation, we require equation (7.26) in integral form. For simplicity, we will assume a footing at ground surface and ignore time effects, so that the coefficients C_1 and C_2 can be omitted and $p_0 = 0$. The equation becomes

$$w = p \int_0^{2B} \frac{I_z}{E_s} \, dz \qquad (7.30)$$

in which both I_z and E_s are functions of depth, I_z having the prescribed shape of Figure 7-4. The obvious case to examine first is that in which E_s is constant, say E_{s0}. Because of the discontinuous nature of I_z, the integration must be performed in two parts. The result is

$$w = \frac{0.6pB}{E_{s0}} \qquad (7.31)$$

If equation (7.29) is substituted for E_{s0} with q_{c0} being the constant cone penetration resistance, the equation becomes

$$w = \frac{0.3pB}{q_{c0}} \qquad (7.32)$$

This result, that the settlement depends linearly on the width of the footing, is, of course, identical with that obtained from the theory of elasticity for a uniformly loaded region on the surface of a homogeneous half-space, equation (7.1), except for the different arrangement of material properties.

A second case to consider is analogous to the one treated in Section 7.2.2 in which the modulus E_s increases linearly with depth from a zero value at the surface, say,

$$E_s = E_{s1}z \tag{7.33}$$

With the substitution of this expression in equation (7.30) and integration, the result is obtained that

$$w = \frac{p}{E_{s1}}(0.8 \log_e 4) = 1.11\frac{p}{E_{s1}} \tag{7.34}$$

Again substituting from equation (7.29) with p_{c1} being the slope of the CPT value with depth gives

$$w = 0.55\frac{p}{q_{c1}} \tag{7.35}$$

which is seen to be independent of footing width and identical in form with equation (7.9). Finally, if, to represent the behavior of sand, approximately, we assume a variation of modulus with the square root of depth

$$E_s = E_{s2}z^{1/2} \tag{7.36}$$

integration of equation (7.30) gives

$$w = 0.75\frac{pB^{1/2}}{E_{s2}} \tag{7.37}$$

or, with substitution of the CPT in the form of equation (7.36),

$$w = 0.38\frac{pB^{1/2}}{q_{c2}} \tag{7.38}$$

a result analogous with that assumed to give equation (7.10).

In uniform deposits of sand it has been reported (29) that the CPT value increases from the surface with depth only to a depth termed the *critical depth* beyond which the value remains essentially constant. The critical depth depends on the relative density of the sand but ranges from 5 to 10 times the diameter of the penetrating object for loose to dense sands.

Evidence for this behavior is not conclusive, because no uniform homogeneous deposits of sands exist in nature and it is difficult and expensive to prepare tanks of uniformly dense sand in the laboratory sufficiently wide and deep to permit definitive tests to be conducted.

If correct, it is a curious result since it indicates that the increasing depth of overburden has no effect on the point resistance. We may conclude that the mechanism of plastic deformation around the penetrometer tip changes from one, near the surface, in which deformation extends to the surface to one, at depth, where volumetric changes in the soil surrounding the tip permit the plastic deformation to proceed in an enclosed region; in this volume, stresses must therefore be essentially independent of the position of the free surface and of the gravitational stress field.

If we accept this hypothesis of a critical depth, the cases of interest in our study of settlement of footings on sand are represented by the uniform profile of q_c with depth as given by equation (7.32) or by the square root variation of q_c with depth. The latter might be reasonably representative of the behavior of soil below a surface footing, whereas the constant q_c with depth model might correspond to the more usual case of a footing embedded to some extent. However, the more usual embedment depths are much less than the critical depth for footings of conventional widths.

The empirical and approximate theoretical relations in this section have been used to estimate settlements and then allowable bearing pressures in small rigid footings on clay and sands. They can also be used to determine the subgrade reaction coefficients that are necessary in the design and analysis of flexible foundation members. However, a description of this process will be deferred until some theoretical studies pertinent to the problem have been described in the following section.

7.3 SETTLEMENT ON NONHOMOGENEOUS HALF-SPACE

Some space was devoted in Chapter 5 to the correspondence of the Winkler model of subgrade response with that of a linearly elastic homogeneous half-space. The difference between the response of the Winkler representation and that of the half-space was observed and various relations were demonstrated. However, the stiffness of soils generally increases with depth and the analytical model should reflect this. This was studied by Gibson (14) who obtained the surprising result that the response is identical to that of a Winkler material. This occurs in the special case in which the elastic modulus (E or G) varies linearly with depth from zero at the surface and the medium is incompressible so that its Poisson's ratio is $\frac{1}{2}$. Gibson showed that, under these conditions, when

$$v = \frac{1}{2} \quad \text{and} \quad E(z) = 3mz \quad \text{or} \quad G(z) = mz \qquad (7.39)$$

since $G = E/2(1 + v)$, where z is distance vertically and m is a constant, the surface settlement w under an area of *any* shape, uniformly loaded by a pressure p, is

$$\begin{cases} w = \dfrac{p}{2m} \text{ within the loaded area} \\ \text{and } w = 0 \text{ outside the loaded area} \end{cases} \qquad (7.40)$$

This implies that if the same pressure is applied symmetrically to a *rigid* foundation slab of any shape, the same settlement will be obtained and, moreover, the pressure distribution between the slab and the underlying continuum will be uniform. The settlement of points originally lying on planes at different depths below the surface is shown in Figure 7-5, which is taken from Gibson's paper.

It follows from equation (7.40) that if the above conditions are met in some soil, a coefficient of subgrade reaction k_0 can be obtained from the relation

$$k_0 = 2m \tag{7.41}$$

Most soil deposits do not behave in this way, but, as Gibson points out, the result may have practical value in the prediction of immediate settlements of loaded areas on uniform, deep beds of normally consolidated clay. This corresponds to the approximate results of Section 7.2.2. In later papers Gibson and his co-workers treated problems in which the inhomogeneous medium is finite in depth (7 and 15) and in which Poisson's ratios other than $\frac{1}{2}$ were considered for the half-space (2 and 16). In the last case, it was found that even a small departure from incompressibility caused a large change in the deflected shape of a uniformly loaded circular area. Clearly, the above identification of the settlements with that of a Winkler material model is restricted only to materials that closely conform to the incompressibility requirement.

From the numerical computations of the behavior of a rigid circular slab performed by Carrier and Christian (8), the region of applicability of a Winkler model can be expressed quantitatively. The soil model is again

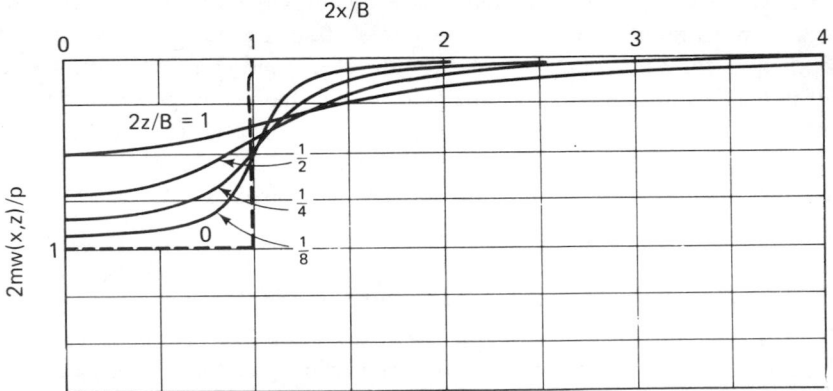

Figure 7-5. Settlement of planes at various depths $2z/B$ due to uniform vertical pressure over infinite strip of width B at the surface (14). The medium has the properties $G = mz$, $v = 0.5$ (Gibson material).

linearly elastic with a modulus E which increases linearly with depth but is nonzero at the surface, so that

$$E = E_0 + dz \qquad (7.42)$$

where d is a constant. When d is zero, the homogeneous soil profile is represented on the one hand, and when E_0 is zero, the extreme case examined by Gibson is recovered on the other. If the foundation slab applying the load to the soil has a diameter or width B, it is convenient to describe the conditions by the ratio E_0/dB. Carrier and Christian find that when this ratio is less than about 0.01 (in the Gibson case, the ratio is zero), a Winkler representation can be used, but the value of the subgrade reaction coefficient depends on Poisson's ratio, as shown in Figure 7-6, where k_0 is proportional to $1/I'$. For $E_0/dB \geq 1.0$, the unmodified Winkler model is inapplicable because the condition of a homogeneous half-space then applies.

To estimate the settlements of footings resting on sand, Terzaghi and Peck (34) suggested the use of equation (7.43):

$$\frac{w}{w_1} = \left(\frac{2B}{B+1}\right)^2 \qquad \text{[equation (7.12)]} \qquad (7.43)$$

where w_1 is the settlement of a square or circular plate of width or diameter $B = 1$ ft on the sand deposit under a given pressure (not load) and w is the

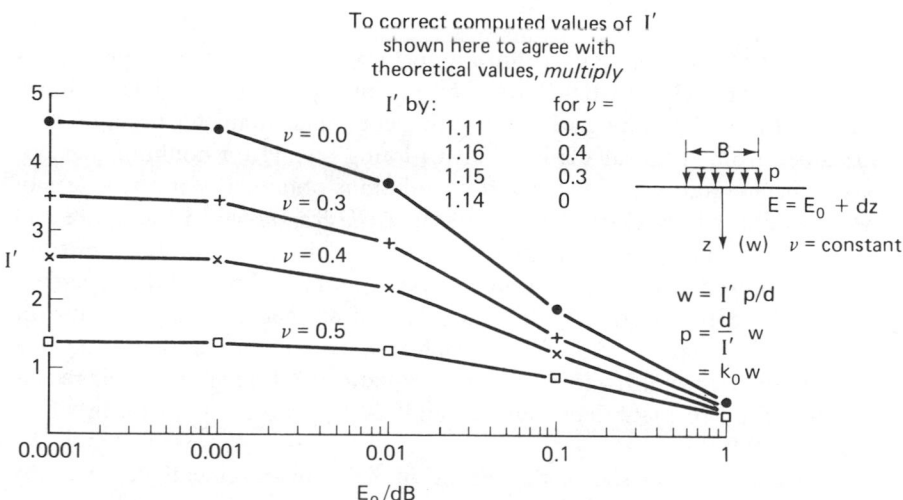

Figure 7-6. Settlement of a smooth rigid circular plate on a nonhomogeneous elastic subgrade by finite element calculation (8).

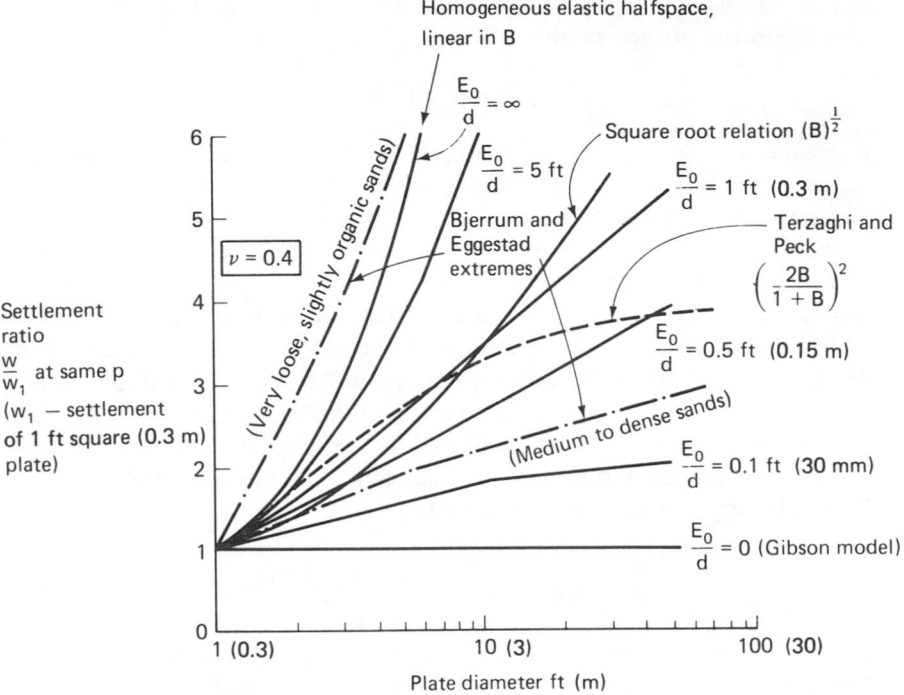

Figure 7-7. Comparison of finite element computations with the results of plate-bearing tests on sand (8).

settlement of the footing of width B in feet (>1 ft) at the same applied pressure. Carrier and Christian have plotted this equation in a figure, shown here as Figure 7-7, along with the settlement versus diameter function they obtained from their calculations for a footing resting on nonhomogeneous soil. This diagram clarifies the various relations obtained. Various values of the modulus of elasticity at the surface, E_0/d, are shown in the figure and range from the extreme of zero (Gibson model except for the Poisson's ratio employed) to the homogeneous case in which the modulus does not vary with depth. In drawing the figure, they used their computed results for a value of Poisson's ratio of 0.4, which they considered characteristic of a medium-dense sand. It can be seen that equation (7.43) matches their results fairly well up to plate diameters of about 20 ft (6 m) for the case in which $E_0/d = 1.0$ ft.

Figure 7-7 also shows the results of field measurements reported by Bjerrum and Eggestad (4) for very loose and medium-dense sands, respectively. It is seen that the very loose sand follows the homogeneous half-space model fairly well, whereas the medium-dense sand results may be represented by the computed result for E_0/d equal to approximately 0.2. In the latter case, a footing with a diameter greater than about 20 ft (6 m) would have a value of E_0/dB equal to or less than 0.01, and the previous criterion of Carrier and Christian indicates that the response of the subgrade soil in such cases could be represented reasonably well by the Winkler model. As was pointed out in the previous section, the settlement of footings in sand calculated by Schmertmann's method could plot anywhere in Figure 7-7, depending on the cone penetration test results observed. If they were relatively uniform with depth, a settlement curve would follow the $E_0/d = \infty$ line. A linear increase with depth would result in a line close to the $E_0/d = 0$ trace; other functions would lie in between. In practice, it is usual to estimate the value of the subgrade coefficient by field load tests, from the results of borings and laboratory tests, or from tables constructed on the basis of past experience. The various methods will be treated next.

7.4   DETERMINATION OF VERTICAL SUBGRADE COEFFICIENTS

The determination of the subgrade reaction coefficient is simpler in the case of slabs, which will be discussed first.

7.4.1   Slabs

In the rare event that the soil is a normally consolidated clay, with test data available on its properties, and *immediate* deflections and stresses are of interest, Gibson's equation (7.41) can be used directly in the determination of k_0. Since there apparently have been no practical tests aimed at assessing the validity of such a procedure, other methods should be used to check the value obtained.

It can be seen from the appropriate equations for the determination of the maximum stress in a slab resting on a Winkler material that the subgrade coefficient k_0 can vary widely without having a great effect on the maximum tensile stress. For pavement design, therefore, the determination of k_0 is not a critical operation. Field tests to determine the coefficient are generally performed with essentially rigid plates in the diameter range of from 12 to 30 in. (0.3 to 0.8 m), since early experiments (Figure 7-8) indicated that the use of plates larger than 30 in. (0.8 m) in diameter effected little change in the value of the coefficient k_0 recorded on uniform soil. In a slab the major proportion of the total reaction occurs below the loading point inside a circle, with a diameter of 4 or 5 times the characteristic length $1/\beta$. Thus, to include possible soil property variations, a logical value of the coefficient of

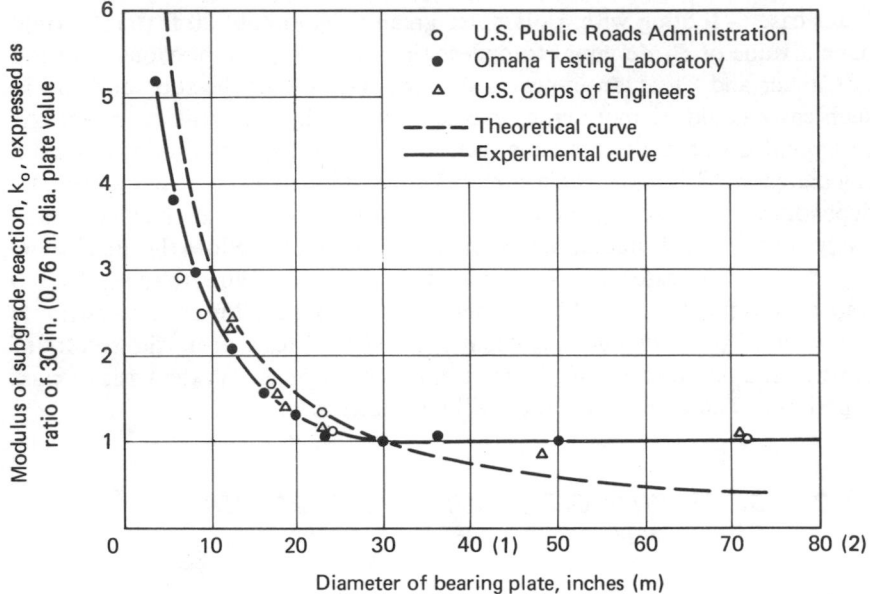

Figure 7-8. Results of bearing tests on circular plates of different diameters in terms of the subgrade reaction coefficient (37). Theoretical curve refers to deflection of uniform elastic half-space under circular load.

subgrade reaction to apply to the design of a particular pavement would be that obtained from a loaded test area of the above diameter. Since for typical concrete pavements and soil conditions, the radius of the area is in the range of 10 to 15 times a pavement thickness of at least 7 or 8 in. (0.2 m), this implies that the coefficient be determined by a load test on a plate of at least several feet (2 m) in diameter. Because of the high loads required, it would obviously be difficult to achieve this in practice. Terzaghi (33) therefore suggested the following procedure. A coefficient k_{01} is measured by a field test on a square or circular plate 1 ft (0.3 m) in diameter. If the subgrade consists of sand, the value of the coefficient k_0 to be used in the design of the slab, beam, or mat is calculated from the relation

$$k_0 = \left(\frac{B+1}{2B}\right)^2 k_{01} \tag{7.44}$$

obtained from equation (7.12) in which B is taken to be the effective diameter of the slab's reaction area. It can be seen from this equation that, for large values of B, k_0 reaches the limiting value of $0.25k_{01}$. Since B cannot be calculated initially, as it, in turn, depends on k_0, it is taken to be 15 times

(Terzaghi gave 14) the *estimated* concrete thickness. The value of k_0 obtained from equation (7.44) is then substituted in equation (5.99) to get the characteristic length $1/\beta$, and this multiplied by 5 gives the diameter of the reaction area to compare with the original estimate of B above. If they differ by more than a factor of 2, the new value of B should be substituted in equation (7.44) and k_0 should be recalculated.

Should the designed slab thickness eventually turn out much different from that assumed, the calculation may have to be repeated. Table 7.2 gives Terzaghi's values for k_{01} on sand for use in preliminary design calculations or where field measurements are unavailable.

TABLE 7.2

Values of Subgrade Reaction Coefficient k_{01} (33) for 1-Ft (0.3 m) Square Area
on Sand Soil (in tons/cu ft; multiply by 0.3 to get values in MN/m³).

Soil Characteristic	Loose	Medium Dense	Dense
Dry or moist sand	20–60	60–300	300–1000
	(30–90)	(90–450)	(450–1500)
Submerged sand	25	80	300
	(35)	(120)	(450)

These values were based on the same original bearing plate data from which the empirical constant in the allowable bearing pressure equation (7.17) was estimated. For a plate width B of 1 ft (0.3 m), it can be seen from that equation that a subgrade reaction coefficient in tons per square foot per inch is given by $N/3$, or in tons per square foot per foot by $4N$, where N is the standard penetration test blow count. However, it was noted in the discussion of Section 7.2.3 that settlements or allowable bearing pressures as given by equation (7.17) had been found to be conservative and it had been suggested that bearing pressures could be increased 50 percent over these values. In the present case, this would imply that a better value for k_{01} for sands could be obtained by the equation

$$k_{01} = 6N \qquad (7.45)$$

for which the blow count is determined at the depth at which the slab is to be placed and corrected for the overburden pressure according to Figure 7-1, and in which k_{01} is in tcf. Since slabs are large in extent, the final value of N for equation (7.45) should be obtained from the zero overburden pressure curve. The modified values of k_{01} are shown in parentheses in Table 7.2.

Soils are nonlinear in their behavior, so that for the Winkler model to hold for a given soil, even approximately, it is necessary to place some restrictions on the load for which the subgrade reaction coefficient is to be determined. To keep to within roughly linear behavior, the stress beneath the structure in the case of a beam or footing should be less than one-half that required to cause failure of the soil. Where a slab is concerned, none of the loads should be greater than one-half that which would cause failure of a circular footing five times the characteristic length in diameter. Under these conditions, therefore, equation (7.44) and Table 7-2 may be employed.

Vesic (36) considers that the equivalent plate diameter of five times the characteristic length, suggested by Terzaghi as being the size for which the subgrade coefficient should be evaluated, is too great. By comparing the settlement of a rigid circular slab on a semi-infinite elastic solid with that of the slab on the Winkler subgrade, Vesic finds that a representative plate diameter is

$$B = \frac{4h}{\pi} \sqrt[3]{\frac{E}{E_s}} \tag{7.46}$$

which is about one-half the diameter suggested by Terzaghi. Equation (7.46) was derived for the case in which the subgrade can be described by a modulus E_s constant to a depth of about 10 characteristic lengths below the surface. If the value of E_s increases, or varies with depth without abrupt changes, equation (7.46) can still be used provided an average value of E_s is taken for the material to a depth of 2.5 characteristic lengths. This enables a representative value of k_0 to be obtained when the modulus generally increases with depth.

In practice, the subgrade modulus of elasticity E_s can be estimated from the results of standard or cone penetration tests as discussed previously, equation (7.29), and can be substituted in equation (7.46) to permit the calculation of k_0. It is preferable because of the variability of soils to combine the results of subsurface tests and the calculation of k_0 from relations such as equations (7.46) and (7.44) with a field plate loading schedule.

When the subgrade consists of *clay*, consolidation will take place if the load is permitted to remain in place. The largest moments and stresses may be developed in a pavement slab at the end of consolidation. Consequently, the coefficient of subgrade reaction to be used in design should be the value obtained from a test carried out long enough for consolidation in the clay to be completed. Moving wheel loads on a pavement will not result in consolidation, and thus the possibility of a stationary load dictates the pavement thickness. Plate load tests are seldom carried out long enough for the consolidation process to go to completion and Terzaghi therefore suggested the values in Table 7.3 for k_{01} for *stiff* clays only (33).

TABLE 7.3

Values of Subgrade Reaction Coefficient for 1-Ft (0.3 m) Square Area, k_{01}, on Stiff Clay Soil.

Soil Characteristic	Stiff	Very Stiff	Hard
Unconfined compressive strength, tons/ft^2 (MN/m^2)	1–2 (0.1–0.2)	2–4 (0.2–0.4)	> 4 (> 0.4)
Value of subgrade reaction coefficient for 1-ft square area, k_{01} tons/cu ft (MN/m^3)	50–100 (15–30)	100–200 (30–60)	> 200 (> 600)

Since stiff clays may be considered to correspond reasonably well to an ideal homogeneous elastic material, in which the settlement under a given load varies directly with the width of the loaded area, it follows that the coefficient of subgrade reaction in this case varies inversely with dimension B, that is to say,

$$k_0 = \frac{1}{B} k_{01} \qquad (7.47)$$

The width B for a slab on stiff clay is calculated as before by using Terzaghi's or Vesic's criteria and any necessary iteration.

For soft to medium-stiff and normally consolidated clays, that is to say, clays with unconfined compressive strength smaller than about 1 ton/sq ft (0.1 MN/m^2), the values of k_{01} that would be obtained in a completely drained loading test are very small. Because of this, Terzaghi recommended that the foundation slab or beam be considered to be rigid for the computation of bending moments. For calculation, the slab would have to be considered finite instead of infinite in size, and the diameter would be 15 times the thickness.

When a raft or mat foundation is subjected to concentrated line or point loads, as in Figure 7-9, the dimensions of the loaded area may be larger or smaller than that of the area of influence of a single load (approximately $5/\beta$ as given earlier). If B, B_1, or B_2 is larger than $5/\beta$, then the value of the subgrade coefficient will be dictated by the dimension $5/\beta$. If the width is smaller, it follows that the calculation of k_0 should be based on the actual width. In the case of unequal column spacing in two directions, as shown in Figure 7-9(b), Terzaghi (33) pointed out that two values of the subgrade reaction coefficient would be required in the calculations. For the calculation of bending moments in the vertical plane through the columns with wider spacing (x-direction), it can be seen that each closely spaced row amounts essentially to a line loading, so that the effective loaded area is a strip of width B_2, on which the appropriate coefficient k_0 should be based. However,

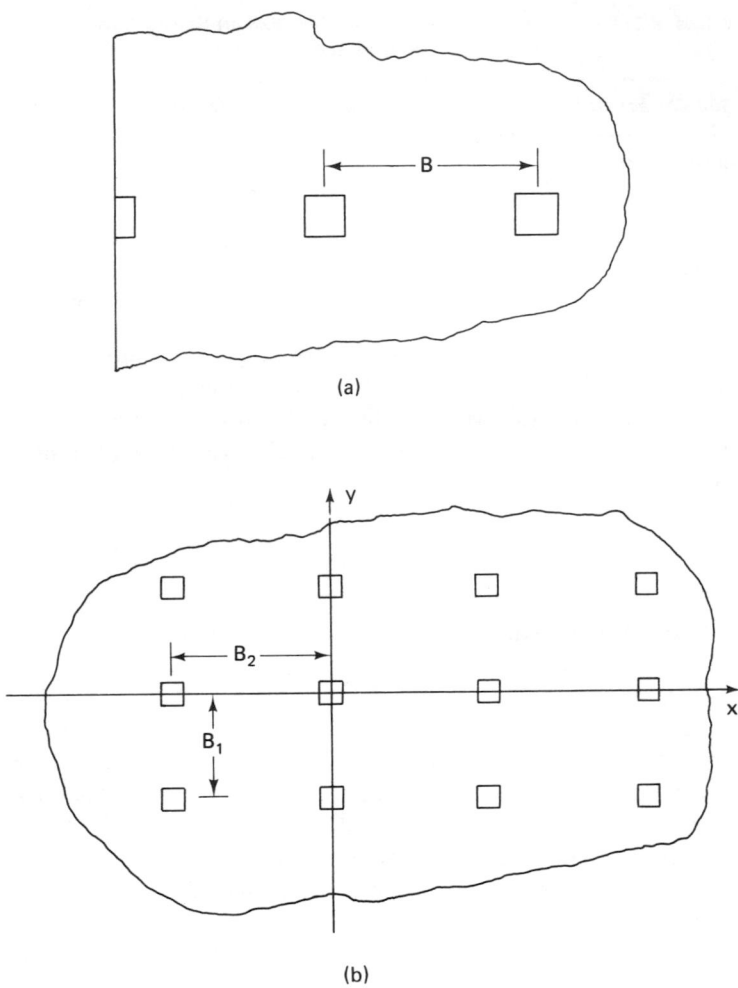

(a)

(b)

Figure 7-9. Lines and arrays of column loads on a slab.

for the bending moments on a cross section in the y-direction, the loaded areas should be considered as rectangles of dimension B_1 by B_2, and k_0 should be computed for this area.

In the case of a mat foundation acted upon by individual concentrated or line loads, the mat will be deflected locally by the loads. However, as a consequence of the overall average loading over the entire mat, displacements will be induced in the subgrade (considered as a half-space) which will give an overall settlement pattern to the mat, depending on the material characteristics and properties and the relative mat/soil stiffness. The local deformations will be superimposed on this. When moments in the mat are computed by using a uniform Winkler subgrade model, only the *local* moments are obtained. Although these may have the dominant values for design purposes, it should be borne in mind that they are superimposed on top of an overall bending moment distribution due to the large-scale response of the supporting medium.

In many cases in practice, the moments generated by the overall pattern of deflections, rather than the local behavior, will dominate in the design of the beam. One way of taking this into account is to permit the coefficient of subgrade reaction to vary over the foundation width to reflect the continuum behavior of the underlying material. That is, for a mat on sand, it would be expected that equal pressures at the center and edges of the foundation will generate greater edge settlements. Thus, the subgrade coefficient could be arranged to diminish from the mat center to the edges. In clay, the opposite effect would be expected. In a detailed computer analysis an appropriate variation of subgrade coefficient based on field studies and experience could be included, but it would be too complicated to do this in a normal Winkler analysis by hand. However, the behavior can be approximately taken into account in the method described by Baker and presented in Section 5.5.

The rest of the procedure for a slab design on stiff clay follows that given previously.

7.4.2 Beams

In the case of very long beams, the width B of the beam is usually known fairly closely at a preliminary stage in design. Then, for a beam on sand, with the value of k_{01} measured by a plate bearing test or obtained from Table 7.2, as before, the coefficient k_0 is calculated from equation (7.44) for the beam width B. To utilize the rigid or flexible beam equations, the value of $k = Bk_0$ is, of course, required.

When the beam rests on stiff clay, k_0 is derived from applying equation (7.47) to the value of k_{01} again obtained from a field test or from Table 7.3.

One further problem remains, that of the beam of finite length L resting on either a sand or clay foundation soil. Since, in the case of sands, as discussed in Section 7.2.3, the settlements of footings of constant width are the same for all footing lengths, it follows that once k_0 has been obtained from equation (7.44), it can be applied to a beam on sand of any length.

When the supporting medium is stiff clay, and the beam has width 1 ft and length L in feet, Terzaghi (33) gave an equation for the subgrade reaction coefficient k_{0L}:

$$k_{0L} = k_{01}\frac{L + 0.5}{1.5L} \tag{7.48}$$

This indicates that the coefficient for a very long 1-ft wide beam is 0.67 of that for the 1-ft square area. Then, for the beam of width B, the coefficient varies, as before, inversely with the width:

$$k_0 = \frac{k_{0L}}{B}(1) \qquad (7.49)$$

or

$$k_0 = \frac{2k_{01}}{3B}(1) \qquad (7.50)$$

for the very long beam. In both the cases of a finite and a long beam resting on a stiff clay, the coefficient used in the solutions for deflection, moment, etc., is $k = k_0 B$. It follows from equations (7.49) and (7.50), therefore, that k is independent of the beam width. Here, a long beam is defined as one whose length is such that λL is greater than about 2.25 (35). Another method of arriving at the coefficient modifying k_{01}/B in equation (7.50) was suggested by Broms (5) based on some work by Vesic (35). Broms writes:

$$k_0 = \frac{\alpha k_{01}(1)}{B} \qquad (7.51)$$

in which the value of α is given by Vesic to be

$$\alpha = 0.52 \sqrt[12]{\frac{kB^4}{EI}} \qquad (7.52)$$

It can be seen that α depends on both soil and beam properties; the effects of these were separated out by Broms in the form

$$\alpha = n_1 n_2 \qquad (7.53)$$

The terms n_1 and n_2 can be related to the unconfined compressive strength of the soil and the beam properties, respectively. Because of the twelfth root, n_1 and n_2 have a rather limited range. For stiff clays, Broms obtained n_1 to be about 0.40; in his paper he works out n_2 for typical piles, which will be discussed in Section 7.5, but here we are concerned with beams, for which n_2 also has a value of about 0.4. The combination gives a value for α for long foundation beams of about 0.16, which is considerably smaller than the Terzaghi value of $\frac{2}{3}$, and appears to be too small, considering that the Terzaghi value generally gives rise to calculated deflections higher than those observed.

The discussion to this point has been concerned with the determination of the subgrade reaction coefficient for horizontal beams and slabs; a reaction coefficient is also important in the study of deflections and bending moments induced in piles by horizontal forces. Although the relations for piles are treated in a later chapter, it is convenient to deal with the question of horizontal subgrade reaction coefficients next. They differ in some respects from the vertical coefficients.

7.5 DETERMINATION OF HORIZONTAL SUBGRADE REACTION COEFFICIENTS

When an individual pile or wall is subjected to lateral loads, the movement of the pile or wall is resisted by the pressures developed between the soil and the pile. The soil behavior may also be represented by a Winkler model, as shown in Chapter 8, so that a lateral or horizontal coefficient of subgrade reaction is required for the calculation of bending moments or deflections in the pile. The application of the Winkler model is more dubious in this case because the loads applied to piles or walls are generally sufficiently high to cause deformation of the soil near the ground surface to failure or at least well into the nonlinear range, even though the actual loads are below the failure load for the structure. This is particularly the case for sands and normally consolidated clays, and it may occur even in overconsolidated clays. However, for preliminary calculations, the Winkler model may be employed with horizontal subgrade reaction coefficients determined in the following way.

If the soil response to a beam resting on its surface is that of a linear spring, then we must represent the behavior of the same soil against an *embedded* beam or pile as, in effect, *two* springs, one at the front and the other at the rear of the pile. This ignores the shearing reactions along the sides. It follows that in an isotropic soil the basic subgrade reaction coefficient k_0 to be used in *pile* studies is *twice* the value that would be employed for the same member acting as a *beam* at the surface of the same soil.

For both sands and clays, the elastic modulus does not vary with horizontal distance from the pile, and as a consequence, the subgrade reaction coefficient varies inversely with the width of the pile in both soils [equation (7.47)], as previously discussed. For stiff clay, the coefficient may be assumed to be constant with depth, and the basic value for a 1-ft (0.3 m) square loaded area can be taken to be twice that obtained by vertical loading tests on the ground surface. It can thus be obtained from Table 7.3 times two. However, tests on overconsolidated London clay have shown that the lateral *modulus* was greater than the vertical modulus by a factor of 1.6. In these circumstances, the values from Table 7.3 may be multiplied *again* by 1.5, say. For a pile of width different from 1 ft (0.3 m), therefore, the horizontal subgrade reaction coefficient is again given by equation (7.47), and it follows that the coefficient $k = k_0 B$ to be used in calculations is independent of the pile width.

In the case of long piles, for which $\lambda L > 2.25$, the subgrade reaction coefficient can be obtained as in Section 7.4.2, by using either the Terzaghi or the Vesic-Broms coefficient. For the typical range of pile dimensions, rather than beams, the coefficient n_2 is of the order of 1.00 to 1.20, so that the value of α is about 0.4 to 0.5 for piles. For short piles ($\lambda L < 2.25$), it is suggested that the method of Section 7.4.2 be employed. A more complicated approach to obtaining the coefficient is described by Broms (5). Again, for a completely imbedded pile, the factor of two is required.

When the lateral loads are applied for long times, clays will consolidate and the pile deformations will increase. Analyses of settlements of both axially and laterally loaded single piles and groups of piles by Poulos (25) indicated that, contrary to the behavior of surface footings under load, the long-term displacement of piles in clays is only 1.1 to 1.25 times the immediate movements. Since displacements are closely related to subgrade reaction coefficients, this suggests that the immediate coefficients given here should be reduced by these numbers to arrive at coefficients for long-term displacement and moment calculations.

In sand, the coefficient in the horizontal direction differs from that in the vertical direction, and it also increases with depth. Including the width effect and assuming a linear variation with depth, we can write the horizontal coefficient of subgrade reaction in a sand for a pile of width B in the following way:

$$k_{0h} = \frac{nz}{B} \tag{7.52}$$

where z is the distance from ground surface and n is a constant, called the *constant of horizontal subgrade reaction* by Terzaghi (33), who has given representative values of the constant as a function of the relative density of the soil. Terzaghi did not discuss the provenance of his values. When the constants suggested by Terzaghi have been used in studies of the lateral deflection of piles in sand (28), it appears that the pile displacements are substantially overestimated. Accordingly, in Table 7.4 Terzaghi's original values have been doubled. As before, for sands, there is no length effect to be taken into account, although this aspect is open to some question here as the soil deformations are not plane strain at the pile top. On the basis of full-scale lateral load tests on piles in medium-dense to dense fine sand below the water table, Reese, Cox, and Koop (28) suggested even higher values for

TABLE 7.4

Values of Constant n for a Pile or Beam 1-Ft (0.3 m) Wide Embedded in Sand
in tons/cu ft; Multiply by 0.3 to Get MN/m^3 Values.

Soil Description	Loose	Medium	Dense
Dry or moist sand	14	42	112
Submerged sand	8 (17)	28 (50)	68 (110)

the parameter n. These are shown in parentheses in Table 7.4. No additional factor of two to account for pile-soil reaction needs to be used with the tabular values. In the field, there is no horizontal equivalent to the plate bearing test at the surface, and values of the coefficient of subgrade reaction for sands (or n-values as in Table 7.4) or clays, if they are to be determined experimentally, must be obtained from test piles with suitable instrumentation. For clays, the horizontal load must be applied long enough to ensure that the clay around the pile is fully consolidated. In reality, in a uniform bed of sand (if such exists) the elastic moduli, and therefore the reaction coefficients, probably increase with depth to a power other than one, possibly in the region of one-half. In that case, a different constant will be required to take the place of n. This problem is discussed in more detail in Chapter 9.

In the cases of soft clay and sands, nonlinear behavior predominates, but field tests will give equivalent secant values of the subgrade reaction coefficient for use in bending moment calculations. However, care must be taken in extrapolating to piles of different dimensions because of the nonlinear nature of the real problem. More discussion of these conditions will be presented in a later section.

In many cases in current practice, piles must be designed taking into account *dynamic* lateral loads due to wind, wave, or earthquake action. In these circumstances, the loads are cyclic in nature, and, of course, no consolidation will take place in a clay subgrade during the time of action of the force. Usually, also, as a consequence of the cyclic loading, the soil properties change with the number of load cycles and it may be necessary to estimate both initial and final values of the subgrade coefficients or equivalent soil behavior. There is some evidence that the coefficient of subgrade reaction decreases by a factor of 3 after about 40 cycles of loading in medium-dense sand with little change thereafter. Reese, Cox, and Koop (28) found smaller changes in their tests. The deterioration is a function of relative density, and it is greatest for loose soils.

In the case of pile groups in both sands and clays at working loads, the group behavior must be obtained by considering the interaction between adjacent piles. This interaction depends on the pile spacing, number of piles, and their length. It is considered in some detail in Chapter 10.

When the pile is axially or torsionally loaded, the soil is principally sheared next to the pile, and the relevant coefficient of soil reaction, k_s or k_θ, is controlled by a soil distortional mechanism different from that of the laterally displaced pile or beam. Tables of values indicated by field experience are not yet available, and rational relations for these coefficients in terms of soil modulus are therefore suggested in Chapter 8.

In the case of sheet pile walls, a guide to the selection of suitable coefficients of subgrade reaction has been given by Terzaghi (33), who distinguishes between the two cases of free earth support and fixed earth support. In the former case, the piles are driven to a relatively shallow depth below the dredge line, and, as a consequence, may move laterally under the soil pressures developed by the dredging process. The portion of the pile to penetration depth D below the dredge line thus moves laterally as a unit, and it behaves in a fashion akin to that of a strip footing of width $2D$. For a stiff clay, the coefficient k can be obtained by considering the previous beam case. However, since there is a free soil surface at the dredge line, the subgrade reaction coefficient in the case of the sheet pile wall is less than that for the beam. Repeating equation (7.50), we have for the *vertical* coefficient of subgrade reaction for an infinite strip footing of width B the expression

$$k_0 = \frac{2k_{01}(1)}{3B} \tag{7.53}$$

where k_{01}, as usual, is the coefficient determined for a 1-ft (0.3 m) square footing at the ground surface under vertical loading. If we replace B by $2D$, we would obtain an equivalent uncorrected expression for the sheet pile:

$$k_0 = \frac{2k_{01}(1)}{3(2D)} \tag{7.54}$$

However, because of the presence of the free surface Terzaghi proposed to reduce this value arbitrarily by rewriting the equation as

$$k_0 = \frac{2k_{01}(1)}{3(3D)}$$

or   (7.55)

$$k_0 = \frac{2k_{01}(1)}{9D}$$

so that the final value of horizontal subgrade reaction coefficient for a sheet pile wall driven D into a stiff clay is two-thirds of the value that would be assigned to the vertical coefficient under a beam on the ground surface, of width $2D$.

For a wall driven a distance D below the dredge line into *sand* under the same assumed conditions of free earth support, Terzaghi proposed the following equation for the horizontal reaction coefficient:

$$k_{0w} = \frac{n_w z}{D} \qquad (7.56)$$

in which n_w is a constant, for which selected values are given in Table 7.5.

TABLE 7.5

Values of Subgrade Constant n_w for Vertical Walls Embedded in Sand
in tons/cu ft (MN/m^3).

Soil Description	Loose	Medium	Dense
Dry or moist sand	2.5 (0.8)	8 (2.4)	20 (6)
Submerged sand	1.6 (0.5)	5 (1.5)	13 (3.9)

It is reasonable to assume that these values are also conservative, and that to obtain realistic values of deflection in the elastic range, they should be multiplied by a factor of 2 or greater.

When the piles are driven farther into the soil, the lowest part of the pile does not deflect under the earth pressure when dredging is carried out, and the condition of *fixed* earth support obtains. The deflected shape of the sheet piling below the dredge line is a damped cosine curve with depth, with the deflection taking place toward the free surface near the dredge level, diminishing to zero at some depth, and then, depending on relative soil and pile stiffnesses and loading, occurring in the opposite direction below this level. For this case, the evaluation of the lateral subgrade reaction coefficients has also been discussed by Terzaghi. The coefficients do not seem to have been much used in practice.

7.6 DISCUSSION

It can be seen from the discussion in previous and subsequent sections that for a uniform, homogeneous linearly elastic subgrade subjected to loads applied through foundation beams and slabs, the calculation of displacements, moments, and stresses involves lengthy computations. When layers of material of different properties are involved, only formal solutions are available. Because of this situation, design engineers turned their attention to the Winkler foundation representation, since it offered simpler (although still complicated) mathematical relations. The real foundation material, soil, does not behave under a foundation in a manner that agrees with either model, although in certain special circumstances agreement may be satisfactorily close.

Because the Winkler approach has been so thoroughly examined, and the literature is full of tabulated values of the functions appearing in the associated mathematics, it has become conventional to identify soil behavior with the Winkler theory. The matching is accomplished through selection of the value of the coefficient of subgrade reaction to suit the particular circumstances of the design under consideration. However, as the idealized model for material behavior and basis for comparison is the linearly elastic half-space, with its impossible mathematics, we have seen that the coefficient is not a property of the subgrade alone, as are the elastic parameters E and v, but depends on the behavior of the beam or slab through which the load is applied. Thus, one cannot select a value for the subgrade coefficient by simply carrying out a load test on the soil in the field and using the coefficient or coefficients thus obtained. In addition, although the basic soil model is linearly elastic, with a modulus which may in certain circumstances vary with depth, this is still a faulty approximation and a better, but still idealized, linear characterization would possess a modulus which depends on the final state of stress at each point. Thus, even a subgrade composed of a homogeneously constituted material (for example, a dry uniform medium sand with the same void ratio at all points) will, under gravity and a small applied surface load, possess properties that vary from point to point. It is the practical recognition of these requirements—that the fictitious property, subgrade reaction coefficient, depends on both loading conditions and material type—which has led to the variety, one might almost say the bewildering variety, of recommendations for its determination presented in Terzaghi's paper, which has been extensively followed in the above discussion.

We conclude that the complexity of a subgrade reaction coefficient that depends on the size and stiffness of the loaded area and on the direction of the loading as well as on the soil material is the penalty paid for the simplification of the calculations. The choice of a coefficient or coefficients in a particular set of circumstances is a good example of the necessity for an engineer to develop the magical quality of judgment based on intuition and experience. This situation is common in many branches of engineering, and it arises, as we have seen, because engineers are better at guessing than at mathematics. In the present case, it is indeed fortunate that the moments and stresses in a beam or slab exhibit such a small sensitivity to the value of subgrade modulus selected and that large factors of safety are readily attainable in most civil engineering structures.

If better models of soil behavior could be used, with acceptable computing times for arriving at the design parameters, the soil behavior at a site would require better characterization. That is to say, at the simplest level, we would have to determine Poisson's ratio and Young's modulus for each soil layer present (and in different directions for anisotropic soils), preferably from field tests. At the next higher descriptive level, we would need to know the variation of these properties with effective stress in the soil. The final state would be a complete nonlinear characterization of the soil both for loading and, in general, for unloading paths. Such characterizations are accompanied by material constants to establish the quantitative material properties in each model. The linearly elastic isotropic material has two such constants; the same material with a pressure dependency of Young's modulus would have three (assuming a Poisson's ratio independent of stress); a second-order nonlinear but still elastic behavior five, and irreversible behavior more still. With the inclusion of anisotropy, the problem of property determination reaches horrifying proportions.

Numerical calculations, such as the finite element approach, are useful for establishing the importance of some of these characteristics in research work. In addition, in practice, their use is likely to spread to cases, for example, in which settlements or stresses require calculation for structures whose design has been established by simpler techniques but which possess complicated boundary conditions whose effects must be assessed. In principle, these finite element techniques can be developed to encompass as difficult material behavior and boundary conditions as can be visualized. For practical problems, however, there is little doubt that they will always be limited to relatively simple material characterizations because of the immense difficulty of determining more complicated soil properties by either in-place or laboratory testing. Even with current and anticipated future developments in both field and laboratory testing, these will be confined to better, more complete, or more extensive determinations of the properties required for simple characterization.

At present, therefore, for design purposes, the choice falls between (a) the hazards associated with subgrade reaction coefficient estimation followed by relatively simple computations (which may be done by computer) and (b) the assessment of more fundamental material properties (E and v) from field and laboratory testing, accompanied by finite element calculations applied to the continuum representation of the problem. Because of the cost of the latter approach and its present unsuitability for design processes, it seems that an engineer's choice for some time to come will devolve on the former method, with, however, an increasing use of computers to perform the detailed calculations.

PROBLEMS

1. How would you obtain the allowable bearing capacity of a strip footing to be founded on a sloping soil deposit? Would you use the same factor of safety against failure as for the same footing on the same soil, with a horizontal surface? How would you estimate the allowable pressure for footing widths for which settlement was the criterion?

2. With an adequate site investigation, what do you think the factor of safety should be in the design of (a) an individual footing, (b) a long footing supporting several columns? Explain briefly.

3. Discuss the advantages and disadvantages of different foundations for a two-story school building in a seismic area. Relatively dense fill exists at the site to a depth of 10 ft (3 m), and is underlain by 30 ft (9.1 m) of sand whose standard penetration resistance averages 10 to 15 blows per foot. Shale bedrock occurs at the 40 ft (12.2 m) depth, and water table is 8 to 10 ft (2.4 to 3.0 m) below ground surface. Include questions of economics.

4. For (a) narrow or small individual footing, (b) wide or mat foundations which consideration, settlement, or failure dominates? Is one or the other more important in different soils?

5. For a particular foundation of any size, how do you estimate what the differential settlement is going to be at adjacent columns for (1) a sand subgrade (2) a clay subgrade?

6. At the site of a proposed nuclear power plant in a desert area, a field *geophysical* (seismic) investigation gave the soil profile and properties in the table. Determine, with an explanation, the value of k_0 you would use in the design of a concrete slab about 4 ft (1.22 m) thick, of dimensions 80 × 120 ft (24 × 37 m) with its base at a depth of about 5 ft (1.5 m). Column and other loads will be applied to the slab to give an average pressure of about 4000 psf (192 kN/m²). The water table is below 225 ft (68.6 m).

	Depth, ft (m)	Unit Weight γ, pcf (kN/m^3)	Poisson's Ratio ν	E, psi × 10^5 (MN/m^2 × 10^3)	G, psi × 10^5 (MN/m^2 × 10^3)
Sand	0–15 (0–4.6)	123 (19)	0.25	0.60 (0.41)	0.24 (0.17)
	15–50 (4.6–15.3)	128 (20)	0.31	0.84 (0.58)	0.32 (0.22)
Clay	50–95 (15.3–29.0)	124 (19.5)	0.34	0.99 (0.68)	0.37 (0.26)
	95–125 (29.0–38.1)	124 (19.5)	0.42	1.29 (0.89)	0.45 (0.31)
	125–175 (38.1–53.4)	128 (20)	0.47	1.59 (1.10)	0.54 (0.37)
	175–225 (53.4–68.6)	130 (20.5)	0.46	2.10 (1.45)	0.71 (0.49)

7. Describe with reference to both field and laboratory tests, a number of methods you might use to obtain the coefficient of subgrade reaction k_0 for use in the analysis of a loaded beam or slab on soil. If you decided to employ a two-parameter (springs and membrane) model in the analysis, how would you arrive at the membrane property, $S(F$ or $FL^{-1})$?

8. For a 1 m (3.3 ft) wide, 0.6 m (1.97 ft) deep concrete beam, spanning 8 m (26.2 ft) between columns, how much differential settlement on a Winkler foundation, $k_0 = 50$ MN/m³ (184 pci), is required to cause tensile cracking in the beam at a tensile stress of 4.5 MN/m² (652 psi)? Assume the slope of deflection is zero at the columns and take the value of EI to be 600 MN-m² (87 k-in.²). How does this estimate of settlement gradient for structural damage to occur compare with the usual 1:300 criterion for architectural distress?

9. Figure P7-9 shows the result of a cone penetration test in sand. Use Schmertmann's method to calculate the settlement of a 1 m (3.3 ft) wide footing with its base at a depth of 3 m (9.8 ft), when it is loaded with an average pressure of 150 kN/m² (3130 psf). The water table is at a depth of 1 m (3.3 ft) below ground surface.

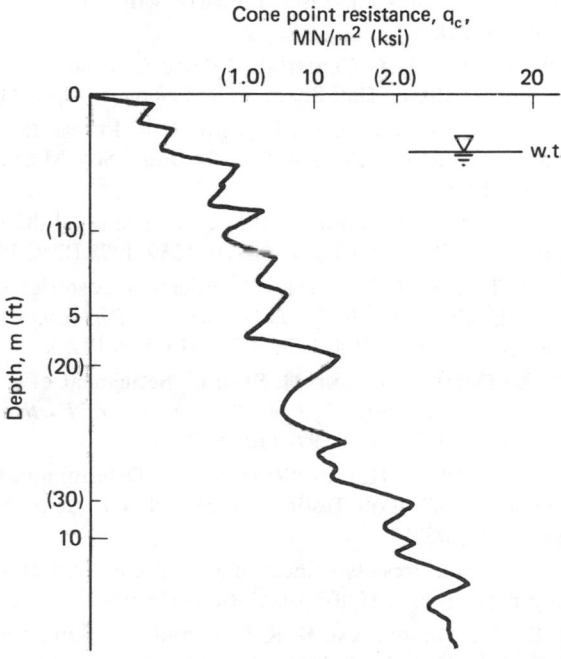

Figure P7-9.

REFERENCES

[1] ALPERSTEIN, R., AND S. A. LEIFER. "Site Investigation with Static Cone Penetrometer," *Proc. ASCE, 102*, GT5, 539–555 (May 1976).

[2] AWOJOBI, A. O., AND R. E. GIBSON. "Plane Strain and Axially Symmetric Problems of a Linearly Non-Homogeneous Elastic Half-Space," *Quart. J. Mechanics and Appl. Mathematics, 26*, 285 (1973).

[3] BAZARAA, A. R. S. "Use of the Standard Penetration Test for Estimating the Settlements of Shallow Foundations on Sand," Ph.D. Thesis, University of Illinois, Urbana, Illinois, 1967.

[4] BJERRUM, L., AND A. EGGESTAD. "Interpretation of Loading Test on Sand," *Proc. Europ. Conf. Soil Mech. and Eng., 1*, 199 (1963).

[5] BROMS, B. "Lateral Resistance of Piles in Cohesive Soils," *Proc. ASCE, 90*, Jour. Soil Mech. Found. Div., SM2, 27 (March 1964).

[6] ———. "Lateral Resistance of Piles in Cohesionless Soils," *Proc. ASCE, 90*, Jour. Soil Mech. Found. Div., SM3 123 (May 1964).

[7] BROWN, P. T., AND R. E. GIBSON. "Surface Settlement of a Deep Elastic Stratum Whose Modulus Increases Linearly with Depth," *Can. Geotech. Journal, 9*, 467 (1972).

[8] CARRIER, W. D., AND J. T. CHRISTIAN. "Rigid Circular Plate Resting on a Non-Homogeneous Elastic Half-Space," *Geotechnique, 23*, 67 (1973).

[9] D'APPOLONIA, D. J., D'APPOLONIA, E., AND R. F. BRISSETTE. "Settlement of Spread Footings on Sand," *Proc. ASCE, 94*, Jour. Soil Mech. Found. Eng., SM3, 735 (May 1968).

[10] D'APPOLONIA, D. J., H. G. POULOS, AND C. C. LADD. "Initial Settlement of Structures in Clay," *Proc. ASCE, 97*, SM10, 1359–1377 (Oct. 1971).

[11] DAVISSON, M. T., AND J. R. SALLEY. "Settlement Histories of Four Large Tanks on Sand," *Proc. ASCE Specialty Conf. on Performance of Earth and Earth-Supported Structures*, Vol. 1, Part 2, 981 (June 1972).

[12] FISCHER, J. A., DETTE, J. T., AND H. SINGH. "Settlement of a Large Mat on Sand," *Proc. ASCE Specialty Conf. on Performance of Earth and Earth-Supported Structures*, Vol. 1, Part 2, 997 (June 1972).

[13] GIBBS, H. J., AND W. G. HOLTZ. "Research on Determining the Density of Sands by Spoon Penetration Testing," *Proc. 4th Int. Conf. Soil Mech. and Found. Eng., 1*, 35 (1957).

[14] GIBSON, R. E. "Some Results Concerning Displacements and Stresses in a Non-Homogeneous Elastic Half-Space," *Geotechnique, 17*, 58 (1967).

[15] GIBSON, R. E., P. T. BROWN, AND K. R. F. ANDREWS. "Some Results Concerning Displacements in a Non-Homogeneous Elastic Layer," *Zeitschr. Ang. Math. u. Phys., 22*, 855 (1971).

[16] GIBSON, R. E., AND G. C. SILLS. "Settlement of a Strip Load on a Non-Homogeneous Orthotropic Incompressible Elastic Half-Space," *Quart. J. Mechanics and Appl. Mathematics, 28*, 233 (1975).

[17] GRANT, R., J. T. CHRISTIAN, AND E. H. VANMARCKE. "Differential Settlement of Buildings," *Proc. ASCE, 100,* GT9, 973–991 (Sept. 1974).

[18] HOLTZ, W. G., AND H. J. GIBBS. "Discussion of Settlement of Spread Footings on Sand," *Proc. ASCE, 95,* Jour. Soil Mech. Found. Eng., SM3, 900 (May 1969).

[19] KOGLER, F. "Discussion of Soil Mechanics Research," *Trans. ASCE, 98,* 299–301 (1933).

[20] MEIGH, A. C., AND I. K. NIXON. "Comparison of In-Situ Tests of Granular Soils," *Proc. 5th Int. Conf. Soil Mech. Found. Eng., 1,* 499 (1961).

[21] MEYERHOF, G. G. "Penetration Tests and Bearing Capacity of Cohesionless Soils," *Proc. ASCE, 82,* Jour. Soil Mech. Found. Div., SM1, 5 (Jan. 1956).

[22] ———. "Shallow Foundations," *Proc. ASCE, 91,* Jour. Soil Mech. Found. Div., SM2, 21–31 (March 1965).

[23] PARRY, R. H. G. "Estimating Foundation Settlements in Sand from Plate Bearing Tests," *Geotechnique, 28,* No. 1, 107–118 (1978).

[24] PECK, R. B., AND A. R. S. BAZARAA. "Discussion of Settlement of Spread Footings on Sand," *Proc. ASCE, 95,* Jour. Soil Mech. Found. Eng., SM3, 905 (May 1969).

[25] POULOS, H. G. "Behavior of Laterally Loaded Piles; II—Pile Groups," *Proc. ASCE, 97,* Jour. Soil Mech. Found. Eng., SM5, 733–751 (May 1971).

[26] ———, AND E. H. DAVIS. *Elastic Solutions for Soil and Rock Mechanics.* New York: John Wiley, 1974.

[27] PRAKASH, S. "Behavior of Pile Groups Subjected to Lateral Loads," Ph.D. Thesis, University of Illinois, Urbana, Illinois, 1962.

[28] REESE, L. C., COX, W. R., AND F. D. KOOP, "Analysis of Laterally-Loaded Piles in Sand," *Proc. 6th Annual Offshore Tech. Conf.,* Paper OTC 2080, 473 (May 1974).

[29] SANGLERAT, G. *The Penetrometer and Soil Exploration.* New York: Elsevier North-Holland, 1972.

[30] SCHMERTMANN, J. H. "Static Cone to Compute Static Settlement Over Sand," *Proc. ASCE, 96,* Jour. Soil Mech. Found. Div., SM3, 1011 (May 1970).

[31] SCHULTZE, E., AND MELZER, K. J. "The Determination of the Density and Modulus of Compressibility of Non-Cohesive Soils by Soundings," *Proc. 6th Int. Conf. Soil Mech. and Found. Eng., 1,* 354 (1965).

[32] SCHULTZE, E., AND E. MENZENBACH. "Standard Penetration Test on Compressibility of Soils," *Proc. 5th Int. Conf. Soil Mech. and Found. Eng., 1,* 527 (1961).

[33] TERZAGHI, K. "Evaluation of Coefficients of Subgrade Reaction," *Geotechnique, 5,* 297 (1955).

[34] ———, AND R. B. PECK. *Soil Mechanics in Engineering Practice.* New York: John Wiley, 1967.

[35] VESIC, A. S. "Bending of Beams Resting on Isotropic Elastic Solid," *Proc. ASCE, 87*, Jour. Eng. Mech. Div., EM2, 35 (April 1961).

[36] ──────. "Slabs on Elastic Subgrade and Winkler's Hypothesis," paper proposed for 8th Int. Conf. Soil Mech. & Found. Engr., Moscow, 1973.

[37] U.K. Dept. of Scientific and Industrial Research, Road Research Laboratory. *Soil Mechanics for Road Engineers.* London: H. M. Stationery Office, 1957.

Pile Analysis: Linearly Elastic Pile and Soil

8

8.1 INTRODUCTION

In the civil engineering field, piles have been a consistently popular topic for study and research. Each year a substantial portion of published papers in geotechnical engineering is still concerned with pile testing and analysis. This is probably the case because, apart from single footings near the ground surface, the pile is the smallest structural unit in foundation engineering on which a field test can be conducted for comparison with analysis. Even so, a loading test on a *fully instrumented* single pile is a formidable task, and a loading test on a group approaches impossibility. The continuing interest in the subject indicates that engineers are still uncomfortable with the analytical techniques available and with their ability to predict the field performance of a pile. In the past decade the development of very large offshore oil drilling structures, including piles whose dimensions are so great as to preclude field test, has generated new concern in field investigation and pile load capacity prediction.

The basic problem of a pile embedded in soil is shown in Figure 8-1 in which various pile top loading conditions are illustrated. We consider first analytical solutions to the problem. Perhaps the most obvious idealization to

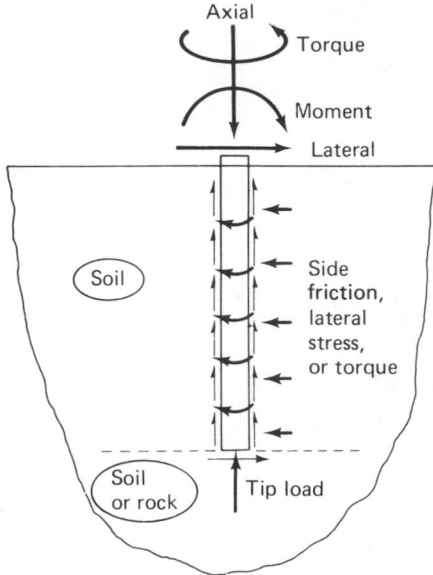

Figure 8-1. Schematic representation of pile and various possible loading conditions.

make is that of considering the pile, which may be a tube, a reinforced concrete rod, or a steel beam of I or H cross section in actuality, to be a linearly elastic rod in a linearly elastic continuum.

Even this superficially simple model presents substantial mathematical difficulties in analysis under any of the loadings of Figure 8-1. Since Mindlin's solution (18) exists for the displacement and stress field in a half-space subjected to a point load below the surface, a model which we might term zeroth-order linearly elastic could represent the load transferred to the medium by the pile as a point load located somewhere in the vicinity of the pile tip. A portion of such a solution, which is Mindlin's point load solution in numerical form, is shown in Figure 8-2(a). The next stage in an approach to realism is to take into account the fact that the load is transferred to the soil from the pile all along its length. A first-order assumption is to represent this by a vertical uniformly distributed line load arranged along the axis of the imaginary pile. A portion of the solution to this problem is shown in Figure 8-2(b) (12). In neither of these solutions does a pile exist; the interactions between an elastic pile and the elastic half-space are omitted. However, we might expect that sufficiently far away from the pile the stresses shown in Figure 8-2(b) would be reasonably close to those which would develop were a pile included. Were Mindlin's solution to be employed in the form of a uniform displacement applied over a circular area below the surface, then the

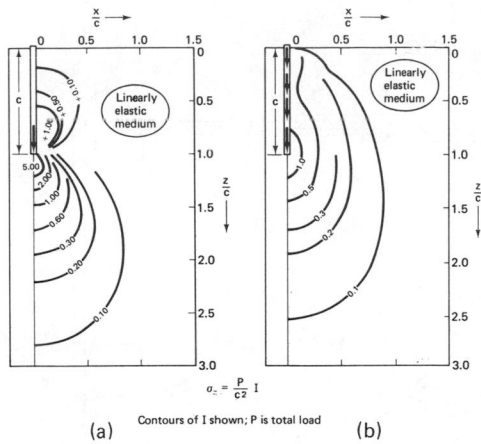

$$\sigma_z = \frac{P}{c^2}\, I$$

(a)   Contours of I shown; P is total load   (b)

Figure 8-2. Analytical continuum representations of pile: (a) pile load applied at tip only (Mindlin case); (b) pile load distributed along length (integrated Mindlin case).

problem of a *rigid* pile in an elastic half-space could be tackled by considering a continuous column of such uniform displacements along the length of the pile.

The next step, however, in the development is to include the pile as an elastic rod. The simplest class of problem for which an exact solution can be obtained for this model is the infinitely long pile subjected to torsion at the top; the solution was obtained by Westmann and is reported in Scott (27). The case of the finite pile under torsion has not apparently been solved.

Next in order of difficulty is the elastic rod, infinite or finite, under axial loading. This problem is difficult if a Poisson's ratio for the pile is included. By setting Poisson's ratio equal to zero, so that the axially loaded pile suffers no distention, Muki and Sternberg were able to solve the problem of the finite pile formally (19). Their solution was evaluated numerically and is shown in Figure 8-3. Unfortunately, results were not computed for values of the ratio of pile to soil shear moduli which are applicable to soil problems (range from 50 to 1000, say).

Finally, there is the case of the pile, finite or infinite, subjected to lateral or moment loading at the top. This problem, in terms of an elastic pile in an elastic medium, does not appear to have been attempted analytically.

Next we come to numerical solutions of pile problems. The most systematic attack has been developed by Poulos, who has used the composite finite element technique described in Chapter 3 to study rigid and compressible piles and pile groups under axial loading (17 and 25), single piles in torsion (24), and single piles and groups under lateral or moment loading at the top (13 and 23). In the case of axial loading, Poulos has extended the

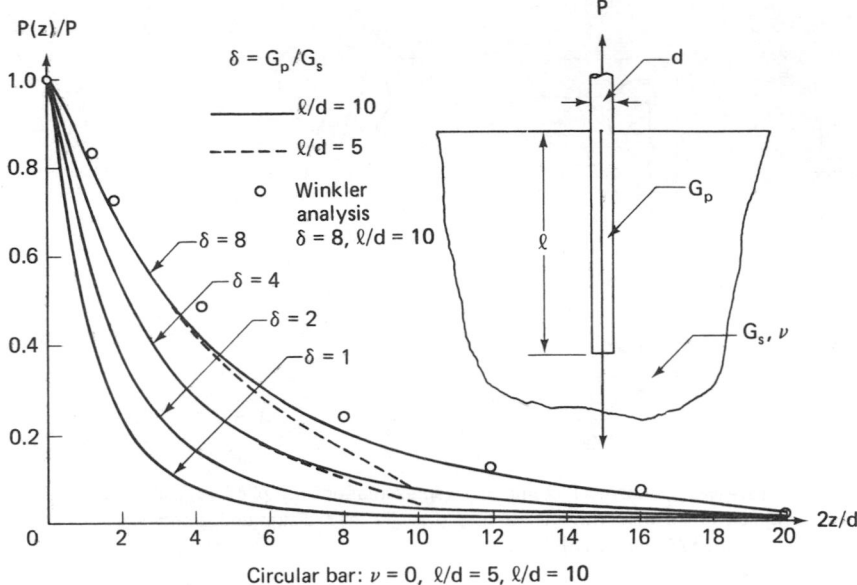

Figure 8-3. Comparison of analytical solution for axially loaded pile in elastic continuum (19) with Winkler representation.

method to account for yielding of the soil along the pile, with consequent slipping of the pile with respect to the soil. In appropriate sections of this and the next chapter we will return to Poulos' results and treat them in more detail. Several investigators have applied finite element techniques (3 and 14) to pile analysis; their results are less general in their application.

It is apparent that a pile under lateral loading is similar in its behavior to a beam loaded transversely, and, consequently, we might envisage representing the elastic response of the soil around the pile by a Winkler or two-parameter model as we have done previously for beams. This technique was employed by Matlock and Reese (16) who developed a series of dimensionless solutions for cases of lateral and moment loads on piles embedded in Winkler foundation soils whose coefficient of subgrade reaction varied in different ways along the pile. These solutions will also be referred to later. Perhaps less obvious but equally valid is the identification of a Winkler model with soil response when the pile loading is axial or torsional. This view has been taken by Scott (26) and Murff (20), has the advantage of mathematical simplicity, and will be considered in the detail in the following sections.

8.2 AXIAL AND TORSIONAL PILE LOADING

It is worth looking at the results of carefully performed field tests on piles to establish the basic aspects of pile-soil interaction which our mathematical models will be required to represent. Suitable tests were carried out at the Arkansas River (2 and 15) test site as part of a U.S. Corps of Engineers program to determine deflection and bearing capacities of proposed pile-supported lock structures. Although various pile types were employed, we will examine the behavior of a 16-in. (0.41 m) diameter, 0.5-in. (13 mm) thick wall, 55 ft (16.8 m) long steel pipe pile driven in a dense fine sand below the water table and subjected to axial loading. The pile was instrumented with strain gauges at various intervals along its length; with them, the axial load in the pile at any section could be determined. In addition, rods were attached to the pile at different depths and extended to the surface in a casing so that they were free of soil friction. These strain rods enabled making accurate measurements of the pile's vertical displacement at different stations along the pile. Figure 8-4 shows the pile test results. Alongside Figure 8-4(b) the small sketch shows the location of strain gauges and rods along the pile (15).

The pile was loaded and unloaded in increments. First, the load was brought up to 50 tons (445 kN), and the pile top deflection and strains were

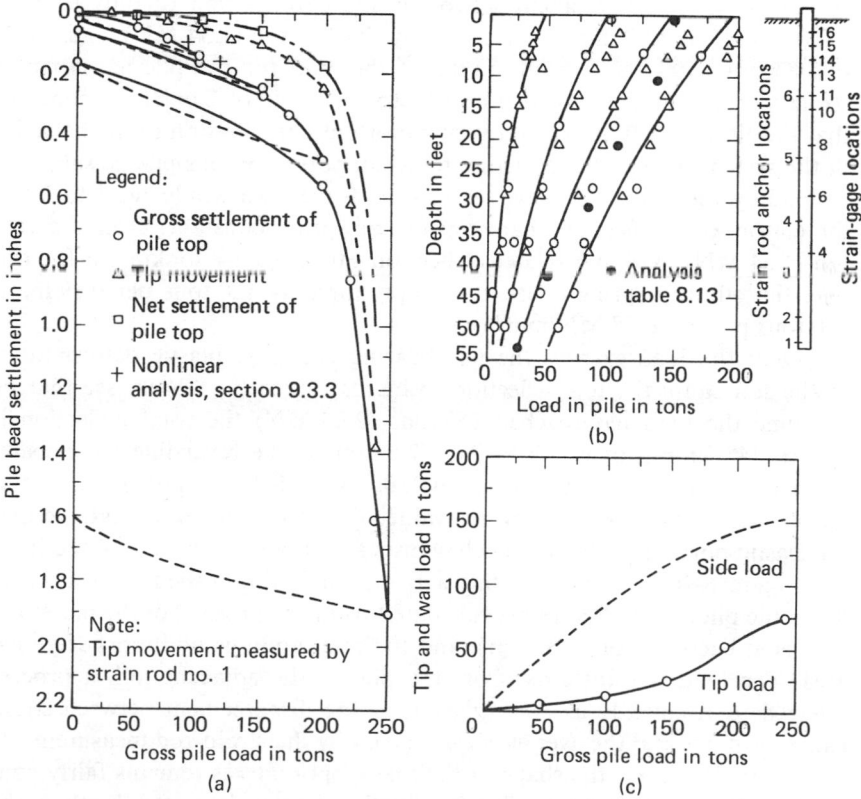

Figure 8-4. Arkansas River (15) field tests of axially loaded pile: (a) pile movement versus load; (b) load distribution in pile; (c) gross load versus tip and wall load.

recorded. It was then unloaded to zero and the rebounds were measured. Next, it was loaded to 100 tons (890 kN), subsequently unloaded, and so on. The resulting load-deflection curve is shown in Figure 8-4(a) in which a number of interesting features is apparent. First, the ultimate bearing capacity of the pile is seen to be slightly less than 200 tons (1.78 MN), say about 185 to 190 tons (1.7 MN), at a top displacement of about 0.4 in. (10 mm). Beyond this value displacement increases rapidly, but stably, with increase of load. Upon unloading, some reversible (elastic) behavior of the pile-soil system is evident. For example, at the beginning of the loading cycle that was taken to 150 tons (1.34 MN), the net top displacement is about 0.03 in. (0.8 mm), which is a residual settlement left by the two previous cycles of load to and from 50 and 100 tons. When 150 tons are resting on the pile, the top displacement is about 0.25 in. (6.3 mm), so that the load has caused a displacement of $(0.25 - 0.03) = 0.22$ in. (5.6 mm). On unloading, the net pile displacement at zero load is about 0.06 in. (1.5 mm), and we see that the elastic part of the behavior is therefore $(0.25 - 0.06) = 0.19$ in. (4.8 mm), while a permanent set, or plastic deformation of only $(0.06 - 0.03) = 0.03$ in. (0.8 mm), was caused by the 150-ton (1.34 MN) load. Since this encompasses the working range of the load for the pile as part of a structure (ultimate load divided by a factor of safety of 2 or 3), we conclude that the elastic portion makes a very important contribution to the behavior of the system. Should the structure to be supported on this pile be subjected to load cycling by wind, wave, or seismic forces, we would need to know the elastic load-deflection characteristic, or spring constant. Over the 150-ton (1.34 MN) load and related deflection range we are looking at for this pile, the pile top spring constant is approximately 150 tons per 0.19 in. = 790 tons per in. (0.28 MN/mm).

Once the load exceeds the pile bearing capacity, plastic deformations in the soil cause the top deflection to increase greatly; we can see that by the time the load has reached 250 tons (2.23 MN) the total deflection is 1.90 in. (48.3 mm), of which 0.30 in. (7.6 mm) is the reversible component. At this level, the top spring constant is $250/0.30 = 830$ tons per in. (0.29 MN/mm) which is close to the previous value. We could, therefore, assume that the elastic portion of the pile's behavior is sensibly independent of the load.

Figure 8-4(b) shows how this active load is diminished with distance down the pile. There figures are calculated from the product of the measured strains at each section of the pile and the pile modulus and cross-sectional area. The load is transferred from the pile to the adjacent soil, a process referred to frequently as *load take-out*. Some disagreement between strain gauge readings and the average strains given by the strain-rod measurements is evident. However, the shape of the load-depth curves remains fairly constant as the load increases. The tip load increases less rapidly than the applied load in the initial stages of loading and more rapidly as failure is approached, as shown by Figure 8-4(c). There is a question about the *shape* of the load-depth curves in Figure 8-4(b), but we will defer discussion of that point until later. Were we interested in the shearing load or stresses along the pile, they would be obtained by calculating or measuring the slope of the load-depth curves in Figure 8-4(b) since the slope indicates the rate of load take-out with distance along the pile. The peak stress in the steel pile is developed at the top, and at a load of 200 tons (1.78 MN) it is about 17,000 psi (117 MN/m²), considerably short of the crushing strength of the material. Thus, usually, the strength of the pile is not a consideration in pile loading. Care must, however, be taken during driving, since use of an inappropriately heavy hammer can cause yield stresses and damage in the pile.

Although the technique appears to be potentially useful and convenient, few torsional pile tests have been done; only one on an uninstrumented pile is reported in the literature (29). The results are shown in Figure 8-5 and will not be discussed further except to remark that they are similar in appearance

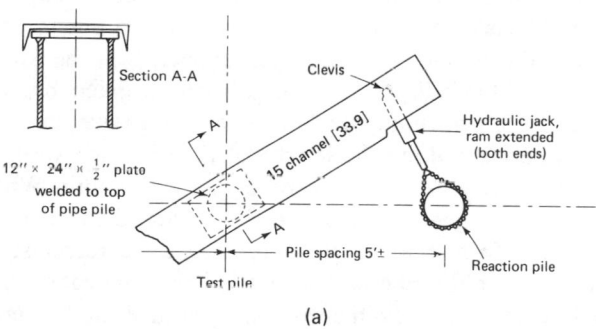

(a)

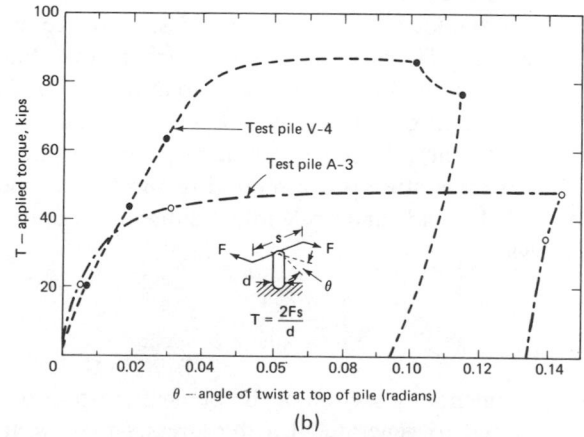

(b)

Figure 8-5. Torsional pile tests (29): (a) test setup; (b) test results.

to the axial pile top load-deflection curves of Figure 8-4(a). The mechanisms of pile-soil interaction that develop are qualitatively similar in the two cases, and the analysis can be handled identically.

We proceed to the analysis of axial and torsional pile-soil interaction using a Winkler representation of the soil's response.

The analysis of the behavior of a single pile under vertical or torsional loads has aspects in common with the deflection of beams supported by soil and loaded laterally. When the pile load is small, the pile and soil displace more or less linearly, and the diminution of load in the pile with depth depends on the relative stiffness of the pile and soil. As the load is increased, the shearing strength of the soil at the pile-soil interface will be exceeded at some point along the length of the pile; from this stage on the displacement or settlement of the top of the pile will increase more rapidly, as slip occurs. Further increase in the load extends the zone where the soil is yielding or slipping next to the pile until yield is eventually reached everywhere along the pile and at its tip, at the maximum or ultimate pile load. We begin by considering only the elastic portion of the response of a pile to an axial load.

In this case, the hypothetical springs connected to the pile reflect the vertical response of the adjacent soil to the load applied by each vertical section of pile, as shown in Figure 8-6. The springs are considered to be distributed continuously along the length of the pile; the spring stiffnesses may be constant or they may vary in some way with depth. Whatever their variation, there will, in general, be one spring attached to the pile base with a stiffness different from the others to represent the response of the soil around the pile tip. If the pile is driven until the tip reaches a very hard stratum or bedrock, this spring will be extremely stiff. With this representation of the foundation, we can now take up typical models of pile behavior.

8.2.1 Rigid pile, axial load

As in the case of the loaded beam, the simplest model to consider is the rigid, axially loaded pile. The analysis is particularly simple, since the entire pile displaces vertically a uniform distance w along its length. The displacement under load, P, applied at the pile top generates shearing stresses, τ, in the soil adjacent to the pile and a pressure, p, at the pile tip. We will assume that both of these stresses are related to the displacement w of the pile relative to the soil by side and tip Winkler subgrade reaction coefficients k_s and k_t as follows:

$$\tau = k_s w \tag{8.1}$$

$$p = k_t w \tag{8.2}$$

Vertical displacement of a section of the pile with respect to the adjacent soil would be expected to generate shearing stresses that would vary with

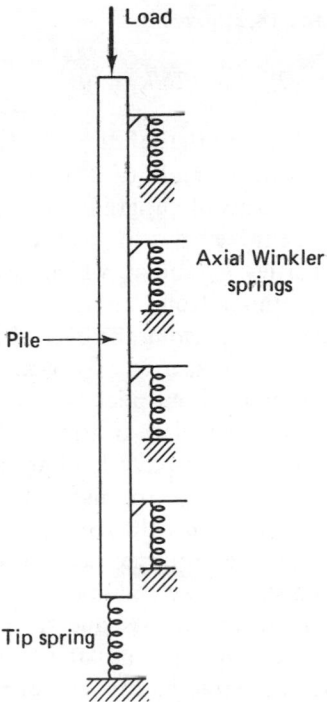

Figure 8-6. Winkler representation of soil response to axially loaded pile.

depth, because of the complicated response of a half-space to the embedded load. However, to retain simplicity in analysis, we will assume that k_s is initially constant along the length of the pile, to represent interaction with the infinite elastic half-space. Because of the stress conditions generated in the half-space by the stresses along the pile surface, the response of the material below the tip of the pile will also be complicated and depend on the pile geometry. Again, here, a constant value of k_t will be assumed.

The force P_s developed along the side of a circular pile of radius a by the pile-soil shearing stresses is given by

$$P_s = 2\pi a l \tau \tag{8.3}$$

and the force acting on the base P_t is

$$P_t = \pi a^2 p \tag{8.4}$$

where l is the pile length.

Thus, the force acting on the pile is the sum of these:

$$P = \pi(2al\tau + a^2 p) \tag{8.5}$$

or using relations (8.1) and (8.2) gives

$$P = \pi w(2alk_s + a^2 k_t) \tag{8.6}$$

This result is in a state akin to the solutions for loaded beams on Winkler subgrades; not much use can be made of it until the material constants k_s and k_t can be identified in terms of subgrade material properties. As in the previous problem, there are three ways to go. We can try to relate k_s and k_t to the soil's elastic properties E_s and v_s which can be obtained, however crudely, by well-known methods from laboratory or field tests on the soil, or we can devise field experiments along the lines of the surface plate loading test, by which k_s and k_t can be measured at a particular site. Lastly, we can obtain the necessary data from actual pile tests. The difficulties in devising an equivalent to the 1-ft diameter plate bearing test look fairly formidable. It will be recalled that the use of a plate loading test at ground surface to determine the Winkler constant k_{01} for use with a beam is not perfect. Corrections must be applied to adjust the measured value to account for the different width and length of the proposed beam as compared to the loading plate. In the present circumstances we would require a test apparatus capable of measuring the vertical displacements due to a vertical shearing stress applied around the periphery of, say, a 1-ft (0.3 m) vertical section of a borehole. Preferably, the test would be carried out at a range of depths over the proposed pile length. At present, there is no standard technique of doing this. The direct shear test is frequently employed to give values for the ultimate shearing resistance of a soil; these are then used in estimates of the ultimate bearing capacity of a pile to be embedded in the soil. Were load and unload tests to be performed in that apparatus, or even, preferably, in the simple shear equipment, the necessary soil property to be used in the assessment of k_s could be obtained as a function of effective stress. In connection with ultimate capacity estimates, Coyle and Sulaiman (8) constructed an apparatus in which a section of model pile passed through a triaxial test cylinder in the axial direction. With it, a controlled lateral soil-pile pressure could be applied and the pile loaded until it slipped. A modulus or subgrade reaction coefficient k_s would also be obtainable from a modification of such a test. These tests are not in standard use, however, and we might therefore consider falling back on the method of relating k_s and k_t to the usual linearly elastic material properties.

At first sight, this does not look too promising, because the advantage of the present analysis lies in its relative simplicity, and yet strictly the identification of the Winkler constants with the elastic properties requires the use of the Mindlin solution to the problem of a point load applied below the surface of an elastic half-space (18). Mindlin's equations are long and their integration to give results for loads applied over circular or peripheral areas is a task well worth avoiding. From the physics of the problem, however, we would expect that the deflection due to a particular load applied over an area of a particular shape and size would diminish as the distance from the surface to the loaded area increases. Thus, even in a homogeneous, linearly elastic half-space the spring constant representing the ratio of load to deflection is not constant with depth. Poulos and Davis (25) have examined the behavior of a rigid pile in an elastic medium through the use of a composite finite element method involving a numerical integration of Mindlin's equations, the result of which is not explicitly available. We will return to their solution after we have considered the determination of k_t.

Suppose we limit ourselves to the consideration of a uniformly loaded circular area of radius a. If it is at the surface, integration of the usual Boussinesq point load equation gives a result for the vertical deflection of the center of the area, w_s:

$$w_s = \frac{2(1 - v_s^2)pa}{E_s} \tag{8.7}$$

where p is the intensity of loading and v_s and E_s are the usual elastic constants for the soil medium. The average deflection is less than this. Another solution which is fairly easy to obtain is available when the loaded area is in the *interior* of an infinite linearly elastic soild. The problem was solved by Kelvin for a point load, and the result may be integrated to give the deflection w_i at the center of a uniformly loaded circular area again:

$$w_i = \frac{(3 - 4v_s)(1 + v_s)pa}{4E_s(1 - v_s)} \tag{8.8}$$

The ratio of the interior to the surface central displacements for the same load and radius of loaded area is obtained from equations (8.7) and (8.8) to be

$$\frac{w_i}{w_s} = \frac{(3 - 4v_s)}{8(1 - v_s)^2} \tag{8.9}$$

For values of v_s ranging from 0 to $\frac{1}{2}$, equation (8.9) shows that the ratio w_i/w_s varies from $\frac{3}{8}$ to $\frac{1}{2}$. On common-sense grounds, we would expect the interior displacement to be about one-half of that at the surface for the same load.

In other words, the *stiffness* of the infinite elastic region under these loading conditions is only about twice as great as that of the semi-infinite region.[1] Looking at the problem of determining k_t, first, we observe that even if by some means we could eliminate the load applied by the pile tip to the elastic half-space, the material below the tip would still deflect down-

---

[1] As usual, we might expect the stiffness to increase from the surface value to the infinite value at a depth of a few loaded-area diameters.

ward because of the displacements induced by the shearing stresses along the side of the pile. Thus, the soil response below the tip would, in effect, be softer than would be indicated by the use of equation (8.8). We will assume, in fact, that the surface expression may be used, equation (8.7). From it we derive

$$k_t = \frac{E_s}{2(1 - v_s^2)a} = \frac{G_s}{(1 - v_s)a} \tag{8.10}$$

To get the spring constant along the side of the pile, we would need the theoretical solution to the displacement at the load, which is produced by a uniform shear loading on the periphery of a short circular cylinder of radius a. As mentioned above, this requires some work on Mindlin's equations; however, we might guess that the final result would have the general form of equation (8.10). In addition, because the loading is closer to a line load around a ring, it might be reasonable to expect that the deflections would be greater, and thus the spring constant smaller, than they would be were the same load distributed over a circular area. Accordingly, we guess that a representation of k_s might be

$$k_s = \frac{G_s}{4(1 - v_s)a} \tag{8.11}$$

We will see shortly how good a guess this is.

With those values for the subgrade reaction coefficients it is possible to calculate the forces acting on the rigid pile. The side force is P_s, from equation (8.3),

$$P_s = \frac{\pi l G_s}{2(1 - v_s)} w \tag{8.12}$$

by substitution from equation (8.11). At the tip of the pile the tip force P_t, from equation (8.4), is

$$P_t = \frac{\pi a G_s}{(1 - v_s)} w \tag{8.13}$$

using equation (8.10) since the displacement is the same all along the pile.

This assumes that the pile base diameter is the same as that of the pile shaft. On occasion, especially with shorter piles or piers, enlarged bases are used. If this is the case, a in equation (8.13) can be replaced with the larger radius and the analysis will proceed in the same way. Summing the forces from equations (8.12) and (8.13) gives the total load P in the pile:

$$P = P_s + P_t = \frac{\pi G_s}{2(1 - v_s)}(l + 2a)w \tag{8.14}$$

from which the displacement can be expressed

$$w = \frac{2(1 - v_s)P}{\pi G_s(l + 2a)} \tag{8.15}$$

In their paper, Poulos and Davis (25) express displacement in terms of the dimensionless parameter wE_sd/P, where d is the pile diameter. Rewriting equation (8.15) in this form gives

$$\frac{wE_sd}{P} = \frac{4(1 - v_s^2)}{\pi\left(\dfrac{l}{d} + 1\right)} \tag{8.16}$$

Another aspect of the pile behavior of interest is the ratio of the load taken by the tip to the total load applied to the pile top. This can be obtained from equations (8.13) and (8.14) in the form

$$\frac{P_t}{P} = \frac{1}{\left(\dfrac{l}{d} + 1\right)} \tag{8.17}$$

Now equations (8.16) and (8.17) can be compared with the results obtained from the more complicated analysis of Poulos and Davis to see if the selected values of k_s and k_t are reasonable. Figures 8-7 and 8-8 show portions of

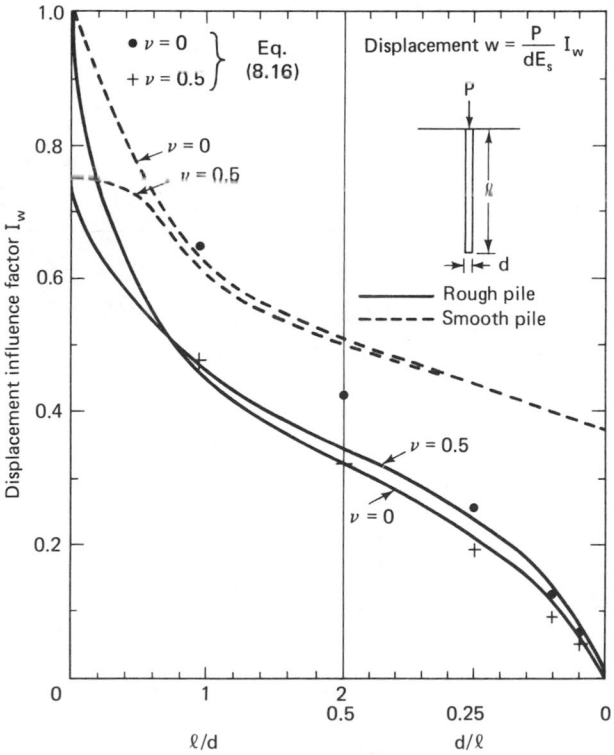

Figure 8-7. Displacement of rigid pile (25).

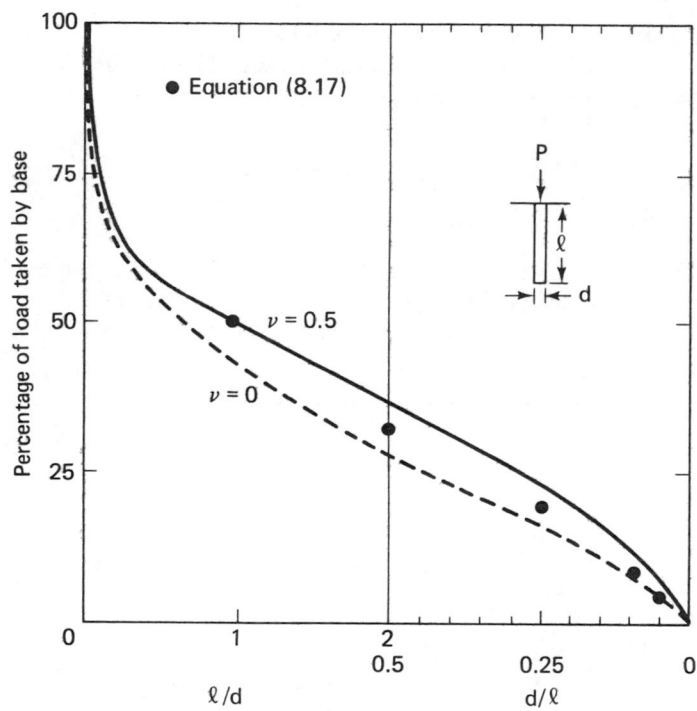

Figure 8-8. Proportion of total load taken by rigid pile tip (25).

diagrams from that analysis in which the exact and approximate results are both plotted.

It can be seen that both approximate dimensionless quantities are in the right range but that they diverge from the better solution at small ratios of l/d. This is to be expected, since the stress field around a short pier is complex, and only for longer piles in proportion to their diameter could we anticipate that an assumption of a uniform subgrade coefficient might be reasonable. For longer piles, the value taken for k_t is relatively unimportant and the tip load of the pile has less effect on its performance.

In the deflection, the effect of Poisson's ratio in the approximate analysis is wrong, since an increase in it diminishes the displacements, whereas in the more correct solution the displacement increases with Poisson's ratio. However, the total effect is relatively small. We might rationalize this result by considering that the Poisson's ratio was probably taken into account more or less correctly in determining k_t, but most likely it plays a less prominent part than we have assigned to it in establishing the k_s relation. The latter could have been expressed in terms of G_s alone.

In general, for longer l/d ratios, say greater than 10, which apply to a large proportion of piles, the approximate analysis appears to supply an adequate estimate of both displacement and tip load in this model. In particular, the small proportion of total load taken by the tip of even a rigid pile, as found by Poulos and Davis, is confirmed. The force in this case can be seen to decrease linearly down the pile.

The model as demonstrated so far has been intended to represent the behavior of a soil whose elastic properties are relatively uniform with depth; this, as usual, would be the case in an overconsolidated clay. In sands, however, the elastic modulus increases with depth, possibly linearly, or in the form of a square root function. Either of these cases can be readily accounted for with this technique. For example, assuming a constant value of Poisson's ratio, a linear function for G_s (or E_s),

$$G_s = G_{s1}z = \left[\frac{E_{s1}z}{2(1 + v_s)}\right] \tag{8.18}$$

and the same relations for k_s and k_t as before, we obtain for the dimensionless settlement of the pile

$$\frac{wE_s(l)d}{P} = \frac{8(1 - v_s^2)}{\pi\left(\dfrac{l}{d} + 2\right)} \tag{8.19}$$

in which $E_s(l)$ is the value of the soil's Young's modulus at the base of the pile. The distribution of force in the pile diminishes parabolically with depth, and, for this case, the tip force is a larger proportion of the applied load. A square root variation of G_s with depth is handled similarly.

There is another physical approach to obtaining the coefficient k_s along the side of the pile. The method to be described will come in handy later. We can consider a horizontal slice of soil of unit thickness, as shown in Figure 8-9, containing the pile as a rigid disk of radius a. To make the conditions definite, we will consider that the soil is influenced by the pile only out to a radius b, at which the displacement is zero. Under axial pile loading, the representative disk displaces in the z-direction. All similar soil and disk slices both above and below the slice in question behave in like fashion. We assume that the deformation of the soil occurs in simple shear,

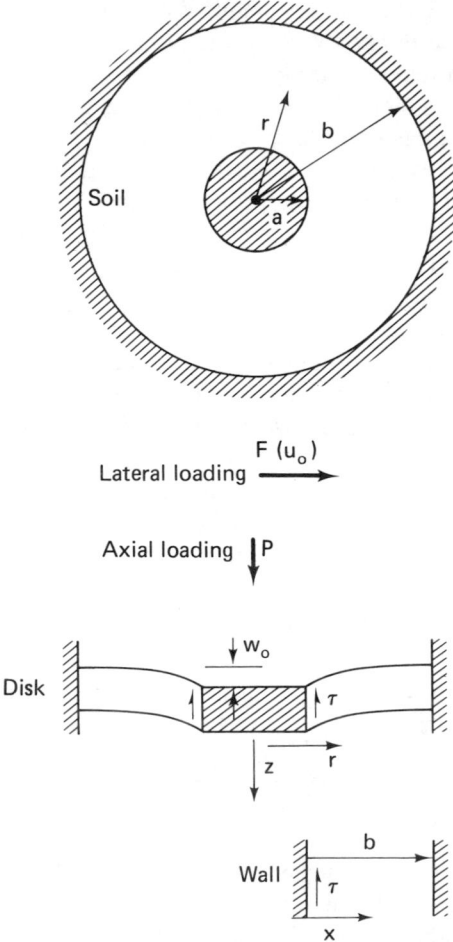

Figure 8-9. Pile section displacement as rigid disk in soil annulus.

so that the equation for vertical displacement is

$$\frac{d^2w}{dr^2} + \frac{1}{r}\frac{dw}{dr} = 0 \tag{8.20}$$

When the displacement of the disk is w_0, the solution to equation (8.20) is

$$w = w_0\left(1 - \frac{\ln\dfrac{r}{a}}{\ln\dfrac{b}{a}}\right) \tag{8.21}$$

From the displacement, the shearing stress at the pile-soil interface is found to be

$$\tau = \frac{Gw_0}{a\ln\dfrac{b}{a}} \tag{8.22}$$

so that the subgrade reaction coefficient k_s is given:

$$k_s = \frac{G}{a\ln\dfrac{b}{a}} \tag{8.23}$$

Comparing this result with equation (8.52), which is later found to give a good approximation to a continuum solution, we find that

$$\frac{b}{a} = 50 \tag{8.24}$$

approximately. Consequently, it appears that the distance between vertically loaded piles for no interaction to take place is about 25 diameters. This is larger than is usually thought to be the case. However, if the figure of 10 diameters, which is sometimes referred to, is employed, then from equation (8.23) we get

$$k_s = \frac{G}{3a} \tag{8.25}$$

nearly, and this value appears to be too large in comparison with the soil continuum results of the next section. Some practical confirmation of the value of 25 diameters appears later when the theoretical behavior of pile groups is compared with field tests.

Occasionally a vertical load may be applied to a wall, which may be considered rigid, embedded in soil. The incompressible pile theory can be used for this plane strain case by taking a unit length (say, 1 ft) of the wall.

The peripheral distance is then the two units of length in contact with the soil. In this case, it seems likely that different, probably lower, values of k_s and k_t should be employed because of the influence of adjacent wall sections on the deformation of the soil. No information appears to be available to aid in the selection of these values, but some guide can be obtained from an analysis along the lines of that given for the axially symmetric case of Figure 8-9. In the right side of the lower part of the figure the distance a now has no meaning, and x is the coordinate axis shown. The dimension b is no longer a radius, but it retains the meaning of the distance at which no vertical displacements occur as a result of the wall movement. With simple shear in the soil, its displacement is now parabolic in shape and the shearing stress, τ, at the wall can be expressed

$$\tau = \frac{2Gw_0}{b} \tag{8.26}$$

so that, in this case,

$$k_s = \frac{2G}{b} \tag{8.27}$$

Computation of k_s depends on an assessment of b, for which, at present, only a guess can be made. In practice, it will probably vary with the depth of the wall, but a distance of about 30 ft (10 m) may be tried.

The reasonably good correspondence of the rigid pile model with a relatively exact solution encourages us to proceed to the analysis of an axially compressible pile along similar lines.

8.2.2 Compressible pile, axial load

As a result of pile compressibility, the force in the pile varies along its length in a fashion different from the previous case but with the decrement in force across a particular vertical section being taken out as before by the soil, as represented by a Winkler spring.

We consider the vertical displacement of the pile in the z-direction to be w. The vertical strain ϵ is therefore, as usual dw/dz, and the stress in the pile at a horizontal cross section is $E\epsilon$ or $E(dw/dz)$; the force F is $EA(dw/dz)$, when A is the cross-sectional area of the pile. For a solid pile, A is the gross cross-sectional area. When a hollow steel tube pile is used, A will be, of course, the cross-sectional area of the pile wall. Later developments in this chapter assume the pile to be solid, but the alterations in the equations to take into account hollow piles, or piles with enlarged bases, are not difficult to make. At a section distant dz, farther down the pile, the force in the pile is $F - dF/dz \cdot dz$. Thus, the difference in the two forces across the section is $dF/dz \cdot dz$. From equilibrium, this must equal the force taken by the soil spring at this section. If, as before, we define the spring constant k of the soil as arising from a stress divided by a displacement, we have

$$\tau = k_s w \tag{8.28}$$

where τ is the vertical shearing stress acting over a circumferential section of the pile dz long, and k_s is the spring constant applying to the soil alongside the pile. The total force F_s on this area is

$$F_s = \tau S\,dz = Sk_s w\,dz \tag{8.29}$$

where S is the pile perimeter. Thus, equating the vertical forces in the pile and soil at the section, we have

$$\frac{dF}{dz} = Sk_s w \tag{8.30}$$

But

$$\frac{dF}{dz} = \frac{d}{dz}\left(EA\frac{dw}{dz}\right) \quad \text{or} \quad EA\frac{d^2w}{dz^2} \tag{8.31}$$

if the pile is uniform along its length, so that, by substituting in equation (8.30) we get the equation

$$EA\frac{d^2w}{dz^2} - Sk_s w = 0 \tag{8.32}$$

or

$$\frac{d^2w}{dz^2} - \lambda^2 w = 0 \tag{8.33}$$

where

$$\lambda^2 = \frac{Sk_s}{EA} \tag{8.34}$$

Equation (8.33), with the appropriate boundary conditions of either displacement or force, describes the behavior of the vertically loaded pile. In dimensionless form, it was used in the numerical examples of Chapter 2. Its solution is

$$w = C_1 e^{-\lambda z} + C_2 e^{\lambda z} \tag{8.35}$$

At the top of the pile, the force applied is P, the axial load. From equation (8.35) the axial load, F, at any section is

$$F = -EA\frac{dw}{dz} = \lambda EA(C_1 e^{-\lambda z} - C_2 e^{\lambda z}) \tag{8.36}$$

At the top, $z = 0$, so that

$$P = \lambda EA(C_1 - C_2)$$

and thus

$$C_1 = \frac{P}{\lambda EA} + C_2 \tag{8.37}$$

At the pile tip, $z = l$, the simplest condition which can apply is zero displacement, and from equation (8.35) we have, for this case,

$$0 = C_1 e^{-\lambda l} + C_2 e^{\lambda l} \tag{8.38}$$

Substituting in equations (8.37) and (8.38), we get

$$C_1 = \frac{P}{\lambda EA} \frac{1}{(1 + e^{-2\lambda l})} \tag{8.39}$$

$$C_2 = -\frac{P}{\lambda EA} \frac{1}{(1 + e^{2\lambda l})} \tag{8.40}$$

An alternative condition, more commonly encountered in practice, at the pile tip permits vertical displacements to occur. In this case, we will require that the force at the pile tip be P_t, so that we get from equation (8.36)

$$P_t = \lambda EA(C_1 e^{-\lambda l} - C_2 e^{\lambda l}) \tag{8.41}$$

for the base. Substituting equation (8.37) in this equation and rearranging, we find eventually that

$$C_1 = \frac{P}{\lambda EA} \left(\frac{\frac{P_t}{P} - e^{\lambda l}}{e^{-\lambda l} - e^{\lambda l}} \right) \tag{8.42}$$

$$C_2 = \frac{P}{\lambda EA} \left(\frac{\frac{P_t}{P} - e^{-\lambda l}}{e^{-\lambda l} - e^{\lambda l}} \right) \tag{8.43}$$

Now we can consider that the deflection of the pile at the tip w_l results from a uniform pressure p applied over the area of the base, which is also supported by a Winkler subgrade of spring constant k_t, so that

$$p = k_t w_l \tag{8.44}$$

or

$$P_t = pA = Ak_t w_l = Ak_t(C_1 e^{-\lambda l} + C_2 e^{\lambda l}) \tag{8.45}$$

using equation (8.35). The term A representing the area of the pile base in this equation can be equal to or larger than the pile area, so that the effect of piles with bulbous tips can be taken into account. Also, for a steel tube pile with closed end, this A is not the same as that used in EA. However, more usually the area of the tip is the same as the pile, and this will be used subsequently. Including equation (8.45) as an expression for P_t in equations (8.42) and (8.43) gives eventually

$$C_1 = \frac{PD_1}{\lambda EA} \tag{8.46}$$

$$C_2 = \frac{PD_2}{\lambda EA} \tag{8.47}$$

where

$$D_1 = \frac{1}{1 + \left(\dfrac{1 - \dfrac{\lambda E}{k_t}}{1 + \dfrac{\lambda E}{k_t}}\right)e^{-2\lambda l}} \tag{8.48}$$

$$D_2 = \frac{1}{1 + \left(\dfrac{1 + \dfrac{\lambda E}{k_t}}{1 - \dfrac{\lambda E}{k_t}}\right)e^{2\lambda l}} \tag{8.49}$$

By substituting from equation (8.34), the term $\lambda E/k_t$ can be represented as

$$\frac{\lambda E}{k_t} = \frac{1}{k_t}\sqrt{\frac{S E k_s}{A}} \tag{8.50}$$

or, for a circular pile,

$$\frac{\lambda E}{k_t} = \frac{2}{k_t}\sqrt{\frac{E k_s}{d}} \tag{8.51}$$

where d is the diameter. A generalized version of these results for different tip conditions is given in Tables 8.11 and 8.12.

A solution to the problem of an axially loaded compressible pile in a certain kind of elastic medium has now been obtained [equations (8.35) and (8.46) through (8.51)]. To make use of the results, it is necessary to introduce equations for k_s and k_t as in the rigid pile solution. In the case of the compressible pile of any normal length-to-diameter ratio (usually greater than 20), the load transferred to the tip is an even smaller proportion of the total load than it is for the rigid pile unless the pile tip tests on, or is embedded in a stiff layer. As a consequence, the selection of a value for k_t is not critical for the uniform material. Only if the soil or rock below the pile tip is much stiffer than the soil surrounding the pile will we need to pick an expression for k_t more carefully. In general, when this happens, equation (8.10) should be employed, but its use leads to unnecessary complexity in the present problem. Instead, we will choose k_t to be equal to k_s (that implies making it too *small*) and we select k_s to be almost the same as that employed previously (equation 8.11) by omitting the term in Poisson's ratio. That is, we take

$$k_t = k_s = \frac{G_s}{4a} \tag{8.52}$$

To evaluate loads and displacements, we need to work out expressions for λ, λl, and $\lambda E / k_t$ for use in various equations. Substituting in equation (8.34), we get

$$\lambda^2 = \frac{E_s}{(1 + v_s)Ed^2} \tag{8.53}$$

Taking the root and multiplying by l gives

$$\lambda l = \left[\frac{E_s}{(1 + v_s)E}\right]^{1/2} \frac{l}{d} \tag{8.54}$$

Using the value obtained for λ and multiplying it by E/k_t, where k_t is given by equation (8.46), gives the following result:

$$\frac{\lambda E}{k_t} = 4\left[\frac{E(1 + v_s)}{E_s}\right]^{1/2} \tag{8.55}$$

From these expressions we can study a number of interesting aspects in the solution.

One variable we will wish to evaluate is the force in the pile as a function of distance down the pile. It is given by equation (8.36), which becomes, after substituting for C_1 and C_2,

$$F = P(D_1 e^{-\lambda z} + D_2 e^{\lambda z}) \tag{8.56}$$

This result can be used to derive the distribution of shearing stress τ between pile and soil, by differentiating the force with respect to distance, z, along the pile and then dividing by the pile perimeter, S:

$$\tau = \frac{-\lambda P}{S}(D_1 e^{-\lambda z} - D_2 e^{\lambda z}) \tag{8.57}$$

The shearing stress can be normalized through dividing it by an average shearing stress P/Sl. Thus,

$$\frac{Sl\tau}{P} = -\lambda(D_1 e^{-\lambda z} - D_2 e^{\lambda z}) \tag{8.58}$$

describes the distribution of shearing stress along the pile. The displacements at the top w_0 and base w_l of the pile can also be written in convenient form. From equation (8.35) and the expressions for the constants we can write

$$w_0 = \frac{P}{\lambda EA}(D_1 - D_2) \tag{8.59}$$

After some rearranging and substituting for λ from equation (8.53), we get

$$\frac{E_s \, dw_0}{P} = \frac{4}{\pi}\left[\frac{E_s(1 + v_s)}{E}\right]^{1/2}(D_1 - D_2) \tag{8.60}$$

It is seen that the dimensionless displacement of the pile top is a function of the ratio of pile to soil properties E_s/E, v_s and the length-to diameter ratio of the pile. Expressed in similar terms, the displacement of the pile tip, w_l, is given by the equation

$$\frac{E_s \, dw_l}{P} = \frac{4}{\pi}\left[\frac{E_s(1 + v_s)}{E}\right]^{1/2}(D_1 e^{-\lambda l} - D_2 e^{\lambda l}) \tag{8.61}$$

It is now time to see how the solution we have obtained compares with a more exact calculation. Using the composite finite element method with 10 elements, and an integration of Mindlin's equations for the influence function, Mattes and Poulos (17) solved the problem of the compressible pile considered here. Their results form a convenient basis for discussion. In their work, Mattes and Poulos, to describe the relative pile/soil stiffness, employ the dimensionless parameter

$$K = \frac{E}{E_s}\frac{A_p}{A_s} \tag{8.62}$$

where A_p is the cross-sectional area of the pile and A_s is the cross-sectional area of soil displaced by the pile. For a solid pile, $A_p/A_s = 1.0$. A particular case is then specified by the pile dimensions, v_s, and K. In one of their sample solutions they calculated the shearing stress variation along a pile with a length-to diameter ratio of 25 for two values of the soil's Poisson's ratio, and K equal to 50 and 5000. The smaller K corresponds to a relatively compressible pile, and the higher K corresponds to a stiff one. Their results are shown In Figure 8-10, which also contains the value of normalized shearing stress calculated from equation (8.52) derived from the Winkler representation of soil support for the case of $v_s = 0$. In carrying out the computations, the pile length-to-diameter ratio was assumed to be 25. All other quantities depend only on the value of K selected and are given in Table 8.1.

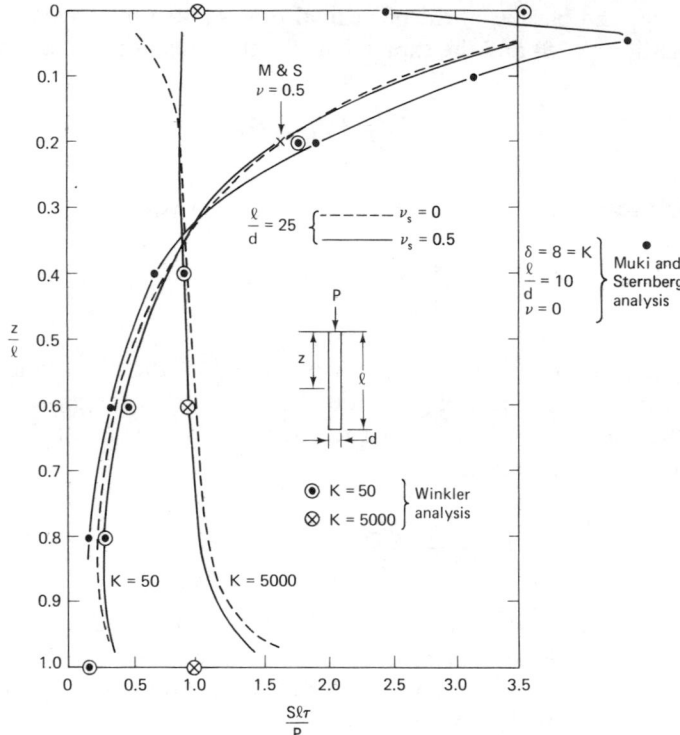

Figure 8-10. Shear stress distribution along single compressible pile, axially loaded (17).

TABLE 8.1

Constants ($\ell/d = 25 : \nu_s = 0$) Required in Evaluation of Equation (8.58)

K	$\lambda\ell$	$\dfrac{\lambda E}{k_t}$	D_1	D_2
50	3.536'	28.284	1.00079	−0.00079
5000	0.354	282.84	1.95755	−0.95755

From these values, the normalized shearing stress was calculated for different distances z along the pile. The results are shown in Table 8.2 as well as in Figure 8-10. Note that the mean shearing stress is less than unity because of the load taken by the pile tip.

Mattes and Poulos also present a diagram, given here as Figure 8-11, showing the dimensionless displacement of the top and base of the $l/d = 25$ pile as a function of K for Poisson's ratios of 0 and 0.5. Equations (8.60)

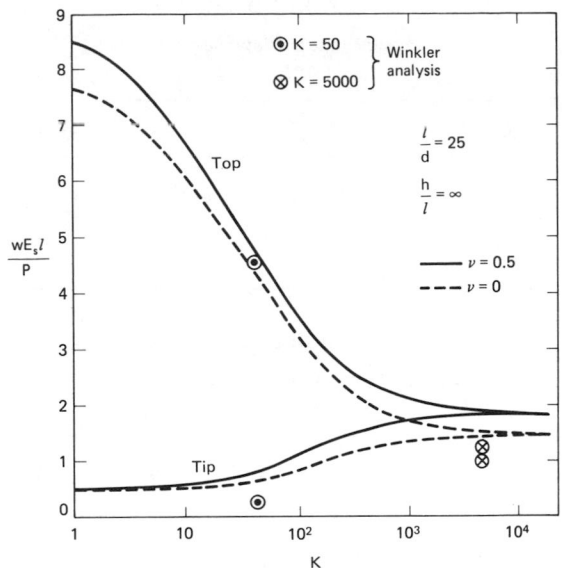

Figure 8-11. Displacement of pile top and tip (17). Single compressible pile.

TABLE 8.2

Values of Normalized Shearing Stress from Equation (8.58)
($\ell/d = 25$; $v_s = 0$; Minus Signs Omitted)

$\dfrac{z}{\ell}$	0	0.2	0.4	0.6	0.8	1.0
$K = 50$						
$\dfrac{S\ell\tau}{P}$	3.542	1.751	0.872	0.448	0.256	0.199
$K = 5000$						
$\dfrac{S\ell\tau}{P}$	1.032	1.008	0.991	0.978	0.971	0.968

and (8.61) were evaluated for $K = 50$ and 5000, $v_s = 0$, and $l/d = 25$ to give the numbers in Table 8.3, which are the four points plotted in Figure 8-11 for comparison with the curves of Mattes and Poulos. The displacements were dimensionalized with respect to l rather than d by multiplying by l/d equal to 25.

Finally, one variable of interest is the variation of the load in the pile with distance down the pile. This can be calculated for the two previous values of K by using the information in Table 8.1 and equation (8.56) to give Table 8.4. The values in parentheses at $z/l = 1.0$ are taken from one of the curves of Mattes and Poulos. A similar calculation for the case $l/d = 10$,

TABLE 8.3

Dimensionless Displacements
$(v_s = 0; \ell/d = 25)$

K	50	5000
$\dfrac{w_0 E_s \ell}{P}$	4.509	1.312
$\dfrac{w_l E_s \ell}{P}$	0.253	1.223

TABLE 8.4

Values of Normalized Force in Pile, F/P
$(\ell/d = 25; v_s = 0)$

$\dfrac{z}{\ell}$	0	0.2	0.4	0.6	0.8	1.0
$F/P \begin{cases} K = 50 \\ K = 5000 \end{cases}$	1.000 1.000	0.492 0.796	0.240 0.597	0.113 0.400	0.046 0.205	0.002(0.01) 0.011(0.04)

$E/E_s = 8$, and $v_s = 0$ was performed in order to compare with the analytical solution of Muki and Sternberg (19). The results are shown plotted in Figure 8-3. Although this is not the most sensitive way to compare the analyses, and the values of the parameters are outside the range of most piles, a fair resemblance is evident.

Having obtained the variables, shearing stress distribution, displacements, and load variation along the pile for comparison with the results of Mattes and Poulos, we must now discuss the results. First, it is obvious that the shearing stress distribution given by this simple analysis is remarkably close to that given by the more complicated calculations. It seems likely that this variable is not very sensitive to the assumptions made, or even to the detailed material properties. The vertical displacement of the pile top is also near that obtained by Mattes and Poulos, especially for softer pile/soil relations. For stiffer piles with respect to the soil, this analysis underestimates the pile settlement in general by about 15 percent. As indicated in Table 8.4, the load carried by the pile base from the simple analysis appears to be overestimated, although the actual percentage of the total load is very small for all pile/soil stiffnesses.

The last result suggests that it might be worthwhile to examine the effect of the base load on the pile settlement. The fact that the base load is high indicates that too high a reaction coefficient (k_t) has been assumed for the material behavior below the pile tip, even though it was deliberately selected to be a relatively low value. The most convenient calculation to make is that based on the assumption that the force P_t at the tip is zero. A solution for this case can be worked out from equations (8.42) and (8.43) or from first principles to give

$$C_1 = \frac{P}{\lambda EA} \frac{e^{\lambda l}}{e^{\lambda l} - e^{-\lambda l}} \tag{8.63}$$

$$C_2 = \frac{P}{\lambda EA} \frac{e^{-\lambda l}}{e^{\lambda l} - e^{-\lambda l}} \tag{8.64}$$

for substitution in equation (8.35) for displacements and in equation (8.36) for force. When the appropriate values are inserted for the pile examined previously, the results given in Table 8.5 appear.

TABLE 8.5

Displacements for Zero Base Load
$(l/d = 25; v_s = 0)$

	K	50	5000
Top	$\dfrac{w_0 E_s l}{P}$	4.509	1.326
Base	$\dfrac{w_l E_s l}{P}$	0.263	1.247

It can be seen by comparison with Table 8.3 that the displacements are little changed by neglecting this effect, so that the major portion of the pile's response to load derives from the forces acting along the side of the pile. In particular, removal of the tip load increases the tip displacement, of course, but it does not bring it up to the value of about 0.6 (dimensionless, as in Table 8.4) found by Mattes and Poulos for the $v_s = 0$, $K = 50$ case (Figure 8-11). For the stiffer pile/soil ratio of 5000, the $P_t = 0$ results are close to those of Mattes and Poulos, but they still are slightly low.

It appears, therefore, that the overall pile behavior is represented quite well with the type of Winkler foundation discussed here, in which the soil reaction coefficients k_s and k_t are given by equation (8.52). Of the two, the value selected for k_s is the more important, and that employed appears to give good results in comparison with the more complex analysis. A slightly smaller value of k_s would give a closer all-around correspondence of displacements without affecting the other variables significantly. In terms of the base load, it appears that k_t as given by equation (8.52) is too high and should be reduced for calculations in which this parameter is of importance. A reduction of k_t to one-half of the value in equation (8.52) would improve the correspondence without adversely affecting the other variables. The displacements are also moved slightly in the right direction by this alteration.

Apart from the probable insensitivity of some of the parameters to the assumptions made for the analysis, the correspondence between the simple and complicated solutions is good enough to prompt the following question: Why is this the case? In the Winkler representation, as usual, a force or stress applied at some point along the surface of the pile produces a deflection at the point of application only. In the elastic half-space an applied load produces deflections all around the load point. We have seen that the behavior of a loaded beam in bending on a Winkler subgrade is somewhat different from the behavior of the same beam on an elastic half-space. Accepting the Mattes and Poulos results as being equivalent to an exact solution, the correspondence in the case of an axially loaded pile seems to be much better. In the case of the beam, we found that a better representation of the continuum subgrade could be achieved by the two-parameter or Pasternak model which permitted deflections adjacent to the loaded area to be taken into account. The physical model of the Pasternak representation included a shear beam or stretched string or membrane at the ground surface in addition to the Winkler springs. The addition of the stretched membrane required a term involving the second derivative of the beam displacement to be brought into the differential equation.

Suppose we attempt to make the same modification in the case of the pile. What would be the result? The stretched string in this case would be vertical, connecting, in effect, the Winkler springs together, so that a load or stress applied at one point would cause an attenuating deflection up and down the vertical column. By choosing a suitable string force (second parameter in the Pasternak model) the pattern of vertical deflections could be made to resemble that in the elastic half-space according to some criterion of identification. Thus, the soil reaction to the pile loading would include a second displacement derivative representing the stretched string behavior in addition to the displacement term describing the Winkler aspect. The second derivative term would be added to equation (8.32), which describes the pile/soil behavior. As a consequence, the resulting equation would contain *two* terms in the second derivative of the displacement. These could be combined into one term with a coefficient describing the contribution of both pile and soil (string) properties, and this coefficient would give rise to a modified value of the parameter λ in the reduced equation, in the form of equation (8.33). This means that the *form* of the equation describing the better representation would be unchanged; it would remain a second-order linear differential equation. The improvement would be reflected only by a change in the coefficient or in the parameter λ. That is an interesting result, because it implies that equation (8.33) as it stands is quite a good description of the physics of pile/soil interaction. Essentially, this is what we found from the numerical examples. It will be recalled that the procedure for arriving at a value of the reaction coefficient k_s, although more or less logical, was quite arbitrary; it turned out to be a reasonably good guess. What probably happened is that it was a less good estimate than it appeared, from the viewpoint of describing the *soil* response in a Winkler model. It was, however, of such a magnitude as to include effectively a contribution of the second parameter of the Pasternak model, so that the numerical results turned out well.

If the pile is embedded in a layered soil profile, the analysis becomes more complicated, but it may still be carried out. For example, if the pile passes through one layer of soil with given elastic properties and is embedded to some length in an underlying layer of different properties, the displacements can be described by two equations of the form of equation (8.35) involving four constants. Consideration of the conditions at the top and base of the pile, and at the interface between the layers where the force in and displacement of the two sections of the pile must match, permits the determination of the constants.

In another example, should the pile be placed in an elastic layer of finite thickness over essentially rigid bedrock (the pile tip does not reach the rock), the present method of analysis can be employed, but the reaction coefficients k_s and k_t must be modified to account for the increase in apparent stiffness by the restraint of the rigid material. In this case, using the composite finite element technique, Poulos and Davis (25) suggest using Steinbrenner's approximation (28) since the problem of stresses and displacements developed by a point load in a finite elastic layer has not yet been solved in convenient form. Steinbrenner made the assumption that the surface displacement in a *finite* layer due to a load at the surface could be obtained from the half-space solution as being equal to the difference in displacements between the surface point and a point vertically below it at a depth equal to the finite layer thickness under the same load. The error involved in this approximation was found by Poulos and Davis to be less than 10 percent except in extreme cases of high Poisson's ratio and thin layers. Steinbrenner's approximation may, therefore, be applied to the determination of the modified values of k_s and k_t in the present formulation.

Poulos and Davis found, using Steinbrenner's assumption, that the distribution of shearing stress along a *rigid* pile was unaffected by the thickness of the finite layer and remained identical to the case of the pile embedded in the semi-infinite medium. In addition, Mattes and Poulos (17) indicate that the settlement of very compressible piles is also largely unaffected by the layer thickness. In such a case, they recommend using the semi-infinite values of their influence factors (or reaction coefficients here). When the value of K gets larger, there is some effect of the layer thickness, but it has significance only when the layer thickness is less than twice the pile length. In general, it appears that for a finite layer thickness of this dimension or greater, the semi-infinite analyses may be used without great error.

By calculating the ratio of the deflection of the top of a flexible pile to that of the top of a rigid pile of the same length, diameter, and loading conditions, a criterion can be determined to decide when an axially loaded pile is rigid. It is found that when this tip load is 0.3 or less of the applied load, the deflection of the flexible pile is less than 5 percent greater than that of the rigid pile when λl is less than 0.4. Thus a pile-soil combination with λl less than this value can be considered to behave as a rigid pile.

Although torsional loading of piles is much less common than either axial or lateral loading, it may occur when a pile group is subjected to off-center lateral loads, and it has intrinsic interest as a potential pile testing technique. The analysis follows that for axial loading almost exactly, except for the determination of the torsional subgrade reaction coefficients, and it can be treated from the points of view of rigid and flexible piles, respectively.

8.2.3   Rigid pile, torsional load

We must first define subgrade reaction coefficients for this case. Rotating the pile causes horizontally directed shearing stresses at the soil-pile periphery and on its base. The relevant subgrade reaction coefficients $k_{\theta s}$ along the side and $k_{\theta t}$ on the base are taken to be described by the equations

$$\tau_s = k_{\theta s} a \theta \tag{8.65}$$

and

$$\tau_t = k_{\theta t} r \theta \tag{8.66}$$

where θ is the angle of rotation of the pile about its vertical axis and r is the radius to the point on the base at which τ_t is measured. In this form, the units of k_θ are the same as in other applications in this book. The problem configuration is shown in Figure 8-12.

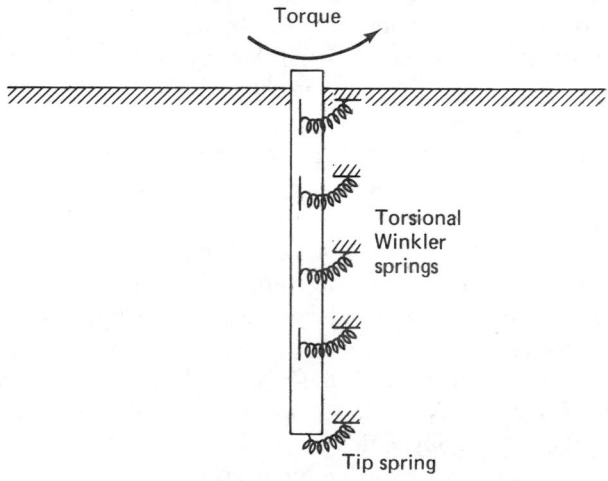

Figure 8-12.   Torsional Winkler model for a pile.

The torque T_s generated around the pile's periphery is

$$T_s = 2\pi a^2 l \tau_s = 2\pi a^3 l k_{\theta s}\theta \tag{8.67}$$

and by the base (from integration of the shearing stress over the base),

$$T_t = \frac{\pi a^4}{2} k_{\theta t}\theta \tag{8.68}$$

when the diameter of the base is the same as that of the pile. Summing equations (8.67) and (8.68), inverting to give θ, and putting in convenient dimensionless form give

$$\frac{G_s d^3 \theta}{T} = \frac{16G_s}{\pi(4lk_{\theta s} + ak_{\theta t})} \tag{8.69}$$

For this analysis, it is convenient and suitable to make the choice

$$k_{\theta s} = k_{\theta t} = \frac{4G_s}{d} \tag{8.70}$$

an expression that will be discussed in the next section. Thus, equation (8.69) becomes

$$\frac{G_s d^3 \theta}{T} = \frac{8}{\pi\left(\dfrac{8l}{d} + 1\right)} \tag{8.71}$$

A comparison of the rotational deflection given by this equation with that calculated by Poulos (24) for the torsion of a rigid pile calculated by the same composite finite element method as before is very favorable as seen in Figure 8-13. Other aspects of torsional pile behavior, for example, when the soil properties change with depth, follow those detailed in Section 8.2.1 on the rigid pile under axial load and will not be discussed further here. We continue to the case of the flexible pile.

8.2.4   Flexible pile; torsional load.

The development of the equation controlling the torsional behavior of a flexible pile follows that of Section 8.2.2 for the pile under axial loading, except for the parametric changes that accompany twisting. Consequently, the equation that results is

$$\frac{d}{dz}\left(GJ \frac{d\theta}{dz}\right) - 2\pi a^3 k_{\theta s}\theta = 0 \tag{8.72}$$

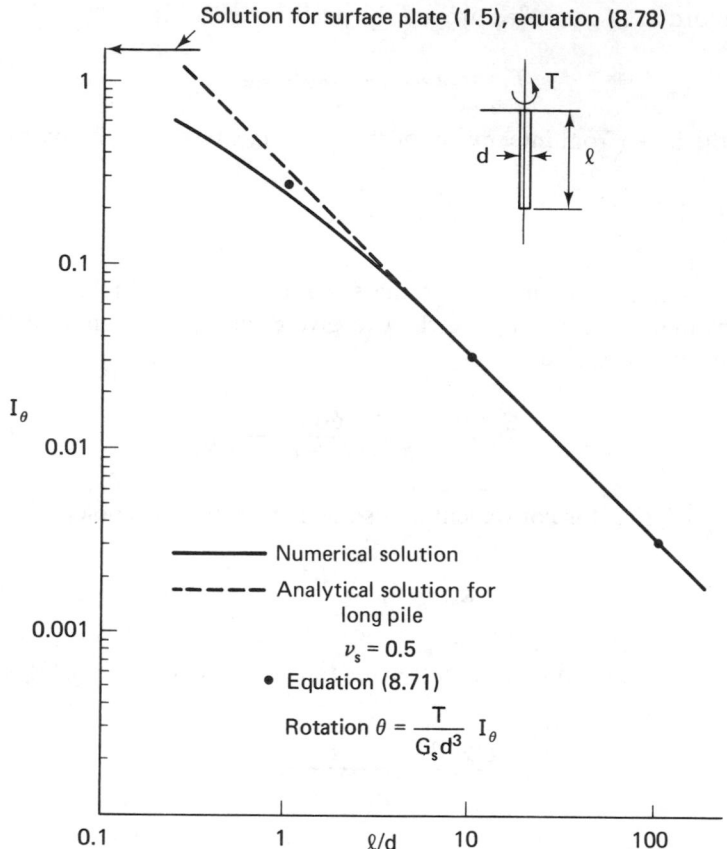

Figure 8-13.   Comparison of solutions (24) for a rigid pile in torsion.

which, if the pile section is uniform along its length, reduces to

$$\frac{d^2\theta}{dz^2} - \lambda^2\theta = 0 \tag{8.73}$$

when

$$\lambda^2 = \frac{2\pi a^3 k_{\theta s}}{GJ} \tag{8.74}$$

if the pile is circular in cross section and if G and J are the shear modulus and polar moment of inertia, respectively. Should the pile be square, or of H or I section, the shearing response of the soil to twisting is more difficult to evaluate, but it is suggested that in this case the dimension a represents an equivalent radius equal to one-half the largest dimension of the cross section.

The polar moment for such a section must, of course, be correctly evaluated. If the pile is a hollow tube, J must be calculated for the tube, and a is the outer radius. In the case that the pile is a solid cylinder, the expression for λ can be simplified to the following:

$$\lambda^2 = \frac{4k_{\theta s}}{Ga} \tag{8.75}$$

It is seen that equation (8.73) is identical in form to equation (8.33) for the axial loading case, and it follows that it has the same solution in terms of θ as the other in terms of w. The expressions for the constants C_1 and C_2 also are the same as before, except for the pile and soil properties, and will not be repeated. All the results can be obtained for the equivalent torsion boundary conditions (zero base rotation, base torque equaling T_t, zero base torque) by replacing the various groups of constants in equations (8.36) through (8.49) and (8.56) through (8.61) as follows:

$$
\begin{array}{cccc}
\text{For} & P & \text{use} & T \text{ (torque)} \\[2mm]
\text{For} & EA & \text{use} & GJ \\[2mm]
\text{For} & \dfrac{\lambda E}{k_t} & \text{use} & \dfrac{\lambda G}{k_{\theta t}}
\end{array}
\tag{8.76}
$$

and the value of λ employed is obtained from equations (8.74) or (8.75) for the torsional case.

The analogous equation to (8.45) for the load at the base in terms of base torque T_t is

$$T_t = \frac{\pi a^4}{2} k_{\theta t}\theta = \frac{\pi a^4}{2} k_{\theta t}(C_1 e^{-\lambda l} + C_2 e^{\lambda l}) \tag{8.77}$$

in which, again, the radius a may be equal to or larger than the pile radius to permit the behavior of piles with enlarged bases to be assessed.

The solution of a particular problem depends, once again, on the selection of values for $k_{\theta s}$ and $k_{\theta t}$. The same difficulties crop up as before, in terms of field tests, but are perhaps intensified because of the lack of consideration of torsional situations in soil mechanics. A field test to obtain $k_{\theta s}$ can be visualized, in which a device could be clamped to a borehole wall and rotated. Alternatively, in some soils the appropriate property might be obtainable from a careful loading and unloading test by using field vane. A determination of the subgrade coefficient $k_{\theta t}$ appears even more difficult; it would require loading a plate axially at depth and then measuring its rotation under applied torque. The alternatives are to elucidate the properties from field torsional tests on piles, which have been rarely attempted, or once again to fall back on theoretical considerations.

The problem of the rotation of a rigid circular plate bonded to the surface of an elastic half-space has been solved (11) and the torque T was found to be related to the rotation θ by the equation

$$T = \frac{16G_s a^3}{3}\theta \tag{8.78}$$

where G_s is the shearing modulus of the soil.

Rotation of the same plate in an infinite elastic space would require twice the torque for the same angular movement, but for the reasons given earlier in the discussion on axial loading, we will confine our attention to the surface relationship. Using the first part of equation (8.77) along with (8.78) enables an expression to be found for $k_{\theta t}$:

$$k_{\theta t} = \frac{32G_s}{3\pi a} \approx 6.8\frac{G_s}{d} \tag{8.79}$$

However, we saw previously that subgrade coefficients established by this connection were generally too stiff; therefore, we will arbitrarily reduce the value of $k_{\theta t}$ thus obtained. In addition, we will assume, as before, that the same description will apply to $k_{\theta s}$, since no equivalent simple analysis yields its relation to the soil properties. In this event, we have

$$k_{\theta t} = k_{\theta s} = \frac{4G_s}{d} \tag{8.80}$$

It may be said that in each of the cases of soil loading by a rigid or flexible member, the pressure or stress applied to the soil can be looked at, from the usual mechanics point of view, as resulting from a modulus times a strain. With G_s being the modulus in equation (8.80) and in other equations like it, and with the strain consisting of displacement divided by distance, it follows that the length $d/4$, for example, in equation (8.80) is a measure of some average distance over which the displacement occurs. An estimate of this distance serves to define the coefficient of subgrade reaction in the various problems. It is the width of the beam in the case of a laterally loaded beam, twice the pile diameter in the axially loaded pile, and one-quarter of the pile diameter here.

In the solutions obtained here for the torsional problem, the Poisson's ratio does not appear. This corresponds to the results of the almost exact analysis of Poulos (24) who found it to have a very small effect. With the equations already set out for the axial loading case, we can now calculate the rotation of the pile top for varying conditions of pile/soil stiffness and pile length/diameter ratios for comparison with the calculations of Poulos. In torsion, Poulos defines a relative pile/soil stiffness parameter K_t by the relation, analogous to equation (8.62) for the axial case,

$$K_t = \frac{GJ}{G_s d^4} \tag{8.81}$$

which, for a solid pile of circular cross section, reduces to

$$K_t = \frac{\pi}{32} \frac{G}{G_s} \tag{8.82}$$

For the conditions shown, the calculated rotation of the pile top is given in Table 8.6 in dimensionless units.

TABLE 8.6

Rotation of Pile Top Under Torsion
($\ell/d = 25$)

K_t	1	100	10,000
$\dfrac{G_s d^3 \theta}{T}$	0.5642	0.0564	0.0135

These results are plotted in Figure 8-14 for comparison with the results of Poulos, and it can be seen that the correspondence is reasonably good.

Another aspect of such a comparison is the distribution of rotation with depth along the pile. Calculations for the present model give the results shown in Table 8.7. These results are plotted in Figure 8-15 in relation to the results of a similar computation performed by Poulos. Figure 8-15 also shows that for stiffness ratio values approaching 10^4, a pile of this l/d ratio behaves as if it were rigid. Table 8.8 shows the values of K_t at which piles of various slenderness ratios can be assumed to be rigid, and the easier equations of the previous section can be employed. By using equations (8.75), (8.80), and (8.82), a simpler criterion can be developed: The pile is stiff if λl is less than 0.4, which is the same value determined for the axially loaded pile.

TABLE 8.7

Distribution of Pile Top Rotation with Depth
($K_t = 100$; $\ell/d = 25$)

$\dfrac{z}{\ell}$	0	0.2	0.4	0.6	0.8	1.0
$\dfrac{G_s d^3 \theta}{T}$	0.0564	0.0233	0.0096	0.0040	0.0018	0.0011

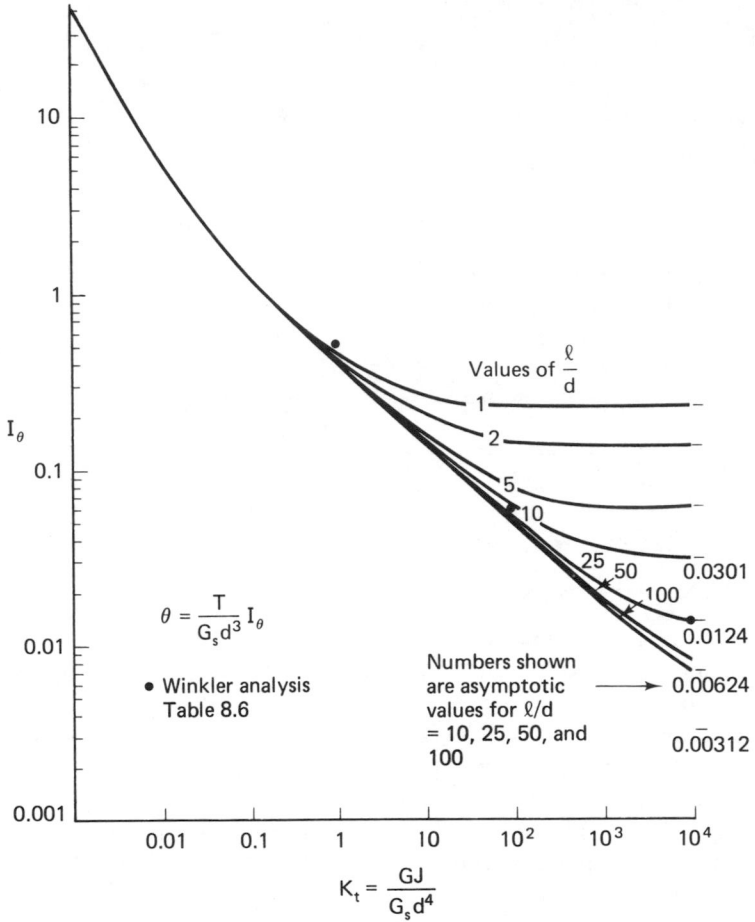

Figure 8-14. Comparison of solutions (24) for a flexible pile in torsion.

TABLE 8.8

Stiffness Ratios at Which Various Piles Are Essentially Rigid

$\dfrac{\ell}{d}$	5	10	25	50
K_t	500	2000	10^4	10^5

The behavior of rigid piles under ground conditions where the soil modulus, and thus the subgrade reaction coefficient increases with depth, has been treated in Section 8.2.1. It remains to consider the effect of this kind

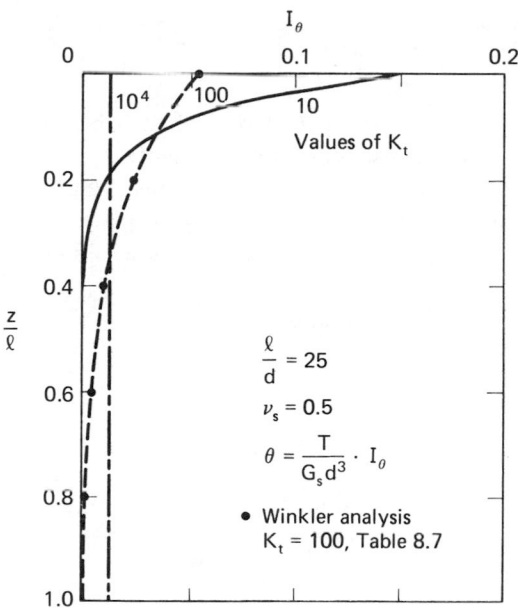

Figure 8-15.   Effect of pile flexibility on rotation distribution along pile; constant G_s.

of profile on the response of flexible piles to load, either axial or torsion. This will be done after we briefly examine an alternative approach to the pile problem.

8.2.5   Complementary formulation for axially loaded pile

The basic equilibrium equation for an element of an axially loaded pile is equation (8.30), which can be rearranged in the following form:

$$\frac{dF}{dz} - Sk_s w = 0 \tag{8.83}$$

Previously, we have used the stress-strain relation of the pile

$$F = EA\frac{dw}{dz} \tag{8.84}$$

and substituted for the gradient of F in (8.83) to give the customary equation in terms of w. Now, instead, we take another derivative in equation (8.83) to give

$$\frac{d^2F}{dz^2} - Sk_s\frac{dw}{dz} = 0 \tag{8.85}$$

and replace dw/dz by the expression (8.84) to get

$$\frac{d^2F}{dz^2} - \frac{Sk_s}{EA}F = 0 \qquad (8.86)$$

This is the equation for an axially loaded, linearly behaving pile in an elastic medium in terms of force instead of displacement.

It is generally more convenient to use this equation instead of the deflection equation when the force, instead of displacement, at the top and bottom of the pile can be specified. When the dimensionless parameters

$$s = \lambda z \quad \text{and} \quad f = \frac{F}{P}\text{(force)} \qquad (8.87)$$

with

$$\lambda^2 = \frac{Sk_s}{EA} \qquad (8.88)$$

as usual are chosen, the dimensionless equation is obtained:

$$\frac{d^2f}{ds^2} - f = 0 \qquad (8.89)$$

Since this has the same form as the displacement equation, the solution is also the same, but it is expressed in terms of force:

$$f = A_1 e^{-\lambda z} + A_2 e^{\lambda z} \qquad (8.90)$$

The solution proceeds as before, but it is obviously more convenient when the boundary force conditions are known, although displacements can also be taken into account.

The usefulness of this development can be demonstrated by a finite difference example. We choose again the axially loaded pile subjected to load P at the top and with zero displacement at the tip. We will select for simplicity (and also because they are used in Section 9.3.4) three elements in the pile, so that $\Delta s = s_l/3 = \lambda l/3$. In consequence, the unknowns in the problem are the forces at points 0, 1, 2, and 3. However, since we have made force dimensionless with respect to applied load, it follows that $f_0 = 1$. This is different from the case in which the problem is formulated in terms of displacements and the applied load is specified. In that case, the requirement at the pile top is that the displacement *gradient* is established. When a coarse network of points is selected, gradient is then applied to the top two points. Since the solution is trying to represent a continuous function, the gradient over this interval is not a good simulation of an actual gradient which occurs at ground surface. On the other hand, the force $f_0 = 1$ *exactly* conforms to the boundary-value requirement.

The displacement at the bottom boundary is fixed at zero. This can be expressed in force terms by the use of equation (8.83), from which it is apparent that at the pile base

$$\frac{df}{ds} = 0 \tag{8.91}$$

The most convenient finite difference way of accomplishing this is to locate a fictitious point $2'$ a distance Δs below the boundary. The force at $2'$, f_2' is given the value f_2, so that the bottom boundary becomes a line of symmetry with zero gradient across it.

From the differential equation (8.89) the finite difference equations can be written for points 1, 2, and 3 as follows:

$$\text{point 1} \quad \frac{f_0 - 2f_1 + f_2}{\Delta s^2} - f_1 = 0 \tag{8.92}$$

$$\text{point 2} \quad \frac{f_1 - 2f_2 + f_3}{\Delta s^2} - f_2 = 0 \tag{8.93}$$

$$\text{point 3} \quad \frac{f_2 - 2f_3 + f_{2'}}{\Delta s^2} - f_3 = 0 \tag{8.94}$$

Substituting $f_0 = 1$ and $f_2' = f_2$, we get

$$-(2 + \Delta s^2)f_1 + f_2 = -1 \tag{8.95}$$
$$f_1 - (2 + \Delta s^2)f_2 + f_3 = 0 \tag{8.96}$$
$$2f_2 - (2 + \Delta s^2)f_3 = 0 \tag{8.97}$$

The solution gives

$$f_2 = \frac{1}{(1 + 4\Delta s^2 + \Delta s^4)} \tag{8.98}$$

$$f_3 = \frac{2f_2}{(2 + \Delta s^2)} \tag{8.99}$$

$$f_1 = \frac{f_2}{(2 + \Delta s^2)} + \frac{1}{(2 + \Delta s^2)} \tag{8.100}$$

Two problems that are studied in other sections are those for which $s_t = \lambda l = 3.3716$ (Section 9.3.4) and $s_t = 2.00$ (Section 2.3.1). The solution from both the finite difference results, equations (8.98) through (8.100), and exact analysis for these values of s_t are shown in Table 8.9. It can be seen that the approximate solutions for only three subdivisions are remarkably good, in contrast to a solution obtained for the same boundary conditions by finite differences based on the displacement formulation.

TABLE 8.9

Force	f_0	f_1	f_2	f_3
s	0	$s_l/3$	$2s_l/3$	s_l
Approximate				
$s_l = 2.00$	1.0000	0.5466	0.3361	0.2750
Exact solution	1.0000	0.5392	0.3271	0.2658
Approximate				
$s_l = 3.3716$	1.0000	0.3465	0.1308	0.0801
Exact solution	1.0000	0.3283	0.1169	0.0692

The same approach can, of course, be applied to the case of the torsionally loaded pile.

8.2.6   Flexible pile; soil modulus increases with depth

Lastly, with respect to both axially and torsionally loaded piles in an elastic medium, the assumption of a soil property constant with depth is only even approximately realistic when the pile is embedded in an overconsolidated clay. For other soils, it would be more reasonable to take the soil property as increasing in some fashion with depth. The two variations that would most commonly apply are a linear or square root variation with distance z. In general, we will postulate

$$k = k_n z^n \tag{8.101}$$

in which the coefficient k is appropriate for either axial or torsional loading, whichever problem is being analyzed. When this general k is included in the differential equation of pile-soil interaction, the latter can be written in the form

$$\frac{d^2 s}{dz^2} - g(z)s = 0 \tag{8.102}$$

in which s and z are now dimensionless deflection and length coordinates obtained by multiplying the actual displacement and distance by the parameter λ, which now depends on the form assumed for k, or $g(z)$. If k is constant with depth, as assumed previously, $g(z) = 1$, and λ is defined by equation (8.34) for the axially loaded pile. Alternatively, if $n = 1$ in equation (8.101), $g(z) = z$ and λ is given by the equation

$$\lambda^3 = \frac{Sk_1}{EA} \tag{8.103}$$

for axial loading, and

$$\lambda^3 = \frac{2\pi a^3 k_{\theta 1}}{GJ} \tag{8.104}$$

in torsion. If the subgrade coefficient increases linearly from a finite value at the surface, as, for example, $k = k_0 + k_1 z$, then the introduction of the new variable $\eta = k_0 + k_1 z$ will permit the results of the straightforward linear case to be applied. After solution, the original variable z can be recovered.

In general, for the exponent n, $g(z) = z^n$ and λ is defined by the relations:

$$\text{(axial)} \qquad \lambda^{n+2} = \frac{Sk_n}{EA} \tag{8.105}$$

$$\text{(torsion)} \qquad \lambda^{n+2} = \frac{2\pi a^3 k_{\theta n}}{GJ} \tag{8.106}$$

The general solution to equation (8.102) is

$$s = A\sqrt{z}\, I_{+1/(n+2)}\left(\frac{2}{n+2} z^{(n+2)/2}\right) + B\sqrt{z}\, I_{-1/(n+2)}\left(\frac{2}{n+2} z^{(n+2)/2}\right) \tag{8.107}$$

where $I_{\pm 1/(n+2)}$ are modified Bessel functions of fractional order, and A and B are constants. In this form, the two components of the solution are not very convenient, since they are both unbounded as z goes to infinity. They behave in the manner of $\sinh z$ (I_+) and $\cosh z$ (I_-) functions, respectively. However, when we have to deal with a very long pile, the mechanics of the problem tell us that all the quantities of displacement, force, etc., will tend to zero with distance down the pile. It is useful in this regard to have a solution that behaves like the exponential functions of equation (8.35), the solution to the problem in which the soil property is uniform with depth. This can be accomplished by rewriting equation (8.107) in the form

$$s = C_1\sqrt{z}\,(I_- - I_+) + C_2\sqrt{z}\,(I_- + I_+) \tag{8.108}$$

in which it is understood that the argument of the I_- and I_+ functions is

$$\left(\frac{2}{n+2} z^{(n+2)/2}\right)$$

and the order is $1/(n+2)$. When written this way, the difference of the Bessel functions in the first parentheses behaves like e^{-z} and vanishes at infinity, whereas the sum, similar to e^z, increases without limit at infinity. Then, as in previous problems involving infinite and semi-infinite beams, we can put C_2 equal to zero in any pile problem in which the pile is very long or very flexible.

Although n can, in principle, be assigned any value, most practical interest is centered in the cases in which it equals unity or the value one-half. In these cases, the functions take the form given in Table 8.10.

TABLE 8.10

Components of Equations 8.107 and 8.108

n	Bessel Function Subscript	Bessel Function Argument
1	$\pm\dfrac{1}{3}$	$\dfrac{2}{3}z^{3/2}$
$\dfrac{1}{2}$	$\pm\dfrac{2}{5}$	$\dfrac{4}{5}z^{5/4}$

When n has the value 1, the solution can be expressed in terms of Airy functions, $Ai(z)$ and $Bi(z)$, which are tabulated (1). Except for multiplicative constants, the two Airy functions are identical to the two variable parts of equation (8.108) when n equals unity. Thus, the solution can be rewritten

$$s = C_1 Ai(z) + C_2 Bi(z) \tag{8.109}$$

in which $Ai(z)$ diminishes and $Bi(z)$ increases with z, as do the first and second groups of functions in equation (8.108). For the case $n = \frac{1}{2}$, it is convenient to make the same abbreviation, so that here

$$s = C_1 Ri(z) + C_2 Si(z) \tag{8.110}$$

where we define

$$Ri(z) = \sqrt{z}\left[I_{-2/5}\left(\frac{4}{5}z^{5/4}\right) - I_{2/5}\left(\frac{4}{5}z^{5/4}\right)\right] \tag{8.111}$$

and

$$Si(z) = \sqrt{z}\left[I_{-2/5}\left(\frac{4}{5}z^{5/4}\right) + I_{2/5}\left(\frac{4}{5}z^{5/4}\right)\right] \tag{8.112}$$

In solving actual problems we also need the first derivatives of all these functions, which will be denoted $Ai'(z)$, $Bi'(z)$, $Ri'(z)$, and $Si'(z)$. The two functions Ri and Si and their derivatives do not appear to be tabulated, and tables and graphs of their values, as well as those of Ai and Bi and their derivatives, are given in Appendix C.

In order to consider in general the various boundary conditions treated previously in the simpler cases in which $g(z) = 1$, it is convenient if the solution to equation (8.102) is written in the general form

$$s = C_1 f_1(z) + C_2 f_2(z) \tag{8.113}$$

in which C_1 and C_2 are constants and $f_1(z)$ and $f_2(z)$ are independent solutions of equation (8.102). The function of $f_1(z)$ will typically be finite at $z = 0$ and decrease to zero as z goes to infinity, whereas $f_2(z)$, also finite at $z = 0$, goes to infinity as z increases. The nature of the functions f_1 and f_2 depends on the variation assumed for the function $g(z)$ in equation (8.102). For the different cases discussed previously, f_1 and f_2 and their first derivatives f_1' and f_2' are given in Table 8.11.

TABLE 8.11

Functions f_1 and f_2 for Given $g(z)$

$g(z)$	$f_1(z)$	$f_2(z)$	$f_1'(z)$	$f_2'(z)$
(a) 1	e^{-z}	e^z	$-e^{-z}$	e^z
(b) z	$Ai(z)$	$Bi(z)$	$Ai'(z)$	$Bi'(z)$
(c) $z^{1/2}$	$Ri(z)$	$Si(z)$	$Ri'(z)$	$Si'(z)$

In equation (8.113) the constants are obtained by applying the boundary conditions. The simplest problem that can be examined is that in which the pile is long and can be considered to extend to infinity in the z-direction. For the semi-infinite pile, the solution f_2 would not be expected to play a part, and thus C_2 must be zero in this case. The other boundary condition occurs at the top of the pile, $z = 0$, and here the *dimensionless* load is taken equal to F. For axial loading, the dimensionless load is P/EA; in torsion it is $T/\lambda GJ$ where P and T are the axial load and torque at any section of the pile. The load is given by the negative gradient of the deflection with respect to depth, so that, here, at $z = 0$,

$$F = -C_1 f_1'(0) \quad \text{or} \quad C_1 = -\frac{F}{f_1'(0)} \tag{8.114}$$

where $f_1'(0)$ is the derivative of s with respect to z at $z = 0$. Thus, the solution for the displacement of the pile is

$$s = -\frac{f_1(z)F}{f_1'(0)} \tag{8.115}$$

and for the dimensionless load R in the pile,

$$R = \frac{f_1'(z)F}{f_1'(0)} \tag{8.116}$$

For a finite pile of dimensionless length d (equal to λl, where l is the pile length), the boundary conditions at the top and bottom of the pile determine the value of both constants C_1 and C_2. There are three conditions of interest: (a) The tip displacement may be zero if the pile is driven to bedrock; (b) the tip load may be taken to be very small or essentially zero if the pile is long and relatively flexible compared to the soil; (c) the tip load may have a finite dimensionless value R_d. The values of the coefficients C_1 and C_2 for these cases are shown in Table 8.12. Here D_1 represents the denominator in case (a), D_2 in case (b) as given in the expression for C_1.

TABLE 8.12

Coefficients C_1 and C_2

Case	C_1	C_2
(a) Tip displacement zero $s(d) = 0$	$\dfrac{f_2(d)F}{f_1(d)f'_2(0) - f_2(d)f'_1(0)}$	$-\dfrac{f_1(d)F}{D_1}$
(b) Tip load zero $R(d) = 0$	$\dfrac{f'_2(d)F}{f'_1(d)f'_2(0) - f'_2(d)f'_1(0)}$	$-\dfrac{f'_2(d)F}{D_2}$
(c) Tip load finite $R(d) = R_d$	$\dfrac{Ff'_2(d) - f'_2(0)R_d}{D_2}$	$\dfrac{-Ff'_1(d) + f'_1(0)R_d}{D_2}$

Although the coefficients appear complicated, inspection shows that they can be evaluated readily by referring to tables or graphs of the appropriate functions at the values corresponding to $z = 0$ (ground surface) and $z = d$ (pile tip).

These results correspond to those previously derived for the axially loaded pile in uniform soil, which can be expressed in the same form. The most difficult case to analyze is (c), in which we have seen it is necessary to substitute the tip load obtained on the basis of the subgrade reaction coefficient times the tip displacement. Rather than doing this to obtain general and complicated equations for C_1 and C_2 as in equations (8.46) through (8.49), it is easier either to carry forward a solution numerically before making the substitution, or to assume a value for R_d which can be checked by calculating the tip displacement subsequently. At the simplest level, use can be made of the practical observation that the tip load is generally in the range of from 0.2 to 0.3 of the applied load for many real piles. In dimensionless terms $R_d = (0.2 \text{ to } 0.3)F$.

It may be helpful at this stage to work through an example of an axially loaded pile.

8.2.7 Analysis of Arkansas River pile test (axial load)

Figure 8-4 shows the results of a pile test carried out under the direction of the U.S. Army Corps of Engineers. The soil at the site was a submerged fine-to-medium sand with a stiffness that increased with depth, as evidenced by the blow counts of standard penetration tests (SPT). The number of blows required for sampling at the pile test site increased from about 20 to 30 near the surface to 100 at a depth of 55 ft (16.8 m). The pile test shown in Figure 8-4 was performed on a 16-in. (0.41 m) diameter, 0.312-in. (7.9 mm) wall thickness, steel pipe pile driven with a Vulcan 140C double-acting steam hammer with a rated energy of 36,000 ft-lb (49 kN-m), to a depth of 52.5 ft (16 m). The number of blows delivered by the hammer per foot of penetration increased fairly uniformly with depth, confirming the SPT picture of soil resistance, to a value of about 45 at the 52.5-ft depth. The total driving time was only a few minutes. Because additional channels were welded to the side of the pile to protect the instrumentation, the cross-sectional area of the pile was somewhat larger than its diameter and wall thickness would indicate. The product EA for the pile is given in the report (15) as 692×10^3 kips (3.08×10^3 MN).

In Figure 8-4(a) the unloading curves show a pile/soil behavior which we will assume is linearly elastic. On unloading from the 150-ton load (1.34 MN), the rebound pile top deflection is about 0.2 in. (5 mm). Using this information and the theoretical developments of the last section, we will analyze the pile to determine what value of axial subgrade coefficient is indicated for the soil at the site. From Figure 8-4(c) it can be seen that, at the 150-ton load, the load transmitted to the pile tip is 25 tons (223 kN); thus, $R_d = 25/150$ or 0.167.

The field evidence indicates that the soil's stiffness increases with depth; for this analysis, we will assume that the subgrade reaction coefficient increases *linearly* with depth, so that n in equation (8.101) equals unity, and we are concerned with the Airy function solution of row (b) of Table 8.11. With the value of $R_d = 0.167$, the tip load is finite, so that the coefficients in the solution are given by case (c) of Table 8.12. To evaluate C_1 and C_2, we need to know the value of the various functions for values of the argument equal to zero (ground surface) and d (pile tip). However d is dimensionless and involves the parameter λ, which, since it includes k_1, equation (8.105), is unknown at the outset. Because of the complexity of the solution, the best approach is a trial-and-error one, beginning with a guess at d (and, hence, through the length of the pile, at λ and k_1). We will assume here that several trial guesses have already been made and that we are coming close to the final result with an estimated value of 1.92 for d. From Appendix C, or tables, (1), we obtain values of the required functions at z equal to zero and d as follows:

Ai(0)	Bi(0)	Ai'(0)	Ai'(1.92)	Bi'(0)	Bi'(1.92)
0.3550	0.6149	−0.2588	−0.0563	0.4483	3.7732

Substituting these values of the derivatives in the expressions for C_1 and C_2 in case (c) of Table 8.12 has the following result:

$$C_1 = 3.8876F \tag{8.117}$$

$$C_2 = 0.0138F \tag{8.118}$$

Employing these coefficients in equation (8.109) for the vertical deflection at $z = 0$ gives

$$s(0) = \lambda w(0) = C_1 Ai(0) + C_2 Bi(0) = 1.3887F \tag{8.119}$$

Substituting for F gives

$$w(0) = \frac{1.3887P}{\lambda EA} \tag{8.120}$$

Since P is 150 tons (1.34 MN), EA is 692×10^3 kips (3.08×10^3 MN), and λ, from our choice of d equal to 1.92, with a pile embedded length of 52.5 ft (16 m), is 3.0476×10^{-3} in.$^{-1}$ (0.12 m^{-1}), we get

$$w(0) = 0.1975 \text{ in. (5.02 mm)} \tag{8.121}$$

This is close enough to the indicated deflection of 0.2 in. (5.08 mm) to terminate the trial-and-error portion of the calculation. We still need to estimate the subgrade reaction coefficient gradient k_1. From equation (8.105) we have

$$\lambda^3 = \frac{Sk_1}{EA} \tag{8.122}$$

or

$$k_1 = \frac{EA\lambda^3}{S} \tag{8.123}$$

The presence of the channels along the outside of the pile to protect the instrumentation increases the perimetral length of the pile to 82 in. in this case, so that, with the other values of the pile constants, and λ, we obtain

$$k_1 = 0.24 \text{ lb/in.}^4 \text{ (2.57 MN/m}^4) \tag{8.124}$$

This value is interpreted as being the increase in the subgrade coefficient (lb/in.^{3}) per inch of depth. From the site investigation work we can consider it to represent the behavior of a medium-dense, fine-to-medium sand below the water table. Whether or not it characterizes all such sands depends on similar evaluations from other pile tests, but, for the present, it may be considered a guide. If we assume the value to be related to a shear modulus G_s of the soil by equation (8.52), then it follows by this model that the shear modulus of the soil varies linearly with depth. At a depth of 25 ft (7.62 m), for example, halfway down the pile, the indicated shear modulus of the soil is 2300 psi (15.9 MN/m²), approximately. Using this value of G_s halfway down the pile as an average, and the pile properties, it is possible to evaluate K (equation 8.62) to be about 600. For the pile's length-to-diameter ratio of 40, it appears to be medium stiff with respect to the soil.

With the value of λ, or d, it is possible to calculate the force in the pile as a function of depth:

$$P(z) = -EA[C_1 Ai'(z) + C_2 Bi'(z)] \qquad (8.125)$$

or, including C_1 and C_2 from equations (8.117) and (8.118),

$$\frac{P(z)}{P} = -[3.8876 Ai'(z) + 0.0138 Bi'(z)] \qquad (8.126)$$

Table 8.13 has been prepared from tabulated values of the derivatives. The calculated load distribution is plotted in Figure 8-4(b), and it can be seen to differ substantially from the shape of the experimentally determined curve, as indicated by the strain gauges. In this instance, the reason for the difference is probably related to the residual strains left in the pile after driving. A discussion of this point is given in Section 9.3.3.

TABLE 8.13

Load Distribution in Pile at 150-Ton (1.34 MN) Axial Load

$\frac{z}{d}$	0	0.2	0.4	0.6	0.8	1.0
z	0	0.384	0.768	1.152	1.536	1.92
P, tons (MN)	150 (1.34)	137 (1.22)	110 (0.98)	79 (0.70)	50 (0.45)	25 (0.22)
Depth, ft (m)	0	10.5 (3.2)	21 (6.4)	31.5 (9.6)	42 (12.8)	52.5 (16)

It is now appropriate to examine the problem of the elastic behavior of an embedded pile subjected to lateral forces and moments at its top.

8.3 LATERAL PILE LOADING

It is helpful in trying to visualize the behavior of piles in lateral loading conditions to examine the behavior of an instrumented real pile under field test conditions. An appropriate test was performed on the same cylindrical pipe pile that was tested axially at the Arkansas River site, and described in Section 8.2. Loads were applied horizontally at the ground line. The load-deflection behavior of the pile is shown in Figure 8-16(a), and the moments

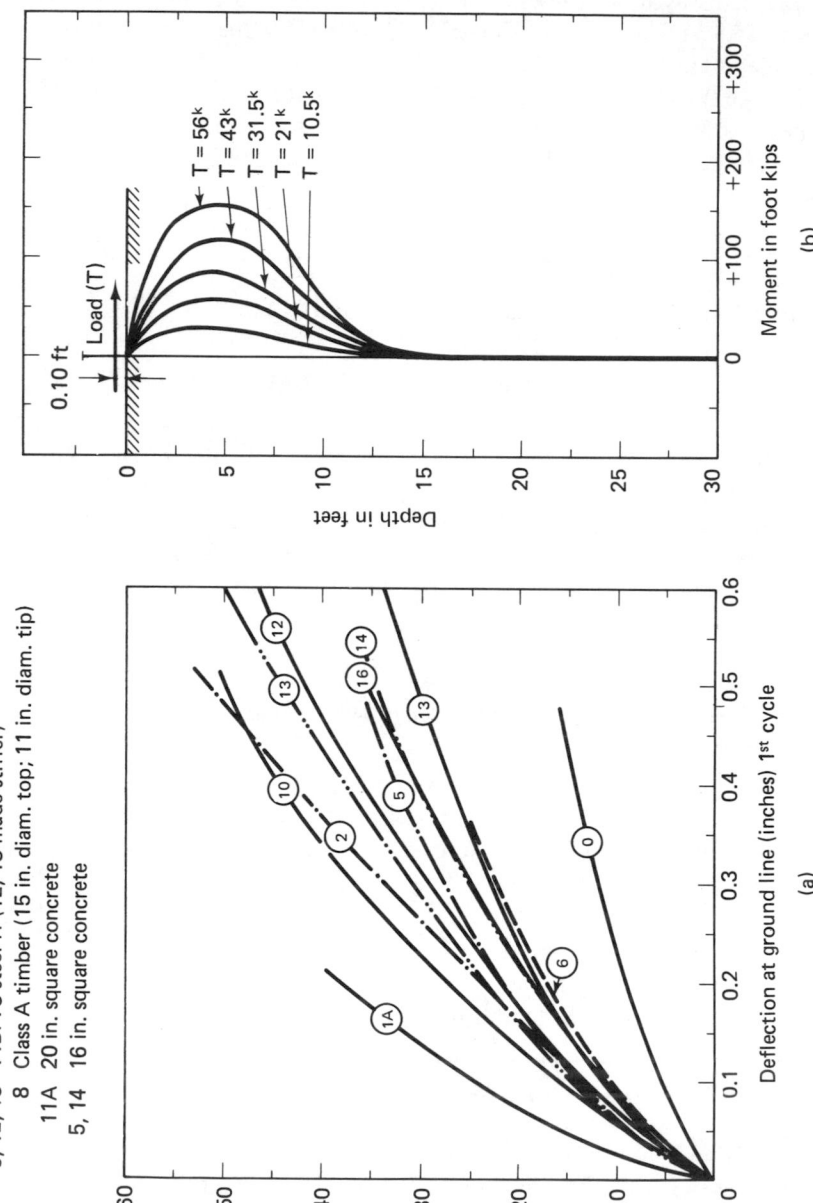

2, 10, 16   16 in. O.D., 0.312 in. wall steel pipe
6, 12, 13   14BP73 steel H (12, 13 made stiffer)
8   Class A timber (15 in. diam. top; 11 in. diam. tip)
11A   20 in. square concrete
5, 14   16 in. square concrete

Load in kips 1st cycle

Deflection at ground line (inches) 1st cycle

(a)

Depth in feet

Moment in foot kips

(b)

Load (T)

0.10 ft

T = 56^k
T = 43^k
T = 31.5^k
T = 21^k
T = 10.5^k

Figure 8-16. Laterally loaded flexible piles, Arkansas River test results (15) : (a) load versus deflection at top for a number of piles; the text refers to pile 2 ; (b) moment versus depth in pile 2 at different loads.

calculated from the strain-gauge data to exist in the pile at the various loads are represented in Figure 8-16(b). Unfortunately, neither the unloading behavior of the pile nor the residual moments left in the pile when the lateral load returned to zero are given in the report. These are important aspects of pile behavior under any loading condition.

In Figure 8-16(a) the line representing the pile behavior is curved all the way and has an appearance very different from the axial load test curve of Figure 8-2(a). In Figure 8-16(a) it would be very difficult to assess a limiting or ultimate lateral load for the pile, a load which although not precisely defined in the axial case was clearer to distinguish. It would be difficult to pick out any portion of the lateral behavior, particularly in the absence of an unloading curve, to which a linearly elastic model could be ascribed. The reason for this is that the soil adjacent to the pile at ground surface yields at the very smallest lateral load. As the load increases, yielding progresses down the pile further. In addition, the surface of the pile opposite to the loading direction separates from the soil and, to some depth, leaves a gap. Some soil inevitably moves or falls into this gap, so that when the load is removed the plastic deformations and the presence of the soil filling behind the pile prevent it from returning to its pristine position. Although this behavior reflects to some extent the presence of the essentially cohesionless sand at the site, it will not be dissimilar to that experienced by a laterally loaded pile in other soils. If this pile-soil system were to be modeled elastically, the model parameter would depend on the load or displacement level specified.

In Figure 8-16(b) the maximum moment in the pile occurs at a depth of from 3 to 5 ft (1 to 1.5 m), and increases, and appears to move down slightly as the load develops. The downward movement of the peak moment is another indication that the behavior is not entirely elastic. As the pile displaces, however, the migration of the maximum moment does not prevent it from increasing, until eventually a load can be applied large enough to yield the pile in bending. This is then the ultimate criterion for failure of a laterally loaded pile, although, in practice, displacement constraints may dominate. Again, this result is different from the mechanism of axially loaded pile failure, which depends entirely on *soil* yielding under normal circumstances.

Lastly, it can be observed for this pile-soil combination in Figure 8-16(b) that the moments have disappeared entirely by a distance of 15 ft (4.6 m) down the 52.5-ft (16 m) embedded pile length. In contrast, the tip load of the axially loaded pile, although a small proportion of the applied load, is frequently large enough to warrant consideration in analysis, and it usually plays an important part at the stage of pile failure. It follows that for normal piles of the class tested as shown in Figure 8-16, the boundary conditions at the pile tip play no part in the pile behavior under lateral load. The pile can be considered to be semi-infinite. It is fortunate that this is the case, because, as we have seen in earlier chapters, beam bending involves a fourth-order differential equation whose solution has four constants to be satisfied by four boundary conditions. The elimination of two of these greatly simplifies the task of analysis. Before we reach this stage, however, we must deal with an important simpler case, which applies when short rigid piers are subjected to lateral loads or moments at or above ground surface.

8.3.1 Rigid pile

The analysis for this problem has been presented in Section 4.3 for a rigid beam subjected to a transverse load and moment arbitrarily placed. In the case of a pile, the load and moment generally act, or are reduced to acting, at the ground surface, and we are, therefore, interested in the special cases of equations (4.62) and (4.63) when $a = 0$. To keep our notation systematic, it is convenient, to consider the vertical, pile-axis coordinate as z and the horizontal displacement caused by moment or lateral load as u. With these conditions, equations (4.62) and (4.63) for deflection under lateral load T_0 and moment M_0, respectively, become

$$u = \frac{2T_0}{kl}\left(2 - 3\frac{z}{l}\right) \tag{8.127}$$

and

$$u = \frac{6M_0}{kl^2}\left(2\frac{z}{l} - 1\right) \tag{8.128}$$

The moment in the beam subjected to load T_0 is given by equation (4.69) with appropriate changes in the coordinate designations; the maximum value is expressed by equation (4.70), at a distance of one-third of the pile length from the point of application. When the pile is loaded by a moment M_0 at $z = 0$, the moment in the beam is

$$M = M_0\left(1 - 4\frac{z^3}{l^3} + \frac{3z^2}{l^2}\right) \tag{8.129}$$

with the maximum, of course, occurring at $z = 0$. In both of these cases of loading, it should be pointed out again that the k employed is equal to k_0 times the pile width.[2] Since the problem solved is linearly elastic, combinations of T_0 and M_0 give deflections and moments which can be obtained by superposition.

For a number of actual soil profiles, a more realistic assumption regarding the subgrade reaction coefficient is to let it vary with depth either linearly or as a square root or other function. The linear case is treated in Chapter 4

---

[2]It will be recollected that k_0 to be used with the pile is twice the value employed with a surface beam, as discussed in Section 7.5.

with the results given in equations (4.72) through (4.82). Because of the linear variation of displacement generated by the rigidity of the beam, it is not difficult (but algebraically tedious) to generate similar equations for other assumed distributions of k with depth. The graphical plots in Chapter 4 illustrate the development of moments in several cases.

A rigid wall totally embedded in the soil is sometimes used as an anchorage for tie-rods of, for example, a sheet-pile retaining wall. In that case, the analysis proceeds as in the case of the rigid beam, but it applies to a portion of the wall and soil of unit length in plane strain. The horizontal coefficient of subgrade reaction appropriate to this case must be selected.

We now turn to study the behavior of flexible piles under lateral and moment loads acting at or near ground surface.

8.3.2 Flexible pile

Here we are concerned with a special case of the general problem discussed in Chapter 5. There are two regions of behavior that require consideration: (a) the general case of the finite pile, which is short enough and stiff enough with respect to the soil conditions that even the tip of the pile undergoes deflections due to the lateral load applied at the top, and (b) the special case of the pile long enough that the deflections, moments, and shears have died out with distance before the tip is reached. In the majority of *piles*, considered as rods of cylindrical, square, or H cross section, with widths from 12 to 72 in. (0.3 to 2 m) and lengths, in proportion, of from 50 to 400 ft (15 to 120 m), the latter situation holds good with respect to lateral loads. Generally, a pile is of intermediate stiffness with respect to axial loads, and it is flexible with respect to horizontal loads. The boundary between a finite and semi-infinite pile from the point of view of lateral loads might, in fact, serve to distinguish piles from *piers*, which are typically of much lower length-to-diameter ratios and relatively stiff with respect to the soil. We will consider the finite pile, loaded by horizontal force or moment at its upper end, first.

When the finite pile is loaded by a lateral force T_0 at its upper end, acting in a clockwise direction, the deflection at that end, u_{0T}, may obtained from the basic solution, or from Hetenyi (13) to be

$$u_{0T} = \frac{2T_0\lambda}{k}\left(\frac{\sinh \lambda l \cosh \lambda l - \sin \lambda l \cos \lambda l}{\sinh^2 \lambda l - \sin^2 \lambda l}\right) \tag{8.130}$$

The expression for the moment M produced is complicated, and it is not possible to obtain the maximum moment in a simplified form; instead, it must be obtained by evaluating the following expression:

$$M = \frac{T_0}{\lambda}\left(\frac{\sinh \lambda l \sin \lambda z \sinh \lambda(l-z) - \sin \lambda l \sinh \lambda z \sin \lambda(l-z)}{\sinh^2 \lambda l - \sin^2 \lambda l}\right) \tag{8.131}$$

Both of these expressions have been evaluated for a variety of beam stiffness, in terms of λl, in Appendix C for the special case in which the load is applied at the end.

If the pile is loaded by a clockwise moment M_0 at its upper end, the deflection at that end, u_{0M}, is obtained from the equation

$$u_{0M} = \frac{2M_0\lambda^2}{k}\left(\frac{\sinh^2 \lambda l + \sin^2 \lambda l}{\sinh^2 \lambda l - \sin^2 \lambda l}\right) \tag{8.132}$$

In this case, the maximum moment occurs at the loaded end, but since combinations of force and moment loading occur in practice, the distribution of moment may be required. It is given by the equation

$$M = \frac{M_0\{\sinh \lambda l[\sinh \lambda(l-z)\cos \lambda z + \cosh \lambda(l-z)\sin \lambda z \\ - \sin \lambda l(\sinh \lambda z \cos \lambda(z-l) + \cosh \lambda z \sin \lambda(z-l)]\}}{\sinh^2 \lambda l - \sin^2 \lambda l} \tag{8.133}$$

The rotation θ_{0M}, at the pile top, caused by the moment M, is also required; it appears in the expression

$$\theta_{0M} = \frac{4M_0\lambda^3}{k}\left(\frac{\sinh \lambda l \cosh \lambda l + \sin \lambda l \cos \lambda l}{\sinh^2 \lambda l - \sin^2 \lambda l}\right) \tag{8.134}$$

The displacement and moment variations, as a graphical expression of equation (8.133) along beams of varying stiffness (λl), are shown in Appendix C for the finite beam or pile.

As in the previous cases of axial load and torsion, this problem has been tackled by Poulos (22) by the same method as before, through the appropriate integration of the Mindlin equation, this time in the lateral direction to simulate the elastic half-space. However, the pile cross section was not represented in this study; instead, the pile was assumed to consist of a thin strip of width d and appropriate stiffness, as expressed by EI. Thus, the shearing stresses developed between a real pile and the medium along the sides at right angles to the direction of displacement were not taken into account. In such an analysis, as in the Winkler model, there are a number of features of interest from a design point of view. The lateral deflection at the top of the pile arising both from applied force and moment is of concern, naturally. In some practical circumstances, however, the pile is loaded some distance above the ground surface, and, at such a load point, the displacements are affected by both the deflection and slope of the pile at ground surface; consequently, the slope must also be known. Lastly, the maximum moment in the pile must be compared, usually with the pile strength, to ensure that it is not exceeded. The deflections and slopes can be related to the loads by dimensionless compliances C as follows:

$$u_{0T} = C_{uT}\frac{T_0}{E_s l}; \qquad u_{0M} = C_{uM}\frac{M_0}{E_s l^2}; \qquad \theta_{0T} = C_{\theta T}\frac{T_0}{E_s l^2}; \qquad \theta_{0M} = C_{\theta M}\frac{M_0}{E_s l^3}$$

$$(8.135)$$

where u_0 and θ_0 are displacement and rotation at ground surface, respectibely, due to horizontal load and moment acting at the surface. The subscripts denote which agency, T_0 or M_0, is causative. It follows from reciprocity that the compliance, C_{uM}, describing the displacement due to a moment, is equal to that, $C_{\theta T}$, giving the rotation caused by a force. Figure 8-17(a) through (c) shows the compliances calculated by Poulos for the three cases. It is common practice to cast the tops of piles arranged in a group into a pile cap to an extent that the top of each pile can be considered to move laterally under horizontal loads without rotation. The pile top or head is said to be *fixed*. Poulos' results (22) for this compliance, C_{uF}, for the lateral displacement due to a horizontal force are shown in Figure 8-17(d). Once again, the parameters in these diagrams are l/d, the length/diameter ratio of the pile, and a relative pile/soil stiffness, K_R, which is expressed by the equation

$$K_R = \frac{EI}{E_s l^4} \tag{8.136}$$

The other variable of interest, the maximum moment in the pile, as obtained in Poulos' analysis, is shown in Figure 8-18, for the pile subjected to lateral load at the top.

As in previous pile problems, the following question arises: Taking the results of Poulos as being essentially exact for the half-space case, how adequate is the Winkler model in this case, considering the convenience of using it? Poulos makes a comparison between the two approaches, basing it on a value of $k(= k_0 d)$ obtained by identifying the deflection of a fixed-head pile translated laterally in the soil (this is equivalent to a rigid beam displaced vertically by a symmetrical loading). However, the value of k_0 arrived at $(0.82 E_s/d)$ is biased by the fact that rigidity is an extreme condition. We can do better, particularly for the more common situation of a relatively flexible or long pile, by returning to the matching considerations of Chapter 5. Using equation (5.57) as a guide, but avoiding its complexity, we choose

$$k_0 = \frac{E_s}{d} \tag{8.137}$$

which has the considerable advantage of being easy to recall. In the Winkler solutions, equations (8.130) and (8.132), the only variable is λl, which

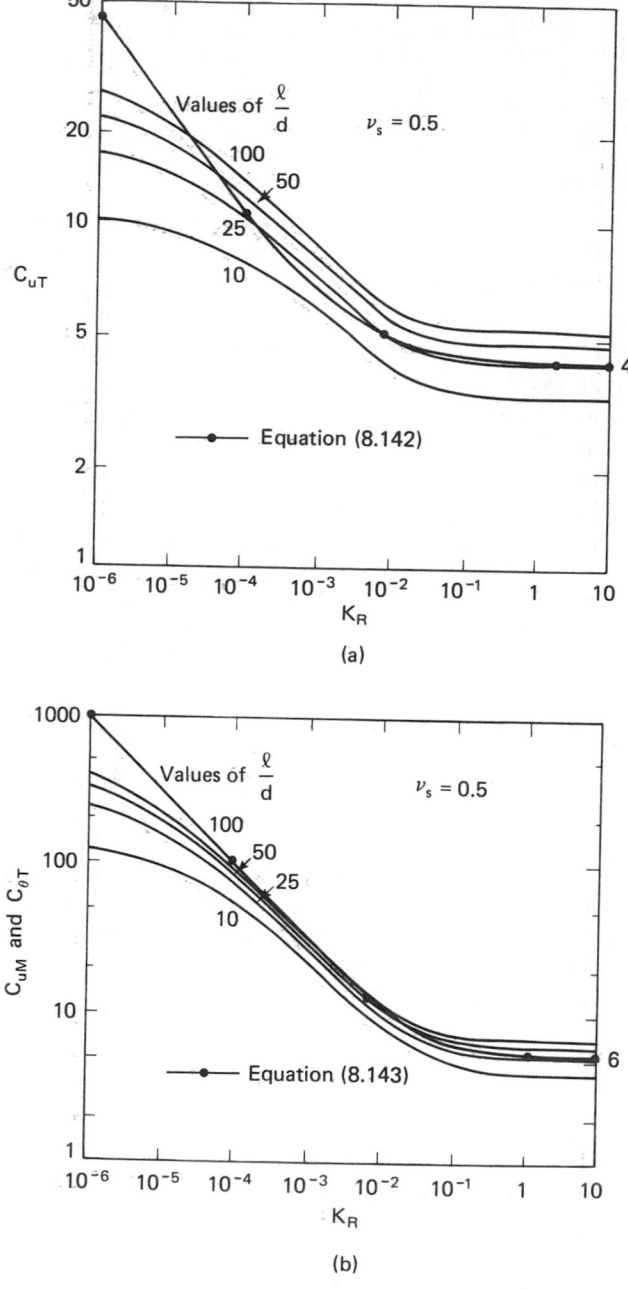

Figure 8-17. Deflection of flexible pile under lateral load as a function of relative pile-soil stiffness and length-to-diameter ratio: (a) free-head pile laterally loaded; horizontal deflection at top; (b) free-head pile; top deflection under moment load or top rotation under lateral load;

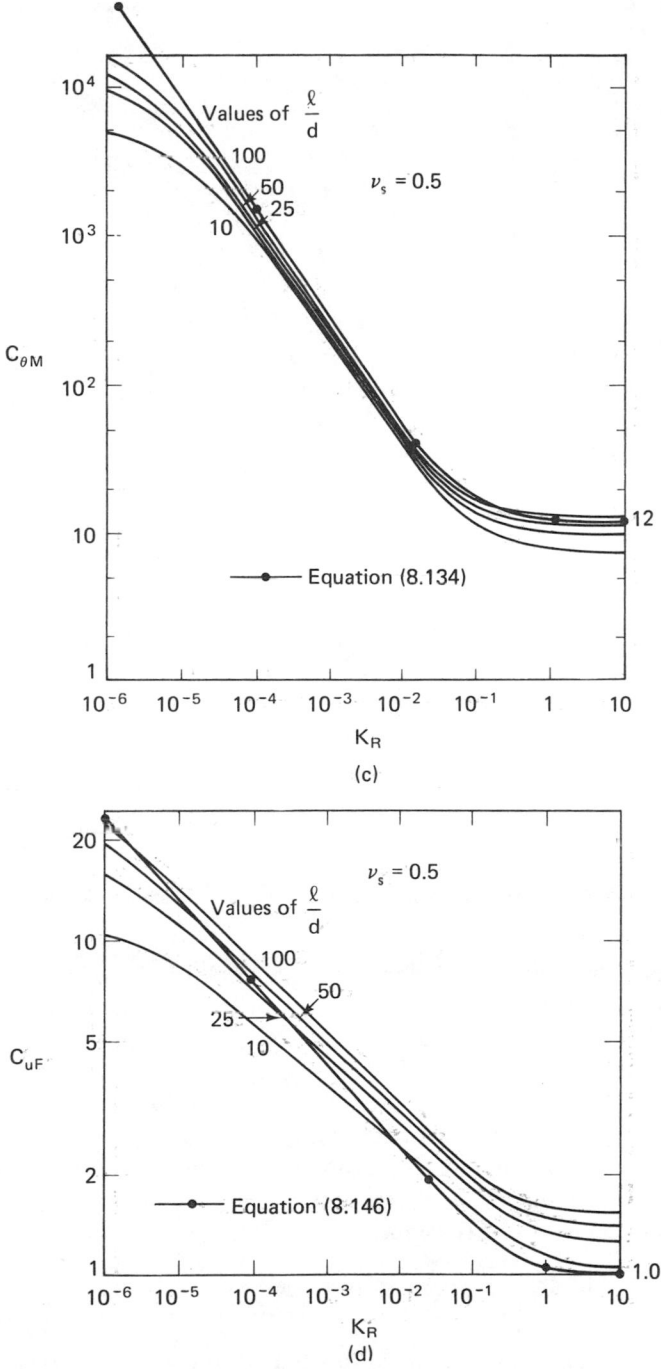

Figure 8-17. (Cont.) (c) free-head pile; top rotation under moment load; (d) fixed-head pile; top deflection under lateral load.

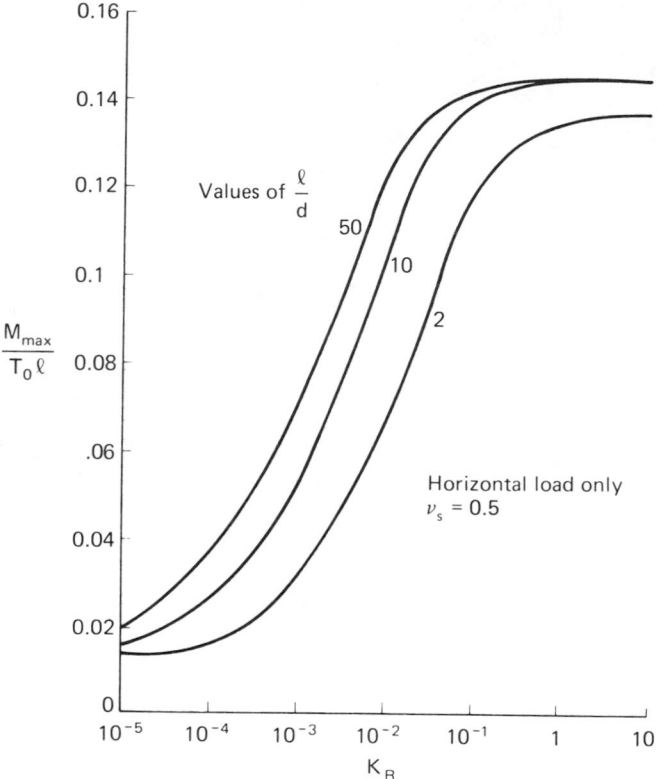

Figure 8-18. Free-head pile under lateral load; maximum moment versus relative pile-soil stiffness and length-to-diameter ratio.

describes, as we have seen before, the relative flexibility of pile and soil; the length-to-diameter ratio does not play a part. Since, as seen in Figures 8-17 and 8-18, the latter ratio does affect the results of Poulos, it follows that the k_0 of equation (8.137) should also be a function of l/d. If we choose to neglect this relation in establishing k_0, there will be one Winkler result to compare with the range of l/d solutions of Poulos. Using equation (8.136), we can express equation (8.130) as an example, in terms of C_{uT} and K_R. The other equations may be developed similarly.

From equation (5.12) we have

$$\lambda^4 l^4 = \frac{k_0 d l^4}{4EI} \tag{8.138}$$

in which equation (8.137) may be substituted to give

$$\lambda l = \left(\frac{1}{4}\right)^{1/4}\left(\frac{E_s l^4}{EI}\right)^{1/4} \tag{8.139}$$

but we can introduce K_R from equation (8.136), so that

$$\lambda l = \frac{1}{\sqrt{2}} \left(\frac{1}{K_R} \right)^{1/4} \tag{8.140}$$

Thus, the function in the right side of equation (8.130) depends only on K_R. As K_R increases, λl decreases, as it should, since a small value of λl represents a stiff pile. We write equation (8.130) as follows:

$$u_{0T} = \frac{2T_0 \lambda l}{kl} f(\lambda l) \tag{8.141}$$

or

$$C_{uT} = \frac{E_s l}{T_0} u_{0T} = \frac{E_s l}{T_0} \frac{2T_0}{kl} \lambda l f(\lambda l)$$

from which, using equation (8.137), we get eventually

$$C_{sT} = 2\lambda l f(\lambda l) \tag{8.142}$$

In which λl can be replaced by K_R from equation (8.140), so that C_{uT} is a function of K_R only. A similar expression can be obtained for the other loading condition, equation (8.132):

$$C_{uM} = 2(\lambda l)^2 g(\lambda l) \quad (= C_{\theta T}) \tag{8.143}$$

in which $g(\lambda l)$ is the trigonometric expression in the right-hand side of equation (8.132). From equation (8.134) the compliance $C_{\theta M}$ is derived:

$$C_{\theta M} = 4(\lambda l)^3 h(\lambda l)$$

in which $h(\lambda l)$ is the function in parentheses on the right-hand side of equation (8.134).

The solution for the displacement u_{0F} of a fixed-head flexible finite pile can be obtained either from first principles or from the result for a finite beam loaded at its midpoint. It is (13)

$$u_{0F} = \frac{T_0 \lambda}{k} \left(\frac{\cosh 2\lambda l + \cos 2\lambda l + 2}{\sinh 2\lambda l + \sin 2\lambda l} \right) \tag{8.144}$$

and the moment required at the pile top to maintain zero slope is

$$M = \frac{T_0}{2\lambda} \left(\frac{\cosh 2\lambda l - \cos 2\lambda l}{\sinh 2\lambda l + \sin 2\lambda l} \right) \tag{8.145}$$

The relevant compliance in this case is

$$C_{uF} = \frac{E_s l u_{0F}}{T_0} = \lambda l j(\lambda l) \tag{8.146}$$

where $j(\lambda l)$ is the expression in parentheses in the right-hand side of equation (8.144). These equations are plotted in Figure 8-17 for comparison with the previous solution. The asymptotic values for rigid beams (small λl) are shown on the right edge of each figure. It is seen that the Winkler solution gives generally reasonable results in the region of pile length-to-diameter ratios of from 25 to 50, which spans the commonly encountered range. This means that the factor of two for the totally embedded pile (beam) is already incorporated in equation (8.137).

A different method of representing lateral pile deflections calculated from a Winkler model was given by Broms (6 and 7).

We proceed to the consideration of semi-infinite piles, for which the solutions for some loading cases of interest were given in Chapter 5. The maximum lateral deflection at the head of the pile resulting from load T_0, for example, is given by equation (5.46), with P_0 replaced by T_0.

Since the dimension l no longer appears in the semi-infinite pile problem, the compliances cannot be specified in their previous form including l. Instead, we define

$$u_{0T} = C_{uT\infty}\frac{T_0\lambda}{E_s}; \qquad u_{0M} = C_{uM\infty}\frac{M_0\lambda^2}{E_s}; \qquad \theta_{0M} = C_{\theta M\infty}\frac{M_0\lambda^3}{E_s} \quad (8.147)$$

By comparing the first expression of (8.147) with equation (8.141), we see that

$$C_{uT\infty} = \frac{C_{uT}}{\lambda l} = 2f(\lambda l)$$

but, from equation (5.46)

$$C_{uT\infty} = 2 \qquad (8.148)$$

which is seen to incorporate the asymptotic form of $f(\lambda l)$. Similarly, for the pile loaded by moment M_0, we can derive the following expression from equation (5.51):

$$C_{\theta T\infty} = \frac{C_{\theta T}}{(\lambda l)^2} = \frac{C_{uM}}{(\lambda l)^2} = 2 \qquad (8.149)$$

From equation (5.48), with $x = 0$, comes the asymptotic result for $4h(\lambda l)$ in the moment-rotation compliance

$$C_{\theta M\infty} = \frac{C_{\theta M}}{(\lambda l)^3} = 4 \qquad (8.150)$$

The displacement of the fixed-head, semi-infinite pile can be derived from the solution to the infinite pile subjected to lateral load, equation (5.16); the resulting asymptotic value for $j(\lambda l)$ in the compliance is

$$C_{uF\infty} = \frac{C_{uF}}{\lambda l} = 1 \qquad (8.151)$$

Through the relation expressed by equation (8.140), these results are plotted on the diagrams of Figure 8-17 and show the range of validity of the assumption of infinity or finiteness for the pile length. For λl greater than about 3, the functions $f(\lambda l)$, $g(\lambda l)$, $h(\lambda l)$ and $j(\lambda l)$ can reasonably be approximated by their asymptotic values in equations (8.148), (8.149), (8.150), and (8.151). In other words, a laterally-loaded pile with $\lambda l > 3$ can be considered to be semi-infinite.

For the pile subjected to lateral load at the top, the maximum moment occurs at a distance of $\lambda z = \pi/4$ from the end, so that the maximum moment is

$$M_{\max} = \frac{T_0}{\lambda} e^{-\pi/4} \sin \pi/4 = 0.3224 \frac{T_0}{\lambda} \tag{8.152}$$

Although the assumption of a soil resistance property uniform with depth simplifies the analysis and may fit an actual soil profile in a few cases, the modulus in most soil profiles increases with depth in some fashion. A convenient idealization for analysis is to assume the subgrade reaction to vary linearly, or with a power function of depth, as in the case of axially loaded piles.

Using equation (5.5), but in which k is now the lateral subgrade reaction coefficient, and varies with distance along the pile, the beam equation becomes, for the pile:

$$EI \frac{d^4 u}{dz^4} + k_n z^n u = 0 \tag{8.153}$$

Defining the usual dimensional parameter λ, we can write the equation in dimensionless form:

$$\frac{d^4 t}{dz^4} + h(z)t = 0 \tag{8.154}$$

in which t and z are dimensionless displacement and distance, respectively, obtained by multiplying the real values by λ, and $h(z)$ represents the function assumed for k. The characteristic λ is given in general by the equation

$$\lambda^{4+n} = \frac{k_n}{EI} \tag{8.155}$$

except when $n = 0$, for which the factor of $\frac{1}{4}$ is included in the right-hand side, to bring the equation into the usual form of equation (5.12), which applies to the constant k solution. With z in dimensionless form, $h(z)$ is expressed by the relation

$$h(z) = z^n \tag{8.156}$$

Other than numerical results, such as those obtained by the finite difference method by Matlock and Reese (16), analytical solutions of equation

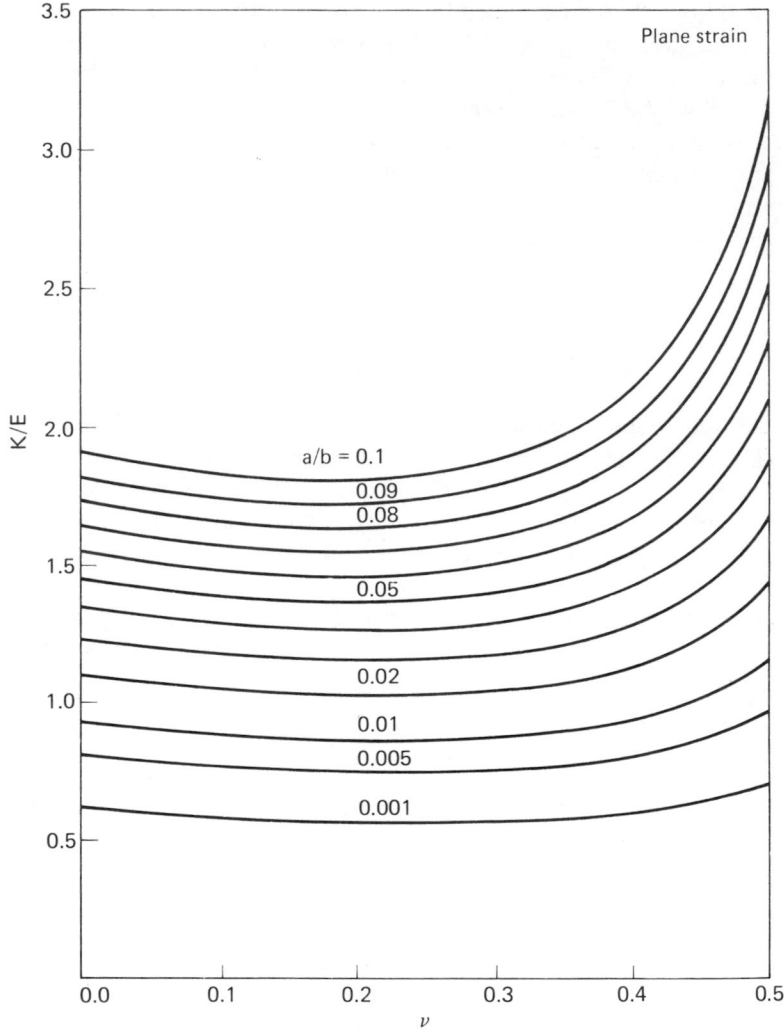

Figure 8-19. Relation among k/E_s, ν_s, and ratio a/b for lateral displacement of rigid disk, radius a, in annulus, outer radius b (5).

(8.154) have not been available. For the case $n = 1$, however, such a result has been obtained by Franklin and Scott (10). It is expressed in the form

$$t = C_1 t_1 + C_2 t_2 + C_3 t_3 + C_4 t_4 \qquad (8.157)$$

in which the t_j are four linear, independent solutions of equation (8.154) established so that t_1 and t_2 are the two functions which decrease exponentially with distance along the pile, as do $e^{-z} \cos z$ and $e^{-z} \sin z$ in the elementary solution, whereas t_3 and t_4 increase exponentially, as do $e^z \cos z$ and $e^z \sin z$. The functions t_j are represented and tabulated in Appendix C. For other values of n, a general solution in terms of four functions has not been obtained, but a limited two-function solution is possible and will be discussed later. Thus, only when $n = 1$ can the case of a finite length of pile be studied by using equation (8.157) and the results in Appendix C.

When the pile is semi-infinite in length, the asymptotic solutions are readily derived by retaining only the terms t_1 and t_2 in equation (8.157), since the deflections, moments, etc., must go to zero as z increases. Using the tabulated values in Appendix C for the functions t_1 and t_2, we find that the displacement at ground surface in this case is given by the equation

$$u_{0T} = 2.429 \frac{T_0}{\lambda^3 EI} = 2.429 \frac{T_0 \lambda^2}{k_1} \tag{8.158}$$

A dimensionless compliance can be constructed in this problem, based on the consideration that both k and E_s increase linearly with z, k_1 and E_{s1} being the constants of proportionality, respectively,

$$C_{0T\infty} = \frac{E_{s1} u_{0T}}{\lambda^2 T_0} = 2.429 \tag{8.159}$$

if it is assumed that the postulate, equation (8.137), holds regardless of depth. Similarly, the other compliances can be obtained to be

$$C_{\theta T\infty} = C_{uM\infty} = \frac{E_{s1} \theta_{0T}}{\lambda^3 T_0} = 1.619 \tag{8.160}$$

$$C_{\theta M\infty} = \frac{E_{s1} \theta_{0M}}{\lambda^4 M_0} = 1.747 \tag{8.161}$$

and

$$C_{uF\infty} = \frac{E_{s1} u_{0F}}{\lambda^2 T_0} = 0.928 \tag{8.162}$$

For problems in which the coefficient of subgrade reaction increases with some power of depth other than unity, a generalized numerical technique has been worked out by Franklin and Scott (10) for the semi-infinite pile. In the particular case of $n = \frac{1}{2}$, the solutions t_1 and t_2 are given graphically and in tabular form in Appendix C. Compliances analogous to those of equations (8.159) through (8.162) can be derived from them.

A result that will be useful in the calculation of the displacements of and loads in pile groups will now be demonstrated. In Section 8.2.1 we showed how the subgrade coefficient k_s could be calculated from the analysis of the axial displacement of a rigid disk embedded in a bounded elastic medium. A similar analysis can be performed for the laterally loaded pile. The geometry is the same as is shown in the top diagram of Figure 8-9, but now the slice of unit thickness is in plane strain in the z-direction (no vertical strains), and the disk is subjected to a lateral force F, causing it to displace u_0 in the same direction. This problem has been solved (4 and 5), and the results can be manipulated (5) into a form convenient for our purposes here. The results are too long to be given in detail, but eventually it can be shown that

$$F = k\left(\frac{a}{b}, E_s, v_s\right)u_0 \qquad (8.163)$$

in which the value of k is indicated to be dependent on the variables shown, including the Young's modulus, E_s, and Poisson's ratio, v_s, of the soil. The relation is shown in Figure 8-19 in which the ordinate is the ratio k/E_s.

Comparing Figure 8-19 with equation (8.137) (with both sides multiplied by d, to give k), which resulted in good correlation of the Winkler model with a numerical continuum analysis, we see that for a typical Poisson's ratio of about 0.3, k/E_s is about unity when $b/a = 50$, as in the case of an axially loaded pile.

8.3.3 Analysis of Arkansas River pile test (lateral load)

The lateral load-deflection and moment relationships for the laterally loaded pile discussed previously are shown in Figure 8-16. They will be used in an illustrative example, although the unloading data that would serve as the basis for an elastic analysis are lacking. In Figure 8-16(a) the load-deflection relation for the pile is not far from a straight line (although it is certainly not elastic); this is unusual in lateral load tests, but it lends a certain validity to our analysis. Since in the moment diagram, Figure 8-16(b), a moment curve has been drawn for a lateral force of 43 kips (191 kN) (T_0), which corresponds to 0.4 in. (10 mm) deflection (u_{0T}) at the ground line, we will correlate with this point. The pile property of interest is EI, given in the report (2) as 2.4×10^{10} lb-in.^{2} (68.9 MN-m^2). As in the previous axial analysis of Section 8.2.7, we will assume here a soil property increasing linearly with depth, and we will attempt to find out what it is numerically.

It is apparent from Figure 8-16(b) that the moments die out long before the pile base at 52.5 ft (16 m) depth is reached. A semi-infinite pile analysis is therefore called for. In the absence of the moment data, we would have had to assume this, and it is always a reasonable first step, at least for piles of the dimensions of the one considered, so long as the soil is not excessively soft.

For the problem thus defined, the equation of interest is (8.158) in which we can substitute the above values to obtain λ^3 and hence λ. The value derived is

$$\lambda = 2.216 \times 10^{-2} \text{ in.}^{-1} \ (0.872 \text{ m}^{-1}) \tag{8.164}$$

Substituting λ into equation (8.155) with $n = 1$ gives

$$k_1 = 128.21 \text{ pci } (34.8 \text{ MN/m}^3) \tag{8.165}$$

Thus, the value of k in this lateral load test increases at this rate with depth. We can consider this value characteristic of a medium-dense sand below the water table. Because of the presence of the protective flanges, the effective diameter of the pile is 24 in., which can be employed with equation (8.165) to give the basic soil lateral subgrade coefficient to be

$$k_{01} = \frac{128.21}{24} = 5.34 \text{ lb/in.}^4 \ (57.1 \text{ MN/m}^4) \tag{8.166}$$

At a distance of 25 ft (7.6 m), approximately halfway down the pile, we can obtain from equation (8.166) the value of k_0:

$$k_0 = 5.34 \times 300 = 1602.63 \text{ pci } (435 \text{ MN/m}^3) \tag{8.167}$$

Lastly, using the relation expressed by equation (8.137), we find that $E_{s1} = k_1 = 128.21$ pci (34.8 MN/m^3), or that, again at a depth of 25 ft (7.6 m), the lateral soil modulus E_s is given by

$$E_s = E_{s1}z = 128.21 \times 300 = 39,000 \text{ psi } (269 \text{ MN/m}^2) \tag{8.168}$$

If we assume a Poisson's ratio of about 0.3, this indicates a value of about 15,000 psi (104 MN/m^2) for G_s at this level, much higher than the 2300 (15.9) we obtained from the axial pile analysis of Section 8.2.7. Although, as we will see from the non-linear analysis of Section 9.4.3, this value is, to some extent, an artifact of the linear analysis, we might conclude that the different mechanisms involved in axial and lateral pile-soil deformations may not lead to consistent estimates of the soil property modulus. Alternatively, evaluation of a soil modulus and use of the relations between modulus and subgrade coefficient suggested in this chapter will not necessarily lead to equally good estimates of axial and lateral pile load-deformation behavior. The results may indicate an anisotropic soil at the site.

It is also worth pointing out that the value of k_1 in equation (8.165) is substantially higher than those quoted for the appropriate soil in Table 7.4. A conclusion is that Terzaghi's suggested values are also conservative in this lateral loading case.

We have now completed an adequate treatment of the *elastic* problem of soil-beam or soil-pile interaction. Although the results given can be applied in a small deflection range of the behavior of real structures, in a large proportion of situations encountered in practice, partial to complete soil yielding adjacent to the beam, slab, or pile will develop. In addition, analysis of possible failure modes of a structure connected to soil demands consideration of yielding of structural elements as well as the soil. In the following chapters the plasticity of soil and structure in such circumstances will be addressed.

PROBLEMS

1. A long hollow tube steel pile 24 inches (0.61 m) in diameter, $\frac{3}{8}$ inch (9.5 mm) wall thickness is driven into uniform clay whose lateral subgrade coefficient is taken to be 50 pci (13.6 MN/m³). A lateral load is to be applied at ground surface. What value would it have just to develop yielding in the steel at 40,000 psi (276 MN/m²)? What would be the pile top deflection at this load? Do you think the *soil* would have yielded at this deflection? What would you use as a working or design load? What are your reasons for the safety factor you selected? (See also Problem 11-5.)

2. List and describe briefly five types of pile. For each pile describe in one or two sentences a situation where the pile has a particular advantage (requires reference to sources other than this book).

3. For a lateral load applied at ground surface to the pile of Problem 1, what are the minimum lengths of the pile which would be considered "long" in a loose, medium, and dense saturated sand? Assume the subgrade reaction coefficient to increase linearly with depth.

4. A dock structure consists of two rows of MZ32 steel sheet piles fixed at their tops to a steel deck as shown in Figure P8-4. For design purposes ship impact is to be considered as a 2-ton (17.8 kN) static load applied to each 1-ft (0.3 m) section. Calculate the maximum moment in the piles and the deflection at the deck for the cases (a) where the deck is rigid, (b) where the deck has the same moment of inertia as the piles.

$$\text{For piles,} \quad \begin{aligned} I &= 220 \text{ in.}^4 \text{ per ft run } (3 \times 10^8 \text{ mm}^4/\text{m}) \\ S &= 38.3 \text{ in.}^3 \text{ per ft run } (1.79 \times 10^6 \text{ mm}^3/\text{m}) \end{aligned}$$

Are the piles adequate for each case?

5. The use of torsion for testing individual piles has been suggested. In what range of pile/soil modulus ratio will the torsional load-deflection measurement at the top of the pile reflect the *soil's* properties? How would you deduce the behavior of the same pile in the same soil under axial load?

6. A soil report by a geotechnical engineering firm recommended the use of reinforced concrete piles to take lateral loads in a structure. The choice of pile size was left to the designer, to be taken from the following table. The table gives the lateral load in kips (1000 lb) (4.5 kN) which would produce $\frac{1}{2}$ inch (13 mm) lateral deflection at the pile top. Because of the loose condition

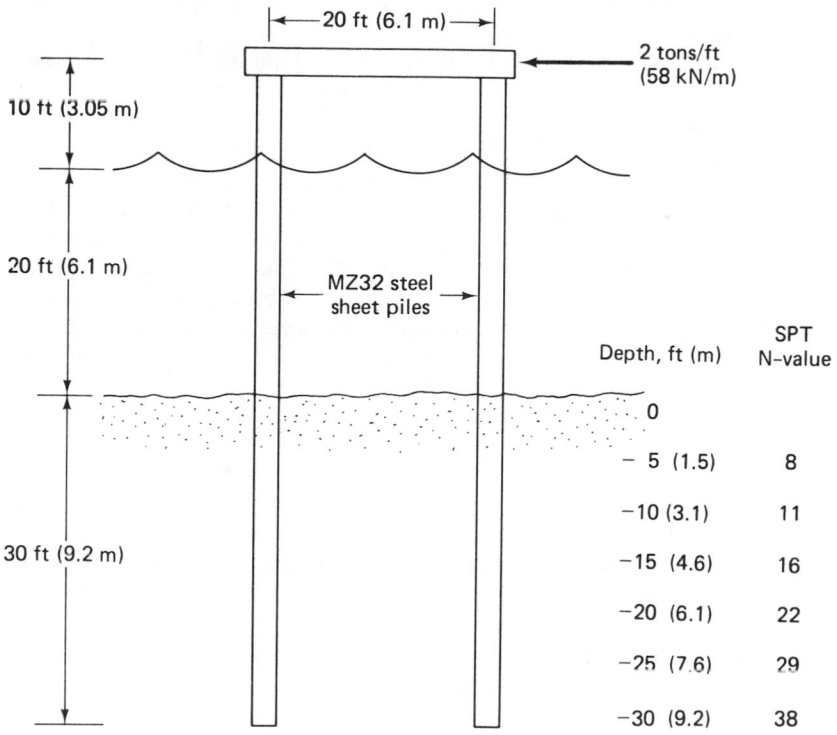

Figure P8-4.

of the top 15 ft (4.6 m) of soil profile, it was also assumed that the top 15 ft (4.6 m) of each pile was not supported by the soil. Two cases of pile restraint at the *top* were considered: (a) fixed (zero slope), (b) free (pinned). Consider the piles to be of *square* cross section.

Reinforced Concrete Pile Width, in. (m)	Lateral Load to Give $\frac{1}{2}$in. Lateral Deflection, kips (kN), at Top of Pile	
	(a) Fixed Head	(b) Free Head
16 (0.41)	2.8 (12.5)	0.7 (3.1)
24 (0.61)	9.6 (42.7)	2.5 (11.1)
36 (0.91)	25.0 (111.3)	6.6 (29.4)

Answer the following questions:

(a) Assume the piles are very long; what value of coefficient of subgrade reaction in a horizontal direction (k) is consistent with the loads given? If the soil into which the piles are to be driven is a sand, do you conclude it is loose, medium dense, or dense?

    (b) Is your answer to (a) changed by considering the piles to be short and rigid?

    (c) For the value of k you get for (a) how long would each pile (width shown in table) have to be, to be considered very long? At what length would it be considered short?

7. (a) For the design of another structure at the site described in Problem 6, the report suggests using a *vertical* coefficient of subgrade reaction of 100 pounds per cubic inch (27.2 MN/m³). Is this consistent with the soil conditions you deduce in Problem 6(a)? [Ignore the top 15 ft (4.6 m) of soil.] (b) Calculate the *vertical* bearing capacity of the 24 inch width pile, taking its length to be 50 ft (15.2 m). (c) Why do you think piles were suggested for the site, considering the information given above the table? For a heavy, important structure, would you suggest any other foundation type? Discuss advantages and disadvantages *briefly*.

8. A structure in a seismic area is to be built on a thick reinforced concrete mat supported on a large number of 16 inch (0.41 m) square concrete piles, fifty feet (15.2 m) long, driven into a medium-dense sand. To help resist possible lateral earthquake loads it is proposed to drive the two outer rows of piles at an angle (batter) of 1 horizontal on 6 vertical into the soil. Calculate how much the horizontal stiffness of a batter pile at the pile cap in FL^{-1} units exceeds that of a vertical pile. Based on the results of your calculations do you consider it would be worth using batter piles? Take the pile $EI = 2 \times 10^{10}$ pound-inches squared (57 MN-m²).

9. Calculate the maximum moment in the Arkansas River test pile of Section 8.3.3. Compare with measured value shown in Fig. 8-16 (b) at 43 kip (191 kN) lateral load and discuss. What lateral load would be required to bring the pile: (a) just to yield in bending at the extreme fibers; (b) to complete yielding, if the soil remained linearly elastic (Winkler)? Are the displacements such that this soil behavior is likely?

10. Set up the equations for an axially-loaded pile which is embedded in a two-layer soil profile. Assume that the tip load is negligible. The upper layer is of depth h, less than the pile length.

11. For the steel pipe pile of the Arkansas River tests, perform an elastic axial load analysis at a load of 150 tons (1.34 MN), assuming that the soil subgrade coefficient is *uniform* with depth and obtain a value for the coefficient. Compare it with the number computed in the analysis in the text.

12. Carry out a lateral load analysis on the Arkansas River pile at the load of 43 kips (191 kN), assuming a lateral subgrade coefficient uniform with depth. Calculate the coefficient required and compare it with the axial value from Problem 11 and the nonuniform value in the text.

13. (a) Calculate the immediate settlement of a 10 m (32.8 ft) long rigid pile 0.6 m (1.97 ft) in diameter embedded in a 20 m (65.6 ft) thick clay layer, under a load of 200 kN (45 k). (b) Calculate the ultimate settlement of the pile after consolidation is complete. The properties of the clay are: $G = 1$ MN/m² (145 psi) $v = 0.3$(drained).

14. If a pile passes through several layers of soil of different properties, a continuous Winkler analysis for either vertical or lateral loads is too tedious to

undertake. In this case a lumped or discrete parameter model can conveniently be used to solve the problem. For this solution, the pile is divided into rigid sections, each of which is attached to its neighbors by springs representing the stiffness of a pile section. Further, each section is connected to the soil by another spring simulating the stiffness of that soil surrounding the pile. By relating the forces and displacements in each pile section, a series of equations results; they can be solved simultaneously. Set up such an equation set for an axially-loaded pile divided into four segments, with a different soil property related to each segment.

REFERENCES

[1] ABRAMOWITZ, M., AND I. A. STEGUN, eds. *Handbook of Mathematical Functions*. New York: Dover, 1965.

[2] ALIZADEH, M., AND M. T. DAVISSON. "Lateral Load Tests on Piles—Arkansas River Project," *Proc. ASCE, 96,* SM5, 1583–1604 (Sept. 1970).

[3] BAGUELIN, F., AND R. FRANK. "Theoretical Studies of Piles Using the Finite Element Method," *I.C.E.(U.K.), Proc. Conf. on Numerical Methods in Offshore Piling,* 65–73 (May 1979).

[4] ———, AND Y. H. SAID. "Theoretical Study of Lateral Reaction Mechanisms of Piles," *Geotechnique, 27,* 405–434 (1977).

[5] BARDET, J.-P. "Determination of a Winkler Modulus for a Rigid Cylinder Embedded in an Elastic Medium," paper submitted to ASCE, Dec. 1979.

[6] BROMS, B. B. "Lateral Resistance of Piles in Cohesive Soils," *Proc. ASCE, 90,* SM2, 27–63 (March 1964).

[7] ———. "Design of Laterally Loaded Piles," *Proc. ASCE, 91,* 3M3, 79–99 (May 1965).

[8] COYLE, H. M., AND I. H. SULAIMAN. "Skin Friction for Steel Piles in Sand," *Proc. ASCE, 93,* SM6, 261–277 (Nov. 1967).

[9] EVANGELISTA, A. "Inclined Piles, Alone and in a Group, Imbedded in an Elastic Medium," *Rivista Italiana di Geotecnica, 10,* No. 3 (July-Sept. 1976) (in Italian).

[10] FRANKLIN, J. N., AND R. F. SCOTT. "Beam Equation with Variable Foundation Coefficient," *Proc. ASCE, 105,* EM5, 811–827 (Oct. 1979).

[11] FREEMAN, N. J., AND L. M. KEER. "Torsion of a Cylindrical Rod Welded to an Elastic Half-Space," *Trans. ASME, J. Appl. Mech.,* 687–692 (Sept. 1967).

[12] GRILLO, O. "Influence Scale and Influence Chart for the Computation of Stresses Due, Respectively, to Surface Point Load and Pile Loads," *Proc. 2nd ICSMFE, 6,* 70–73 (1948).

[13] HETENYI, M. *Beams on Elastic Foundation.* Ann Arbor: University of Michigan Press, 1946.

[14] HOLLOWAY, D. M., G. W. CLOUGH, AND A. S. VESIC. "Mechanisms of Pile-Soil Interaction in Cohesionless Soil," Duke University Soil Mechanics Series No. 39 (1975).

[15] MANSUR, C. I., AND A. H. HUNTER. "Pile Tests—Arkansas River Project," *Proc. ASCE*, *96*, SM5, 1545–1582 (Sept. 1970).

[16] MATLOCK, H., AND L. C. REESE. "Generalized Solutions for Laterally-Loaded Piles," *Proc. ASCE*, *86*, SM5, 63–91 (Dec. 1960).

[17] MATTES, N. S., AND H. G. POULOS. "The Settlement of a Single Compressible Pile," *Proc. ASCE*, *95*, SM1, 189–207 (Jan. 1969).

[18] MINDLIN, R. D. "Force at a Point in the Interior of a Semi-Infinite Solid," *Physics*, *7*, 195–202 (1936).

[19] MUKI, R., AND E. STERNBERG. "Elasto-Static Load-Transfer to a Half-Space from a Partially-Embedded Axially-Loaded Rod," *Int. J. Solids and Structures*, *6*, 69–90 (1970).

[20] MURFF, J. D. "Response of Axially Loaded Piles," *Proc. ASCE*, *101*, 356–360 (March 1975).

[21] POULOS, H. G. "Analysis of the Settlement of Pile Groups," *Geotechnique*, *18*, 449–471 (1968).

[22] ———. "Behavior of Laterally Loaded Piles: I. Single Piles," *Proc. ASCE*, *97*, SM5, 711–731 (May 1971).

[23] ———. "Behavior of Laterally Loaded Piles: II. Pile Groups," *Proc. ASCE*, *97*, SM5, 733–751 (May 1971).

[24] ———. "Torsional Response of Piles," *Proc. ASCE*, *101*, GT10, 1019–1035 (Oct. 1975).

[25] ———, AND E. H. DAVIS. "The Settlement Behavior of Single, Axially-Loaded Incompressible Piles and Piers," *Geotechnique*, *18*, 351–371 (1968).

[26] SCOTT, R. F. Class notes, California Institute of Technology, 1968.

[27] ———. "The Freezing Process and Mechanics of Frozen Ground," U.S. Army CRREL, Cold Regions Science and Engineering, Monograph II-D1 (Oct. 1969).

[28] STEINBRENNER, W. "Tables for Settlement Calculations," *Die Strasse*, *1*, 121–124 (1934) (in German).

[29] STOLL, U. W. "Torque Shear Test for Cylindrical Friction Piles," *Civil Engineering*, 63–65 (April 1972).

Pile Analysis
Including Yielding   9

9.1 INTRODUCTION

In the analyses of piles under axial, torsional, and lateral loads described in the previous chapter, both pile and soil were considered to behave linearly elastically. However, in all the analyses, regardless of the soil model adopted, it was found that fairly high stresses could be generated in the soil adjacent to the top of the pile even by relatively low loads. In practice, under most conditions, these stresses would exceed the shearing strength of the soil, and for an analysis to be realistic, this effect should be taken into account. When the loading is axial, the stresses in the pile itself are fairly low at normal working loads, so that *pile* yielding in this mode is seldom a question. For the similar case from the analysis point of view, as treated here—torsion— it *is* possible to stress a pile to yield at loads that might be in a working range from the soil stressing point of view, and this must be considered in such a loading condition. In the case of a laterally loaded pile, development of yield moments in the pile itself may be the primary consideration in establishing the load range of the pile, and it must be included in the analysis.

Yielding in soil is a complex phenomenon, the object of a considerable research effort in soil mechanics. We will not treat it in detail here but will refer the reader to a summary of yield and plasticity information given in Appendix B. Our consideration of the yield stresses in the soil adjacent to a pile will be simplified in a manner analogous to that in which we handled elastic soil behavior in Chapter 8, with the addition of plasticity, but the results are sufficiently realistic to describe a number of practical circumstances of interest. However, some preliminary aspects of yielding are worth a little more discussion.

The stress-strain behavior of an elasto-plastic element, say steel (although the argument applies also to soil), under homogeneous stressing conditions in, for example, a tension or compression test, is frequently taken to be as shown in Figure 9-1(a). A portion of the behavior of OA is taken to be linear and reversible, whether this is real or imagined. At A nonlinear partially irreversible behavior due to slips between or in the basic building blocks (crystals or grains) of the material develops and by point B this is complete; the element yields at constant load or stress. After point A unloading and reloading occurs along paths such as DE, in the vicinity of D at least, more or less parallel to OA. Since this behavior is complicated, for analysis purposes it is usually idealized to that shown in Figure 9-1(b).

If a structural component, such as a beam, is made up of elements with the idealized behavior of Figure 9-1(b), then its behavior in bending (or torsion, or combined loads) follows the curve of Figure 9-1(c). Up to point A in Figure 9-1(c) all the elements of the component are stressed linearly elastically; at A the most highly stressed elements reach the yield condition of point A in Figure 9-1(b). However, since the other elements are still behaving linearly elastically, the generalized load-strain curve does not become horizontal at A in Figure 9-1(c). To achieve this, the load (moment, torque) must be increased until all elements of the beam reach the state represented by point A in Figure 9-1(b). The beam, however, is only a component of a still larger structural assemblage, for whose analysis the behavior of Figure 9-1(c) is again too complicated. Consequently, the generalized load-strain curve of Figure 9-1(c) is idealized back to the shape of Figure 9-1(b) as in Figure 9-1(d); now the load ordinate is moment, torque, or axial load, and the strain abscissa is hinge angle, twist, or axial displacement.

With a return to the idealized behavior shown by Figure 9-1(b) or (d), but with different slope and yield quantities associated with different members, analysis of the structural assemblage proceeds. It will give rise to an assemblage response as illustrated in Figure 9-1(e). The initial straight line behavior OA is terminated by yielding of a structural component at the load corresponding to point A. The load can be increased, with that component yielding along the path represented by AC in Figure 9-1(d), until another component yields at the load corresponding to point B. When all the components have reached yield, the path CD is followed. Were this assemblage part of a total structure, its behavior again would be simplified to the basic

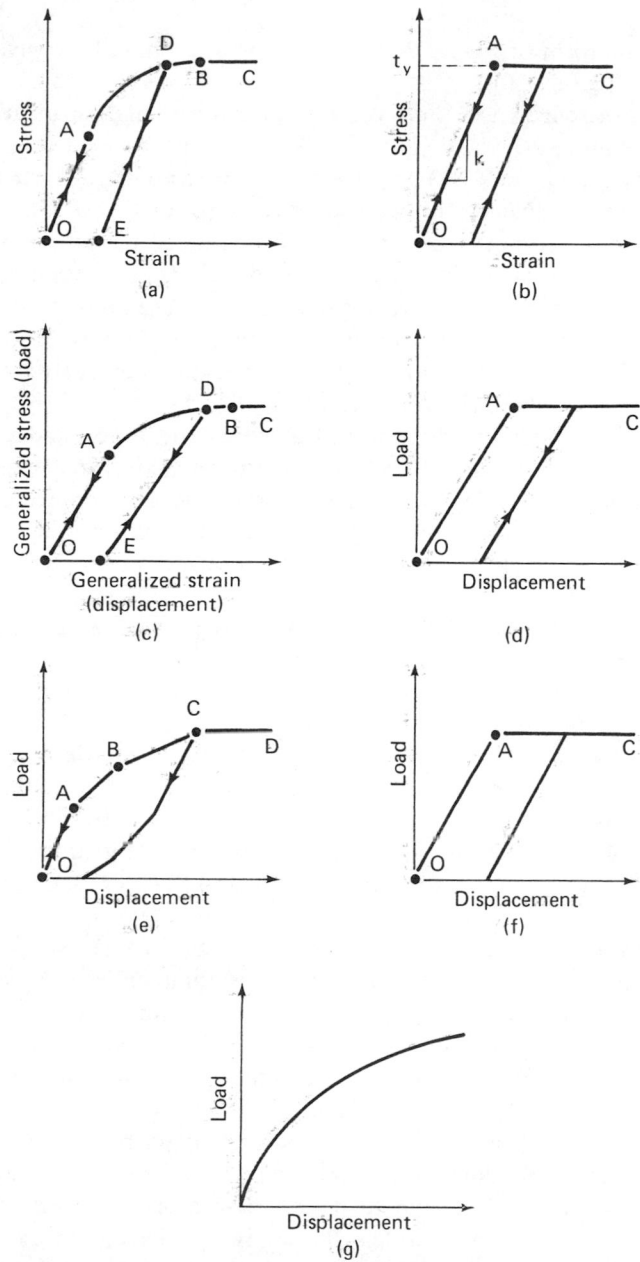

Figure 9-1. Sequence of yielding in structural elements, components, and assemblages: (a) element behavior, somewhat idealized; (b) element behavior, more idealized; (c) component behavior; (d) component behavior, idealized; (e) assemblage behavior; (f) assemblage behavior idealized. (g) structure behavior.

elastoplastic form of Figure 9-1(f) for analysis, eventually to give the result shown in Figure 9-1(g).

It is apparent that in the succession of analyses and idealizations represented by Figure 9-1, the details of the representation of element or component behavior, Figure 9-1(a) or (b), are of little or no importance to the final structural load-displacement curve, Figure 9-1(e) or (g). In deriving Figure 9-1(g), it does not matter how the curve is assumed to run from A to B in the element or component Figure 9-1(a) or (c). It is of somewhat greater importance where the point A is located in the idealizations of Figure 9-1(b) or 1(d), but even this is not crucial to the final result. Only in special structures (5) where every element may achieve yield simultaneously is the detailed performance of components significant.

When the behavior of the top of a pile is concerned under axial, torsional, or lateral loading, these considerations apply. Soil elements yield first at the top of the pile; as the load increases, yielding progresses down the pile. Soil elements are successively involved. It follows that the intricate details of the soil behavior may not be of concern in the analysis. This point will be examined in the cases studied. The soil behavior will be represented by curves of the idealized shape of Figure 9-1(b), (d), or (f). Only when a very rigid pile is loaded axially will almost all soil elements reach yield more or less simultaneously.

Once again, it is convenient to base our analysis on the Winkler foundation model. Although in its original form this representation accounted for soil behavior by a linear spring, it is not necessarily restricted to that function, and behaviors such as those shown in Figure 9-1(a) or (b) can be employed. In addition to yielding, the spring element can be extended to include time-dependent properties by the addition of dashpots or general nonlinear behavior.

The more complicated models generally require the use of computers for analysis and are not adapted to simple preliminary design calculations. As described in Chapter 8, linear Winkler spring representations were employed for the soil in pile analyses by Matlock and Reese (10). Subsequently the same authors and co-workers used a Winkler model of the type shown in Figure 9-1(a) (8, 9, 16, and 17) while that illustrated in Figure 9-1(b) was included in computational methods described by Murff (13).

To derive the nonlinear relation,[Figure 9-1(a)] for use in their analyses of laterally-loaded piles from field and laboratory soil test results, the American Petroleum Institute (1), Reese (16), and Reese, Cox, and Koop (17) suggest a detailed step-by-step procedure of some complexity. For the reasons suggested above, this level of complication is not considered essential to the development of a workable analysis technique for piles. However, because of its application in practice, the technique is set out in Section 9.5.

The approach described here thus employs a Winkler model in the form of Figure 9-1(b).

9.2 SOIL BEHAVIOR

If a relation of the form of Figure 9-1(b) is to be used to represent soil behavior, it requires that two soil properties be identified: a spring or elasticity constant k and a yield value t_y. These are related to the element behavior of Figure 9-1(a) as obtained from soil tests and the pile width. The relationship of the spring constant to soil and pile properties has been dealt with in Chapter 8; a more detailed discussion of the yield value is deferred until later. In this section, for the present, it will be assumed that these properties can be assumed or extracted from the data available. Attention will therefore be directed to the ways in which the properties may vary with depth. The pile cross section is taken here to be constant with length, but, if required, a variation can be incorporated in the analysis.

In the following discussion the parameters k and t_y may apply to axially, torsionally, or laterally loaded pile conditions. The pile dimension is included in these values; that is to say, k, here, equals Sk in Chapter 8. The simplest model of a soil profile is that in which both soil properties, $k(FL^{-2})$ and $t_y(FL^{-1})$, are constant with depth, as shown in Figure 9-2(a). This might fit the behavior of certain overconsolidated clays. As the pile deflects, the soil resistance at any depth is linear up to the value of deflection, s_* (axial, torsional, or lateral), given by the ratio t_y/k.

For higher deflections, the soil resistance is constant. Since both properties are constant with depth, s_* is also constant with depth. Because a pile in all deformational modes is loaded at the top, the deflection at the top is greater than that at depth and will reach the value s_* first. Up to the point at which s_* is attained at zero depth, the total load-deformation curve is linear; thereafter it exhibits softening and generally follows a curve of the type shown as OAB in Figure 9-1(c). The extent to which it flattens out after B depends on the mode of loading and the pile and soil properties.

For a first attempt at a more complicated profile, it might be postulated that one of the properties, k or t_y, remains constant with depth, while the other varies linearly. However, this behavior would not logically fit a realistic soil profile. If the stiffness increases with depth, so, too, should the strength. For any reasonable model of soil behavior varying with depth, it would be expected that both properties would change; the first obvious choice would be to make the variation linear, as in Figure 9-2(b), with zero values at the surface. Such a profile might be considered to represent idealized sand or normally consolidated clay behavior. Thus, $t_y = t_{y1}z$, $k = k_1z$. The value of deflection s_* at which linear behavior terminates is again given

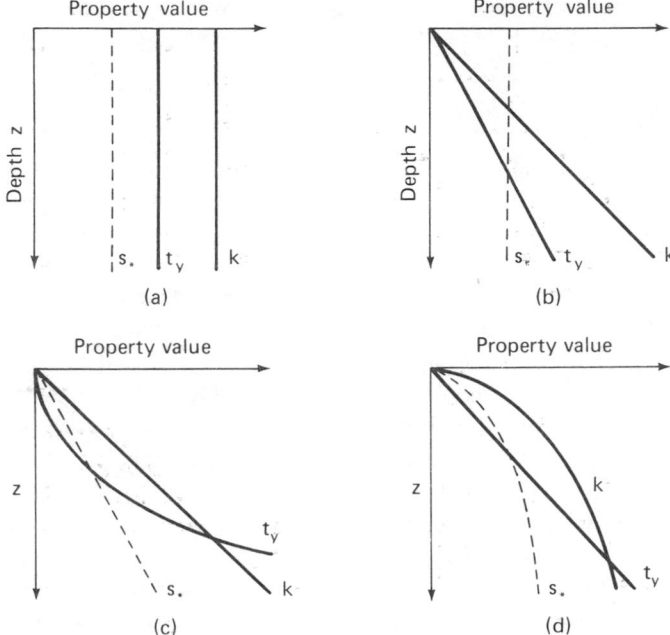

Figure 9-2. Variation of yield stress, t_y, coefficient of horizontal subgrade reaction, k, and yield deflection, s_*, with depth: (a) properties constant with depth; (b) properties linear with depth; (c) t_y parabolic, k linear with depth; (d) t_y linear, k varying with one-half power of depth.

by the ratio t_y/k, which, with the linear variation, leads once more to s_* constant with depth and equal to t_{y1}/k_1.

Nevertheless, although this model appears reasonable, the implication of the mathematics here hardly corresponds to intuition. These linear relations require that a pile embedded in a material with zero shear strength at ground surface can be deflected a finite distance s_* at the ground surface with a linearly elastic relation between load and pile top deflection up to this deflection before nonlinear plastic effects come into play. It must be pointed out that there is no mathematical difficulty in this. Although the displacement is finite, the stress developed in the soil at the ground surface is zero because the constant k is also assumed to be zero there. However, experimental studies of pile behavior in sands, for example, for which this model might be expected to be reasonable, indicate that irreversible deformations are caused at and near ground surface at all values of pile load. In parallel it is observed that for piles in sand no linear portion of behavior such as OA in Figure 9-1(c) develops. This implies that s_* is zero at ground surface. The consequence is that the variation of either t_y or k must be different from linear with depth. Either or both variations may be nonlinear, but they cannot follow the same power function, or the deflection s_* will again be constant with depth. For s_* to be zero at ground surface, the yield stress must be a higher power function of depth than the spring constant. Although they do not discuss this question in their paper in which they analyze lateral load tests on piles in sand, Reese, Cox, and Koop (17) suggest a mechanism for yielding of the soil near the top of a pile which gives rise to a yield pressure increasing as the square of the depth. In this case, with k constant or varying with a power of depth less than 2, the limiting deflection $s_* = u_*$ (lateral load case) is zero at the ground surface. A quadratic variation of t_y may be possible, for lateral loading, but it is not easy to see how this could develop for vertical or torsional loading. In the latter cases, the yield load would be more likely to be related to the lateral (horizontal) soil stress which is presumably close to linear with depth in many practical situations.

With respect to the spring constant, a number of soil investigations (18) have indicated Young's or the shear modulus of sand to vary as the one-half or one-third power of the overburden pressure. It is reasonable to expect that the spring constant k might correspond to this behavior. Thus, for example, a linear variation of the yield load t_y with depth, and variation of k with the one-half power of depth, would give a limiting deflection s_* also changing with the one-half power of depth and zero at ground surface. In general, as will be seen below, the complexity of the solutions is not affected by different or complicated variations of the yield load but is increased to some extent by more involved spring constant functions of depth. The identification of these properties for specific soils will be discussed later.

9.3 AXIAL AND TORSIONAL PILE LOADING

In the axial pile tests performed at the Arkansas River site, and described in Section 8.2, plastic yielding developed in the soil adjacent to the pile or at the pile-soil interface as the load was increased. The load-deflection behavior at the pile top reflected this yielding; an analysis of this pile test will be presented later. The response of the Arkansas River test piles also depended on the axial flexibility of the piles with respect to the soil, but it is more appropriate to begin a general elastoplastic analysis by considering the pile to be rigid.

9.3.1 Rigid pile; axial or torsional load

Once again, as in Section 8.2.1, the basis for the analysis is pile displacement constant along its length. As we have seen in the previous section, yielding of the soil can affect the pile response in two ways. Either the variations with depth of both the yield stress and the spring constant are similar, so that the vertical displacement $w_*(= s_*)$ at which yielding is initiated is constant with depth [Figure 9-2(a) and (b)], or the depth variations differ in such a way that $w_*(s_*)$ increases with depth and may or may not be zero at ground surface [Figure 9-2(c) and (d)].

In the first case, at an elementary level, the calculations are particularly simple: the load-deflection relation is linear, following equation (8.6), up to the value of load P_*, at which the vertical displacement of the entire pile reaches w_*, at which yield commences all along the pile. Equation (8.6) therefore also relates P_* and w_*. The load is constant for subsequent displacements, so that the load-displacement relation is similar to the idealized behavior of Figure 9-1(b). This presupposes that the displacement required to yield the soil below the tip of the pile is the same as that generating yield along the side. If the tip displacement at yield, w_{*t}, which may be of the order of 5 percent of the pile diameter, is larger than that along the pile perimeter, then the pile load-displacement behavior is still linear up to displacement w_*, but the load is not constant thereafter. Instead, a further straight-line portion of the load-displacement curve follows yielding along the side and is caused by the additional (assumed) elastic behavior (as shown in Figure 9-3) of the soil below the tip. The load versus pile displacement relation has the appearance of segments OAB of Figure 9-1(e), and it is flat thereafter. In this case, the point B occurs at a displacement of w_{*t}, and the total yield load taken by the pile is given by the sum of the load required to cause yielding along the side of the pile and the tip yield load. The tip yield load may be obtained by bearing capacity computations following Terzaghi, as they have been modified by Meyerhof (12) and Vesic (21 and 22). It generally makes a contribution of less than 30 percent of the total ultimate load that the pile can bear, unless the tip is driven into a denser, stronger material.

If the elastic and plastic soil properties are such that w_* increases with depth, then a deflection $w > w_*(z_0)$ at depth z_0 will develop yielding in the

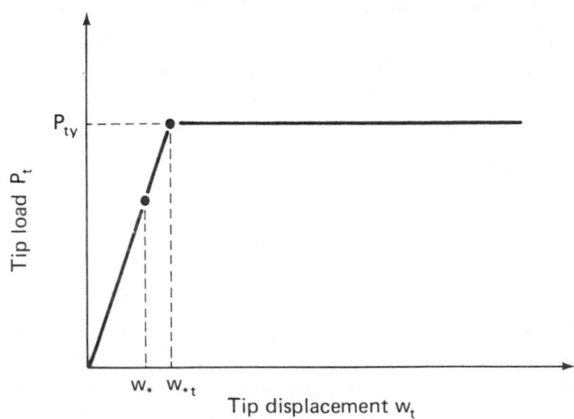

Figure 9-3. Relation assumed between pile tip load and tip displacement.

soil above z_0, while the soil below z_0 is still in the elastic range. The load at the pile top for this deflection is obtained from the sum of the soil yield stresses around the pile perimeter to z_0 and the force to cause the elastic deflection below.

9.3.2 Compressible pile; axial or torsional load

As the load on the pile is increased, the shear strength of the soil at the top of the pile will be exceeded first in all the cases under study. At greater loads, a zone of yielding soil will extend to a depth z_0, below which the elastic behavior described previously still holds. Ultimately, the yielded region will extend all the way to the tip of the pile; when this occurs and yielding develops below the tip, the load capacity of the pile is exceeded.

The yield stress t_y in the soil at the soil-pile interface is specified as one of the functions of depth illustrated in Figure 9-2. If it is put in dimensionless form, $h(z)$, the equation for the pile in the *yielded* zone is given by the equation

$$\frac{d^2s}{dz^2} - \alpha h(z) = 0 \tag{9.1}$$

Here again, the dimensionless displacement s and distance z are obtained by multiplying the real dimensional values by the constant λ representing the properties of both pile and soil in the elastic solution, as given by equation (8.34) or by equations (8.87) and (8.88) in general. In equation (9.1) α is a constant depending on the numerical value selected for the yield properties of the soil and the pile and soil elastic properties. For different variations of t_y with depth, α is given in Table 9.1.

In Table 9.1, S is the pile perimeter, D is its diameter or equivalent diameter in the case of a square pile, and λ is determined for the particular variation of k with depth selected for the unyielded portion of the pile, $z > z_0$, according to equation (8.105).

TABLE 9.1

Relation Between t_y and α

Equation for Soil Yield Stress	$h(z)$	Value of α	
		Axial Loading	Torsional Loading
$t_y = t_y$	1	$\dfrac{t_y}{\lambda EA}$	$\dfrac{S\,Dt_y}{2\lambda^2 GJ}$
$t_y = t_{y1}z$	z	$\dfrac{t_{y1}}{\lambda^2 EA}$	$\dfrac{S\,Dt_{y1}}{2\lambda^3 GJ}$
$t_y = t_{y2}z^2$	z^2	$\dfrac{t_{y2}}{\lambda^3 EA}$	$\dfrac{S\,Dt_{y2}}{2\lambda^4 GJ}$

Since now the function $h(z)$ is specified in a given problem, the displacement gradient is obtained by integrating equation (9.1) once, and the displacement is obtained by integrating it twice. Two constants, as usual, will be introduced in the integration process; one will be determined by the boundary condition at $z = 0$, the other by the boundary condition where the yielded region solution must coincide with the elastic region solution. In particular, for the case in which $h(z) = z^r$, we have

$$\frac{ds}{dz} = \frac{\alpha z^{r+1}}{(r+1)} + C_3 \tag{9.2}$$

and

$$s = \frac{\alpha z^{r+2}}{(r+1)(r+2)} + C_3 z + C_4 \tag{9.3}$$

Taking the dimensionless applied load F as defined in Chapter 8, we find that

$$C_3 = -F \tag{9.4}$$

and, thus, substituting in equation (9.3), we have

$$s = \frac{\alpha z^{r+2}}{(r+1)(r+2)} - Fz + C_4 \tag{9.5}$$

This leaves one constant C_4 still to be determined. However, at the depth z_0, the displacement is that value holding at the limit of elastic behavior, in other words, s_*, (dimensionless). Thus, from equation (9.5), we get

$$s_* = \frac{\alpha z_0^{r+2}}{(r+1)(r+2)} - Fz_0 + C_4$$

or
$$\tag{9.6}$$

$$C_4 = s_* - \frac{\alpha z_0^{r+2}}{(r+1)(r+2)} + Fz_0$$

At this stage, we should pause and consider the problem. Normally, in a mechanics problem like this one, we are given or we know the value of the applied load F and would like to determine quantities like the deflection at the top of the pile, distribution of load down the pile, and so on. If we follow this reasoning in the present case, then we quickly find that, for a given load, z_0 is one of the unknowns in the problem and must be determined. However, on substitution of known quantities in equations such as (9.6), z_0 appears in at least quadratic form, which is inconvenient for solution. In the elastic equations to be introduced later, z_0 appears in the argument of functions which may be quite complicated. A solution can therefore be much more speedily obtained if, in fact, z_0 is taken to be known from the beginning and used to calculate, eventually, the load F, employed in the equations as an unknown quantity. This is somewhat inconvenient from the point of view of the final result (that is, instead of calculating deflections, etc., for loads of 10, 20, 30, . . . kips, we will obtain loads and deflections for the depth of the plastic zone of 10, 20, 30, . . . ft), but it greatly facilitates the numerical work. In fact, when lateral loads on piles are involved, the solution must be obtained this way, as we will discuss later.

Reconsidering equation (9.6) from this viewpoint—that z_0 is specified—we see that F is the fundamental unknown to be determined at this stage. We now introduce the elastic solution which holds in the depth range $z_0 < z < d$, equation (8.113):

$$s = C_1 f_1(z) + C_2 f_2(z) \qquad (8.113)$$

which introduces further constants C_1 and C_2. The unknowns are now three: C_1, C_2, and the load F. To obtain them we apply the displacement and force requirements at $z = z_0$ and the force or displacement condition at the pile tip $z = d$. Considering displacement first, we have, from equation (8.113),

$$s_* = C_1 f_1(z_0) + C_2 f_2(z_0) \qquad (9.7)$$

since, as we have seen, the displacement at this level is s_*. The force $F(z_0)$ at $z = z_0$ is given by

$$F(z_0) = -C_1 f_1'(z_0) - C_2 f_2'(z_0) \qquad (9.8)$$

but this must be equal to the force left in the pile at this level from the yielding region, which is given by equation (9.2) as

$$F(z_0) = F - \frac{\alpha z_0^{r+1}}{r+1} \qquad (9.9)$$

Setting equal the two values of force in the pile at z_0 in equations (9.8) and (9.9) gives

$$F - \frac{\alpha z_0^{r+1}}{r+1} = -C_1 f_1'(z_0) - C_2 f_2'(z_0) \qquad (9.10)$$

The final equation for the system is given by the tip conditions, which are handled in the same way as in Section 8.2.6. In particular, if the dimensionless tip load is taken to be R_d, which is assumed to be known, the last equation is

$$R_d = -C_1 f_1'(d) - C_2 f_2'(d) \qquad (9.11)$$

Thus, when z_0 is selected, the unknown quantities are C_1, C_2, C_3, C_4, and F, which are given by equations (9.4), (9.6), (9.7), (9.10), and (9.11). These may be rewritten in the following matrix form in which the columns represent the coefficients of the unknowns $C_1, \ldots F$, and the last column is the right-hand side. The equation numbers are given on the right.

$$
\begin{bmatrix}
+f_1(z_0) + f_2(z_0) & 0 & 0 & 0 \\
+f_1'(d) + f_2'(d) & 0 & 0 & 0 \\
+f_1'(z_0) + f_2'(z_0) & 0 & 0 & +1.0 \\
0 & 0 & +1.0 & 0 & +1.0 \\
0 & 0 & 0 & -1.0 & +z_0
\end{bmatrix}
\begin{pmatrix} C_1 \\ C_2 \\ C_3 \\ C_4 \\ F \end{pmatrix}
=
\begin{cases}
+s_* \\
-R_d \\
+\dfrac{\alpha z_0^{r+1}}{r+1} \\
0 \\
\dfrac{\alpha z_0^{r+2}}{(r+1)(r+2)} - s_*
\end{cases}
\quad
\begin{matrix}
(9.7) \\ (9.11) \\ (9.10) \\ (9.4) \\ (9.6)
\end{matrix}
$$

$$(9.12)$$

In practice, these equations may be solved simultaneously on a computer or on certain makes of "pocket" calculators. In a hand or other calculator calculation, however, it is more convenient to solve the first two equations, (9.7) and (9.11), simultaneously for C_1 and C_2, which can be substituted in the third equation, (9.10), to give F. The dimensionless force F is then inserted in the last two equations, (9.4) and (9.6), to give the coefficients C_3 and C_4. If, as suggested in Section 8.2.6, the tip load, R_d, is chosen to be a constant proportion, say 0.2 to 0.3, of the applied load F, then the second equation, (9.11), is modified by replacing the zero in the F column by the proportion selected (positive), and the $-R_d$ on the right side is changed to zero. This requires that the first three equations be solved simultaneously for C_1, C_2, and F. It can be seen from equation (9.5) that the dimensionless deflection $s(0)$ at the surface $z = 0$ is equal to C_4. The equations are, of course, somewhat simplified when $h(z) = 1$ in the case in which $r = 0$.

Thus, by choosing a succession of values of the yield depth, z_0, from zero to the full length of the pile, the history of applied load versus displacement can be traced out, up to the point at which yielding has progressed all the way down the sides of the pile. Further axial displacement, utilizing the assumed ideal elasto-plastic tip load-displacement relationship, brings the soil below the tip to yield. This adds a small linear portion to the last point on the top load-displacement curve, due both to the small further elastic displacement of the tip and the slight additional compression of the pile under the increased load in it. In the next section we will show the numerical analysis of a test pile as an example.

9.3.3 Analysis of Arkansas River pile test (axial load)

For a numerical example we will consider the Arkansas River pile test described in general in Section 8.2.7. Before proceeding with the analysis here, we must discuss the test a little more.

Strain gauges were installed on the pile at 2-ft (0.6 m) intervals to a depth of 19 ft (5.8 m), except for the top 5 ft (1.5 m) where there were none. Thereafter they were at 4-ft (1.2 m) intervals to 35 ft (10.7 m); below that level they were at 5-ft (1.5 m) intervals to the pile tip. Six strain rods were also anchored to the pile at varying spacings in the range of from 4 to 10 ft (1.2 to 3 m), so that the pile's vertical displacement could be measured at various depths. Average strains over the intervals between the rod anchors were calculated. A complete description of the instrumentation is given in the paper (7).

Test pile 2 was loaded axially to failure and the strain gauge and strain rod readings were used to calculate the load distribution in the pile at various pile head load levels [50, 100, 150, and 200 tons (0.45, 0.89, 1.34, and 1.78 MN)]. During the loading process the pile was apparently loaded to 50 tons (0.45 MN), unloaded to zero, loaded to 100 tons (0.89 MN), unloaded to zero, and in similar fashion to 150 tons (1.34 MN) and 200 tons (1.78 MN). The paper does not discuss this and it does not say at what stage the strains were recorded. Presumably, they were measured at 50, 100, 150, and 200 tons on first loading to each of those values. In that event, the strains recorded at 200 tons (1.78 MN) included any residual strains caused by (a) driving and (b) subsequent load and unload cycles. Later analysis in the paper indicates that substantial residual strains were generated by the driving process. It is reasonable to assume that they were also modified by the load cycling. Graphs of pile load versus movement, load distribution along the pile, and relations among gross, tip, and wall loads are shown in Figure 8-4 for test pile 2.

The effect of the load cycling was to increase the settlement of the pile top at any load compared to the displacement that would have occurred at that load arrived at by a monotonic load increase. It will be assumed therefore in the analysis here that somewhat smaller vertical displacements than those recorded in Figure 8-4(a) should be taken since the effect of cycling the load is not included in the method employed here. The load distribution in the pile in Figure 8-4(b) is given by the strains recorded by both strain rods and strain gauges. Evidently, more reliance has been placed on the rod readings, since, for the most part, the curve in the figure has been passed through them; at the top of the pile the strain gauge readings have been ignored. It is difficult to understand the solid line-load distribution curves of Figure 8-4(b) at all loads. The saturated sand of the test site would be expected to sustain only very small effective stresses near the top of the pile, with the stresses increasing with depth. Thus, it would be anticipated that, when the gross load in the pile was substantial, the shearing stress between pile and sand at the top of the pile would be at the yield or shear strength value for the pile-to-sand friction coefficient.

Near the top the soil's *shearing strength* would be expected to vary approximately linearly with depth. Thus, the *force* transferred out of the pile and into the soil in at least the top few feet (1 meter) would vary parabolically with depth. This implies that a vertical tangent should exist to the curves in Figure 8-4(b) at ground surface. No matter what the complexity of the soil-pile stress relations at depth, the shearing strength of the sand at ground surface must be very small, at a level insufficient to support the gradients of the load distribution curve at ground surface shown in Figure 8-4(b). The shearing strength of the sand cannot be essentially uniform with depth, as indicated by the top 10 or 20 ft (3 or 6 m) of the load-distribution curves. However, although these questions may be raised, in order to perform the analysis, it is still necessary to use the basic data given in Figure 8-4.

Some assumptions are required for the computations to proceed. The shearing strength for the sand and the Winkler spring coefficient will both be taken to be linear with depth. Thus, in equations (8.102) and (9.1), $g(z)$ and $h(z)$ are both equal to z. Use of a linear equation for both relations implies a dimensional value of $s_*, = t_{y1}/k_1$, constant with depth. We proceed to establish this value from the results.

From Figure 8-4(c) the tip load on the average over the load range except near the peak is about 0.2 of the gross load. We will assume this ratio to hold for all the calculations. It is also assumed that at 200 tons (1.78 MN) the soil around the pile tip is close to yielding and that the frictional resistance along the pile wall is fully mobilized. Since 140 tons (1.25 MN) [200 tons minus a 60-ton (0.53 MN) tip load] are removed by pile wall friction at the 200-ton load, the average shearing resistance along the pile is 424 lb/in. (74.3 kN/m) (the perimeter is included in all the calculations). With the assumption of a linear shearing resistance with depth, the shearing resistance at the base of the pile is then 848 lb/in. (148.6 kN/m) and the *slope* of this shearing resistance (t_{y1}) is 1.285 lb/in.² (8.9 kN/m²).

The difficulty with the calculation at this stage is that a value of k is required for the soil surrounding the pile. Since this is assumed here to be a linear function of depth, the value of k_1 must be arrived at. However, before any calculations of pile load-deflection relations can proceed, the value of λ, (equation 8.105), which depends on k_1, is needed. In practice, a value of λ might be guessed at, load-settlement calculations made with results that would be compared to the measurements, λ modified, and the computations repeated. The comparison of analysis with measurements thus requires a trial-and-error solution. This was done in this case, but only the final results are presented.

Examination of the load-movement curve for the pile in Figure 8-4(a) indicates that if an initially purely elastic portion of behavior is present, it is very small (that is to say, our model is incorrect). But, if it is assumed that the response could reasonably be taken as linear up to a load of about 60 tons (0.53 MN), then the deflection, 0.075 in. (1.9 mm), at about this value represents w_*. We take w_* therefore to be 0.075 in. From the value for t_y obtained above, it follows that $k_1 = 17.13$ lb/in.^{3} (4.65 MN/m^3). To compare this value with the value of 0.24 lb/in.^{4} (2.57 MN/m^3) calculated in Section 8.2.7 from the unloading path, we must divide 17.13 by the perimeter, 82 in. (2.08 m), to get 0.21 lb/in.^{4} (2.24 MN/m^4). From equation (8.105), $\lambda^3 = k_1/EA$, so that $\lambda = 2.91 \times 10^{-3}$ in.$^{-1}$ (0.115 m^{-1}) and the dimensionless length of the pile, $d = 1.833$. With these values, the complete load-settlement behavior of the pile can be calculated from the analysis, with an assumption for the tip load.

For the assumed linear variation of both t_y and k with depth, the pile top will displace downward a distance s_* before yielding begins in the soil. This elastic portion of the load-settlement relation is calculated first. From Table 8.11 for this case, f_1 and f_2 are $Ai(z)$ and $Bi(z)$, respectively, and equation (8.113) gives, in dimensionless form, the top settlement

$$s(0) = C_1 Ai(0) + C_2 Bi(0) \tag{9.13}$$

so that the constants C_1 and C_2 in Table 8.12 must be evaluated. They involve $Ai'(0)$, $Ai'(d)$, $Bi'(0)$, and $Bi'(d)$ where d is the dimensionless pile length. Assuming $R_d = 0.2F$, the values of the constants are as follows:

$$Ai(0) = 0.3550 \quad Ai'(0) = -0.2588 \quad Ai'(1.833) = -0.0658 \quad C_1 = 3.8950F$$
$$Bi(0) = 0.6149 \quad Bi'(0) = 0.4483 \quad Bi'(1.833) = 3.1481 \quad C_2 = 0.0179F$$

Thus, from equation (9.13) and the constants, we get

$$s(0) = 1.3937F \tag{9.14}$$

Since $F = P/EA$, and the dimensional deflection at the pile top $w(0) = s(0)/\lambda$, the relation between pile top deflection, $w(0)$, in the elastic range and load is given by

$$P = 1.445 \times 10^6 w(0) \tag{9.15}$$

where P is in pounds and $w(0)$ is in inches.

To find the load at which plastic behavior begins at the top of the pile, we equate the deflection in the elastic range to w_*, 0.075 in. (1.9 mm); thus, the limiting value of P is 1.0837×10^5 lb or 54.18 tons (0.482 MN). This corresponds reasonably well with the assumption about the limit of elastic behavior.

For the plastic deformation portion of the pile response to load, it will be assumed that yielding of the soil (slip) has occurred next to the pile to a dimensionless depth z_0. The load transfer and shearing stress distribution above this depth are known, and from z_0 to d the pile-soil behavior is assumed to be still elastic. In the latter region the soil spring constant will be given by the same equation as before. For the elastic portion of the pile-soil behavior below z_0, the same equations from Table 8.12 hold for C_1 and C_2, except for the replacement of $Ai(0)$, $Bi(0)$, $Ai'(0)$, and $Bi'(0)$ with $Ai(z_0)$, $Bi(z_0)$, etc. When the constants are determined for a particular value of z_0, the displacement at z_0 is found and equated to s_*. This gives the value of the load P, from which the pile top displacement can be found.

Choosing the value $z_0 = 1.00$ and with R_d/F still equal to 0.2, we can follow the calculations through by using the following values from Appendix C:

$$Ai(z_0) = 0.13529 \qquad Ai'(z_0) = -0.15915$$

$$Bi(z_0) = 1.20742 \qquad Bi'(z_0) = 0.93244$$

When these are substituted in the matrix equation, (9.12), or in the separate equations, along with the value for α, 2.1929×10^{-4}, derived from the yield and pile properties, the solution is found to be: top deflection, $w(0) = 0.168$ in. (4.3 mm) at a load, P of 118.6 tons (1.06 MN). Similar calculations for $z_0 = 0.5$ and 1.5 gave, respectively, $w(0) = 0.112$ in. (2.8 mm) at $P = 80.6$ tons (0.72 MN) and $w(0) = 0.218$ in. (5.5 mm) at $P = 152.6$ tons (1.36 MN). These values are plotted in Figure 8-4(a) and may be compared with the measured pile behavior. The deflections are seen to be somewhat smaller than those observed, which were the result of cyclic loading. Other values of t_{y1} and k_1 or other variations of yield and elastic soil reaction with depth may be tried to give a better fit, if desired.

There is one more point to be made concerning pile behavior under axial or torsional loads. The point also applies to laterally loaded piles, as discussed in the next section, and has important consequences for field tests. When the threshold deformation, s_*, is exceeded during pile loading, plastic deformations occur between pile and soil above the level at which s_* is reached. The load-displacement behavior is nonlinear beyond this point, in the present model, and, when the load is removed, residual displacements and stresses are left in the pile. They form a self-equilibrating system with no load at the pile top. In particular, when s_* is zero at ground surface, no part of the load-pile top displacement relation is linear, and even the slightest load will cause some small irreversible component of displacement.

Before a field test can commence, the instrumented pile must be installed in the soil. From the point of view of analysis, it is difficult to do this satisfactorily by any method. But, since the test is usually intended to indicate the behavior of future piles that will be driven at the same site to support a structure, the test pile is commonly thumped into the ground with the same hammer used for driving the construction piles. Such a process involves, of course, continuous plastic deformation of the soil surrounding the pile, until it is at its design depth. It also subjects the instrumentation to very high stresses; failures are likely. When pile-driving ceases, residual stresses are left in the pile, whether its future utility lies in axial, torsional, or lateral load resistance. However, the strain gauges are installed on the pile before driving, and it is difficult to obtain true zeros on these at ground surface. In addition, driving and the temperature conditions in the driven pile will always change them. Consequently, it is common practice to take strain gauge readings *after* the pile has been driven and to consider that these are zero strains, or are base values, for the measured quantities under load. If the post-driving residual strains are large, it will be seen that this can lead to a distorted view of the pattern of load take-out in the pile at subsequent load levels. It is probably the reason for the anomalies in the shape of the load versus distance-down-the-pile curves in Figure 8-4(b). Among other things, the use of these incremental values in place of the real, unknown total values would give a misleading indication of the pile-soil shearing stress at which yielding occurred. The stress indicated by the incremental strain gauge readings would be different from that actually developed in the soil at the recorded load.

9.3.4 Nonlinear approximate analysis

Another method of including nonlinear soil behavior in pile analysis, with possibly more realistic soil behavior, is to employ an approximate analysis based on the residual methods described in Chapter 2. A convenient and fairly realistic representation of certain soil behavior adjacent to a pile is given by an exponential function in the following form, in terms of shearing stress τ versus displacement w:

$$\tau = B(1 - e^{-Cw}) \qquad (9.16)$$

where B and C are soil properties that can also be made functions of depth, z, if desired. At very large displacements $\tau = B$, which is therefore the shearing strength of the soil. The constant C expresses the variation of stress with displacement. The rate of change of stress with displacement is

$$\frac{d\tau}{dw} = BCe^{-Cw} \qquad (9.17)$$

and at zero displacement

$$\frac{d\tau}{dw} = BC \qquad (9.18)$$

which can therefore be taken to be the initial subgrade reaction modulus defined in the same way as in the linear model.

With the usual definitions of terms, the nonlinear equation describing the axial displacement of a pile is

$$EA\frac{d^2w}{dz^2} - SB(1 - e^{-Cw}) = 0 \tag{9.19}$$

in which, as before, it is convenient to make the dimensionless substitutions

$$s = \lambda z \quad \text{and} \quad \psi = \frac{\lambda EA}{P}w \tag{9.20}$$

Here λ is not defined the same as formerly; it is obtained after substitution of equation (9.20) in equation (9.19). Equation (9.19) becomes

$$\frac{d^2\psi}{ds^2} - \frac{SB}{\lambda P}[1 - e^{-CP\psi/\lambda EA}] = 0 \tag{9.21}$$

and, accordingly, we select

$$\lambda = \frac{SB}{P} \tag{9.22}$$

to give equation (9.21) in the form

$$\frac{d^2\psi}{ds^2} - (1 - e^{-D\psi}) = 0 \tag{9.23}$$

where

$$D = \frac{CP}{\lambda EA} = \frac{P^2C}{SEAB} \tag{9.24}$$

In contrast to the linear form of the equation, version (9.23) contains a constant D which depends on the load and material properties. Thus, for each value of load to be applied, it must be solved anew for one soil and pile. Before solving an example problem, therefore, numbers have to be selected for these parameters. Here we will choose the soil properties B and C to be 10 psi (69 kN/m²) and 0.5 in.$^{-1}$ (0.2 mm^{-1}), respectively. Substituting in equation (9.18) makes the initial tangent subgrade reaction coefficient equal to 50 pci (13.6 MN/m³). The values also mean that the soil resistance equals $(1 - e^{-1}) = 0.632$ of its shear strength at a relative pile-soil deflection of 0.2 in. (5.1 mm). We will take the load to be 400 tons (3.56 MN), so that, from equation (9.22), $\lambda = 9.425 \times 10^{-4}$ in.$^{-1}$ (3.71 $\times$ 10^{-2} m^{-1}). It is desirable to choose an example in which the pile is fairly flexible in the axial direction, so that the nonlinear soil behavior will clearly emerge in the calculation. For convenience in the calculation, we will assume a value of $s_l = \lambda l = 1.5$, so that l in this case is 1591.5 in. or 133 ft (40 m). The pile is therefore quite long. Finally, we select a pile of 24-in. (0.61 m) diameter, 0.375-in. (9.5 mm) wall, approximately, so that $EA = 8.4 \times 10^8$ lb (3740 MN), with perimeter $S = \pi \times 24$ in. (1.92 m). The exponent D in dimensionless equation (9.23) has therefore the value 5.0524.

As in the examples in Chapter 2, we choose a polynomial representation for pile deflection ψ:

$$\psi = a_0 + a_1 s + a_2 s^2 \qquad (9.25)$$

At the top of the pile the applied load is P, which is given by the expression in dimensions

$$P = -EA\left(\frac{dw}{dz}\right)_0 \qquad (9.26)$$

Substituting from the dimensionless expression (9.20), it appears that the dimensionless pile top requirement is

$$\left(\frac{d\psi}{ds}\right)_0 = -1 \qquad (9.27)$$

So that, by differentiating (9.25) and making $s = 0$, we have

$$a_1 = -1 \qquad (9.28)$$

the first control over the undetermined coefficients. At the base of the pile, for comparison with linear solutions obtained formerly, the displacement is again taken to be zero, and we have

$$a_0 + a_1 s_l + a_2 s_l^2 = 0$$

or, substituting for s_l,

$$a_0 + 1.5a_1 + 2.25a_2 = 0 \qquad (9.29)$$

the second condition on the coefficients. With three unknown coefficients, another equation is required. It can be obtained by any of the residual methods discussed previously. We will do a simple collocation here, making the residual zero at $s = 1.0$. The residual is

$$R = 2a_2 - [1 - e^{-5.0524(a_0 + a_1 s + a_2 s^2)}] \qquad (9.30)$$

Substituting $s = 1$ and equating to zero yield

$$e^{-5.0524(a_0 + a_1 + a_2)} + 2a_2 = 1 \qquad (9.31)$$

The three equations, (9.28), (9.29), and (9.31), can then be solved for the coefficients a_0, a_1 and a_2. It is most convenient to substitute $a_1 = -1$ from

(9.28) in (9.29), from which a_0 can be determined in terms of a_2 for insertion in equation (9.31). The result is an equation in a_0:

$$e^{-0.5614(5a_0-3)} - \frac{8}{9}a_0 + \frac{1}{3} = 0 \tag{9.32}$$

The zeros of the left side can be obtained by a programmable calculator or other computer or simply by trial and error to give $a_0 = 0.8832$. The last coefficient is therefore $a_2 = 0.2741$, and the polynomial approximation is

$$\psi = 0.8832 - s + 0.2741s^2 \tag{9.33}$$

At $s = 0$, $\psi_0 = 0.8832$, and the vertical settlement of the top of the pile under the 400-ton (3.56 MN) load is

$$w_0 = \frac{P\psi_0}{\lambda EA} = 0.8925 \text{ in. (22.7 mm)} \tag{9.34}$$

The same process can be repeated to give the pile top deflection under other loads. At 100 and 250 tons (0.89 and 2.23 MN), for example, the settlements are 0.1097 and 0.4268 in. (2.8 and 10.8 mm), respectively. It is also of interest to solve the linear problem by the same approximate method, when the linear soil reaction coefficient is taken at the same value, 50 pci (13.6 MN/m³) as the initial tangent value of the nonlinear behavior. This calculation results in the value of 1.1918 for a_0 for the same pile and the same boundary conditions. The pile top spring constant is 746.5 tons/in. (262 MN/m) for the linear case. Lastly, the linear problem can also be solved, of course, exactly by using the coefficients from equations (8.39) and (8.40). In that case, the pile top spring constant is 891.9 tons/in. (313 MN/m). Here, $\lambda l = 3.3716$.

All these results, linear and nonlinear, are shown in Figure 9-4. Over the range of loads considered, from 0 to 400 tons (3.56 MN), the effect of the nonlinear soil behavior is evident. At the latter load level, which is high for a pile of this cross section, the stress in the pile steel at the top is about 28,000 psi (193 MN/m²), and with the nonlinear soil behavior, the stress in the soil adjacent to the top of the pile is about 99 percent of its shear strength. In fact, the soil is stressed to 90 percent or greater of its shear strength at the 400-ton load to a depth $s = 0.487$, or in about the top one-third of its depth.

A nonlinear solution can also be developed through the use of finite differences. It is of value to illustrate the technique in tackling the same problem as we studied above by the collocation method. For simplicity of exposition, we divide the pile into three segments, Δs, of equal length $s_l/3$, as in Section 8.2.5. This gives four points on the pile at which the displacement is to be calculated. The displacements, from the pile top down, are then

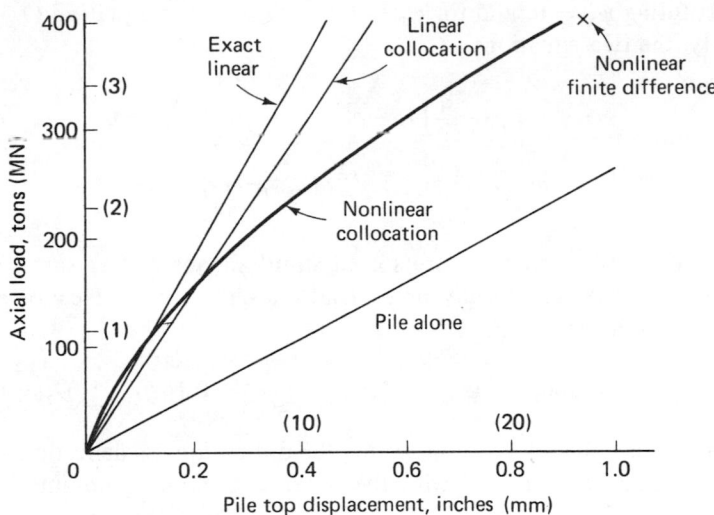

Figure 9-4. Linear and nonlinear analyses of pile displacement under axial loading.

ψ_0, ψ_1, ψ_2, and ψ_3. Once again we apply the boundary conditions first. At the base, of course, the displacement is zero; so

$$\psi_3 = 0 \tag{9.35}$$

At the top, the condition (9.27) holds, and, translated into finite difference terms, this becomes

$$\frac{\psi_1 - \psi_0}{\Delta s} = -1$$

or

$$\psi_1 - \psi_0 = \frac{S_l}{3} \tag{9.36}$$

Since there are three unknowns (ψ_0, ψ_1, and ψ_2) to be determined, two more equations are needed. They come about from the finite difference form of equation (9.23) applied to the central points 1 and 2 in turn, as follows:

At point 1:

$$\frac{\psi_0 + \psi_2 - 2\psi_1}{(\Delta s)^2} - (1 - e^{-D\psi_1}) = 0 \tag{9.37}$$

At point 2:

$$\frac{\psi_1 + \psi_3 - 2\psi_2}{(\Delta s)^2} - (1 - e^{-D\psi_2}) = 0 \tag{9.38}$$

Substituting $\psi_3 = 0$, and for ψ_0 in terms of ψ_1 from equation (9.35), gives, finally, the two equations

$$-\psi_1 - \frac{s_l^2}{9}\left(1 - e^{-D\psi_1}\right) + \psi_2 + \frac{s_l}{3} = 0 \tag{9.39}$$

$$\psi_1 - 2\psi_2 - \frac{s_l^2}{9}(1 - e^{-D\psi_2}) = 0 \tag{9.40}$$

When these two equations are solved simultaneously after substitution for s_l and D, either graphically or by trial-and-error or by Newton's method (6), the results are

$$\psi_0 = 0.9266; \quad \psi_1 = 0.4266; \quad \psi_2 = 0.1476; \quad \psi_3 = 0 \quad (9.41)$$

in which the value of ψ_0, representing the dimensionless deflection at ground surface, can be compared with the other approximate nonlinear solution value of 0.8832.

9.4 LATERALLY LOADED PILE

In the case of the pile under axial loading, the yielding of the pile material itself under load is almost never a consideration in the analysis. Under lateral loading conditions, however, the eventual criterion for the ultimate load *is* determined by the maximum moment developed in the pile and the strength properties of the pile. Conversely, working loads may be controlled by pile deflection, commonly taken not to exceed 0.75 in. (20 mm). Consequently, in this circumstance, there are three régimes of behavior of interest: elastic response of soil and pile, yielding of the soil, and ultimately, yielding of the pile.

The field behavior of a pile under lateral load in the Arkansas River tests was described in Section 8.3, and the load-pile top deflection relation was seen to be highly nonlinear with no linear portion of the response evident either in the observations or justified by the physical mechanism at work. It was also pointed out in that section that both deflections and moments in a normal pile in the field are essentially zero at a depth of from 20 to 30 ft from the point of load application. This was advantageous in the subsequent analysis, since it meant that conditions at a pile tip, in general, have little or no effect on the response of the pile top or the moments developed. Only two solutions of the equation describing the elastic portion of pile behavior are therefore generally required.

On occasion, however, extremely stiff, and generally short caissons or piers may be subjected to lateral loads and moments. If they may be considered stiff enough to be rigid by the criterion of Chapter 4, they represent one extreme case of the behavior of piles under lateral loads and will be considered first. The situation of a moderately stiff pile whose tip conditions do influence moments and deflections results in an analysis that requires more algebra than that for the relatively flexible pile, since all four of the elastic zone solutions are required, but it is not inherently more difficult. Since, in practice, it is less common than either the rigid or the flexible pile, detailed development of the solutions will be omitted here. The solutions can be obtained by an extension of the methods described.

Although for axial and torsional pile loading the yield stress between soil and pile is fairly obviously related to the shearing strength of the soil, or the shearing or frictional resistance at the pile-soil interface, the situation is different in the case of lateral loading (4). Here the pile deflects sideways, and the ultimate resistance to the movement involves both the soil's shearing strength and geometrical considerations. Calculation of the yield stress against the pile requires the assumption of a plastic mechanism in the soil surrounding the pile; the mechanism will be somewhat different for circular and square piles. Near the ground surface both physical intuition and observation indicate that the mechanism will be three-dimensional, since vertical movements of the free surface occur in addition to the horizontal displacements. At depths greater than several pile diameters from the surface the mechanism will become essentially two-dimensional, with all displacements occurring in the horizontal plane. The solution of such a plasticity problem is not easy and is preferably avoided, although an attempt has been made by Reese, Cox, and Koop (17) to treat it approximately.

Although such studies are worthwhile, a more pragmatic approach for pile engineering design and analysis purposes is to assume various relations between depth and a simple yield stress acting on the pile, without consideration of the geometry of plastic deformation, and to compare the results obtained from analysis using these relations and the behavior of field test piles in various soils. From the fits of the analytical solutions to the field tests, both qualitative and quantitative properties can be obtained and matched to different soil conditions. This will be discussed further with reference to the example calculation at the end of the chapter.

For piles under lateral loading, the same considerations apply to the relation between yield stress and subgrade reaction coefficient variation with depth, with respect to the threshold deflection s_*, as have been described in terms of axial and torsional pile loading. The greatest deflections are generated at ground surface, and diminish with depth. Where the deflection exceeds s_*, the soil yields and imposes a soil yield pressure on the section of pile above; below, the soil behavior is still elastic. If the soil properties are such that s_* is finite at the surface and is either constant or increases with depth, then the pile lateral load versus top deflection behavior will be elastic up to the point where the top deflection reaches s_*. Beyond that point only a portion of the deflection will be elastic. If s_* is zero at ground surface, which is a logical deduction for sands, then the load-deflection curve will include plastic deformation and plastic energy dissipation from the beginning. On unloading from any load the pile will exhibit a residual displacement or set. It follows therefore, under these conditions, that, on removal of load, residual moments will also be left in the pile. This has important consequences for pile behavior, particularly if cyclic or cyclic reversal of load applications is to be taken into account.

9.4.1 Rigid pile

If the soil is assumed to have yield and elastic property variations with depth such that the threshold displacement, s_*, is finite at the pile top, then the pile load-deflection-moment-depth relations are calculated on an elastic soil basis, as in Sections 4.3 and 8.3.1, when the top displacement is smaller than this value. When it is larger, then the pile movement is resisted by the yield pressure down to the depth at which s_* is reached and by the elastic property of the subgrade reaction coefficient below this depth.

For illustration, the analysis will be made for the case in which the yield stress, $p_y(FL^{-1})$, and the subgrade reaction coefficient, $k(FL^{-2})$, are both constant with depth. Both these parameters include the pile width. In this case, the displacement takes place in the horizontal or x-direction and is designated, as usual, u. Consequently,

$$u_* = \frac{p_y}{k} \qquad (9.42)$$

The lateral displacement, u, of the pile under horizontal load T_0 at the ground surface is a linear function of depth, z:

$$u = az + b \qquad (9.43)$$

where a and b are constants. We will consider the load to be great enough that the pile deflection exceeds u_* from the surface down to a depth z_0, at which the deflection is just equal to u_*. Thus, in equation (9.43) we have

$$u_* = az_0 + b \qquad (9.44)$$

This is the first equation in the unknowns of the problem. The second equation is obtained, as usual, from the condition of horizontal force equilibrium:

$$T_0 - p_y z_0 - k\int_{z_0}^{l} (az + b)\, dz = 0 \qquad (9.45)$$

where l is the length of the pile. Moment equilibrium forms the third equation; we will take moments about the top of the pile to get

$$\frac{p_y z_0^2}{2} + k \int_{z_0}^{l} (az + b)z \, dz = 0 \tag{9.46}$$

Normally, as in Chapters 4 and 8, we would take the unknowns of the problem to be a, b, and z_0 under the given load T_0. However, we can see that the same problem arises here as in the axial loading case; namely, z_0 appears in the equations in polynomial form. Solution for z_0 is therefore complicated in general. The equations are all linear in a, b, and T_0, though; therefore, if we take these three parameters as the unknowns, they can be obtained for given values of z_0 and the other parameters l, k, and p_y. Performing the analysis in the present case gives, eventually,

$$a = -3u_* \left[\frac{l^2}{(l - z_0)^2 (2l + z_0)} \right] \tag{9.47}$$

$$b = u_* \left[\frac{2l^3 + z_0^3}{(l - z_0)^2 (2l + z_0)} \right] \tag{9.48}$$

and

$$T_0 = \frac{p_y l}{2} \frac{(l + 2z_0)}{(2l + z_0)} \tag{9.49}$$

From the constants the deflection of the pile can be obtained by using equation (9.43). Substituting in this equation for the deflection at the tip of the pile shows that, as in the elastic problem, the tip displacement is negative. That is, the tip displaces in the direction opposite to the top. At some distance down the pile for a particular value of z_0, the pile displacement is zero. Since from equation (9.43) b is the deflection at ground surface, u_0, z_0 could be obtained in terms of T_0 from equation (9.49) and substituted in equation (9.48) to give an equation for u_0 in terms of T_0 independent of z_0. It is apparent that such an operation will be time-consuming, and, for a practical problem, not worth the trouble, since for given real values of the parameters, u_0 and T_0 can be calculated for a range of values of z_0 and subsequently plotted against each other.

If z_0 is set at zero, the expression for T_0 becomes

$$T_0 = \frac{p_y l}{4} \tag{9.50}$$

But this is the condition at the end of the elastic range, so that T_0 would not be expected to depend on p_y. The latter can be removed from equations (9.49) and (9.50) through the use of equation (9.42) to give

$$T_0 = \frac{k u_* l}{4} \tag{9.51}$$

Now, since $z_0 = 0$, the deflection of the top of the pile is just equal to u_*, so that

$$T_0 = \frac{ku_0 l}{4} \tag{9.52}$$

which is now identical to the result in equation (8.127) when z there is taken equal to zero.

The moment in the pile at a distance z down the pile from the ground surface is given by the equations

$$0 < z < z_0 \qquad M = T_0 z - \frac{p_y z^2}{2} \tag{9.53}$$

$$z_0 < z < l \qquad M = T_0 z - p_y z_0 \left(z - \frac{z_0}{2} \right) - k \int_{z0}^{z} (as + b)(z - s)\, ds \tag{9.54}$$

Although the latter equation may be evaluated in general form, it is easier to do this for a specific problem after a and b have been determined numerically.

At first sight, it might seem in this problem that z_0 could be increased in increments, causing soil yielding to progress steadily down the pile with correspondingly waxing load and displacement. This all depends, of course, on the assumption that the *pile* does not yield. The displacement does increase under load, but what happens is that at a particular value of load, the *tip* displacement (in the negative direction) also reaches the threshold displacement, $-u_*$. By substituting the constants a and b from equations (9.47) and (9.48), the length l in the right-hand side of equation (9.43), and $-u_*$ for the displacement on the left side, we can determine the value of z_0 at which this happens. Following the algebra through, we arrive at the following cubic equation for z_0:

$$2\left(\frac{z_0}{l}\right)^3 - 3\left(\frac{z_0}{l}\right) + 1 = 0 \tag{9.55}$$

The only root of this equation in the interval 0 to 1 is

$$\frac{z_0}{l} = \frac{\sqrt{3} - 1}{2} \approx 0.366 \tag{9.56}$$

which may be substituted in equations (9.48) and (9.49) to find the top deflection and load at which the tip deflection reaches the threshold. They are

$$T_0 = (\sqrt{3} - 1)\frac{p_y l}{2} \approx 0.366 p_y l \tag{9.57}$$

and

$$u_0 = \frac{1}{3}(3 + 2\sqrt{3})u_* \approx 2.155 u_* \tag{9.58}$$

so that the top of the pile moves over a considerable distance after yielding begins there, before the soil goes plastic at the pile tip. Now that we have developed yield at the tip, it is necessary to assume a plastic zone propagating up the pile from the tip. We will call the distance z_1, from the top of the pile to this depth and pursue the analysis further for $z_0 > 0.366l$. The distance z_1 becomes another unknown in the problem, and it is described by adding one further equation to the system:

$$-u_* = az_1 + b \tag{9.59}$$

Equation (9.44) remains the same as before, but equations (9.45) and (9.46) require modification to the following forms:

Horizontal equilibrium:

$$T_0 - p_y z_0 - k \int_{z_0}^{z_1} (az + b)\, dz + p_y(l - z_1) = 0 \tag{9.60}$$

Moment equilibrium about the pile top:

$$\frac{p_y z_0^2}{2} + k \int_{z_0}^{z_1} (az + b)z\, dz - p_y \frac{(l^2 - z_1^2)}{2} = 0 \tag{9.61}$$

The elastic region of behavior now extends only from z_0 to z_1, and this effect is represented by the changes in the upper limit of the integral terms in these equations. The last terms account for the horizontal force and moment generated by the yield pressure p_y acting from z_1 to l at the bottom of the pile. Equations (9.44), (9.59), (9.60), and (9.61) now must be solved simultaneously to give a, b, T_0, and z_1 in terms of the given quantities. The algebra is heavier than before, but the result is:

$$a = -2u_* \frac{\left(z_0 + \dfrac{1}{\sqrt{3}}\sqrt{2l^2 - z_0^2}\right)}{(l^2 - 2z_0^2)} \tag{9.62}$$

$$b = u_* \frac{\left(l^2 + \dfrac{2z_0}{\sqrt{3}}\sqrt{2l^2 - z_0^2}\right)}{(l^2 - 2z_0^2)} \tag{9.63}$$

$$\frac{z_1}{l} = \frac{1 - \left(\dfrac{z_0}{l}\right)^2 + \dfrac{1}{\sqrt{3}}\left(\dfrac{z_0}{l}\right)\sqrt{2 - \left(\dfrac{z_0}{l}\right)^2}}{\left(\dfrac{z_0}{i}\right) + \dfrac{1}{\sqrt{3}}\sqrt{2 - \left(\dfrac{z_0}{l}\right)^2}} \tag{9.64}$$

$$T_0 = p_y l \left\{ \frac{\left[\dfrac{1}{\sqrt{3}}\sqrt{2 - \left(\dfrac{z_0}{l}\right)^2}\left(\dfrac{2z_0}{l} - 1\right) + 1 - \dfrac{z_0}{l}\right]}{\left[\left(\dfrac{z_0}{l}\right) + \dfrac{1}{\sqrt{3}}\sqrt{2 - \left(\dfrac{z_0}{l}\right)^2}\right]} \right\} \tag{9.65}$$

When the value of z_0/l from equation (9.56) is substituted in these equations, it is found, correctly, that $z_1/l = 1.0$ (no yielding at the tip), and the values of T_0 and u_0 in equations (9.57) and (9.58) are recovered.

Once again, provided the pile does not yield, then, as T_0 is increased, the deflection goes up, the upper plastic zone extends down the pile, and the lower one works its way up from the tip. The limit to this process is reached when the intervening elastic region vanishes and the load is sustained by opposing plastic zones. When this occurs, z_0 and z_1 become equal at a value that can be determined from first principles from the limiting forms of equations (9.60) and (9.61) or by replacing z_1/l in the left side of equation (9.64) by z_0/l and solving for z_0/l. Either method gives the result

$$\frac{z_0}{l} = \frac{1}{\sqrt{2}} \approx 0.707 \tag{9.66}$$

at an ultimate load of

$$T_{0u} = p_y l(\sqrt{2} - 1) \approx 0.414\, p_y l \tag{9.67}$$

These last two results can be obtained, of course, much more simply from the horizontal force and moment equilibrium equations than by the tortuous route we have taken. At the ultimate load the pile top deflection is infinite. At this stage the maximum or ultimate moment M_u occurs in the pile. It can be obtained readily from the rectangular distribution of yield pressure on the pile and is given by the expression

$$M_u = \frac{T_{0u}^2}{2p_y} \tag{9.68}$$

and occurs at the depth

$$z_u = \frac{T_{0u}}{p_y} \tag{9.69}$$

For the stage before this, the moments can be obtained from an equation analogous to equation (9.54). As in that former case, it is easier to evaluate the moments specifically for a given problem than to write a general expression.

The analysis above was performed for the condition where both the yield stress, p_y, and the subgrade reaction coefficient, k, were constant along the pile length; this resulted in u_*, the threshold deflection, also being constant. These conditions might reasonably hold in practice for a caisson installed in an overconsolidated clay. In this, the simplest of circumstances, the mechanisms and the analysis turned out to be perhaps surprisingly complicated, although general analytical results were developed.

In other cases of p_y and k variation with depth, analysis is still possible, if desired, to obtain general expressions for deflection, load, and moment, but the equations obtained are even clumsier than those for the simple prior case. In practice, it is easier for specific cases to adopt the following general approach, since the numerical information is known or assumed in advance.

It is assumed that the yield stress and subgrade reaction coefficients can be expressed in power form:

$$p_y = p_{yq} z^q \tag{9.70}$$

$$k = k_r z^r \tag{9.71}$$

so that

$$u_* = \frac{p_y}{k} = \frac{p_{yq}}{k_r} z^{q-r} \tag{9.72}$$

In practice, the yield exponent q will always be greater than or equal to the subgrade elastic exponent r, so that the threshold deflection is either constant or increases with depth. Since the pile deflects as a rigid body, the fundamental equations employed for all problems, before the tip deflection reaches the local value of u_*, are as follows:

Threshold deflection:

$$u_* = \frac{p_{yq}}{k_r} z^{q-r} = az_0 + b \tag{9.73}$$

Horizontal force equilibrium:

$$T_0 - \int_0^{z_0} p_{yq} z^q \, dz - \int_{z_0}^{l} k_r z^r (az + b) \, dz = 0 \tag{9.74}$$

which may be integrated to give

$$T_0 - \frac{p_{yq}}{(q+1)} z_0^{q+1} - \frac{k_r}{(r+2)} (l^{r+2} - z_0^{r+2})a - \frac{k_r}{(r+1)} (l^{r+1} - z_0^{r+1})b = 0 \tag{9.75}$$

Moment equilibrium about top of pile:

$$\int_0^{z_0} p_{yq} z^q z \, dz + \int_{z_0}^{l} k_r z^r (az + b) z \, dz = 0 \tag{9.76}$$

which may be integrated to give

$$\frac{p_{yq}}{(q+2)} z_0^{q+2} + \frac{k_r}{(r+3)} (l^{r+3} - z_0^{r+3})a + \frac{k_r}{(r+2)} (l^{r+2} - z_0^{r+2})b = 0 \tag{9.77}$$

Since, as before, we are assuming that the calculations will be pursued for a given value of z_0, the unknowns are a, b (equal to the top deflection, u_0), and T_0, so that it is convenient to assemble the equations in matrix form for solution. We put them in the order (9.73), (9.77), and (9.75).

$$
\begin{bmatrix}
z_0 & 1.0 & 0 \\
\dfrac{k_r}{(r+3)}(l^{r+3}-z_0^{r+3}) & \dfrac{k_r}{(r+2)}(l^{r+2}-z_0^{r+2}) & 0 \\
-\dfrac{k_r}{(r+2)}(l^{r+2}-z_0^{r+2}) & -\dfrac{k_r}{(r+1)}(l^{r+1}-z_0^{r+1}) & +1.0
\end{bmatrix}
\begin{Bmatrix} a \\ b \\ T_0 \end{Bmatrix}
=
\begin{Bmatrix}
p_{yq}z_0^{q-r} \\
-\dfrac{p_{yq}}{(q+2)}z_0^{q+2} \\
\dfrac{p_{yq}}{(q+1)}z_0^{q+1}
\end{Bmatrix}
$$

$$(9.78)$$

If the solution is being obtained by hand, numerical values can be worked out for the coefficients for a particular value of z_0 and the first two equations solved simultaneously for a and b, since they do not contain T_0. This follows the analysis given for the simpler case. The resulting numbers for a and b are then substituted in the third equation to give T_0. If a computer or a programmable calculator is employed, it is more convenient to solve all three equations simultaneously to obtain values for a, b, and T_0 with one entry to the program.

Once the other quantities have been determined, the moment in the pile is obtained from the following expressions:

$$ 0 < z < z_0 \qquad M = T_0 z - \int_0^z p_{yq}s^q(z-s)\,ds \qquad (9.79) $$

which may be integrated to give

$$ 0 < z < z_0 \qquad M = T_0 z - \frac{p_{yq}}{(q+1)(q+2)}z^{q+2} \qquad (9.80) $$

and

$$ z_0 < z < l \qquad M = T_0 z - \int_0^{z_0} p_{yq}s^q(z-s)\,ds - \int_{z_0}^z k_r s^r(as+b)(z-s)\,ds $$

$$(9.81)$$

whose integration is most conveniently performed for specific values of q and r.

The final problem to be considered for a rigid pile is that of the application of a moment M_0 to the pile top. Returning, first, to the simplest case in which yield pressure, p_y, and the subgrade coefficient, k, are constant with depth, we can see that there is a certain simplification in this situation. Because no net lateral force is applied to the pile, the point of zero deflection is always halfway down the pile, which therefore deflects equal amounts top and bottom in opposite directions. The elastic case before the limiting deflection u_* is attained has been treated in Chapters 4 and 8, and the deflections u_* and $-u_*$ are reached simultaneously at the top and bottom of the pile, so that here z_1, the distance from the surface to the level at which $-u_*$ is attained, is given by the equation

$$z_1 = l - z_0 \qquad (9.82)$$

With deflection zero at the midpoint of the pile, the horizontal equilibrium of force is satisfied, and the constants in the pile deflection equation are related:

$$b = -\frac{al}{2} \qquad (9.83)$$

This time, consideration of moment equilibrium in the beam shows that z_0 appears in quadratic form, from which it can be evaluated specifically to give

$$z_0 = \frac{1}{2}\left[l - \sqrt{3l^2 - \frac{12M_0}{p_y}} \right] \qquad (9.84)$$

Clearly, when the radical is equal to l, $z_0 = 0$, and this defines the end of purely elastic behavior. From equation (9.84) the deflection at this stage is

$$u_0 = u_* = \frac{6M_0}{kl^2} \qquad (9.85)$$

which is the result obtained also from equation (8.128) when z is made equal to zero therein. The deflection is related to z_0 by the equation

$$u_* = az_0 + b \qquad (9.86)$$

but b and a are connected by equation (9.84), so that

$$a = -\frac{2u_*}{(l - 2z_0)} \qquad (9.87)$$

$$b = \frac{u_* l}{(l - 2z_0)} \qquad (9.88)$$

Finally, z_0 reaches its maximum value $l/2$ when the radical in equation (9.84) has zero value, that is, when M_0 reaches its ultimate value M_{0u}, where

$$M_{0u} = \frac{p_y l^2}{4} \qquad (9.89)$$

Because the inclusion of plastic behavior renders the problem nonlinear, the case of combined lateral load and moment at the end of a pile cannot be treated by superposing the two individual solutions. When a moment must be taken into account, it can be expressed in terms of the load T_0 in the following way:

$$M_0 = T_0 h \qquad (9.90)$$

where h is a length. If the pile actually sticks up above the ground surface and T_0 acts at a point above the ground, then h is the real distance from the point of action of T_0 to ground level. If the moment at the pile top acts for some other reason, h may be a fictitious length introduced to simplify the analysis. To solve the problem including force and moment, the distance h is introduced as another coefficient in the matrix of equation (9.78). It will be recalled that the second line of the matrix was obtained from moment equilibrium of the pile about the point $z = 0$. This is the only place the presence of an acting moment has an effect. Thus, the zero in the third place on the second row of the matrix is replaced by h. Now the full set of equations must be solved to obtain a, b, and T.

Although it would be possible to construct the solution to the complete problem in the second stage, after the deflection $-u_*$ had been exceeded at the bottom of the pile, it becomes complicated in the general case. Albeit tedious, the appropriate matrix equation is not difficult to establish in a particular problem when values of q and r have already been selected. In practice, with soil properties increasing with depth, it is most likely that the maximum moment in the pile will be the criterion dictating the maximum load that can be applied before the soil at the pile tip yields. In that case, the solutions already given, equations (9.78), (9.80), and (9.81), are the relevant expressions. With the plastic behavior assumed for the soil, the load-deflection curve of a rigid pile does become horizontal at an ultimate load, although this is probably of academic interest only. We will see in the next section that this is not the case for the long flexible pile under lateral load and/or moment.

9.4.2 Flexible pile

Because of the fourth-order nature of the behavior of a flexible pile, there are four solutions, controlled by four constants, which contribute to the pile's deflection, as we have seen in previous chapters. These constants are obtained from the boundary conditions. If the conditions at the tip of the pile play a part in its behavior, the problem becomes more complex. However, as discussed in Chapter 8, for most long piles, as opposed to caissons, in practice, deflections, moments, and shears disappear at a distance down the pile only a fraction of the pile length. Therefore, it is convenient and practical to ignore the condition at the tip and to include in the elastic part of the pile-soil behavior only those two elastic solutions that diminish in an exponential fashion with length.

As in the case of the rigid pile, we assume power-law variations of yield pressure p_y and subgrade reaction coefficient k with depth, equations (9.70) and (9.71). In the range of plastic soil behavior to depth z_0, therefore, the equation controlling the pile deflection is

$$EI\frac{d^4u}{dz^4} + p_{yq}z^q = 0 \tag{9.91}$$

and in the elastic region below this $z_0 < z < l$ the equation is

$$EI\frac{d^4u}{dz^4} + k_r z^r u = 0 \tag{9.92}$$

Equation (9.91) can be integrated successively four times to give

$$S = EI\frac{d^3u}{dz^3} = -\frac{p_{yq}}{(q+1)}z^{q+1} + EIC_1 \tag{9.93}$$

$$M = EI\frac{d^2u}{dz^2} = -\frac{p_{yq}}{(q+2)(q+1)}z^{q+2} + EIC_1 z + EIC_2 \tag{9.94}$$

$$\frac{du}{dz} = -\frac{p_{yq}}{EI(q+3)(q+2)(q+1)}z^{q+3} + C_1\frac{z^2}{2} + C_2 z + C_3 \tag{9.95}$$

$$u = -\frac{p_{yq}}{EI(q+4)(q+3)(q+2)(q+1)}z^{q+4} + \frac{C_1 z^3}{6} + \frac{C_2 z^2}{2} + C_3 z + C_4 \tag{9.96}$$

Two of the constants can be immediately determined from the boundary conditions at $z = 0$. If the applied load consists of a force T_0 only, and the moment is zero, then $C_2 = 0$ and $C_1 = T_0/EI$. When only a moment, M_0, loads the pile, $C_1 = 0$ and $C_2 = M_0/EI$. In the event both are acting, $C_1 = T_0/EI$ and $C_2 = M_0/EI$. In the above equations these constants can be replaced by the loads. Equations (9.93) through (9.96) hold down to the depth z_0, at which elastic behavior takes over. We will wish to make sure that all the variables on the left side of equations (9.93) through (9.96) are matched to the same variables determined from the elastic solution at $z = z_0$. The first set of values is obtained by replacing z by z_0 in these equations.

$$u(z_0) = -\frac{p_{yq}}{EI(q+4)(q+3)(q+2)(q+1)}z_0^{q+4} + \frac{T_0 z_0^3}{6EI} + \frac{M_0 z_0^2}{2EI} + C_3 z_0 + C_4 \tag{9.97}$$

$$\frac{du}{dz}(z_0) = -\frac{p_{yq}}{EI(q+3)(q+2)(q+1)}z_0^{q+3} + \frac{T_0 z_0^2}{2EI} + \frac{M_0 z_0}{EI} + C_3 \tag{9.98}$$

$$M(z_0) = -\frac{p_{yq}}{(q+2)(q+1)}z_0^{q+2} + T_0 z_0 + M_0 \tag{9.99}$$

$$S(z_0) = -\frac{p_{yq}}{(q+1)}z_0^{q+1} + T_0 \tag{9.100}$$

In the elastic region $z_0 < z < l$ the applicable equation is equation (9.92), whose solution is, for the case in which the conditions at the tip of the pile have no effect on the variables near the pile top,

$$u = A_1 u_1(\lambda z) + A_2 u_2(\lambda z) \tag{9.101}$$

where λ is given by equation (8.155) and u_1 and u_2 are the two solutions that diminish as (λz) increases. From this result, the other variables in the elastic solution can be obtained by differentiation.

$$u' = \frac{du}{dz} = \lambda(A_1 u_1' + A_2 u_2') \tag{9.102}$$

$$M = EIu'' = EI\lambda^2(A_1 u_1'' + A_2 u_2'') \tag{9.103}$$

$$S = EIu''' = EI\lambda^3(A_1 u_1''' + A_2 u_2''') \tag{9.104}$$

where the primes indicate differentiation. Specifically, the values $u(z_0)$, $u'(z_0)$ $M(z_0)$, and $S(z_0)$ are obtained by substituting (λz_0) in the above equations and obtaining the values $u_1(\lambda z_0)$, $u_1'(\lambda z_0)$, etc., $\ldots$, $u_2(\lambda z_0)$, $u_2'(\lambda z_0)$, $\ldots$, etc.

When z_0 has been selected, as before, the unknowns in the problem are A_1, A_2, C_3, C_4, T_0, and M_0. However, M_0 as in the previous problem studied, can be replaced by a moment considered to be generated by T_0 in the form $T_0 h$, where h is known, and the variables reduce to five. To describe them, we have the four equations obtained by equating expressions (9.97) through (9.100) with the four equations (9.101) through (9.104), couched in terms of λz_0. One more equation is required; it is the one developed from the condition that the displacement is known, u_*, at $z = z_0$. We can therefore set up the equations in the usual matrix form, reordering them somewhat for convenience.

$$
\begin{bmatrix}
u_1(\lambda z_0) & u_2(\lambda z_0) & 0 & 0 & 0 \\[2mm]
u_1''(\lambda z_0) & u_2''(\lambda z_0) & \dfrac{-(z_0 + h)}{EI\lambda^2} & 0 & 0 \\[3mm]
u_1'''(\lambda z_0) & u_2'''(\lambda z_0) & -\dfrac{1}{EI\lambda^3} & 0 & 0 \\[3mm]
u_1'(\lambda z_0) & u_2'(\lambda z_0) & -\dfrac{z_0(z_0 + 2h)}{2EI\lambda} & -\dfrac{1}{\lambda} & 0 \\[3mm]
0 & 0 & \dfrac{z_0^2(z_0 + 3h)}{6EI} & z_0 & 1
\end{bmatrix}
\begin{Bmatrix}
A_1 \\[2mm] A_2 \\[2mm] T_0 \\[2mm] C_3 \\[2mm] C_4
\end{Bmatrix}
=
$$

$$
=
\begin{Bmatrix}
\dfrac{p_{yq}}{k_r} z_0^{q-r} \\[4mm]
-\dfrac{p_{yq} z_0^{q+2}}{EI\lambda^2 (q + 2)(q + 1)} \\[4mm]
-\dfrac{p_{yq} z_0^{q+1}}{EI\lambda^3 (q + 1)} \\[4mm]
-\dfrac{p_{yq} z_0^{q+3}}{EI\lambda (q + 3)(q + 2)(q + 1)} \\[4mm]
\dfrac{p_{yq} z_0^{q+4}}{EI (q + 4)(q + 3)(q + 2)(q + 1)} + \dfrac{p_{yq}}{k_r} z_0^{q-r}
\end{Bmatrix}
\tag{9.105}
$$

The first equation comes from setting the elastic region displacement equal to u_* at z_0, the last equation from doing the same for the plastic region displacement. The middle three equations are obtained from matching the two sets of solutions. For manual solution it may be observed that the first three equations do not contain C_3 and C_4; they may be solved for A_1, A_2, and T_0. Substitution in the fourth and then in the fifth equation gives C_3 and then C_4, which is equal to the pile displacement at the surface, $z = 0$, according to equation (9.96). If a computer or programmable calculator is available, it is more convenient to solve the five equations simultaneously, since this takes only one setup for the problem.

For the properties selected for soil and pile, a value of λ is calculated first. Then for a chosen plastic depth z_0, λz_0 is obtained. Depending on the variation arrived at for k, values for the functions u_1, u_2, and their derivatives are found in the tables of Appendix C. All of the other numbers in equation (9.105) are calculated from the given or assumed problem parameters. The moment in the pile in the plastic region is calculated from equation (9.99) with the addition of $M_0(T_0 h)$ if required. In the elastic region the moment is given by equation (9.103), where u_1'' and u_2'' are determined from the tables for the selected λz, and A_1 and A_2 come from the solution of the set of simultaneous equations. In the case of a long pile loaded laterally, deflection decreases with distance down the pile, so that, for usual distributions of p_y and k staying at least constant or increasing with depth, there is no likelihood of the threshold displacement being exceeded at any depth other than by propagation of yielding from the surface down. The complexities introduced by yielding of the soil around the pile tip are therefore avoided for the long pile.

An analysis similar to the above has been described by Van Leyden (20). The method will be illustrated in the following section by a worked example.

9.4.3 Analysis of Mustang Island pile

A series of tests was carried out on the lateral loading of two piles driven at a site on Mustang Island, Texas. The tests are described in papers by Cox, Reese, and Grubbs (2) and Reese, Cox, and Koop (17) from which the following description is derived.

In these field tests the soil at the site consisted primarily of a dense uniform fine sand, with a water content of about 25 percent and a relative density in the range of from 80 to 100 percent estimated from Standard Penetration Test blow counts to a depth of 80 ft. In the top 10 ft (3 m) the relative density was somewhat lower. The sand profile was interrupted by a clay layer from a depth of 40 to 50 ft (12 to 15 m); this would have had no effect on the lateral load-deflection behavior of the piles, however. For the tests, the water table was above ground surface. Here and there in the profile there was evidence of clay or silt soil in the sand.

The two test piles were cylindrical A-53, grade B, seamless steel tubes, 24-in. (0.61 m) outside diameter, 0.375-in. (9.5 mm) wall thickness, and 79 ft (24 m) long. The bottom 38 ft (11.6 m) of the piles were not instrumented, since only lateral load tests were planned; inside the next upper 31 ft (9.4 m), 34 active and 6 dummy strain gauges were installed in 17 pairs, and the top 10 ft (3 m), protruding above ground surface, was again free of instrumentation. The embedded length was therefore 69 ft (21 m). Installation of the piles proceeded by driving the ungauged 38-ft (11.6 m) length open-ended until the top was at ground level and then clearing the soil out before welding on the 31-ft (9.4 m) long instrumented section that was sealed with diaphragms both top and bottom to protect the strain gauges from moisture. After the two sections were welded, driving was continued; when 69 ft (21 m) of penetration had been reached the top 10-ft (3 m) section was bolted on. Driving records for the two piles were very similar. The distance between the centers of the two test piles was 24 ft (7.3 m). Between them was installed a reaction frame for the horizontal load. It consisted of four 14WF78 piles 6 ft (1.8 m) on axes, driven to 20-ft (6.1 m) penetration, centered midway between the two test piles, and connected together at the top by the reaction frame.

A hydraulic ram reacting against the frame was used to load the piles at the flange level 1 ft (0.3 m) above ground surface. Load was measured by a load cell. Displacements were measured at two points on the unloaded uninstrumented 10-ft (3 m) section above ground, so that displacement and rotation could both be obtained. The load, acting 1 ft (0.3 m) above the ground line, generates a moment in the pile at ground surface, and the pile deflection and moments are greater than would have been the case had the same lateral load been applied actually at ground surface.

The two test piles were subjected to static and cyclic lateral loads. The results of the static load tests on pile 1 are reproduced in Figure 9-5. Figure 9-5 (a) is in terms of displacements and Figure 9-5 (b) is in terms of moments. In the cyclic tests, a greater load (by a factor of 3 or 4) was applied in one direction (major) than the other to simulate the effect of wave forces on an offshore structure, and cycles at increasing load levels were carried out. Unfortunately, the residual displacement and moments remaining in the piles at the end of the first cycle at zero load are not shown.

The product of modulus, E, and moment of inertia, I, of the test piles, EI, was approximately 6.0×10^{10} lb-in.^{2} (172 MN-m^2) as measured; the maximum stresses in the piles were limited to 27,000 psi (186 MN/m^2) compared to a measured yield stress of about 40,000 psi (276 MN/m^2). For this yield stress, the limiting elastic moment in the pile is 6.7×10^6 lb-in. (757 kN-m); the highest moment generated in either of the piles during the tests was about 4.5×10^6 lb-in. (509 kN-m). By cone penetration tests a significant increase in soil density was observed to result from pile driving,

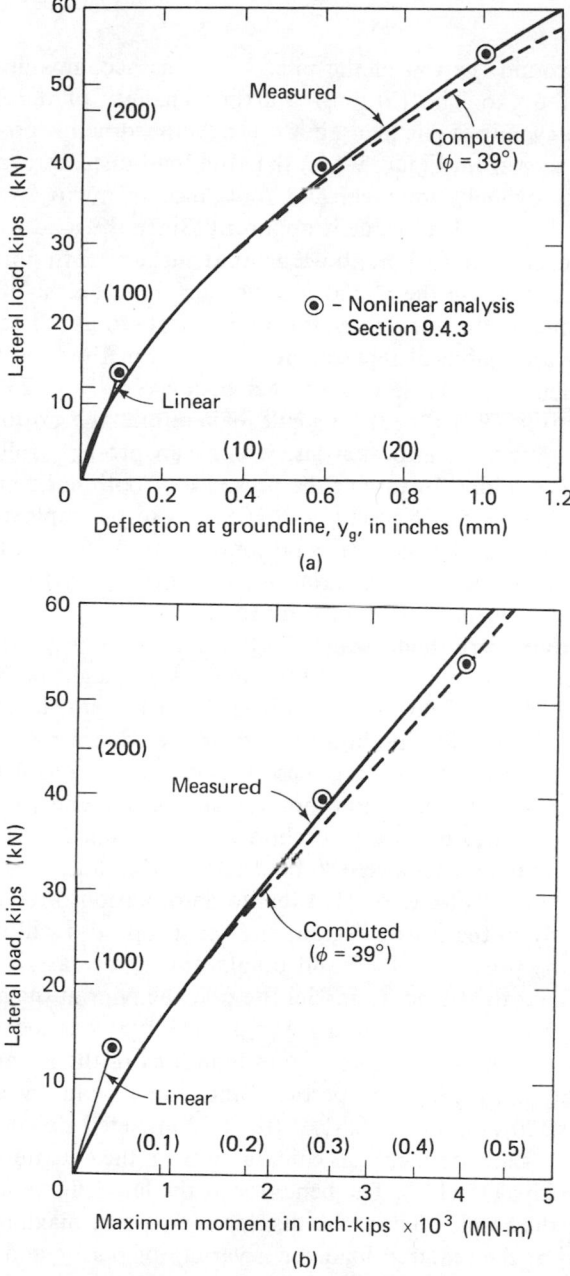

Figure 9-5. Comparison of measured behavior of field pile at Mustang Island, Texas (2) and analytical results: (a) lateral load versus ground line deflection; (b) lateral load versus maximum moment. Circles are the results of the computations described here; dashed lines show behavior computed from *p-y* analysis.

particularly around the top of the pile. This was accompanied by surface settlements of up to 2 in. (51 mm). Further changes of density occurred because of the cycling of pile loads, but this developed below ground surface.

It can be seen from Figure 9-5(a) that the load-displacement relation at the pile top is strongly nonlinear and that, moreover, no approach to an asymptotic or limiting load value is apparent. Since the load was applied to the pile at a level 12 in. (0.3 m) above ground surface, both horizontal force and a moment acted on the pile at ground surface. Because of the amount of data available on the test results, it is appropriate to use these field studies as a basis for an analytical representation by the methods of this chapter.

The moments were close to zero at a distance of from 25 to 30 ft (7.5 to 9 m) down the 69-ft (21 m) long pile. Therefore, the bottom boundary conditions do not enter into analysis, which can properly follow the lines of the previous section. Because of the presence of both horizontal load and moment, the full matrix of equation (9.105) has to be employed. There are two areas of choice in the analysis as presented: (a) the *form* of the functions describing the plastic and elastic variations with depth, and (b) the numerical values required to identify them quantitatively.

Several trials were made which will not be described here, but they served to determine that a good fit to the test results might be obtained with a model in which the yield pressure varied linearly with depth in the soil, and the subgrade reaction modulus varied as the square root of depth. To obtain quantitative values defining these variations, two engineering criteria of fit were selected: (a) maximum moment in the pile and (b) deflection at ground surface. It was necessary to choose a particular load for which the analytical model parameters were to be selected; this was taken as 42.5 kips (189 kN). It is likely, of course, that in the deformation of soil around the pile, particularly in the plastic region, the soil properties will not stay constant during the process of load and displacement increase. Consequently, it was decided not to attempt to model the pile behavior at the highest loads.

To begin the analysis, the matrix equation (9.105) was solved for several values of z_0 for each of a number of combinations of the linear plastic, p_{y1}, and square root elastic, $k_{0.5}$, properties. Since the z_0 values were selected at fixed intervals [20, 40, 60 in. (0.51, 1.02, 1.52 m), etc.] down the pile, the load required to generate each z_0 is calculated from the equations and cannot be set at 42.5 kips (189 kN). The behavior at the latter figure must be interpolated from the results. Thus, the displacements and maximum moments were obtained at the required load for a variety of plastic and elastic parameter numbers. Interpolating among these gave the values of p_{y1} and $k_{0.5}$ which, with the given pile properties, would develop the observed ground-level displacement and maximum moment in the pile at the 42.5-kip (189 kN) load.

For the test results shown in Figure 9-5, this technique gave the numbers $p_{y1} = 13.3$ psi (92 kN/m²) and $k_{0.5} = 340$ lb/in.$^{5/2}$ (1.47 MN/m$^{5/2}$). For these values, the load-displacement curve and moments at different load values are also shown in Figure 9-5. At a depth of 25 ft (7.6 m) down the pile the subgrade reaction modulus from the square root variation assumed is 5890 psi (40.6 MN/m²), and therefore the basic lateral subgrade coefficient, k_0, is 5890/24 = 245 pci (66.5 MN/m³). Using equation (8.137) indicates that the lateral elastic modulus E_s for this soil is 5890 psi (40.6 MN/m²), which, with a Poisson's ratio of 0.3, indicates a shearing modulus, G_s, of 2270 psi (15.7 MN/m²). This is considerably lower than the value obtained by *elastic* analysis of the Arkansas River lateral load test, described in Section 8.3.3, and it is more closely compatible with the value indicated by the axial load analysis.

9.5 *p-y* CURVES

Many offshore oil drilling and production platforms are pinned to the ocean floor by piles, which can reach very substantial dimensions. Because the lateral loads due to wave, and, depending on location, seismic action, are important to the stability of these structures, the oil industry has put a considerable effort into determining the factors that influence lateral pile-soil interaction. Many field tests and experiments have been conducted on full-size, although at the lower end of the diameter range, piles (2, 7, 8, 11, and 17). Unfortunately, many of the detailed reports have remained confidential, and the papers which have appeared tend to summarize the results or conclusions. From the studies, partly analytical and partly empirical analysis procedures have been developed (1, 7, 8, and 17) which are now widely employed in the petroleum industry for calculating the lateral movements of and forces in piles. Because of their practical interest, the methods will be briefly described here. They make an interesting comparison with the previous discussion of pile behavior under lateral loads.

In practice, the problem is immensely complex, not only because of the soil behavior but also because of the uncertainty of the loading conditions. However, there are three régimes of loading of concern: (a) the pile behavior under single, short-term but not dynamic loading (although dynamic loading is, of course, a concern under seismic conditions); (b) response to cyclic, static loads whose amplitude, in general, will increase over some period of time, reach a maximum, and then decrease, with random variations in amplitude at any time; and (c) behavior under loads following such a cyclic sequence. Condition (a) for an offshore platform may arise from ship forces, but it is, in any event, a necessary prerequisite to defining the behavior under the other load systems. Storm wave forces are responsible for the cyclic loading of regime (b), whereas the application of loads to the structure following a storm forms the realistic basis for assessing the response to condition (c).

Realistically, but unfortunately, soil properties do not remain constant under repeated or successive straining of the material. Original structure and fabric are disturbed or even destroyed, pore pressures are generated and dissipate, and permanent (until the next loading) distortions are imposed. The soil parameters of stiffness and strength continually change during cyclic loading, and then they are modified again by consolidation after loading ceases. Understanding and explanation of these responses, through, for example, plasticity theory, are only in infancy (14, 15, and 19), and resort must be made to empirical or semi-empirical correlations based on tests. The disadvantage with these procedures is that they are not general; an investigator derives a method that describes the test pile and soil behavior reasonably satisfactorily. There is no assurance that a much larger pile in a different soil will follow the same path. However, the following procedures, with modifications that accumulate in time, are widely used. In the usually employed nomenclature, which appears to date from the 1950's, the symbol p is used to denote the pressure acting on the pile in units of FL^{-2}, although in the final curves it usually represents the reaction force per unit length of the pile, FL^{-1}, and the pile deflection is given by the symbol y. The design procedure requires the construction of pressure-deflection—"p-y"—curves at different depths along the pile; these are then employed in computer calculations of the pile response under lateral load. Methods for determining p-y curves for clays and sands have been published and will be described. Although in this book, consistent with usage in solid mechanics, the lateral displacement is taken as u, and y is a coordinate direction, the original symbols p and y will be adhered to in this section because of their practical connotation.

9.5.1 Soft clay

The basic reference for this section is a paper by Matlock (8) which summarizes a great deal of experimental work on which the semi-empirical analysis procedure is based. The analysis begins with the development of static p-y curves at various depths in the soil profile. Construction of these will be illustrated with respect to a typical curve, as shown in Figure 9-6(a), in which the ordinate is the ratio of pile-soil interaction pressure, p, to the ultimate pressure, p_u, and the abscissa is a normalized lateral deflection y/y_c. The ultimate pressure is established by conventional bearing capacity theory:

$$p_u = N_c \cdot c \qquad (9.106)$$

in which N_c is the bearing capacity factor for soil possessing cohesion c alone. Cohesion may be determined by field or laboratory investigations in the preferred order: (a) in situ vane shear tests; (b) confined (triaxial) com-

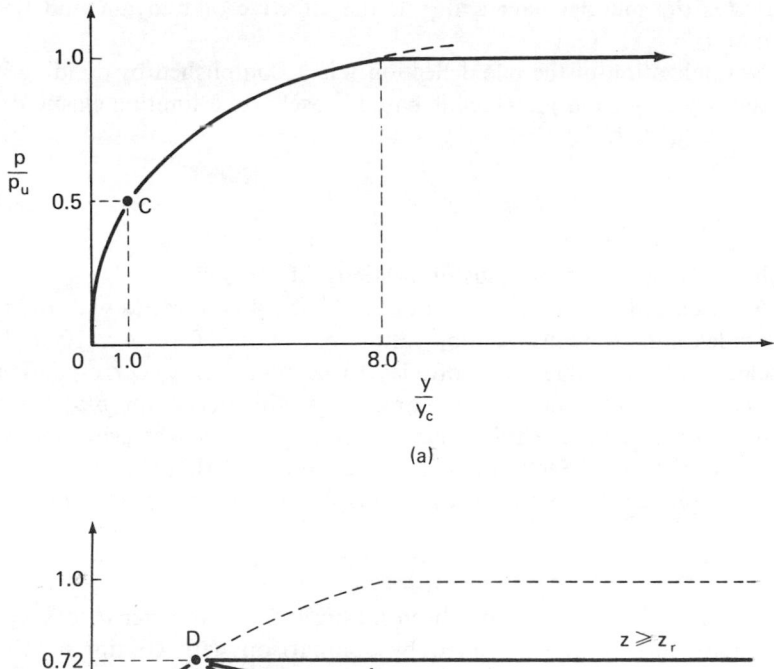

Figure 9-6. *p-y* curve for soft clay : (a) static loading ; (b) cyclic loading (8).

pression tests; (c) miniature vane tests on undisturbed tube samples; (4) unconfined compression tests.

The first complication is that, because the confinement of the soil changes from the surface down, N_c is not independent of depth. It has been found from pile tests that N_c should be taken at the value 9 below a particular depth z_r and to vary linearly from 3 to 9 from the surface to this depth, which is given for homogeneous soft clays by the equation

$$\frac{z_r}{d} = \frac{6}{\dfrac{\gamma' d}{c} + 0.5} \tag{9.107}$$

where d is the pile diameter and γ' is the effective or buoyant unit weight of the soil.

Normalization of the pile deflection y is accomplished by dividing it by an arbitrary deflection y_c, which is based loosely on a limiting elastic deformation of the soil:

$$y_c = \frac{2.5c}{E} d \tag{9.108}$$

in which E is a secant modulus of elasticity of the soil.

As discussed in Chapter 7, the ratio E/c for a clay varies with the plasticity index and overconsolidation ratio from about 50 to over 1000 (3). If we select a typical value for a soft clay to be 100, then y_c corresponds to a displacement of 0.025 of the pile diameter. At this deflection, $p/p_u = 0.5$. It is interesting to compare this value to what we would get using the plane strain disk theory of Section 8.3.2. There we found that the reaction on a section of pile could be related to the displacement by the relation

$$pd = ky \tag{9.109}$$

where k is a stiffness and p has been multiplied by diameter d to keep the dimensions correct. It was found by comparison with continuum theory that k was approximately equal to E. If p is replaced by $0.5p_u$, at which y_c is to be determined, and p_u is given by equation (9.106), then equation (9.109) on rearranging gives

$$y_c = 0.5N_c \frac{cd}{E} \tag{9.110}$$

which, over the range from 3 to 9 for N_c, gives values not dissimilar to that given by equation (9.108). Depending on the viewpoint, this confirms one approach or the other.

Now that point C (Figure 9-6) has been established, it remains to choose a pressure-displacement relation to describe the response below yield. Matlock selects a cubic for this portion of the curve

$$\frac{p}{p_u} = 0.5\left(\frac{y}{y_c}\right)^{1/3} \tag{9.111}$$

which intersects the constant-level yield line $p/p_u = 1.0$ at $y/y_c = 8$. This curve has the disadvantage of being tangential to the pressure axis at the origin, so that the initial soil stiffness is infinite. In numerical analysis the curve is approximated by a piecewise linear function. If the first segment is given a finite slope, the difficulty is avoided. Another approach, which is also convenient for analysis, would be to represent the pressure-displacement curve by an exponential relation

$$\frac{p}{p_u} = 1 - e^{-B(y/d)} \tag{9.112}$$

in which y is referred directly to the pile diameter and B is a constant. The latter can be arrived at by making the initial, finite slope of the curve equal to the pile/soil stiffness k/d. The initial slope is given by the derivative of equation (9.112) at the origin:

$$\left(\frac{dp}{dy}\right)_0 = \frac{p_u B}{d} \tag{9.113}$$

which, when equated to the stiffness, gives

$$B = \frac{k}{p_u} \tag{9.114}$$

or

$$B \simeq \frac{E}{N_c c} \tag{9.115}$$

At depth, and with the value of $E/c = 100$ as before, B is then approximately equal to 10. This approach requires an estimate of the yield pressure p_u and the subgrade stiffness k at each required depth.

However, with the relations (9.106) through (9.111) and the *p-y* curves they define, Matlock was able to represent quite well the deflection and moment distribution curves in a field static lateral load test on a hollow, circular 12.75-in. (0.32 m) diameter pile. The calculational method was, of course, derived from this test sequence. Some independent corroboration was obtained from another instrumented pile test (11), but some of the fitting parameters required alteration to achieve the fit.

The next practical question concerns the cyclic behavior of the pile. With lateral loads coming from wave forces, the number of cycles is large and can be considered to be infinite effectively. Matlock's tests indicate that no soil deterioration occurs during cycling to stresses or displacements less than those indicated by point D in Figure 9-6(b) which duplicates the static curve of Figure 9-6(a). The cyclic resistance reaches a peak at this point, which occurs at a dimensionless displacement $y/y_c = 3$ and stress $p/p_u = 0.72$. At larger displacements, the soil resistance at the ground surface diminishes, until it reaches zero at a value of $y/y_c = 15$. At this and larger displacements, and at greater depths but less than z_r, the soil resistance is constant at

$$\frac{p}{p_u} = 0.72\frac{z}{z_r} \tag{9.116}$$

When $z/z_r > 1$, p/p_u has the value 0.72. At displacements $3 < y/y_c < 15$ the pressure displacement function is linear, as shown in Figure 9-6(b) between the values at $y/y_c = 3$ and $y/y_c = 15$.

The experiments showed that in the soft clay, deflections of the pile greater than $y/y_c = 3$ develop a cavity in the clay next to the pile. After the cavity is formed, the clay resistance to movement is zero until the pile deflects sufficiently to make contact with the soil again. The cavity does not close up after cycling is over, so that on subsequent loading the pile will encounter no soil resistance up to the previous maximum displacement, less some elastic rebound. The post-cycling load-deflection path is shown in Figure 9-6(b) for clay at some depth $z/z_r < 1$. The reversible portion AB of the path is given the slope OC. On the negative side of the axes, for the reverse cycle, the same path is traced for cyclic loading.

No experimental data are yet available with which to check this analysis technique on overconsolidated, stiff, or fissured clays.

9.5.2 Sand

As in the case of clay, the approach for sands also requires the construction of p-y curves which are based on a few field tests, chief among them, the Mustang Island tests performed by Cox, Reese, and Grubbs (2) and analyzed by Reese, Cox, and Koop (17). The following instructions come from the latter paper and are based essentially entirely on the one field test series, although, in other tests, the investigators had gained insight into the mechanisms involved.

The procedure begins again with the development of the ultimate sand resistance against lateral movement of the pile. Because of the complex three-dimensional deformation of the soil near ground surface, a complexity that also extended to the plane strain plastic distortion of the sand at depth, the investigators performed approximate upper bound analyses of both mechanisms. Only their results will be presented here. At shallow depths, z, the ultimate resistance p_{us}, is given by the expression

$$\frac{p_{us}d}{\gamma' z^2} = A\left[\frac{K_0 \tan\phi \sin\beta}{\tan(\beta - \phi)\cos\alpha} + \frac{\tan\beta}{\tan(\beta - \phi)}\left(\frac{d}{z} + \tan\beta \tan\alpha\right)\right.$$
$$\left. + K_0 \tan\beta\,(\tan\phi \sin\beta - \tan\alpha) - \tan^2(\beta - 2\alpha)\right] \quad (9.117)$$

and at greater depths, the yield pressure p_{ud} is

$$\frac{p_{ud}}{\gamma' z} = A\{[\tan^2(\beta - 2\alpha)](\tan^8\beta - 1) + K_0 \tan\phi \tan^4\beta\} \quad (9.118)$$

where K_0 is the at-rest earth pressure coefficient, ϕ is the angle of internal friction, β is $45° + \phi/2$, α is $\phi/2$, and A is an empirical coefficient obtained from the test results as a function of depth. It is illustrated in Figure 9-7. The p/p_u versus, in this case, y/d curve has the same general shape as the clay function, but it is constructed differently. It is shown in Figure 9-8, which indicates the three points on the curve that determine it for a particular depth.

As before, the first of these is the beginning of the constant $p/p_u = 1$ yield portion. It is considered to begin at the displacement indicated at point C:

$$\frac{y_u}{d} = \frac{3}{80} \tag{9.119}$$

Point C is joined by a straight line to point B, where

$$\frac{p_B}{p_u} = \frac{B}{A}; \qquad \frac{y_B}{d} = \frac{1}{60} \tag{9.120}$$

in which B is another empirical adjustment factor derived from the Mustang Island tests and is shown in Figure 9-9. The third point is A, which lies on a straight line through the origin, representing a linearly elastic response of the soil. For point A, the coordinates are derived through the relations

$$\frac{y_A}{d} = \left(\frac{p_B}{k_1 z}\right)^{m/m-1} \left(\frac{d}{y_B}\right)^{1/m-1} \tag{9.121}$$

$$p_A = \frac{z}{d} n y_A \tag{9.122}$$

where

$$m = \frac{p_B (y_u - y_B)}{y_B (p_u - p_B)} \tag{9.123}$$

and n is the constant of horizontal subgrade reaction. For submerged sand, it can be obtained from the values in parentheses in Table 7.4. The grouping nz/d in equation (9.122) represents the coefficient of subgrade reaction as indicated by equation (7.52).

Finally, the curve joining points A and B is a parabola with equation

$$\frac{p}{p_u} = \frac{p_B}{p_u} \left(\frac{y}{y_B}\right)^{1/m} = \frac{B}{A} \left(\frac{y}{y_B}\right)^{1/m} \tag{9.124}$$

Reese, Cox, and Koop (17) note that in some sands the properties selected will result in a displacement at A greater than that calculated for point B. In that case, point B and the parabola should be omitted.

In Figure 9.5(a) and (b) the dashed curves labeled "computed ($\phi = 39°$)" are those obtained by computer calculation based on the *p-y* relationships

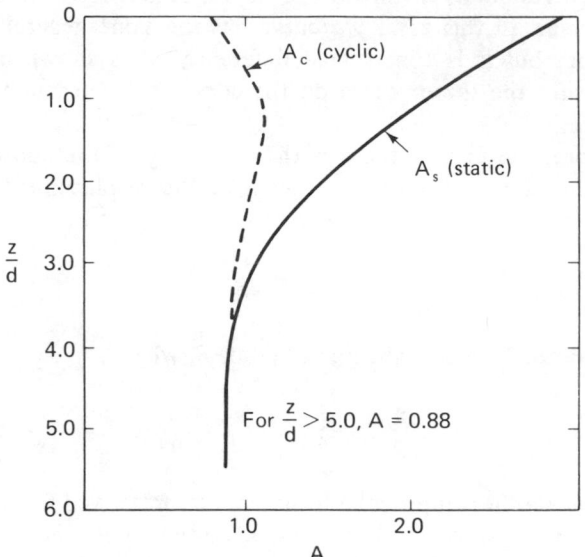

Figure 9-7. Empirical coefficient A for ultimate soil resistance versus depth (17).

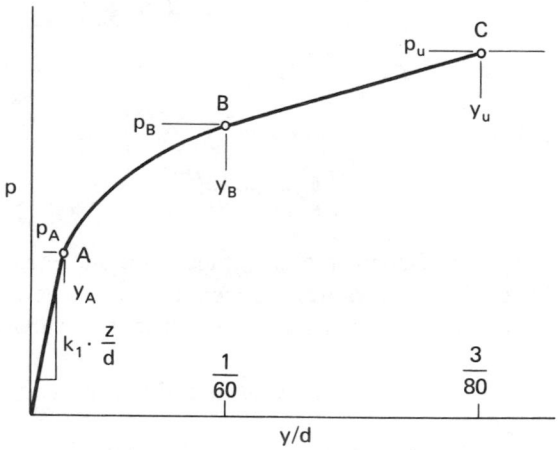

Figure 9-8. p-y curve for sand (17).

derived according to the above procedure. They may be compared with the other curves calculated analytically in Section 9.4.3 and the measured experimental values.

Further field testing is required to substantiate this calculational procedure in circumstances of other piles and granular soils.

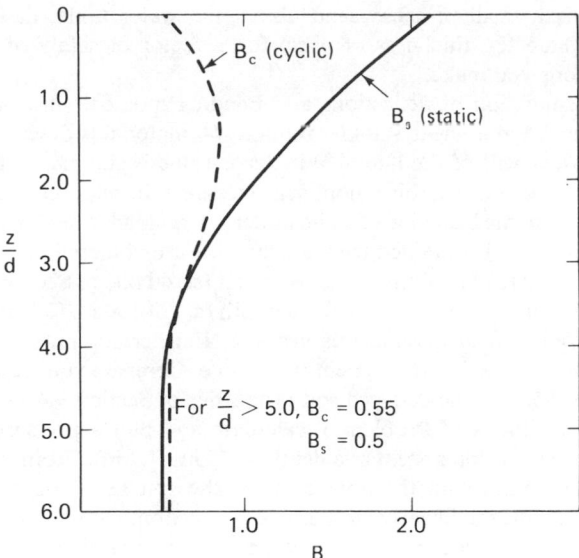

Figure 9-9. Empirical coefficient *B* for soil resistance versus depth (17).

PROBLEMS

1. Anchors for the tie-rods of a sheet pile wall are to be constructed as shown in Figure P9-1. Each tie-rod [spaced at 5 ft (1.5 m) intervals] exerts a force in the horizontal direction of 100,000 lb (445 kN) on the anchor. Assuming that

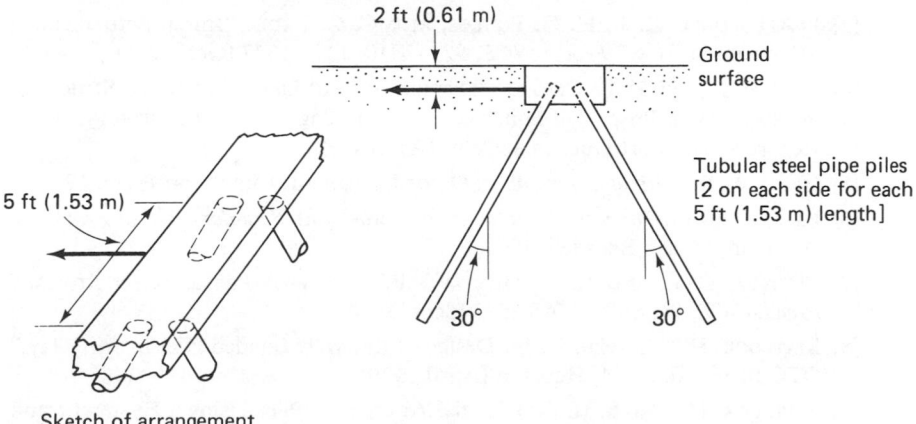

Figure P9-1.

the soil is a medium-dense sand above the water table, design the anchor (length, diameter, thickness of piles) for a factor of safety of 1.5. State any assumptions you make.

2. The determination of deflections and bending moments in a laterally-loaded pile imbedded in a linear Winkler foundation material is described in Chapter 8. The assessment of the lateral Winkler k-value is summarized in Chapter 7. However, real soils exhibit a nonlinear stress-strain relation. Discuss how you would analyze the behavior of a pile under lateral load in such a real soil, where the soil property is obtained from a triaxial test or otherwise.

3. For the lateral load analysis of the Mustang Island pile of Section 9.4.3, assume that yield of the soil has occurred to a depth (z_0) of 100 in. (2.5 m) and calculate the load required to develop this yielding. Further compute the pile top displacement and maximum moment in the pile. Compare your results with those of Figure 9-5. Use the deduced soil properties of Section 9.4.3.

4. For the conditions of Problem 3, calculate and plot a pressure-displacement curve for the pile-soil system at a depth of 50 in. (1.3 m). From the information in Section 9.5.2 obtain the p-y curve for the system at the same depth and compare it with the linearly elastic-plastic function. Comment.

5. How would you modify the lumped-parameter analysis of Problem 8-14 to account for yielding of the soil around the pile as the load is increased?

REFERENCES

[1] AMERICAN PETROLEUM INSTITUTE. *Recommended Practice for Planning, Designing, and Constructing Fixed Offshore Platforms*, RP2A 10th ed. Dallas (March 1979).

[2] Cox, W. R., L. C. REESE, AND B. R. GRUBBS. "Field Testing of Laterally Loaded Piles in Sand," Offshore Technology Conference Paper No. 2079, Houston (May 1974).

[3] D'APPOLONIA, D. J., H. G. POULOS, AND C. C. LADD. "Initial Settlement of Structures in Clay," *Proc. ASCE, 97*, SM10, 1359–1377 (Oct. 1971).

[4] GILL, H. L., AND K. R. DEMARS. "Displacement of Laterally Loaded Structures in Nonlinearly Resposive Soil," Naval Civil Engineering Lavoratory, Tech. Report R-670, Port Hueneme, Calif. (April 1970).

[5] HEMP, W. S. *Optimum Structures*. Oxford, England: Clarendon Press, 1973.

[6] KELLER, H. B., *Numerical Methods for Two-Point Boundary Value Problems*. Waltham, Mass.: Blaisdell, 1968.

[7] MANSUR, C. I., AND A. H. HUNTER. "Pile Tests—Arkansas River Project," *Proc. ASCE, 96*, SM5, 1545–1582 (Sept. 1970).

[8] MATLOCK, H. "Correlations for Design of Laterally Loaded Piles in Soft Clay," OTC Paper No. 1204, Houston (April 1970).

[9] MATLOCK, H., AND S. H. FOO. "Axial Analysis of Piles Using a Hysteretic and Degrading Soil Model," *I.C.E. (U.K.) Proc. Conf. on Numerical Methods in Offshore Piling*, 99–105 (May 1979).

[10] MATLOCK, H., AND L. C. REESE. "Generalized Solutions for Laterally Loaded Piles," *Proc. ASCE, 86*, SM5, 673–694 (Oct. 1961).

[11] MATLOCK, H., AND R. T. TUCKER. "Lateral Loading of an Instrumented Pile at Sabine, Texas," report to Shell Development Co., Austin (1961).

[12] MEYERHOF, G. G. "The Ultimate Bearing Capacity of Foundations," *Geotechnique, 2*, 301 (1951).

[13] MURFF, J. D. "Response of Axially Loaded Piles," *Proc. ASCE, 101*, 356–360 (March 1975).

[14] PREVOST, J-H. "Plasticity Theory for Soil Stress-Strain Behavior," *Proc. ASCE, 104*, EM4, 1174–1194 (Oct. 1978).

[15] RANDOLPH, M. R., AND C. P. WROTH. "Analysis of Deformation of Vertically Loaded Piles," *Proc. ASCE, 104*, GT12, 1465–1488 (Dec. 1978).

[16] REESE, L. C. "Laterally Loaded Piles: Program Documentation," *Proc. ASCE, 103*, GT4, 287–305 (April 1977).

[17] ——, COX, W. R., AND F. D. KOOP. "Analysis of Laterally Loaded Piles in Sand," OTC Paper No. 2080, Houston (May 1974).

[18] RICHART, F. E., J. R. HALL, AND R. D. WOODS. *Vibrations of Soils and Foundations.* Englewood Cliffs, N.J.: Prentice-Hall, 1970.

[19] VALANIS, K. C. "A Theory of Viscoplasticity without a Yield Surface," *Archiwum Mechaniki Stosowanej, 23*, No. 4, 517–533 (1971).

[20] VAN LEYDEN, W. "Plastic-Elastic Analysis of Laterally Loaded Free-Standing Piles," *de Ingenieur, Bouw-en Waterbouwkunde, 9*, B101–B111 (June 1971).

[21] VESIC, A. S. "Expansion of Cavities in Infinite Soil Mass," *Proc. ASCE, 98*, SM3, 265–290 (March 1972).

[22] ——. "Design of Pile Foundations," U.S. Transportation Research Board, NCHRP Synthesis 42, Washington, D.C., 1977.

Pile Groups under Combined Loads

10

In previous sections we have considered the behavior of individual piles under a variety of loading conditions when the soil response has been assumed to be either completely elastic or to involve plastic deformations. Although single piles are occasionally required to resist loads, more usually piles function in a group. The consequence is that the response of each pile is modified by the stress conditions imposed on the soil by other members of the group. Group behavior is very complex and, apart from the soil conditions, is generally dependent on the pile spacing, the length and relative stiffness of the piles, and the number of piles in the group, unless the total is very large.

The failure of pile groups under axial loading has been extensively examined by analysis (10, 16, and 20) and model studies (17 and 19), but it has not been so thoroughly investigated at full scale because of the magnitude of the loads required. A major difficulty in the assessment of pile group performance is the disturbance of the soil caused by the number of piles driven. Sometimes the soil, such as a sensitive clay, degrades as a result of multiple pile driving and sometimes it is rendered denser and stronger, as, for example, when it consists of a loose to medium-dense sand. In the former case, the bearing capacity per pile in a group is smaller than that of the individual pile; in the latter, the group pile capacity can be much greater than the individual. The term applied to the difference in capacity between the individual pile and the same pile in a group is *group efficiency*; it is expressed as the ratio of the bearing capacity of the group pile to that of the single pile in the same soil. A number of formulas have been adduced (10) to represent the group efficiency as a function of the ratio of pile spacing (center-to-center) to pile diameter and the number of piles in the group. Pile length is not usually included. The formulas have been compared with the behavior of model pile groups in clay (20). At pile spacings of 8 diameters or more the piles behave essentially as individuals, whereas at the usual distances of piles in a group—$2\frac{1}{2}$ to 3 diameters—a considerable amount of interaction takes place. At failure at the smaller spacings of 2 diameters or less an entire pile group in a clay behaves as a unit and exhibits a bearing capacity failure as if it were a single deep foundation.

The model studies show that during axial loading of a pile group connected at the pile tops by a rigid pile cap the load, at conventional spacings, is not equally distributed among the piles. As the total load is increased, corner piles take a greater proportion of it than piles along the edges, and the latter in turn bear more load than piles in the center of the group. Failure levels are reached by the corner piles first, followed by edge and center piles. At failure of the whole group, the loads taken by edge and interior piles are still smaller than those borne by the corners.

Questions still remain regarding the satisfactory estimation of failure loads on single piles and pile groups. At building sites resort is always made, after all the analyses, to field pile tests as a final criterion. This is, however, not possible for the large piles used to support steel offshore oil platforms. These piles have calculated capacities of 10,000 tons (89 MN) or greater. No test loading method has been devised. For such piles, and for other shorter ones, capacities are estimated on the basis of side friction and point bearing. The shearing strength of the soil is obtained either from laboratory tests or samples or from field vane, standard penetration, or cone tests.

It is always difficult in these calculations to determine a suitable value of the lateral effective stress that the soil will bring to bear on the pile wall some time after driving. As the pile tip penetrates the soil it pushes the soil aside to an extent depending on whether the pile is solid or a hollow tube. After some distance of penetration, hollow tube piles generally plug and thereafter behave as solid piles. In effect, the soil along the line of the pile's motion is pushed aside in a mechanism similar to that developed by an expanding cavity. It is thereafter strongly sheared as the pile wall slides by. The result is a very disturbed zone out to a distance of about a pile diameter. Depending on the initial soil condition, a saturated soil will develop pore pressures during the penetration process. If they are positive, pile driving is eased, but pile resistance immediately after driving is small. With subsequent dissipation of pore pressure, effective stresses build up against the pile walls, and the resistance increases with time. In clays, full pile capacity may not be reached for months after driving. The difficulty in analysis is that the soil is so disturbed that its stress condition at various stages in this process cannot be accurately determined; in particular, the final effective lateral stress and shear strength developed against the pile may vary widely from soil to soil. Attempts have been made to simulate the effect of pile driving by analysis of an expanding cylindrical cavity in a soil with a selected constitutive behavior such as represented by the Camclay model developed in Cambridge University research (16), *cap* models (4), or other representations. Sufficiently variable behavior is indicated by pile installations in different soils that it is apparent that much more work needs to be done in this area. Analyses must be accompanied by field tests on real piles with total pressure and pore water pressure measurements at the pile wall, and at various radial distances, as functions of time during and after driving. A further complication is posed by some soils, such as those of calcareous origin, whose grains break down during the mechanical stresses of driving. Thus, in them, not only are the prior material properties altered by pile installation, but the soil material itself is also rendered into a different form of altered grain size.

Most aspects of failure of piles and pile groups are extensively discussed in the literature (10 and 21); here we will confine our attention to deformational behavior.

10.1 DISPLACEMENT OF PILE GROUPS; AXIAL LOADS

At working loads on a pile group, where the material response can be considered to be mostly in the elastic range, we are concerned with the displacement of the pile top under the various loading conditions to which it may be subjected. Therefore, we need to know the spring constants at the pile top for axial, lateral, moment, and combined loads as in the case of fixed-head displacement. These quantities have been obtained in previous sections under the assumption that the soil behavior can be represented by a Winkler model. Should data warrant tackling the additional complexity of the two-parameter model, similar quantities can be obtained in the case of lateral loads.

In these analyses the soil property appears in the form of the coefficient of subgrade reaction k_0, which has been expressed previously in terms of the soil's elastic modulus, or has been obtained from field tests, tables based on them, or experience with pile performance. It is apparent that when a number of piles is present in a group, the individual pile behavior should no longer be based on the single-pile value of k_0. The simplest addition that can be made to the analysis is to modify the k_0-value to reflect the presence of companion piles. Presumably, the modification will differ depending on whether the loading is axial or lateral. In the case of torsion applied to a pile group, the individual pile response will be that of a laterally loaded pile. In each case, the effect of adjacent piles is to lower the reaction coefficient, since the soil at the location of the neighbor piles is already being moved by those piles, and the response of a single representative pile in the group is therefore more compliant than it would be alone.

10.1.1 Large group

Predicted behavior can be based on analysis, reinforced in part by observation, although studies of pile group deflections under measured loads are rare. If a subgrade reaction coefficient method is being used, since it is convenient for preliminary investigations, the first question to arise concerns the value of k_0 to use for a pile group. For axial loads, the antiplane analysis of Section 8.2.1 can be used in conjunction with reciprocity to give an approximate analysis of the relation of pile group response to that of a single pile. Considering that a pile in the center of a zone of influence, again taken to be 50 pile radii in radius (25 diameters) and unit thickness, is subjected to an axial force F, then vertical displacements, $w(r/a)$ as a function of dimensionless radius r/a, where a is the pile radius, are given by the equation

$$w\left(\frac{r}{a}\right) = w_0\left[1 - \frac{\ln\left(\frac{r}{a}\right)}{\ln 50}\right] \tag{10.1}$$

in which w_0 is the displacement developed at the pile by the load F. Thus, we can write that load F at the central location 0 causes a vertical deflection w_{10} at location 1, distant r/a away, where w_{10} is given by equation (10.1). From reciprocity, it follows that load F at location 1 will give a deflection w_{01} in a homogeneous medium at location 0 and that w_{10} equals w_{01}. At a central location 0 the total deflection w due to loads F applied to all piles of a group at various distances will be given by the sum of all the deflections caused at 0, or

$$w = w_0\left[1 + \sum_{1}^{n}\left(1 - \frac{\ln\left(\frac{r_n}{a}\right)}{\ln 50}\right)\right] \tag{10.2}$$

where r_n is the radius to the nth pile.

Now for the single pile under load, we can write the load-deflection relation for a section at arbitrary depth in the usual way:

$$F = kw_0 \tag{10.3}$$

where F is the force per unit length of the pile. Similarly, for the same pile as a central member of a group of n piles, subjected to force F on each pile,

$$F = k_n w \tag{10.4}$$

where k_n is the modified subgrade coefficient due to the group. The displacements w_0 and w from these two relations can be substituted into equation (10.2) to give the expression

$$\frac{k_n}{k} = \frac{1}{1 + \sum\limits_{1}^{n}\left[1 - \dfrac{\ln\left(\dfrac{r_n}{a}\right)}{\ln 50}\right]} \tag{10.5}$$

in which the summation is taken for all piles out to the radius ratio, here 50, at which no further deflection is taken to be caused by load on the central pile.

Equation (10.5) gives the effective k_n of a group of n piles in terms of the k of a single pile, when all are subjected to the same load. The value of k_n can be used to give the deflection of a fairly large group of piles as a result of uniform loading. The summation depends on the pile spacing. If, for example, the piles are located 25 diameters apart, then in this model there is no interaction and the reaction coefficient is that of the single pile. For different spacings, there will be various numbers of piles participating inside the influence zone. From the above model we can construct Table 10.1. The results are also plotted in Figure 10-1.

TABLE 10.1

Axial Loading; Pile Group Subgrade Reaction Coefficients
($b/a = 50$)

Spacing Pile Diameters	Number of Piles Involved	Ratio $\dfrac{k_n}{k}$
25	1	1
20	5	0.8142
15	9	0.5916
10	21	0.3361
5	69	0.0976
4	121	0.0630
3	221	0.0357
2.5	305	0.0249

In this analysis, as applied to a real pile group, it is assumed that the major portion of the resistance to pile deflection comes from side friction and that the contribution of the tip resistance is relatively small, or alternatively, behaves the same as the side resistance. The ratios come from the analysis of an elastic disk of unit thickness and diameter taken to be 50 times the pile diameter, so that the value of k of a single pile is obtained by multiplying the right side of equation (8.23), by the pile perimeter, $2\pi a$

$$k = \frac{2\pi G}{\ln 50} = 1.61G \tag{10.6}$$

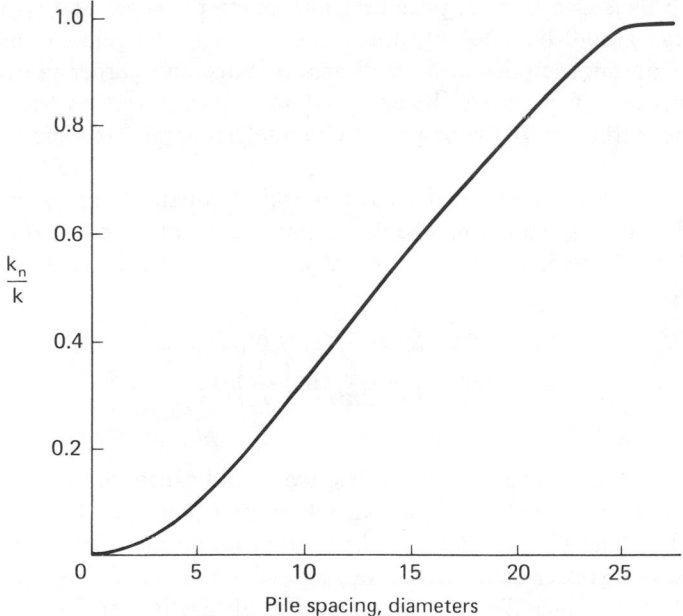

Figure 10-1. Axial loading; large pile group. Ratio of subgrade reaction coefficient of group pile to that of single pile.

If the soil has a modulus G which increases with depth, or if the soil is stratified with different moduli in the various layers, then k varies with depth in the same way as G according to equation (10.6). The axial load-deflection relation for the pile top is obtained by calculations based on the equations of Section 8.2.2. For a large group of piles embedded in the same nonhomogeneous medium, the pile responses will be the same at any one depth, and k_n at that depth or in the layer is modified from the single-pile value at the same depth according to Table 10.1, as a function of the pile spacing. Such large groups of piles are encountered in the foundations of nuclear reactors, substantial buildings, docks, or grain elevators, for example.

10.1.2 Small group

However, more commonly, smaller numbers of piles occur in a group supporting a single column or component of a large structure. Generally, their tops are tied together in a single reinforced concrete block several feet thick, and the piles are embedded to a depth of at least a foot (0.3 m) in the block, depending on the lateral load. Groups of 4, 9, 16, or 25 piles in square arrays are common. In this case, the applied vertical column load, through the rigid pile cap, causes all piles to displace through the same vertical distance. Since the group is now finite, the different piles experience differing degrees of interaction with the other piles, and thus take up different axial

loadings. This is also true for piles near and along the edges and corners of
a very large group, but they are less important to the overall foundation
response than are the piles of a small group. Edge and corner piles of the
large group can, if necessary, be analyzed in the same way as the smaller
group. The method will now be given. The analysis applies to rigid or fairly
stiff piles.

Again, for axial loading, the basic working equation comes from the
centrally loaded disk solution. The deflection at a point at radius r/a, $w(r/a)$
due to a load F applied to a rigid centrally located disk (pile cross section),
radius a, is

$$w\left(\frac{r}{a}\right) = \frac{F}{2\pi G} \ln\left(\frac{\dfrac{b}{a}}{\dfrac{r}{a}}\right) \qquad (10.7)$$

In general, in an expression such as this, we would cancel out the radius a,
but here we have to bear in mind that when r/a equals or exceeds b/a, we
have decided that $w(r/a)$ will be zero. Contributions do not exist from loaded
disks at this or greater distance.

We choose a nine-pile (3×3) group for illustration of the approach,
as shown in Figure 10-2 in which the pile numbering system is also given.
Two closely adjacent piles are spaced s apart, where s is the number of pile
diameters between pile centers. The radial distance between these two piles

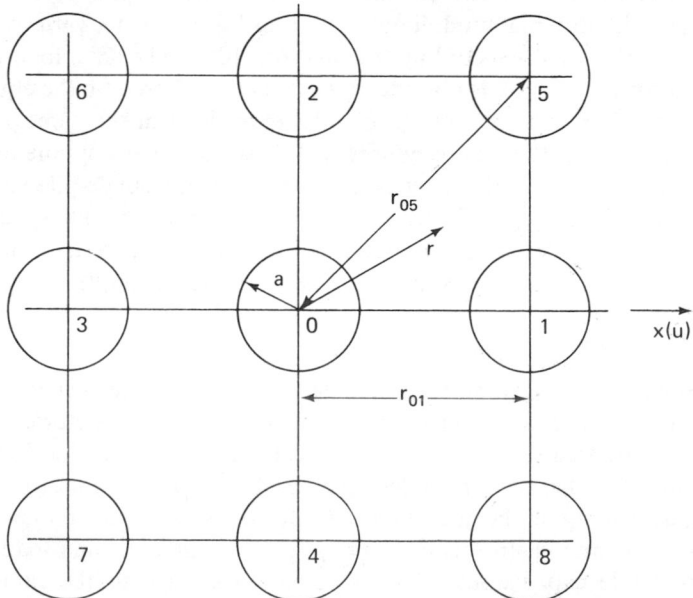

Figure 10-2. Nine-pile array.

is thus $2s$. We consider that the pile sections, subjected to equal vertical displacement, generate different loads. In homogeneous material, by symmetry, there will be three loads: F_0 at the center, F_1 at the center of each edge, and F_s at each corner.

From vertical equilibrium, we get

$$F_0 + 4F_1 + 4F_s = F \tag{10.8}$$

where F is the total axial load applied to the group. For convenience, we normalize by F and rewrite equation (10.8):

$$F_0 + 4F_1 + 4F_s = 1 \tag{10.9}$$

where F_0 now equals F_0/F, etc. Using equation (10.7), we see that the deflections of the three typical pile sections are

$$w_0 = \frac{1}{2\pi G}\left[F_0 \ln\left(\frac{b}{\frac{a}{1}}\right) + 4F_1 \ln\left(\frac{b}{\frac{a}{2s}}\right) + 4F_s \ln\left(\frac{b}{\frac{a}{2\sqrt{2}\,s}}\right) \right] \tag{10.10}$$

$$w_1 = \frac{1}{2\pi G}\left[F_1 \ln\left(\frac{b}{\frac{a}{1}}\right) + F_0 \ln\left(\frac{b}{\frac{a}{2s}}\right) + 2F_s \ln\left(\frac{b}{\frac{a}{2s}}\right) \right.$$
$$\left. + 2F_1 \ln\left(\frac{b}{\frac{a}{2\sqrt{2}\,s}}\right) + 2F_s \ln\left(\frac{b}{\frac{a}{2\sqrt{5}\,s}}\right) + F_1 \ln\left(\frac{b}{\frac{a}{4s}}\right) \right] \tag{10.11}$$

$$w_s = \frac{1}{2\pi G}\left[F_s \ln\left(\frac{b}{\frac{a}{1}}\right) + 2F_1 \ln\left(\frac{b}{\frac{a}{2s}}\right) + F_0 \ln\left(\frac{b}{\frac{a}{2\sqrt{2}\,s}}\right) + 2F_s \ln\left(\frac{b}{\frac{a}{4s}}\right) \right.$$
$$\left. + 2F_1 \ln\left(\frac{b}{\frac{a}{2\sqrt{5}\,s}}\right) + F_s \ln\left(\frac{b}{\frac{a}{4\sqrt{2}\,s}}\right) \right] \tag{10.12}$$

But since these deflections must all be the same,[1] we can write

$$w_0 - w_1 = F_0\left[\ln\left(\frac{b}{\frac{a}{1}}\right) - \ln\left(\frac{b}{\frac{a}{2s}}\right) \right] + F_1\left[4\ln\left(\frac{b}{\frac{a}{2s}}\right) - \ln\left(\frac{b}{\frac{a}{1}}\right) - 2\ln\left(\frac{b}{\frac{a}{2\sqrt{2}\,s}}\right) \right.$$
$$\left. - \ln\left(\frac{b}{\frac{a}{4s}}\right) \right] + F_s\left[4\ln\left(\frac{b}{\frac{a}{2\sqrt{2}\,s}}\right) - 2\ln\left(\frac{b}{\frac{a}{2s}}\right) - 2\ln\left(\frac{b}{\frac{a}{2\sqrt{5}\,s}}\right) \right] = 0$$
$$\tag{10.13}$$

---

[1]Or very nearly so. They will be slightly different at a particular depth below the soil surface due to the different loads in the piles, but the major component of deflection is due to the soil's response, so long as the piles are considered to be fairly stiff.

$$w_0 - w_s = F_0 \left[\ln\left(\frac{\frac{b}{a}}{1}\right) - \ln\left(\frac{\frac{b}{a}}{2\sqrt{2}\,s}\right) \right] + F_1 \left[4\ln\left(\frac{\frac{b}{a}}{2s}\right) - 2\ln\left(\frac{\frac{b}{a}}{2s}\right) \right.$$

$$\left. - 2\ln\left(\frac{\frac{b}{a}}{2\sqrt{5}\,s}\right) \right] + F_s \left[4\ln\left(\frac{\frac{b}{a}}{2\sqrt{2}\,s}\right) - \ln\left(\frac{\frac{b}{a}}{1}\right) - 2\ln\left(\frac{\frac{b}{a}}{4s}\right) \right.$$

$$\left. - \ln\left(\frac{\frac{b}{a}}{4\sqrt{2}\,s}\right) \right] = 0 \tag{10.14}$$

In these two equations $2\pi G$ has been canceled out and all terms have been divided by F to give the normalized forces as in equation (10.9). Only terms are to be included in which the spacing between the two piles involved is less than b/a. If, for a given pile spacing, all terms are to be included, or if the appropriate ones are left out, then $\ln(b/a)$ can be canceled from the remainder to give a simpler form for the equations. If, for example, all terms are valid, equations (10.13) and (10.14) become

$$F_0 \ln 2s + F_1 \ln\left(\frac{2}{s}\right) + F_s \ln\left(\frac{5}{4}\right) = 0 \tag{10.15}$$

and

$$F_0 \ln 2\sqrt{2}\,s + F_1 \ln 5 + F_s \ln\left(\frac{\sqrt{2}}{s}\right) = 0 \tag{10.16}$$

These two equations, (10.13) and (10.14) or (10.15) and (10.16), together with equation (10.9), can be solved for the three unknowns F_0, F_1, and F_s for the value of s selected. When a solution is obtained, the dimensionless displacement $w' = 2\pi Gw/F$ of the pile disk group can be calculated from any one of equations (10.10), (10.11), or (10.12). The subgrade reaction coefficient for each pile is computed by dividing the dimensionless force in the pile by the displacement w', thus

$$k_i = \frac{2\pi G F_i}{w'} \tag{10.17}$$

However, from equation (10.7) the coefficient k for a solitary pile can be written:

$$k = \frac{2\pi G}{\ln\left(\frac{b}{a}\right)} \tag{10.18}$$

so that, dividing equation (10.17) by equation (10.18) the ratio of the individual reaction coefficients can be obtained, as for the case of the large pile

group

$$\frac{k_i}{k} = \frac{F_i}{w'} \ln \left(\frac{b}{a}\right) \qquad (10.19)$$

In the present case of the nine-pile group, the values of F_i, w', and k_i/k for different spacings have been calculated and are given in Table 10.2.

TABLE 10.2

Displacement, Forces, and Relative Stiffnesses in a Nine-pile Element Group

Spacing Diameters, s	F_0	F_1	F_s	w'	Group Displacement Efficiency or $\frac{k_n}{k}$	$\frac{k_0}{k}$	$\frac{k_1}{k}$	$\frac{k_s}{k}$
2	−0.0298	0.0721	0.1854	2.2274	0.195	−0.0524	0.1266	0.3256
3	−0.0024	0.0820	0.1686	1.8822	0.231	−0.0051	0.1705	0.3504
4	0.0116	0.0864	0.1607	1.6338	0.266	0.0277	0.2069	0.3848
5	0.0203	0.0890	0.1560	1.4399	0.302	0.0551	0.2417	0.4237
6	0.0264	0.0907	0.1527	1.2810	0.339	0.0806	0.2770	0.4663
8	0.0345	0.0929	0.1484	1.0294	0.422	0.1312	0.3697	0.5641
10	0.0417	0.0973	0.1423	0.8439	0.515	0.1932	0.4510	0.6596

Several observations can be made from these results. First, at close pile element spacings, $s = 2$ and 3, the center pile element is subjected to tension; that is to say, the soil around it is tending to drag it down, so that, rather than resisting the applied load, it is acting in the same direction. If this is correct, the center pile could be removed with the result that the overall stiffness of the pile system would be slightly increased. In this case, the apparent stiffness of this element is negative. It is interesting to note that in one of the pile *bearing capacity* efficiency formulas, Feld's rule, the efficiency is calculated by subtracting $\frac{1}{16}$ of the single pile bearing capacity from each pile for every near neighbor it has (that is, every neighbor at s and $s\sqrt{2}$). When this is applied to a 3×3 pile group, it turns out that the center pile may be removed without affecting the efficiency or the total bearing capacity of the group. Feld's rule makes no allowance for pile spacing, since it was formulated for pile groups at usual spacings—$2\frac{1}{2}$ to $3\frac{1}{2}$ diameters.

Second, the corner pile element takes the greatest load and has the largest relative stiffness. Consequently, as the overall load is applied, the force builds up most rapidly in the corner element, then in the side center element, and most slowly at the center. Finally, for this model, the piles would have to be spaced at 25 diameters or greater spacings for there to be no interaction between elements. In this case, the load on each pile would be $F/9$ and each element would have a single-pile stiffness according to equation (10.18).

An obvious question at this stage is: How good is this model in representing the behavior of a nine-pile group of fairly stiff piles? Although model test data are available (Whitaker) and confirm the conclusions that the corner piles are stiffer than the edge center piles which are stiffer than the center pile, it is more satisfactory for numerical comparisons to use real, full-scale test results. One of the few systematic field pile group tests performed is reported by Berezantzev (2) and discussed by Leonards (11). The field tests were performed on piles 10.6 in. (0.27 m) in diameter and 18 ft (5.5 m) long embedded in a medium-dense silty sand. Axial load-deflection tests were carried out on single piles and square groups of 4, 9, 16, and 25 piles at various spacings ranging from 3 to 6 diameters. Data in the original paper are sparse, in that, for example, the pile material is not mentioned. If it is assumed that they were lightly reinforced concrete piles, then with the soil properties indicated, it can be calculated that the piles were quite stiff ($\lambda l \sim 0.5$). Wooden piles, however, would have been less stiff, with a λl of about 1+. For the sake of analysis, the piles are assumed stiff. Although small, compared to many construction piles, the piles are much larger than model scale, and they could be used realistically in practice. In the form of the original paper, the data are inconvenient for use here, but Leonards has redrawn them in the form of Figure 10-3 which we can employ.

For analysis, it will be assumed that the soil in the Russian tests was uniform across the site. Non-uniformity in depth does not affect the results. With stiff piles, the vertical deflection at the pile top is directly proportional to the magnitude of the subgrade reaction coefficient k acting along the pile sides. Consequently, the relative deflections indicated by the preceding analysis can be employed in comparison with the relative deflections recorded at the same load on a field pile group at, say, different spacings, even though the value of k and its distribution in the Russian tests are not known.

One of the first points to arise concerns the selection of the limiting boundary of the single-pile influence zone; this was chosen at 25-pile diameters based on the previous analysis of the value required to give a reasonable relation for k based on the soil property G. From the field tests we can check this by comparing the deflection of a single pile at a given load with the deflection of a nine-pile group at the same axial load per pile. Since we are concerned here, of course, with the elastic part of the behavior, not the response at failure, we take our data from the initial part of the load-deflection curves of Figure 10-3. From Figure 10-3(a) we can extrapolate the initial, more or less linear, part of the single-pile curve to a load of 35 tons. The displacement appears to be about 2.2 mm. A similar extrapolation of the behavior of the nine-pile group at 3-diameter spacing to 35 tons (312 kN) gives about 10 mm. Thus, Berezantzev obtained a ratio, at the same

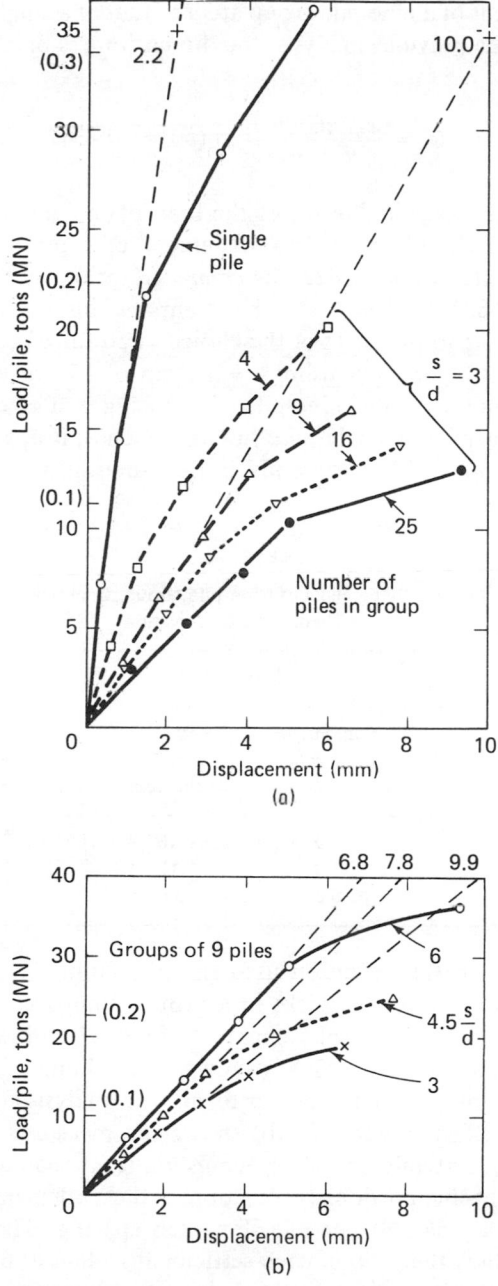

Figure 10-3. Displacement versus pile load: (a) single pile and various groups at 3*d* spacing in silty sand; (b) nine-pile group at various spacings [After Leonards (11) and Berezantzev (2)].

load, of settlement of a nine-pile group at $3d$ to that of a single pile of $10.0/2.2$ $= 4.55$. From the previous analysis, the dimensionless displacement w' of a single pile is

$$w' = \frac{w_0 2\pi G}{F} = \frac{1}{9} \ln (50) = 0.4347 \qquad (10.20)$$

since a single pile takes in effect $\frac{1}{9}$ of the load of the nine-pile group. Since from Table 10.2 the dimensionless settlement of a nine-pile group at $3d$ spacing is calculated to be 1.8822, the *computed* ratio of group to single-pile settlement is $1.8822/0.435 = 4.33$. This compares quite favorably with the observed value. As to the effect of the choice of 50 for b/a, we can solve the same problem as before, but using a b/a ratio of 15, for example. In that case, it turns out that the nine-pile $3d$ spacing settlement is only 2.30 times the settlement of a single pile. In consequence, it appears that the influence distance $b/a = 50$ corresponds, in this analysis also, reasonably well to practice.

TABLE 10.3

Relative Settlements of Nine-pile Group, Comparison
of Theory and Observation

	Measured		
Pile Spacing, s	Settlement at 40 tons (356 kN) mm	Ratio to s = 6 Settlement	Calculated Ratio to s = 6 Settlement
3	9.9	1.46	1.47
4.5	7.8	1.15	1.20
6	6.8	1.00	1.00

Analogously to the definition used for pile failure, we can establish a *displacement efficiency* for the disks in a group, compared to a single disk, by dividing the single disk displacement w' of 0.4347 by the group displacement w' at different spacings. This is shown in a column of Table 10.2.

A second point to arise is whether or not the analysis correctly accounts for the spacing effect. Figure 10-3(b) shows the measured displacement of the nine-pile group at different pile spacings in the Russian tests. Again using the initial straight-line portion, we can obtain the settlements, this time at a constant 40 tons (356 kN) per pile, for each spacing. They are shown in Table 10.3 in which the ratio of each settlement to that at $6d$ spacing is also given. From the column of calculated dimensionless group settlements in Table 10.2, the same ratio can be obtained and displayed in Table 10.3 for comparison. The last two columns show again a reasonable correspondence. A more interesting comparison could have been made had the individual pile loads been recorded during the Russian experiments. These results are also generally confirmed by Whitaker's tests in model pile groups (20).

The method described in this section can be used to obtain the behavior of piles attached to rigid caps in other groupings or configurations, symmetrical ($4 \times 4, 5 \times 5$, etc.) or otherwise ($2 \times 3, 3 \times 8$, etc.). It is worth inserting a cautionary remark here. The driving of a number of piles, particularly at close spacings into certain soils, can effect substantial changes in the soil's properties. Loose sands may be densified. Indeed, piles are often used for this purpose. Sensitive clays may be disturbed with a considerable loss of strength. In these circumstances, the change in material properties will affect the response of the pile group to load. In a loose sand the densification effect may be such that the subgrade reaction coefficients for a pile group are as great as or greater than that for a single pile instead of smaller. The settlement of the group may not be as much greater than that of a single pile as is illustrated by the analysis here. In sensitive clays the opposite effect may occur. In addition, the analysis obviously does not hold for piles that gain the major portion of their resistance from tip embedment in a strong or firm bearing layer.

10.2 DISPLACEMENT OF PILE GROUPS; LATERAL LOADS

Here again we turn our attention to the value of, in this case, the horizontal subgrade reaction coefficient as it is affected by closely adjacent piles. The basic solution of interest is that for the laterally loaded rigid disk (pile section) in a bounded elastic medium, as used in Section 8.3.2 for the determination of the reaction coefficient in terms of the soil properties. The solution is much more complicated than in the axially loaded disk problem, and it also involves two components of displacement. However, the principle in arriving at an answer is the same as before—reciprocal relations between adjacent pile loads and displacements. We have to make a decision again as to the effective relative interaction distance (b/a) between piles. In Section 8.3.2 good correlation between Winkler and elastic half-space solutions was again found when the radius to the outer rigid boundary was taken at 25 pile diameters ($b/a = 50$), and we will use that result here. Lateral deflection of a disk in a plane strain elastic medium is therefore considered to induce no displacement of another disk 25 or more diameters away.

10.2.1 Large group

For a large group of n piles connected by a rigid pile cap, each will be subjected to the same lateral force and displacement at the top, and it is required to find from the disk analysis the horizontal subgrade coefficient that can be used to give the pile deflections, moments, and shears as a function of the pile spacing. The argument is the same as that of the previous section, except that the disk displacement function can no longer be included explicitly, and the analysis leads to the following equation for the ratio of the group horizontal subgrade coefficient k_{hn} to the single pile coefficient k_{h1}:

$$\frac{k_{hn}}{k_{h1}} = \frac{u_1}{\sum\limits_{n=1}^{j=n} u_j} \tag{10.21}$$

where u_j is the horizontal disk displacement and the summation is taken over all the points representing pile locations within the influence zone of the loaded disks. It is convenient in the solution to take the disk displacement as unity. In this solution, as in the previous solution for axial loading, the assumption is made that the deflection of adjacent piles (disks) can be taken to be that of a point in the elastic medium located at the position of the pile center. The very complex problem of the interaction of a group of finite rigid disks in an elastic plane is not undertaken. Compared to the other simplifications made in the problem, the effect of this assumption, at the disk spacings considered, is negligible. One further point is that, with this geometry, the displacement of a disk in the x-direction in general causes displacements of adjacent disks in both x- and y-directions (or r- and θ-directions). We are only interested in the x-component for all piles, since they are considered to move as a unit.

Here, Poisson's ratio v also enters the problem; the calculations have been performed for $v = 0.3$, although other values can be used, and the results are presented in Table 10.4 and Figure 10-4. Since the interaction distance is again 50, the number of piles involved at each spacing is the same as in the axial loading case. To illustrate the effect of the assumption of a particular value for the interaction distance b/a, the calculation was repeated for $b/a = 20$ (10 diameters) with the result shown in Table 10.5 and Figure

TABLE 10.4

Lateral Loading; Large Pile Group, Horizontal Subgrade
Reaction Coefficients (b/a = 50)

Spacing Pile Diameters	Number of Piles Involved	Ratio $\frac{k_{hn}}{k_{h1}}$
25	1	1
18.75	5	0.8551
12.5	9	0.5794
10	21	0.4230
7.5	37	0.2674
5	69	0.1283
2.5	305	0.0326

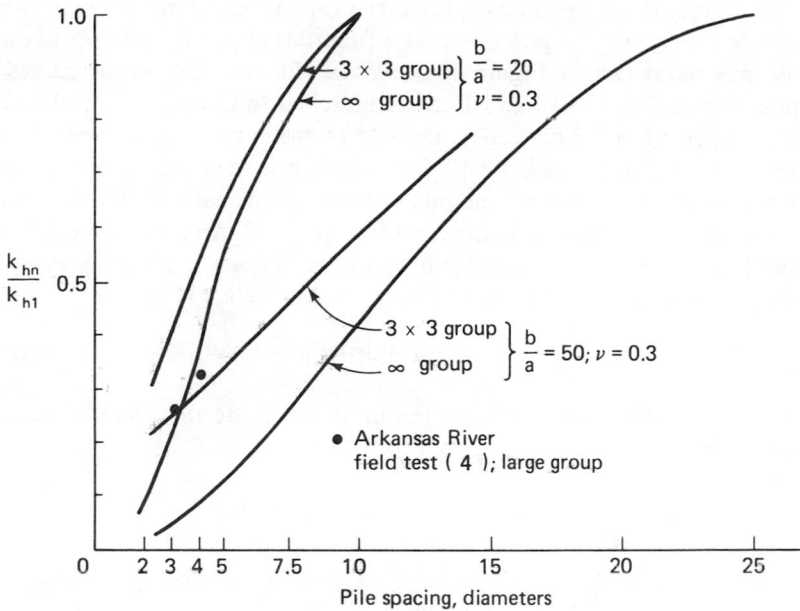

Figure 10-4. Ratio of lateral subgrade coefficient for group to that of a single pile.

TABLE 10.5

Lateral Loading; Large Pile Group, Horizontal Subgrade
Reaction Coefficients ($b/a = 20$)

Spacing Pile Diameters	Number of Piles Involved	Ratio $\dfrac{k_{hn}}{k_{h1}}$
10	1	1
7.5	5	0.8146
5	9	0.5065
4	21	0.3533
3	37	0.2138
2	69	0.0989

10-4. Because of the different interaction distance the number of piles affected at each spacing is different from before.

For the lateral loading case, the ratio of k_{hn}/k_{h1} is invariant with respect to the direction of loading. In this analysis the value of k_{h1} for a particular layer of soil is related to the Young's modulus of the soil, when Poisson's ratio is 0.3, by the expression

$$k_{h1} = 1.04E \qquad\qquad (10.22)$$

according to Figure 8-19.

Based on model experiments, Prakash (15) has related the average lateral subgrade coefficient of a pile group as a function of pile spacing to that of a single pile as shown in Figure 10-5. When the two theoretical curves of Figure 10-4 or Tables 10.4 and 10.5 are matched to the experimental results at a pile spacing of 7.5 diameters, the rest of the curves can be plotted. The comparison is shown in Figure 10-5 in which it is seen that it is difficult to distinguish between the two bounding criteria on this basis, although a somewhat better correspondence is apparent for $b/a = 20$. An asymptotic value of about 20 also seems more plausible in this case. If $b/a = 20$ is an appropriate limiting influence ratio, the corresponding value for k_{h1} is

$$k_{h1} = 1.40E \tag{10.23}$$

In practice, further field tests are required to decide the correct value or range of values.

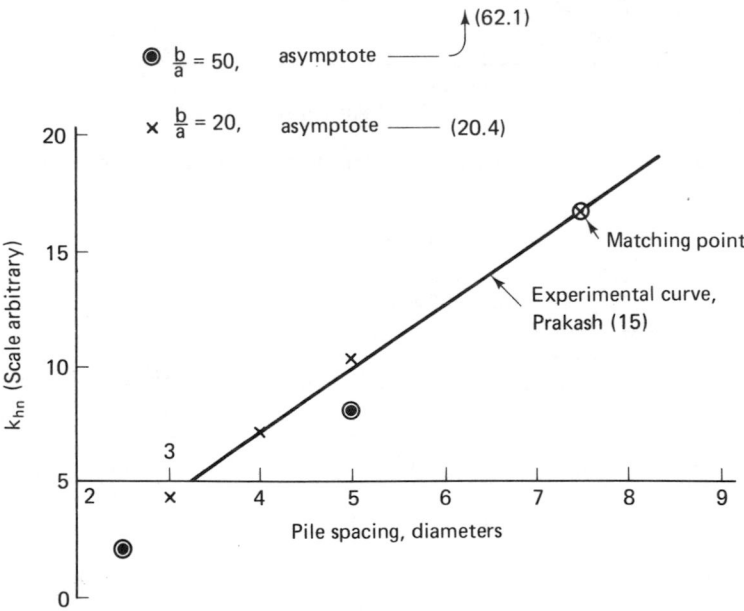

Figure 10-5. Comparison of group lateral subgrade coefficient with experimental results.

10.2.2 Small group

When the pile group is small, in a square or other array, and the piles are contained at the top in a rigid pile cap, the lateral forces vary from pile to pile, as in the axial loading case. The equivalent plane strain problem with which the lateral subgrade reaction coefficients can be established consists again of a system of disks in an elastic medium. If the piles are stiff, all the disks are constrained to undergo the same displacement u in one direction as a consequence of a total force applied to the whole group. This is also taken to be approximately true if the piles are flexible. It is again convenient to take a 3×3 group to illustrate the method. The array is shown in Figure 10-2. There is less symmetry here than in the axial loading problem and it can be seen (or shown) that different unknown pile forces occur in piles 0, 1, 2, and 5, since the forces in piles 1 and 3 are the same, as are those in piles 2 and 4 and in 5, 6, 7, and 8.

We apply a force T in the x-direction to the whole array and it results in a displacement u. The analysis begins with a calculation of the displacement at each of the reference disks 0, 1, 2, and 5 as a result of forces T_0, T_1, T_2, and T_5 applied to the surrounding elements. Thus,

$$u_0 = C_{00}T_0 + 2C_{01}T_1 + 2C_{02}T_2 + 4C_{05}T_5 \tag{10.24}$$

$$u_1 = C_{10}T_0 + (C_{11} + C_{13})T_1 + 2C_{12}T_2 + 2C_{15}T_5 + 2C_{16}T_6 \tag{10.25}$$

etc., where the C_{ij} are compliances relating the x-displacement at a given disk i to the x-force at disk j. The compliances for a given b/a ratio and pile spacing can be established from the r- and θ-displacements of the basic disk solution of Section 8.3.2, since

$$u = u_r \cos \theta + u_\theta \sin \theta \tag{10.26}$$

In this case, the compliances are a function of both the x- and y-coordinates of a disk. It should be pointed out that a force applied to the group in the x-direction (or any other) will develop a reaction force in each disk, in general, oriented at some angle to the x-direction; we are only concerned here with the x-components, however. An individual disk force is not developed only in the x-direction because a displacement of a single central disk causes displacements at adjacent points in both x- and y- (r- and θ-) directions and movement in the y-direction is restrained by the pile cap.

The rigidity of the pile cap requires that the differences $(u_0 - u_1)$, $(u_0 - u_2)$, and $(u_0 - u_5)$ all equal zero. Thus, three equations of the four required can be formulated:

$$u_0 - u_1 = (C_{00} - C_{01})T_0 + (2C_{01} - C_{11} - C_{13})T_1$$
$$+ 2(C_{02} - C_{12})T_2 + 2(2C_{05} - C_{15} - C_{16})T_5 = 0 \tag{10.27}$$

etc. The fourth equation comes from the equilibrium requirement that the sum of the pile forces equal the applied force T. If we normalize by dividing by T, we get the equation

$$T_0 + 2T_1 + 2T_2 + 4T_5 = 1 \tag{10.28}$$

In the previous equation of the group represented by equation (10.27) the values of T_j can similarly be put in dimensionless form. Simultaneous solution of the three equations of the type of (10.27) and equation (10.28) gives the load ratios in each pile. For the case $b/a = 50$ and $v = 0.3$, this has been done for the nine-disk array at various spacings. The results are presented in Table 10.6.

TABLE 10.6

Lateral Forces, Displacements, and Relative Stiffness in a Nine-pile Element Group
($b/a = 50$, $v = 0.3$)

Spacing, s	T_0	T_1	T_2	T_5	u'	Group Displacement Efficiency or $\dfrac{k_{hn}}{k_{h1}}$ ave	$\dfrac{k_{h0}}{k_h}$	$\dfrac{k_{h1}}{k_h}$	$\dfrac{k_{h2}}{k_h}$	$\dfrac{k_{h5}}{k_h}$
2.5	−0.0179	0.0924	0.0660	0.1753	0.1782	0.230	[−0.0370]	0.1912	0.1365	0.3627
4	0.0133	0.0939	0.0809	0.1593	0.1382	0.297	0.0355	0.2505	0.2158	0.4250
5	0.0244	0.0963	0.0851	0.1532	0.1197	0.343	0.0752	0.2966	0.2621	0.4719
7.5	0.0432	0.1024	0.0891	0.1434	0.0883	0.464	0.1804	0.4276	0.3720	0.5988
10	0.0576	0.1107	0.0904	0.1351	0.0691	0.593	0.3073	0.5906	0.4823	0.7208

A similar table can be prepared for any other b/a ratio or array geometry. It is interesting to note again the negative value of T_0 at the closest spacing of 2.5 diameters; it indicates that the central disk is being displaced by the soil rather than vice versa. Disks 5 at the corners take the greatest share of the load, followed by 1 and 2 at the front and edge center, respectively. As the spacing increases, the load is shared more equally.

Taking values from Table 10.6 at the various spacings, equation (10.24) [or equation (10.25) or any other displacement expression] can be used to calculate the displacement of the group, which can be expressed in the form

$$u = \frac{u'T}{G} \tag{10.29}$$

where T is the total load per unit thickness of the plane strain section, G is the shearing modulus of the soil, and u' is a dimensionless coefficient. Values of u' are shown in Table 10.6 for the case represented. Equation (10.22) gives the value of k_{h1} for a solitary disk in terms of the soil modulus E. Using G instead for $v = 0.3$ gives

$$k_{h1} = 2.7213G \tag{10.30}$$

so that in terms of the load $T(1)$ applied to a single disk the equation

$$T(1) = 2.7123Gu \tag{10.31}$$

holds. If this disk forms one of a group of nine disks so widely spaced that no interaction occurs, the force applied to each will be the same and equal to $T/9$. Equation (10.31) becomes

$$\frac{T}{9} = 2.7123Gu \tag{10.32}$$

which upon rearranging gives

$$u' = \frac{Gu}{T} = 0.0410 \tag{10.33}$$

the asymptotic value of u' for this problem.

It remains to calculate the relative values of the horizontal subgrade reaction coefficients for the disks at various spacings. For a particular separation distance, equation (10.29) gives the displacement of the group. With the T_j of Table 10.6 representing the ratios of load in each disk at a particular spacing to the total load, the value of subgrade coefficient for any one disk becomes

$$k_{hj} = \frac{GT_j}{u'} \tag{10.34}$$

Dividing this by equation (10.30) gives the relative subgrade coefficient for any one disk:

$$\frac{k_{hj}}{k_{h1}} = \frac{0.3687T_j}{u'} \tag{10.35}$$

For the spacings and conditions of Table 10.6, the relevant values of the relative coefficients have been calculated and are shown in the table. A lateral displacement efficiency or ratio k_{hn}/k_{h1} calculated as for vertically loaded disks is also shown in Table 10.6. As indicated earlier, the analytical model may give more realistic values of horizontal subgrade coefficient when the influence zone radius, b/a, is taken at 20 rather than 50. For Poisson's ratio $v = 0.3$ again, values as in Table 10.6 are given for $b/a = 20$ in Table 10.7.

In this case, for a single pile,

$$k_{h1} = 3.6353G \tag{10.36}$$

so that

$$u' = 0.0306 \tag{10.37}$$

TABLE 10.7

Lateral Forces, Displacements, and Relative Stiffness in a Nine-pile Element Group
($b/a = 20$, $v = 0.3$)

Spacing, s	T_0	T_1	T_2	T_5	u'	Displacement Efficiency or $\frac{k_{hn}}{k_{h1}}$ ave	$\frac{k_{h0}}{k_{hn}}$	$\frac{k_{h1}}{k_h}$	$\frac{k_{h2}}{k_h}$	$\frac{k_{h5}}{k_h}$
2.5	−0.0104	0.1015	0.0612	0.1712	0.0897	0.341	[−0.0319]	0.3113	0.1877	0.5250
3	0.0066	0.1050	0.0681	0.1618	0.0764	0.401	0.0238	0.3781	0.2452	0.5826
4	0.0320	0.1148	0.0757	0.1467	0.0579	0.528	0.1520	0.5454	0.3596	0.6970
5	0.0575	0.1193	0.0812	0.1354	0.0471	0.650	0.3358	0.6967	0.4742	0.7908
7.5	0.1019	0.1145	0.1029	0.1158	0.0351	0.872	0.7986	0.8973	0.8064	0.9075
10	0.1111	0.1111	0.1111	0.1111	0.0306	1.000	1.0000	1.0000	1.0000	1.0000

which is the asymptotic value of u' here. For this value of b/a, a disk spacing of 10 diameters corresponds effectively to disks at infinite spacing; therefore, the values T_0, T_1, etc., at $s = 10$ are those of an assembly of nine disks behaving as single entities. The group ratio, k_{hn}/k_{h1}, or displacement efficiency, is shown in Figure 10-4 for both Tables 10.6 and 10.7 for comparison with the results of the infinite group analysis. It can be seen that the ratio increases for smaller pile groups, which are therefore effectively stiffer than an infinite group. This follows from the increasing proportion of edge and corner pile contribution in the smaller group.

The literature yields few instances of tests of pile groups subjected to lateral loads. One case is described by Kim, Brungraber, and Kindig (9) who carried out lateral loading tests on groups of six 10 BP 42 37-ft (11.3 m) long steel H-piles at two different spacings of 3 ft (0.91 m) and 4 ft (1.22 m) ($s = 3.6$ and 4.8). The piles were driven to refusal on limestone through a soft to medium stiff silty clay to clay soil, and the six piles in each group were connected at their tops by a rigid concrete cap. The piles can be considered relatively stiff. Loading was applied in the direction of the three-pile line. Lateral deflection of the cap of the more closely spaced group was greater than that of the other group, as expected; the ratio of the two deflections at the same load was about 1.7.

For stiff piles embedded in a homogeneous medium in the 3×3 array for which Table 10.6 was calculated, the relative deflection of pile groups at different spacings is closely approximated by the ratio of the dimensionless quantities u'. For flexible piles, the relative deflections vary with the ratio of the u' to the 0.75 power. In the nine-pile group represented in Table 10.7 the ratio of the deflections at spacings of $3.6d$ and $4.8d$ comes to approximately

1.3. Figure 10-4 indicates that a theoretical analysis of the six-pile group would give a greater deflection ratio, perhaps in the range of 1.5. If the analysis is accepted, it is possible that the clay soil of the test site was softened more by the disturbance caused by the group at 3-ft (0.91 m) spacing than by the group at 4-ft (1.22 m) spacing. Alternatively, an increased ratio would be given by a smaller interaction distance b/a, but this seems implausible.

For assistance in the design of pile foundations in sand for the locks and dams of the Arkansas River Navigation Project, Davisson and Salley (3) conducted extensive lateral load tests on model pile groups in sand. The different groups included from 100 to 150 piles at spacings of from 3 to 4 diameters; batter piles formed members of some groups. In the analysis it was assumed that the soil support could be represented as a Winkler foundation with a horizontal subgrade coefficient increasing linearly with depth, so that

$$k_h = k_{h1}z \tag{10.38}$$

In this case, the length parameter λ for analysis is given by the equation

$$\lambda^5 = \frac{k_{h1}}{EI} \tag{10.39}$$

Davisson and Salley found from their tests that

$$\frac{\lambda_n}{\lambda_1} = 1.3 \quad \text{at 3 diameter spacing}$$

$$= 1.25 \quad \text{at 4 diameter spacing} \tag{10.40}$$

where λ_n was the average λ describing group behavior and λ_1 was obtained from a single-pile test. Thus, from equation (10.39) it follows that

$$\frac{k_{hn}}{k_{h1}} = \left(\frac{1}{1.3}\right)^5 = 0.2693 \text{ at } 3d$$

$$= \left(\frac{1}{1.25}\right)^5 = 0.3277 \text{ at } 4d \tag{10.41}$$

When these two values are compared with the analytical results of Figure 10-4, it is seen that the correspondence with the $b/a = 20$ infinite group curve is quite good. Although the tests were on model piles and the nonlinear behavior of the soil affected the upper portions of the piles, it appears from both examples in clay and sand that the analysis offers at least some guide to the selection of horizontal subgrade coefficients for groups based on the behavior of single test piles under lateral load.

In their tests on both model and full-scale piles, Davisson and Salley found that the effect of cycling the load was to increase the deflection of pile groups by 70 to 90 percent after about 100 cycles. They suggest that it is conservative to assume that the lateral deflection computed for the group based on a single-pile test, loaded once, will be doubled under repeated cyclic loading. Deflections, even in sand, also increase with duration of loading. Long-term displacements of a group may be 30 percent greater than those observed in short-term tests.

In Section 5.2.1 it was pointed out that the deflection of a beam or pile depends almost directly on the reciprocal of the subgrade reaction coefficient and is therefore quite sensitive to the value selected. But the moment in and the resultant pressure acting on the laterally loaded beam or pile are much less dependent on the value of subgrade coefficient. It follows that the variation of subgrade reaction coefficient caused by the interaction of adjacent piles will have corresponding effects. This has also been pointed out by Poulos (13) who notes from his studies that pressure distribution and moments in piles and pile groups are almost unaffected by interaction but that displacements and rotations are strongly influenced, as compared to single piles under the same load. Because of the interdependence of pile numbers and spacings, Poulos also notes that the displacement of a pile group at a given load is dependent on the breadth of the group rather than on the number of piles. This is true for both axially and laterally loaded arrangements. In consequence, in those practical circumstances in which deflection instead of ultimate load is the determining factor in design, economies can be achieved by using smaller numbers of piles at larger spacings.

From his analysis, Poulos finds that for both axially and laterally loaded pile groups in clays, 80 to 90 percent of the final displacement occurs immediately upon loading. This parallels the previous findings for piles in sands. Although these conclusions, strictly speaking, develop from analyses of the response of idealized piles in homogeneous media, they probably also hold in the case of soils whose properties change with depth.

10.3 ECCENTRIC AND MOMENT LOADS ON PILE CAP

The results of the preceding sections can be used to give values of subgrade reaction coefficients based on soil properties as functions of pile spacing. The examples considered pile groups subjected to vertical and lateral loads disposed symmetrically with respect to the group. The arrangement of the piles themselves was also symmetric. In practice, however, a pile group is frequently arranged to resist vertical, horizontal, and moment loads applied to the pile cap, and the group may include both vertical and inclined (*batter*) piles. In those conditions the vertical and lateral subgrade reaction coefficients and their variation with depth are known or assumed for each pile, and a calculation is required to give the distribution of pile top forces and the components of displacement of the pile cap. The cap is almost always taken to be rigid, and the computation proceeds in the following way.

With selected pile dimensions and sectional properties and the known subgrade reaction, expressions can be formulated for the displacement response of each pile to axial, lateral, and moment loads applied to the pile top. The sums of these unknown load components must be equal to the applied force components at the pile cap. Further, with a rigid pile cap, the pile top displacements are related to the movements of the cap. Thus, the conditions of equilibrium (balance of forces), compatibility, or consistency of deformations (pile-cap geometry) and the constitution of the system (pile-soil response) are all accounted for. In the literature the relevant equations have been set up and solved by a variety of numerical and graphical means (1, 5, 7, 12, 13, and 19), most of which have been developed to simplify either the problem representation or its solution or to formulate it in terms familiar to structural engineers. Along these lines, the mystery of the response of piles embedded in soil has been resolved by simulating the embedded pile by means of a free-standing, fixed-ended pile of fictitious length (5 and 12). The fictitious length is different for axial and various conditions of lateral loading. Fixity of the pile at the cap has been bypassed by the use of dummy piles (1 and 7). For wooden piles, shrinkage leads, in many cases, to pile-top conditions less than completely fixed, and possibly even pinned. Steel or concrete piles with adequate embedment length in the pile cap can almost always be considered fixed at the upper end.

Since, in previous chapters, a variety of solutions has been given for the stiffness or compliance of a pile top under different loading conditions, it is convenient to dispense with diverse artifices and utilize these pile responses directly. If the pile cap is attached to the piles some distance above the soil surface, as is the case for bridge or dock piers, the pile stiffness at the cap can be obtained by a suitable combination of responses of the embedded and free-standing portions. Alternatively, when the pile cap is embedded, the stiffness due to the pile cap-soil interaction must be added to the system. The first requirement, therefore, is to obtain stiffnesses at the pile top for the various conditions of top displacement pertinent to the cap movement. It will be considered here that the pile is completely immersed in the soil and that the cap is close to but does not interact with the ground.

10.3.1 Pile stiffness

If the pile is pinned at the top, the displacements of interest are axial along the pile and lateral transverse to the pile's axis. For a pile fixed in the pile cap, its response to axial, transverse, and rotational movements is required in the calculations.

Axial displacement. From Section 8.2, the axial displacement e at the top of the pile[2] under an axial force P can be expressed

$$e = \frac{P}{\lambda EA}[f(\lambda l)] \tag{10.42}$$

where the function $f(\lambda l)$ depends on the pile tip conditions and the assumed distribution of soil subgrade coefficient along the pile. The force P is positive in compression. Rearranging gives

$$P = \frac{\lambda EA}{f(\lambda l)} e = K_{Pe}e \tag{10.43}$$

where

$$K_{Pe} = \frac{\lambda EA}{f(\lambda l)} \tag{10.44}$$

In these equations, λ is the usual characteristic distance parameter and EA and l are the pile properties and length, respectively. In the case of a semi-infinite pile in uniform soil, $f(\lambda l) = 1$, for example, and

$$K_{Pe\infty} = \lambda EA \tag{10.45}$$

Lateral displacement without rotation. Figure 10-6(a) illustrates the lateral movement, s, of a fixed-ended pile. Such a displacement generates both a moment, M, and a lateral force, T, at the pile top. The solution is obtained either from the case of a finite beam loaded at the center or from first principles:

$$T = K_{Ts}s \tag{10.46}$$

where, for a soil with uniform properties,

$$K_{Ts} = \frac{k}{\lambda}\left(\frac{\sinh 2\lambda l + \sin 2\lambda l}{\cosh 2\lambda l + \cos 2\lambda l + 2}\right) \tag{10.47}$$

and

$$M = K_{Ms}s \tag{10.48}$$

where

$$K_{Ms} = \frac{k}{2\lambda^2}\left(\frac{\cosh 2\lambda l - \cos 2\lambda l}{\cosh 2\lambda l + \cos 2\lambda l + 2}\right) \tag{10.49}$$

---

[2]We reserve the displacement components u, v, and w for the pile cap.

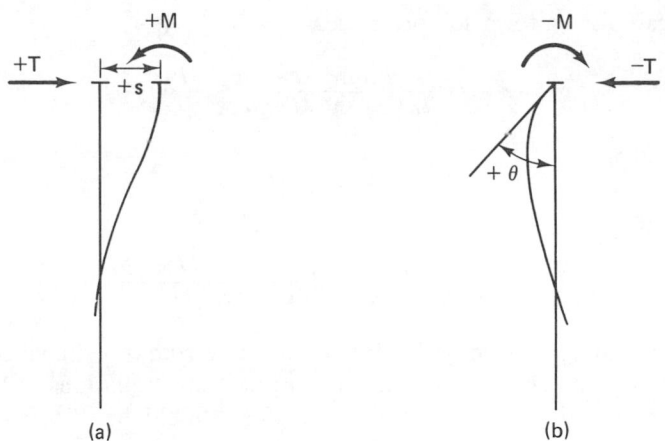

Figure 10-6. Moment and lateral force convention: (a) displacement without rotation; (b) rotation without displacement.

When the pile is semi-infinite, the stiffnesses become

$$K_{Ts\infty} = \frac{k}{\lambda} \tag{10.50}$$

and

$$K_{Ms\infty} = \frac{k}{2\lambda^2} \tag{10.51}$$

in a uniform soil.

In these equations the force T and moment M are both positive as shown in Figure 10-6(a) for a positive displacement s. If the soil reaction coefficient varies linearly along the length of the pile, the values of K_{Ts} and K_{Ms} can be determined from the tables of Appendix C. For soil with properties varying in some other manner (say, as the square root of depth), the values can still be obtained from Appendix C but only for the case of a semi-infinite pile, unless a numerical beam solution is performed for the individual case. In many cases of fairly long piles in practice, the assumption of a semi-infinite pile gives reasonable values of the K's required.

Rotation without displacement. The rotation θ of the top of a pile without lateral displacement is shown in Figure 10-6(b). By appropriately summing the solutions for a finite beam subjected to a lateral force and a moment at the end, the force required to prevent displacement can be determined:

$$T = -K_{T\theta}\theta \tag{10.52}$$

where, again for uniform soil properties,

$$K_{T\theta} = \frac{k}{2\lambda^2}\left(\frac{\sinh^2 \lambda l + \sin^2 \lambda l}{\cosh^2 \lambda l + \cos^2 \lambda l}\right) \tag{10.53}$$

and

$$M = -K_{M\theta}\theta \tag{10.54}$$

where

$$K_{M\theta} = \frac{k}{2\lambda^3}\left(\frac{\sinh \lambda l \cosh \lambda l - \sin \lambda l \cos \lambda l}{\cosh^2 \lambda l + \cos^2 \lambda l}\right) \tag{10.55}$$

Here both T and M are negative for the positive rotation shown in Figure 10-6(b). It can be shown that the two expressions, equations (10.49) and (10.53), are identical to one another, as they should be. The asymptotic values for the semi-infinite pile in uniform soil are

$$K_{T\theta\infty} = \frac{k}{2\lambda^2} \tag{10.56}$$

$$K_{M\theta\infty} = \frac{k}{2\lambda^3} \tag{10.57}$$

These values of K can be related to the compliances C of Section 8.3.2 and Figure 8-17.

For non-uniform properties, the appropriate K values in this case can also be derived from Appendix C. Now we can proceed to relate the displacements of a pile to those of the cap.

10.3.2 Pile to cap compatibility

In Figure 10-7 is shown in two dimensions a cap supported on a group of piles. It can represent a slice through a long dock or pier structure, for example. The cap is subjected to forces as shown and displaces u, w, and β in the given coordinate system. Each pile is constrained to undergo axial, lateral-without-rotational, and rotational-without-lateral displacement by each of the components of cap movement. By establishing the geometry of deformation, we can relate pile to cap displacements. It is convenient to place the origin of the coordinate system in the center of the cap, because many pile arrangements are symmetrical about the center, with the consequence that some of the terms are zero in the subsequent equations. The geometrical requirements will be expressed in terms of Figure 10-7, but they can be developed for any geometry of cap or pile connection.

The cap undergoes positive translations of u and w in the x- and z-directions and a positive rotation β about the origin, as shown in Figure 10-7. Each of these displacements, in turn, will be referred first to the movements of the pile, second to the forces generated in the pile by these movements, and third to the reaction components in the x-, z-, and moment directions on

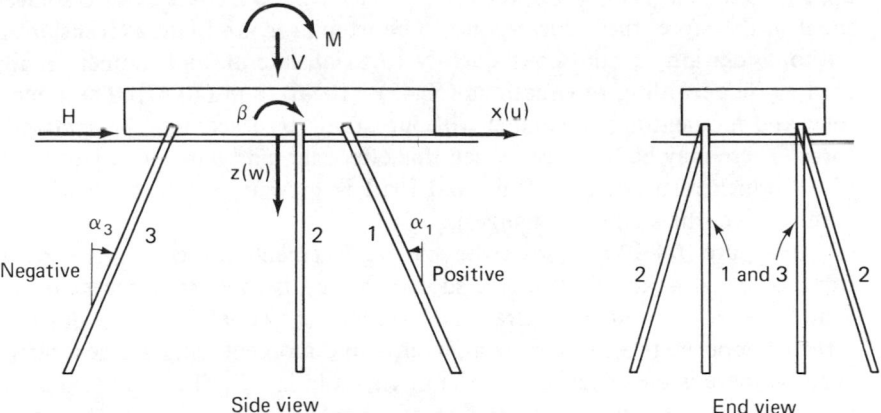

Figure 10-7. Pile-supported cap.

the pile cap. Figure 10-8 shows the geometry of the displacements for a typical pile, number j with positive slope α_j. Horizontal displacement, u, is indicated in Figure 10-8(a).

At the cap the attachment point of the pile is displaced from P to P'; the components of displacement parallel and perpendicular to the pile length

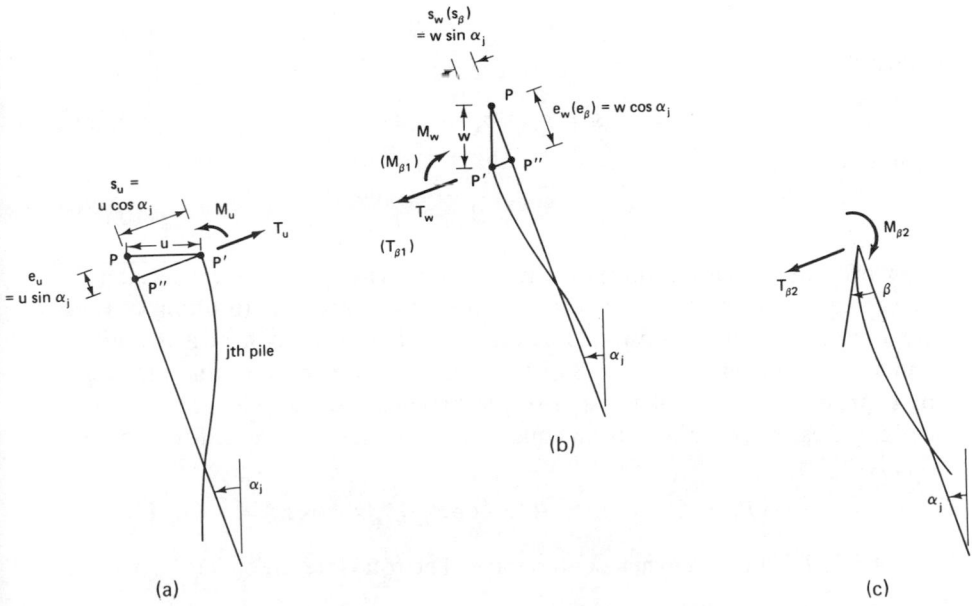

Figure 10-8. Pile displacements occasioned by cap movement: (a) cap horizontal translation u; (b) cap vertical translation w; (c) cap rotation β.

are PP'' and $P''P'$, equal, respectively, to $u \sin \alpha_j$ and $u \cos \alpha_j$. For displacement u, therefore, these correspond to shortening e_u and lateral translation without rotation, s_u, and consequently to axial, lateral, and moment loads in the pile according to equations (10.43), (10.46), and (10.47). The lateral load and moment are calculated without taking into account the axial pile force. There may be occasions when this effect should be included [equation (5.81) which also applies when axial load is present; when the load S is compressive, the sign of S changes].

In Figure 10-8(b) are shown the pile displacements caused by the vertical cap movement w. In this case, the shortening, e_w, is equal to $w \cos \alpha_j$, in the same sense as e_u, and the lateral displacement, s_w, equals $w \sin \alpha_j$, but it is directed opposite to s_u. Again, axial, lateral, and moment loads are generated. Finally, there is the effect on the pile of cap rotation β. There are two components here: (a) a vertical translation of the pile head by amount βx_j, where x_j is the coordinate of the top of the pile, and (b) a rotation of the pile top without translation. Since the first of these takes place exactly the same as the vertical movement w, with the replacement of w in Figure 10-8(b) by βx_j, the pile displacements can be expressed similarly. Rotation β alone is shown in Figure 10-8(c); it develops moment and lateral force in the pile but no axial component.

In sum, the total displacements of the pile can be expressed as follows:

Axial:

$$e = e_u + e_w + e_\beta \tag{10.58}$$

Lateral:

$$s = s_u - s_w - s_\beta \tag{10.59}$$

Rotation:

$$\theta = \beta \tag{10.60}$$

in which the subscripts indicate the causative cap displacements. Lateral movement is a displacement without rotation; rotation due to tilting of the cap occurs without translation (when the translation component is included under s_β). Following Figure 10-8, the right-hand sides of these equations can be expressed in terms of the cap displacements and the pile slope. Thus, making these substitutions, from equation (10.59) the total axial force in the pile is

$$P_j = K_{Pej}(u \sin \alpha_j + w \cos \alpha_j + \beta x_j \cos \alpha_j) \tag{10.61}$$

Again P is positive when compressive. The total lateral force is

$$T_j = K_{Tsj}(u \cos \alpha_j - w \sin \alpha_j - \beta x_j \sin \alpha_j) - K_{T\theta j}\beta \tag{10.62}$$

in which the last term comes from the rotation. Lastly, we have the total pile moment

$$M_j = K_{Ms_j}(u \cos \alpha_j - w \sin \alpha_j - \beta x_j \sin \alpha_j) - K_{M\theta_j}\beta \qquad (10.63)$$

In equations (10.62) and (10.63) the positive directions of action on the pile are still as shown in Figure 10-6(a). For subsequent calculations it is convenient to express these in terms of forces directed along the coordinate axes, as follows:

$$X_j = -P_j \sin \alpha_j - T_j \cos \alpha_j \qquad (10.64)$$
$$Z_j = -P_j \cos \alpha_j + T_j \sin \alpha_j \qquad (10.65)$$

where the positive force directions are now those of the cap (Figure 10-1). From equations (10.64) and (10.65), using equations (10.61) and (10.62), we get

$$X_j = -K_{Pe_j}[u \sin^2 \alpha_j + (w + \beta x_j) \sin \alpha_j \cos \alpha_j]$$
$$\quad -K_{Ts_j}[u \cos^2 \alpha_j - (w + \beta x_j) \sin \alpha_j \cos \alpha_j] + K_{T\theta_j} \beta \cos \alpha_j \qquad (10.66)$$
$$Z_j = -K_{Pe_j}[u \sin \alpha_j \cos \alpha_j + (w + \beta x_j) \cos^2 \alpha_j]$$
$$\quad + K_{Ts_j}[u \sin \alpha_j \cos \alpha_j - (w + \beta x_j) \sin^2 \alpha_j] - K_{T\theta_j}\beta \sin \alpha_j \qquad (10.67)$$

with the moment in equation (10.63) remaining the same, since the positive moment direction on the pile gives a positive resulting moment on the cap. These are the forces acting on the *pile* during the required movement of the cap. Taking them in the negative direction gives the forces resisting the cap displacement. Thus, the equilibrium equations can be formulated for the cap by equating to zero the sum of all components of the cap force in each direction in turn. All *n* piles are taken into account in the summation. The equilibrium equations (10.68) through (10.70) follow; they include the components of equations (10.66) and (10.67) in the horizontal and vertical directions and the moment from equation (10.63). However, to the moments contributed from each pile head must be added the moment applied to the cap by the *Z* component of force in each pile. This gives the third term in equation (10.70).

Horizontal:

$$H + \sum_1^n X_j = 0 \qquad (10.68)$$

Vertical:

$$V + \sum_1^n Z_j = 0 \qquad (10.69)$$

Moment:

$$M + \sum_{1}^{n} M_j + \sum_{1}^{n} x_j Z_j = 0 \qquad (10.70)$$

When the piles are disposed symmetrically about the coordinate origin and have the same stiffnesses for the various component forces, the terms $\sum \sin \alpha_j \cos \alpha_j$ and $\sum x_j \cos^2 \alpha_j$ vanish. For small batter angles, expressions containing $\sin^2 \alpha_j$ can also be ignored.

The system of equations (10.68) through (10.70) contains as unknowns the displacements u, w, and β of the cap, for which the set is solved. When they are obtained, they may be substituted into equations (10.61) through (10.63) to give the forces on each pile.

A method of analyzing cap displacements and pile stresses when the cap forces are known therefore proceeds as follows. One or more single piles is tested under axial and lateral loading at the site. Representatives of all piles of differing sections should be tested. It is generally not possible to perform a test where moment is applied to the pile top, but a fixed-head lateral load test may be possible. For a linear analysis, and by using the methods of Chapter 8, vertical and horizontal subgrade coefficients are calculated from the surface loads and deflections. The coefficients may be selected as functions (e.g., linear) of depth, if the tables of Appendix C are used. Since, especially for lateral loads, the soil behavior is usually nonlinear, even for small deflections, the subgrade reaction coefficients must be selected for the range of estimated deflections of the pile under the working cap loads.

If subsequent analysis indicates that cap and pile displacements will fall outside this range, a new subgrade coefficient should be selected and the computation performed again. One or two iterations will usually be sufficient. Such a linearization procedure is normally adequate for conservative design, and it requires less labor than a nonlinear analysis. If, in some circumstances, an estimate of the nonlinear behavior is needed, then the methods of Chapter 9 can be adapted to the results of the pile load test to give a nonlinear response function at the pile top. These functions are inserted in modified forms of equations (10.68) through (10.70), which are then solved for increasing values of cap loads to give a nonlinear cap load-displacement behavior.

The selection of subgrade coefficients from the test results should be made conservatively to reflect soil response and profile differences in various locations as indicated by boring and soil tests, as well as the pile tests. If the piles are to be widely spaced laterally, on the order of 8 pile diameters or greater, under the cap, no interaction between piles need be taken into account. But when they are spaced at distances of only a few pile diameters, interaction will occur, and the calculated subgrade reaction coefficients, and therefore pile K's, must be modified by the methods indicated earlier in this section. For a very large pile group, of say more than 50 piles in a more or less square array or of more than 7 rows of piles in a rectangular arrangement, the group can be considered infinite, and the subgrade reaction coefficient and K-values for all piles can be adjusted downward accordingly. For small groups of up to 50 piles, this method can also be used by employing the group deflection efficiency calculated as in Table 10.7 for the example nine-pile array, and it will give reasonable values for the *cap* displacement. But, as shown by the analysis of the small group, the reactions of individual piles vary widely from corner to center positions under these conditions; therefore, the loads in individual piles will be calculated incorrectly if an average subgrade coefficient is employed. Some piles may be overstressed under actual loading conditions. If the individual behavior is to be approximately accounted for, then subgrade reaction coefficients and K-values must be assessed for each pile in a small group and included in the summation equations. When these have been solved, the displacements can be entered, as before, into the individual pile force expressions to give the single pile forces. The individual pile subgrade coefficients can be calculated by the methods described in this section and illustrated by the nine-pile example.

If batter piles are used, their axial stiffnesses K_{P_e} can frequently be taken as those of individual piles, particularly if they are sloping out of the plane of Figure 10-7, for example, in a small group. However, the lateral response develops mostly in the top 10-to 20-pile diameters, and some average spacing must be assumed in the calculation of the appropriate K's to represent the increasing distance between piles. Although it would be desirable to perform tests on a battered pile to obtain its behavior as different from that of a vertical pile, particularly in the lateral direction, it is difficult in practice to do this, and the response at small batter angles can be taken to be the same as that of a vertical pile.

Sometimes, as pointed out by Francis (5), a batter pile in a two-dimensional array slopes out of the plane of the figure; for example, Figure 10-7 might represent a six-pile group, with piles 2 battered as shown in the end view. In this circumstance, the angle α of the batter pile should be taken as zero in the summation of horizontal forces, equations (10.66) and (10.68), but it should be taken as its real, out-of-plane value in the calculation of vertical forces, equations (10.67) and (10.69), and moments arising from these vertical forces in the third term of equation (10.70). Francis, however, does not do this in the worked example of his paper.

If the appropriate expressions are used to replace equations (10.61) through (10.70), the analysis can be extended to handle three-dimensional geometries and load conditions, but, of course, the analysis becomes considerably more complicated. An efficient tabulation scheme or computer algorithm must be established.

The analysis will now be illustrated by a two-dimensional example.

The concrete cap of Figure 10-9 is a plane section of a large pile cap running perpendicular to the paper and subjected to lateral, vertical, and moment loads, of which the proportion taken by the cap and a three-pile group as shown is given in the figure. The piles both in the direction of the long axis of the cap and in the plane of the figure are widely spaced and may be assumed to act as individuals. All piles are made of reinforced concrete and have the same dimensions: 14 in. (0.36 m) square by 50 ft (15.2 m) long. Each has a moment of inertia, I, and area A, respectively, of 4000 in.[4] $(1.67 \times 10^{-3}$ m$^4)$ and 250 in.^{2} (0.16 m^2), including equivalent amounts for the steel, and, with a Young's modulus, E of the concrete, of 3×10^6 psi $(2.1 \times 10^4$ MN/m$^2)$, the products of EI and EA are 1.2×10^{10} lb-in.^{2} (34.4 MN-m^2) and 7.5×10^8 lb $(3.34 \times 10^3$ MN). At the site the soil is taken to be uniform with depth; the values of k_s, the side resistance, and k_h, the lateral resistance, are determined to be 2000 (13.8 MN/m^2) and 1000 psi (6.9 MN/m^2), respectively. It will further be assumed that 0.2 of the axial

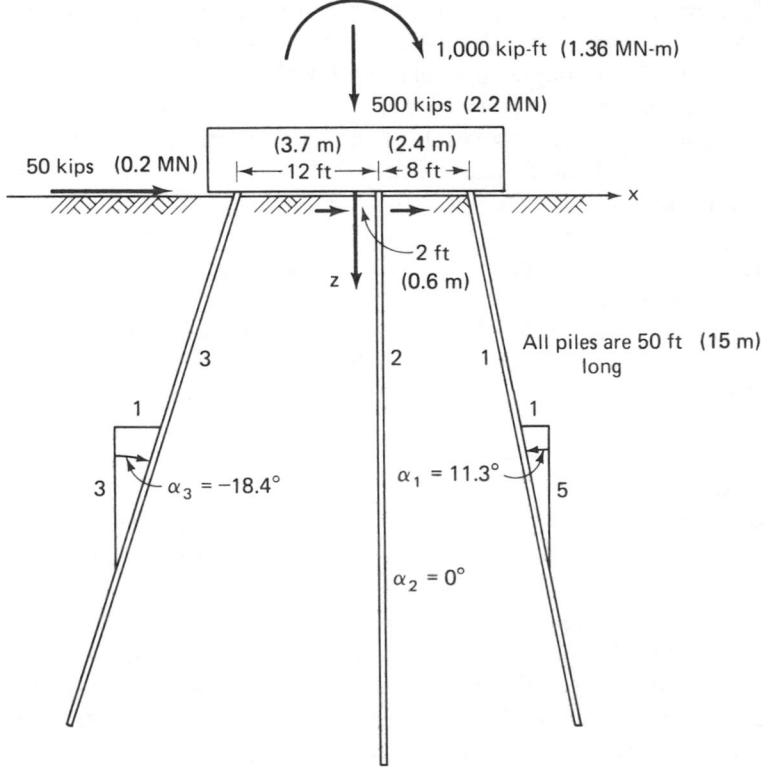

Figure 10-9. Cap and pile arrangement in example problem.

load in each pile reaches the pile tip and that the pile cap is close to ground surface but does not interact with the soil. The calculations follow.

It is first necessary to calculate the axial, λ_a, and lateral, λ_h, values of the characteristic length parameter. We find that

$$\lambda_a = 1.633 \times 10^{-3} \text{ in.}^{-1} \ (6.43 \times 10^{-2} \text{ m}^{-1});$$

$$\lambda_h = 1.201 \times 10^{-2} \text{ in.}^{-1} \ (0.473 \text{ m}^{-1})$$

from which

$$\lambda_a l = 0.980; \qquad \lambda_h l = 7.21$$

where l is the pile length. Thus, the piles are reasonably stiff in the axial direction but flexible laterally with respect to the soil. Little error will be introduced by assuming the piles to be semi-infinite in length under lateral loads. This will frequently be the case in practice. For axial loading, and using equations (8.42), (8.43), and (8.35), we obtain the function $f(\lambda_a l)$ of equation (10.44) to be

$$f(\lambda_a l) = \left(\frac{0.4 - e^{\lambda l} - e^{-\lambda l}}{e^{-\lambda l} - e^{\lambda l}}\right) = 1.1532 \tag{10.71}$$

so that, from equation (10.44)

$$K_{P_e} - \frac{\lambda EA}{f(\lambda_a l)} = 1.0620 \times 10^6 \text{ lb/in. (186 MN/m)} \tag{10.72}$$

In this calculation, as in subsequent ones, the spring constants will be derived as if the piles were vertically embedded in the soil.

For the semi-infinite pile case, the other spring stiffnesses required for lateral loading follow from equations (10.50), (10.51), (10.56), and (10.57):

$$K_{Ts} = \frac{k_h}{\lambda_h} = 8.3264 \times 10^4 \text{ lb/in. (14.6 MN/m);}$$

$$K_{T\theta} = K_{Ms} = \frac{k_h}{2\lambda_h^2} = 3.4664 \times 10^6 \text{ lb (15.4 MN):} \tag{10.73}$$

$$K_{m\theta} = \frac{k_h}{2\lambda_h^3} = 2.8863 \times 10^8 \text{ lb-in. (32.6 MN-m)}$$

Substituting these values and the geometrical properties of the pile arrangement into the summations, equations (10.68) through (10.70) of the terms, equations (10.66), (10.67), and (10.63) gives the following set of equations:

$$3.8535 \times 10^5 u - 1.0541 \times 10^5 w + 4.7666 \times 10^7 \beta = H$$

$$-1.0541 \times 10^5 u + 3.0504 \times 10^6 w + 3.2295 \times 10^7 \beta = V \tag{10.74}$$

$$4.7666 \times 10^7 u + 3.2295 \times 10^7 w + 3.0537 \times 10^{10} \beta = M$$

in which the numerical coefficients form a stiffness matrix. For example, in the first of these equations we see from equations (10.66) and (10.68) that the coefficient of u, say k_{11}, is given by

$$k_{11} = K_{Pe} \sum \sin^2 \alpha_j + K_{Ts} \sum \cos^2 \alpha_j \qquad (10.75)$$

Here $\sum \sin^2 \alpha_j = 0.1385$ and $\sum \cos^2 \alpha_j = 2.8615$, and when these are multiplied by the spring constants in turn, the result, 3.8535×10^5, is obtained. The second coefficient is

$$k_{12} = (K_{Pe} - K_{Ts}) \sum \sin \alpha_j \cos \alpha_j \qquad (10.76)$$

in which the angle summation comes out to be -0.1077, and the coefficient, therefore, is -1.0541×10^5. In the third term we have the coefficient

$$k_{13} = (K_{Pe} - K_{Ts}) \sum x_j \sin \alpha_j \cos \alpha_j - K_{T\theta} \sum \cos \alpha_j \qquad (10.77)$$

In this equation $\sum x_j \sin \alpha_j \cos \alpha_j = 59.0760$, $\sum \cos \alpha_j = 2.9293$, and multiplication by the stiffnesses and summing give $k_{13} = 4.7666 \times 10^7$. The other coefficients follow similarly. The check on the matrix of coefficients is that it is symmetrical, as it should be for this structure. In general, the matrix will be symmetrical unless piles with batters out of the plane of the figure are included. In the latter case, the pile contributions appear differently in the various equations.

When the right-hand loads are replaced by their design values from Figure 10-9 and the simultaneous equations are solved, the displacements

$$u = 0.1863 \text{ in. } (4.7 \text{ mm}); \qquad w = 0.1712 \text{ in. } (4.3 \text{ mm});$$

$$\text{and } \beta = -7.90 \times 10^{-5} \text{ rads}$$

are obtained. With these values, the axial, lateral, and moment forces in each pile can be calculated from equations (10.61) through (10.63), with the following results:

Force	Pile		
	1	2	3
Axial P, lb (kN)	2.0723×10^5	1.7980×10^5	1.1944×10^5
	(922)	(800)	(532)
Transverse T, lb (kN)	1.2852×10^4	1.5794×10^4	1.9755×10^4
	(57)	(70)	(88)
Moment M, lb-in. (kN-m)	5.4644×10^5	6.6894×10^5	8.3382×10^5
	(62)	(76)	(94)

The sign convention of Figure 10-7 is employed. From equations (10.64) and (10.65) the horizontal and vertical components of force at the top of each pile should be computed as a check on the arithmetic. When this is done, we get

$$\sum X_j = -50.009 \text{ kips } (-222.54 \text{ kN})$$
$$\sum Z_j = -500.05 \text{ kips } (-2.2252 \text{ MN})$$
$$\sum M_j + \sum x_j Z_j = -12,001.8 \text{ kip-in. } (-1.3566 \text{ MN-m})$$

which when compared with the loads represent an acceptable level of accuracy in the numerical work.

In the context of a real structure the calculated displacements are acceptable, but the axial and lateral loads would have to be examined in relation to the axial bearing capacity and yield bending moment in the piles, respectively, to ensure that the behavior was in the elastic range for which the properties were assumed.

Sometimes the pile deflections and forces are calculated ignoring the contribution of the flexural components to the equations. In effect, the piles are considered to be pinned at both ends and each is given the calculated axial stiffness obtained earlier. Care must be taken when this is done, however, since it is necessary that the pinned-ended pile group form a structurally stable structure. For example, if in the pile group in Figure 10-9 the batter piles were equally spaced about the center vertical pile and had the same slope, lateral and moment restraint would vanish if they were pinned at their ends. Elimination of the flexural stiffness terms, which supply this support, would give a set of equations whose solution would have no meaning (except for vertical load only). This is not always obvious. In the worked example being considered, for instance, the piles are not *quite* symmetrically disposed. They form a just barely stable system. If, therefore, only axial force terms are retained in the equation, the applied forces will give rise to very large values of the deflections u, w, and β, since these can occur with only small length changes in the individual piles.

Only a small additional numerical complication enters if the piles are of different cross sections and lengths, since the various stiffnesses must then be included inside the summation terms. The same consideration applies if the pile group is small and closely spaced, so that different values of k_s and k_h must be applied to various piles according to the requirements investigated earlier. Since the number of equations to be solved is only three, they may be solved, for example, by Newton's method (8) even if nonlinear pile-soil reactions are introduced along the lines of Chapter 9. The equations then become nonlinear functions of the variables u, w, and β. An example of nonlinear equations solved by this method appears in Section 9.3.4.

PROBLEMS

1. A friction pile foundation in a deep deposit of soft clay consists of many piles spaced at 2.5 diameters in both directions. Axial loading tests run on single piles at this site indicate failure under a 60-ton (534 kN) load and a settlement of 0.4 in. (10 mm) under a 15-ton (134 kN) load. The actual average load per pile (total load ÷ no. of piles) that will be caused by the building will be 15 tons (134 kN). Give your ideas on whether or not every pile will carry approximately 15 tons (134 kN) and whether or not the average settlement of the structure will be about 0.4 in. (10 mm). What will be the proportion of the short-term settlement to the settlement, several years after construction?

2. Figure P10-2 indicates the behavior of a group of piles in clay at two different pile spacings. At spacing $2d$, the corner piles took 90% of the load supported by a single pile in the same soil at failure; at the $4d$ spacing, the corner piles took the same load as a single pile at failure. Compare the loads taken by the two groups of piles at $\frac{1}{2}$ the failure load, and at the failure load. What are the "efficiencies" of the two groups at these two loads? How do the safety factors of individual piles in the group compare with the group safety factor at $\frac{1}{2}$ the group failure load? Is the situation unsatisfactory for any reason? How do the observed efficiencies compare with those given by the two formulas at $\frac{1}{2}$ and total failure loads? Based on the observations, or otherwise, have you any suggestions for improving the design calculations for such pile groups? You may suggest experiments, if desirable.

3. The rigid concrete block shown in Figure P10-3 is cast on top of the double row of steel sheet piles embedded in sand, and is subjected to a lateral load. Calculate the force per running foot and the maximum moment per running foot about an axis perpendicular to the paper in each pile row, by any method you please. How much lateral deflection would you estimate to take place? What would be the maximum (ultimate) lateral force taken by the system?

4. What difference do you think it would make to the plots of percentage load on each pile versus percentage load in Figure P10-2 if the piles were: (a) very soft; (b) very stiff, axially, compared to the soil. Illustrate and explain.

5. Suppose you are going to perform some laboratory experiments on model piles in groups, but you are going to apply a *horizontal* load to the pile cap.
   (a) Draw a diagram of efficiency versus pile spacing for this case for a group of, say, 3 × 3 piles. Explain your curve.
   (b) Draw a diagram of horizontal load on each pile versus load on the group and discuss.
   (c) To represent a realistic range of full-scale pile/soil stiffness, would $\frac{1}{8}$ in. (3.2 mm) diameter brass rods used by researchers for axial load studies in clay be satisfactory? (Make some guesses and perform some calculations.)

6. Discuss (a) the relative elastic displacements of a single pile and a pile group (say 4 × 4) under the same single pile load (piles the same size); (b) the ultimate loads of a single pile and a pile group. The piles are embedded in a medium-stiff clay and the group is loaded through a rigid pile cap. Take the problem

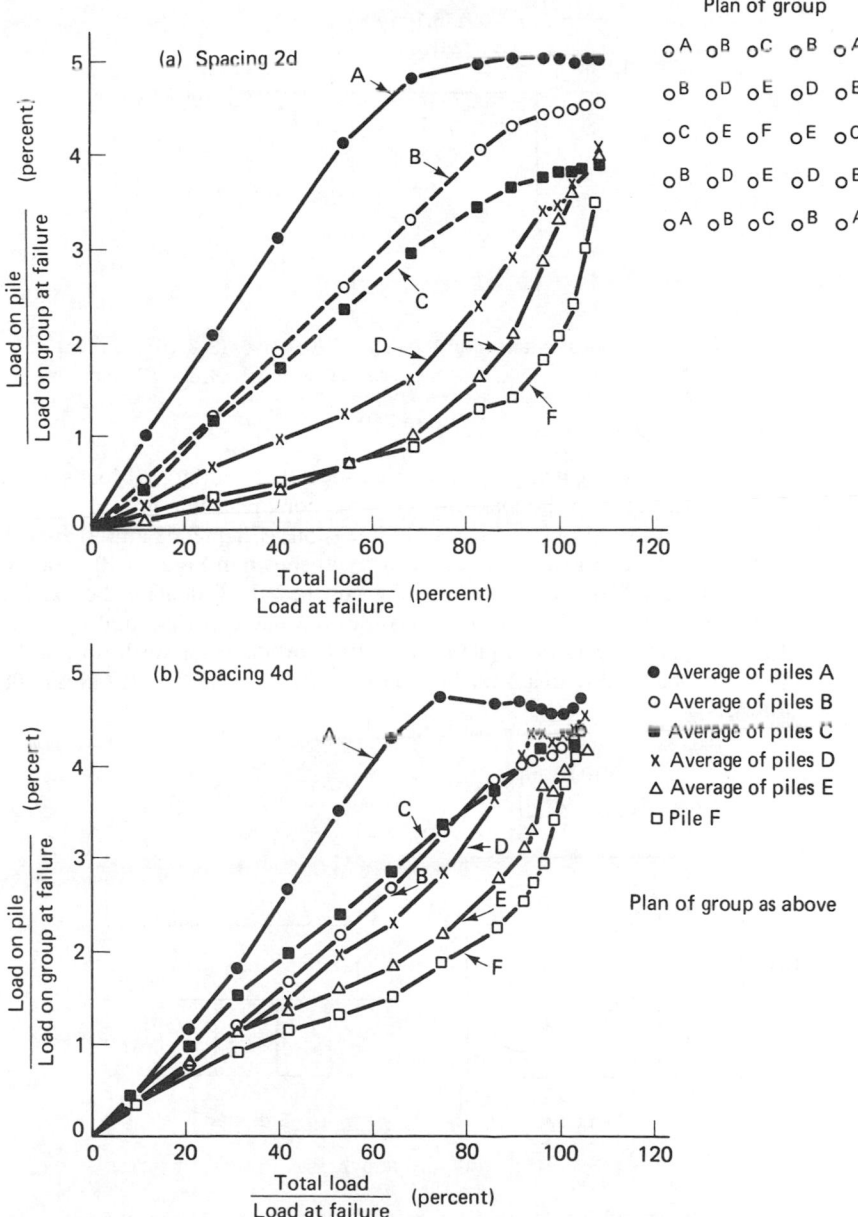

Figure P10-2. (After Reference 20.)

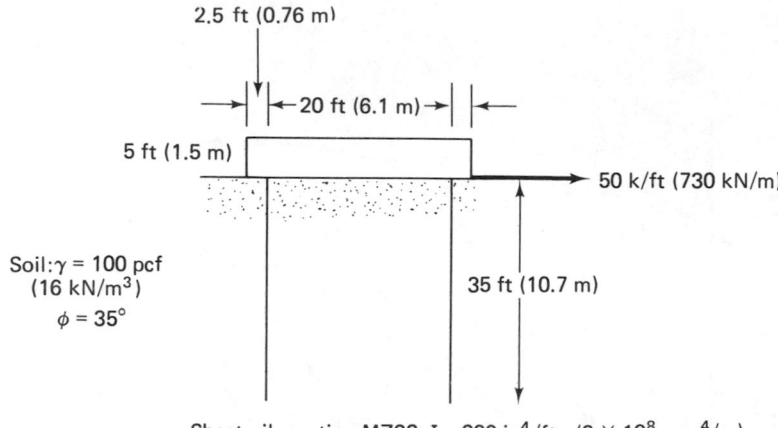

2.5 ft (0.76 m)

20 ft (6.1 m)

5 ft (1.5 m)

50 k/ft (730 kN/m)

Soil: γ = 100 pcf
(16 kN/m³)
ϕ = 35°

35 ft (10.7 m)

Sheet pile section MZ32, I = 220 in.⁴/ft   (3 × 10⁸ mm⁴/m)
S = 38.3 in.³/ft   (1.8 × 10⁶ mm³/m)

Figure P10-3.

in 2 parts: (1) vertical loads and displacements; (2) horizontal loads and displacements. Mention the load distribution among piles in a group.

7. Four piles of the Arkansas River test dimensions (Chapters 8 and 9) are driven in the soil of the test site, in a square array as shown in Figure P10-7. (a) What would the axial load capacity of the group be? (b) Calculate the axial load settlement relation of the group. Consider the pile cap rigid and explain the basis for your calculations. Pile is 16 in. (0.41 m) diameter steel tube, 0.312 in. (7.9 mm) thick wall, area 23.86 in.² (0.0154 m²), $EA = 692 \times 10^3$ kips (3.08 × 10³ MN).

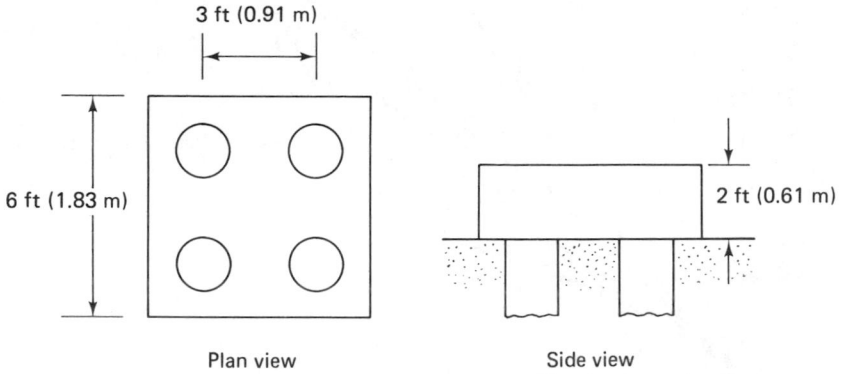

3 ft (0.91 m)

6 ft (1.83 m)

2 ft (0.61 m)

Plan view   Side view

Figure P10-7.

8. Use the tables of Appendix C to calculate the four pile top stiffnesses for a very long (semi-infinite) pile embedded in a soil whose subgrade coefficient

increases: (a) linearly with depth from zero at the surface, and (b) with the square root of depth.
9. From the table in the sample exercises of Section 10.4, calculate the axial stresses in the various piles, and the magnitudes and depths of the maximum moments. Finally, compute the maximum extreme fiber stresses. Are they too high?

REFERENCES

[1] ANDERSEN, P. *Substructure Analysis and Design*, 2nd ed. New York: Ronald Press, 1956.

[2] BEREZANTZEV, V. G. "Load-Bearing Capacity and Deformation of Piled Foundations," *Proc. 5th ICSMFE, 2*, 11–15 (1961).

[3] DAVISSON, M. T., AND J. R. SALLEY. "Model Study of Laterally Loaded Piles," *Proc. ASCE, 96*, SM5, 1605–1627 (Sept. 1970).

[4] DIMAGGIO, F. L., AND I. S. SANDLER. "Material Model for Granular Soils," *Proc. ASCE, 97*, EM3, 935–950 (June 1971).

[5] FRANCIS, A. J. "Analysis of Pile Groups with Flexural Resistance," *Proc. ASCE, 90*, SM3, 1–32 (May 1964).

[6] HAIN, S. J., AND I. K. LEE. "The Analysis of Flexible Raft-Pile Systems," *Geotechnique, 28*, No. 1, 65–83 (1978).

[7] HRENNIKOFF, A. "Analysis of Pile Foundations with Batter Piles," *Trans. ASCE, 115*, 351–381 (1950).

[8] KELLER, H. B. "Numerical Methods for Two-Point Boundary Value Problems," Waltham, Mass.: Blaisdell, 1968.

[9] KIM, J. B., R. J. BRUNGRABER, AND C. H. KINDIG. "Lateral Load Tests on Full-Scale Pile Groups," *ASCE PES, 1*, Part 2, 1105–1133 (1972).

[10] LEONARDS, G. A., ed. *Foundation Engineering*. New York: McGraw-Hill, 1962.

[11] ———. "Settlement of Pile Foundations in Granular Soil," *ASCE PES, 1*, Part 2, 1169–1184 (1972).

[12] O'CONNELL, W. J. L., AND P. M. QUINLAN. "Design of Pile Groups," *Trans. Inst. Civ. Engrs. of Ireland, 81*, No. 1, 1–19 (Nov. 1954).

[13] POULOS, H. G. "Behavior of Laterally Loaded Piles: II Pile Groups," *Proc. ASCE, 97*, SM5, 733–751 (May 1971).

[14] ———. "Estimation of Pile Group Settlements," University of Sydney, School of Civil Engineering, Research Report No. R288 (May 1976).

[15] PRAKASH, S. "Behavior of Pile Groups Subjected to Lateral Load," Ph.D. Thesis, University of Illinois, 1962.

[16] SCHOFIELD, A. N., AND C. P. WROTH. *Critical State Soil Mechanics*. New York: McGraw-Hill, 1968.

[17] TEJCHMAN, A. F. "Model Investigations of Pile Groups in Sand," *Proc. ASCE, 99*, SM2, 199–217 (Feb. 1973).

[18] TERZAGHI, K., AND R. B. PECK. *Soil Mechanics in Engineering Practice*, 2nd ed. New York: John Wiley, 1967.

[19] VETTER, C. P. "Design of Pile Foundations," *Trans. ASCE, 104*, 758–811 (1939).

[20] WHITAKER, T. *The Design of Piled Foundations*. London: Pergamon Press, 1970.

[21] WINTERKORN, H. F., AND F. Y. FANG, eds. *Foundation Engineering Handbook*. New York: Van Nostrand Reinhold, 1975.

Limit Design   *11*

11.1 INTRODUCTION; UPPER AND LOWER BOUNDS

In the discussion so far the basis for engineering design has been stress (the maximum tensile stress on the underside of a slab, for example) which can be considered in terms of moment in a slab, beam, or wall. Since the principal engineering structural materials, steel and concrete, can be considered to a good approximation to behave linearly up to a certain stress level, above which yielding takes place, linear analyses model the material behavior reasonably well until at some point in the structural arrangement the limiting stress is reached. Because unrestrained deformation or flow of the material does not occur until the limiting stress is exceeded by a substantial amount, a structure has a certain safety reserve if the limiting stress is nowhere exceeded in it. (We are not concerned with problems of instability here.) The linear analyses investigated up to this point hold good for this region of structural behavior, and they constitute *lower bound* estimates of the safety of a structure.

As the load or some portion of load on a structure is increased, a limiting stress will first be reached in the extreme fiber at some section. However, before a beam, slab, or other structural member undergoes substantial

425

deformations under applied load it is necessary not only to exceed the limiting stress in the extreme fibers but also to reach the yield stress throughout the entire depth at a particular section of the beam. When this occurs, a *plastic hinge* is formed at this point in the beam or slab, and relative rotation of the beam on each side of the hinge occurs with no increase of moment at that point. At this time the yield stress may also have been reached at other sections, and other plastic hinges may be in the process of developing. If the formation of one or more plastic hinges results in the development of a collapse "mechanism," the *structure* will collapse at this stage of loading.

These stages in the deformation of a structure under increasing load up to the point of collapse have been described by two methods of analysis. The first concerns itself with solely linear behavior and thus is terminated at the point in loading where the first fibers at some section reach the yield stress. This method has been described in Chapters 4, 5, 6, and 8. The second method involves considerations of fully plastic behavior of members or sections of members and will be examined in the following pages.

The implications of the development of plastic behavior in structural components have been studied in detail since the 1950's, and these studies have led to methods referred to as *limit analysis*. We will begin by summarizing some of the features associated with plastic behavior of structures and described in Appendix B.

The yield surface of an ideal, elasto-plastic material is a surface in principal stress space that is convex outward. It is described mathematically by a yield function, f, in terms of stresses, σ_{ij}:

$$f(\sigma_{ij}) = k^2 \tag{11.1}$$

say, in which k is a constant. The material cannot exist at stress states where f exceeds k^2, and it is in an elastic state when f is less than k^2. Since, in plasticity, we are generally interested in work and dissipation relations, the state of strain is usually described in terms of a strain rate. The plastic strain rate is assumed to be derivable from a plastic potential function. The plastic potential is identified with the yield function f, so that plastic strain rates are obtained from the relation

$$\dot{\epsilon}_{ij}^p = d\lambda \frac{\partial f}{\partial \sigma_{ij}} \tag{11.2}$$

in which $d\lambda$ is a positive scalar factor of proportionality. This equation is called a *flow rule*. Since when an element of material reaches a yield state of stress it deforms continuously, we are interested in the relative magnitudes of the components of the plastic strain rate at this stress state. Equation (11.2) requires the plastic strain increment vector to be normal to the yield surface at the point representing the yield stress state. In practice, of course, materials do not exhibit perfectly plastic behavior, so that, in general, they do not continue to flow at the yield stress. Instead, their strains are limited as a consequence of strain-hardening or for geometrical reasons.

Since it is only possible to obtain exact elasto-plastic solutions for relatively simple problems, an approach that has developed in both continuum and structural mechanics has been to consider the conditions necessary to cause collapse to take place. In a continuum this implies the calculation of the stresses at which unlimited plastic deformations develop; in a structure it is the load at which the structure would deform continuously. In both these cases, strain-hardening or other effects will generally limit the displacements or deformations. To assist in assessing the calculated collapse loads, three theorems have been postulated (1 and 2).

Theorem 1: Only plastic strain increments take place at the stress or load at which failure occurs (here "failure" is equivalent to collapse or plastic flow); the stresses remain constant.

This theorem demonstrates that in areas of the body which are still behaving elastically there are no elastic strain *increments* while yielding goes on. These parts may be considered to be rigid.

Theorem 2: If any system of stresses can be found in internal equilibrium under a load condition P_l, which satisfies the other stress boundary conditions and which does not exceed the yield condition anywhere in the body, then the load P_l must be less than or equal to the collapse load.

Theorem 2 places a lower limit or limits, depending on how many solutions are attempted, on the collapse load. It is therefore referred to as the *lower bound theorem*. It is important to note that it applies to *any* set of stresses meeting the above requirement; it is not necessary that the stress system be an exact solution of the linearly elastic problem for the body under the loading and boundary conditions. Thus, in establishing a lower bound we concentrate our attention on the questions of equilibrium and the yield condition without considering how the body deforms.

Theorem 3: Choose a possible collapse mechanism of deformation or displacement of the body and calculate a load or load system P_u required to operate the mechanism by equating the rate of work done by the external loads to the internal rate of energy dissipation. The load P_u will be either greater than or equal to the collapse load.

Approaching the problem in this way, we will find different values for the load P_u depending on the collapse mechanism postulated. Theorem 3 tells us that the exact (ideal) collapse load will be smaller than any value of P_u

we obtain. The theorem is therefore termed the *upper bound theorem*. In this case, a load is arrived at by considering a plausible failure mechanism and an associated energy relationship, so that the required approach is different from that involved with lower bound estimates.

In any particular problem we can get an idea of how closely our solutions approach the exact answer by examining the spread between the lower bound and upper bound loads. If they are widely separated, this is an indication that either our stress distribution in the lower bound solution is far from the one that will obtain near collapse or we have not made a good guess at the possible collapse mechanism for the upper bound solution. Possibly both considerations may apply. In this case, different upper and lower bound solutions may be sought to reduce the difference between them. It is obvious that we are interested in the highest lower bound load and the lowest upper bound load, since the lower bound load is a conservative value for structural design purposes and the upper bound load is unconservative.

So far the exposition has been based on continuum considerations, but qualitatively the same theorems and results apply to a variety of structural configurations provided we replace the continuum stresses and strains with *stress resultants* and *generalized strains* for the structure. For beams and columns, the stress resultants are moments, shears, and axial forces and the respective generalized strains are rotations and shearing and axial displacements. Stress space is replaced by stress-resultant space in which the yield surface is again convex outward, and the same form of flow rule holds with respect to the plastic potential in terms of stress resultants. Thus, the generalized plastic strain rates in the coincident generalized strain rate space are normal to the stress-resultant yield surface.

For a beam, a two-dimensional stress-resultant or load space may involve axial load as one axis and torsional moment as another [as in Appendix B, Figure B.8(b)] or moment and axial load. The load space frequently encountered for plates has axes of radial and tangential moments. In the latter case, the axes of the coincident generalized strain rate space are the corresponding rotation rates. For the example of a footing (or tractor tread) resting on soil described briefly in Appendix B (Figure B.9), the three-dimensional load space consists of axes of vertical load, horizontal load, and moment, and the corresponding displacement rates form the axes of the strain rate space. Because the geometry of the collapse mechanism is fundamental to the analysis of the upper bound loads, it is necessary to confirm that the generalized plastic strain rates corresponding to a particular collapse mechanism obey the normality requirement with respect to the yield criterion at each point on the yield surface which describes the stress-resultant or load state. This will be illustrated later by an example of a slab on an elastic subgrade material.

In the following problems and worked examples the analysis of the upper bound collapse mechanism invokes a displacement Δ of some part of the system. Other displacements or rotations are expressed in terms of this Δ through the geometry of the assumed mechanism. The upper bound theorem is expressed in terms of *rate* of work and *rate* of dissipation, and it effectively defines a calculation of the amount of work done or energy dissipated in a virtual, small displacement $\dot{\Delta}$ added to Δ without changing the geometry of the mechanism. Thus, all the calculations should be written in terms of *rate* and $\dot{\Delta}$. However, for convenience, and since the mathematics is not changed by the substitution, the expressions will be written in abbreviated form as "work done" and "energy dissipated" by using the virtual displacement in the form Δ instead of $\dot{\Delta}$.

11.2 BEAMS

Figure 11-1 gives an example of a simply supported beam. Here a conventional linearly elastic analysis indicates the relation between load P and, say, the deflection at the midpoint of the beam in terms of the beam properties. After the limit stress is reached (in the case illustrated, in the extreme fibers at the beam midpoint), the elastic analysis can no longer be used to obtain deflections. At this stage the normal stress distribution through the depth of the beam is as shown by the line AOA' in Figure 11-1(b). (It is assumed that the beam is homogeneous and that it possesses yield stresses equal in both tension and compression. Other conditions do not affect the conclusions of the analysis.) If we consider the beam to be a continuum, then the load P at which this state of stress is reached under the load constitutes a first lower bound. When P is increased above this value, more of the beam at the top and bottom surfaces reaches the yield stress and the stress distribution at some later stage of loading might be represented by the line BOB' in Figure 11-1(b). The simple elastic beam theory no longer can give the load-deflection relation for the beam; however, the limit of the beam's resistance to deflection has not been reached. This limit will be attained when the stress in the beam is at the yield value on both sides of the centerline (or neutral axis) as shown by the distribution $DACOC'A'D'$ in Figure 11-1(b). At this stage in the stressing this section of the beam is exerting its maximum resistance to bending, and it is at its yield moment M_p. It will be understood that this gross beam behavior depends on the behavior of the beam material, which must exhibit a stable (ideally elastic-plastic, here) stress-strain characteristic. If the behavior is unstable (brittle, for example), the full yield moment M_p cannot be developed.

So far we have discussed the details, from a continuum mechanics point of view, of how the beam reaches its yield moment. In developing this picture of the behavior, we would have reference to the yield surface for the beam *material*, expressed in terms of the stresses, and we would identify strain rates

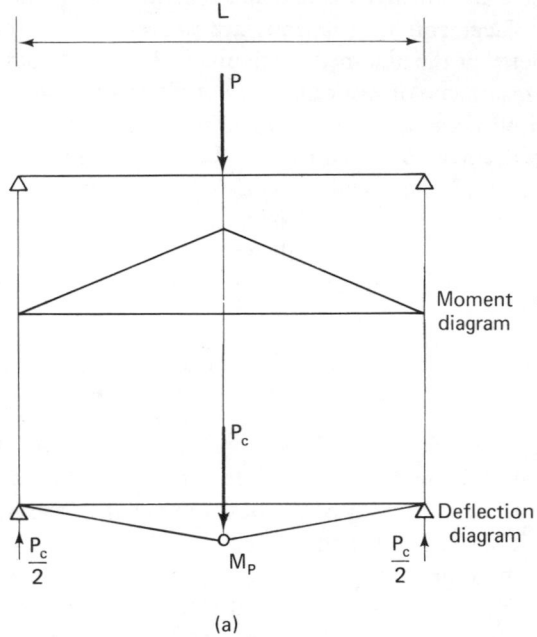

(a)

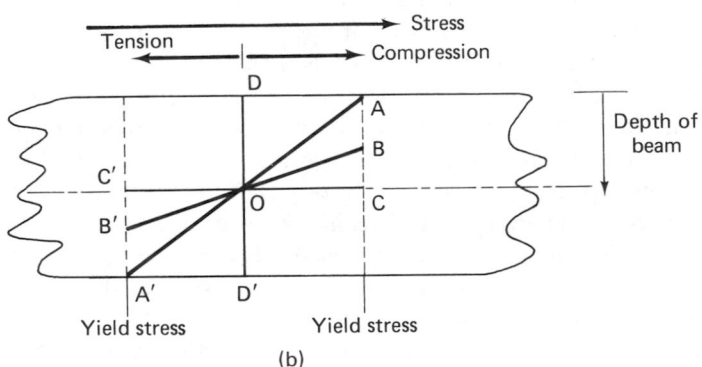

(b)

Figure 11-1. Collapse load on simply supported beam: (a) moment and deflection; (b) stresses at center section of beam.

superimposed on the stress space and yield surface. Replacing stress by moment and strain rates by rotation gives the generalized coordinates referred to earlier. The picture of the beam's stress-strain behavior is simplified to that of a moment-rotation relation. For an ideal elasto-plastic material behavior, the moment-rotation relationship is, in general, a complicated function, which can, in turn, be simplified also to an elasto-plastic representation as described in Chapter 9. The moment at a point in the beam is now taken to vary linearly with rotation at the point up to the yield moment, at which it remains constant, while the rotation continues. Instead of the yield stress, we now take the yield moment M_p as the variable of importance in beam-loading problems. The yield moment will generally be determined directly by experiment, or it may be calculated from the material property and beam geometry. Returning to the example of Figure 11-1(a), we can calculate the relation between the load and the yield moment M_p for the beam from equilibrium considerations. This maximum moment will occur in this example at the center of the beam. When ultimately a hinge develops there, the beam has a three-hinge mechanism (two at the two simple supports and one at the middle) and therefore fails. The moments are, of course, zero at the supports and the center moment is M_p.

The limit theorems of plasticity described earlier guide us in calculating lower and upper bounds for the load at collapse. If we take the lower bound first and use a continuum approach to the beam's behavior, in which the criterion of yield is applied to the elements or fibers of the beam, we have to determine the load at which the extreme fiber reaches its yield stress. For a load at the center of the beam, the moment at the center is $PL/4$. Elementary beam theory relates the extreme fiber stress, σ, to the moment

$$\sigma = \frac{Md}{2I}$$

or, at yield

$$M_y = \frac{2I\sigma_y}{d}$$

where d is the beam depth, I is its moment of inertia, and σ_y is the yield value of stress. Inserting P gives a lower bound value of P, P_{l1}:

$$P_{l1} = \frac{8I\sigma_y}{dL}$$

However, it is clear that this essentially microscopic approach to the problem gives an extreme lower bound, since only the boundary fibers are barely brought to yield.

We can replace the analysis by one incorporating the generalized stress coordinates in which the yield condition is referred to section moment rather than fiber stress. We make the assumption by the process described at the beginning of Chapter 9 that the moment-rotation (or curvature) relation for a point on the beam is linear up to some yield moment M_p at which continuous rotation yielding occurs in the beam. By comparing the rectangular stress distribution in the beam when M_p is reached with the triangular one with extreme fibers only at σ_y, we see that M_p equals $1.5M_y$.

Therefore, for a second, higher lower bound, using our idealized elastic-plastic moment-rotation behavior, we are required to apply a load P_l such that equilibrium is satisfied and the yield or plastic moment M_p is not exceeded anywhere in the beam. This condition is met if we apply a load P_l so that M_p is just attained at the center of the beam. From equilibrium the reactions at each support are equal to $P_l/2$, so that we find

$$P_l = \frac{4M_p}{L} \tag{11.3}$$

Because of the relation between M_y and M_p, this value of P_l is 50 percent higher than P_{l1}. The second limit theorem then states that P_l is less than or equal to the load P_c required to cause collapse of the system.

Application of the upper bound theorem requires calculating the rate at which energy is dissipated in the plastic hinge and equating this to the rate at which work is being done by an applied load P_u. However, in beams, the deflection from the initial configuration necessary to develop plastic hinges is small, and it is therefore convenient to write a virtual work equation in terms of a small arbitrary deflection instead of a rate of work expression. In the present case, we will assume that the beam deflects Δ under the load P_u which is at least sufficient to generate the yield moment M_p at the beam center. The (rate of) work done by the load is then $P_u \Delta (P_u \dot{\Delta})$. The (rate of) rotation at the center hinge is $4\Delta/L(4\dot{\Delta}/L)$, and thus the (rate of) energy dissipated at the hinge is $4\Delta M_p/L(4\dot{\Delta}M_p/L)$. Equating the two expressions gives

$$P_u = \frac{4M_p}{L}$$

which the upper bound theorem states is either equal to or greater than the collapse load. In this case, the lower and upper bound values of P coincide, so that the exact collapse load P_c for this problem has been obtained: $P_l = P_u = P_c$.

This technique offers a different choice of design method from that used previously and based on the elastic analysis. In the latter method, a limiting stress for the material is found, usually from building codes or design manuals, and this is divided by a safety factor on stress to give a working stress. Then the structure is so proportioned that the design load will produce this working stress in the structure. The calculation of the design load is based on a linearly elastic analysis. An economical design is one in which the design load will produce the working stress simultaneously at a number of points in the structure. However, as we have seen, when ductile materials are used, the safety factor against the actual collapse of the structure is much greater than the one used to reduce the limit stress to the working stress. An increase of the load to that of the design value times the stress safety factor will deflect the structure, still linearly, until the limit stress is reached but only in the extreme fibers of certain sections. A further increase is needed until the yield stress is reached in these sections, and more load must be added until a plastic hinge is formed, generally at one section only. For common structural arrangements, a further increase is required until a collapse mechanism develops and the structure falls down.

Since the intermediate stages between the end of the linear range and collapse cannot readily be calculated, an alternative design approach is to calculate the collapse load for the structure and apply a factor of safety based on *load* to this (*load factor*) to arrive at the design load. Since the early 1950's this approach has been widely developed and is described in detail for steel frame structures in many papers and a number of books (1, 5, and 11). When this technique is employed, a check must be made of the deflections developed when the *design* load is applied. These may be unacceptably high for some structures or structural components. They may also be difficult to calculate.

Although there were initial doubts about the ability of reinforced concrete to sustain large enough moments without failure (ductility) at initially developing plastic hinges, so as to permit other hinges to form as the load increases, the use of ultimate load analysis in reinforced concrete design has been growing (6, 7, and 12). To permit initially forming hinges to maintain their moment through the deformations necessary for the formation of later hinges, the concrete must be underreinforced. The application of this method to problems of foundation engineering seems to have been limited to date. We will proceed to consider some aspects of design by analysis of ultimate load.

First, to enlarge on the calculational techniques, let us consider that the beam of Figure 11-1(a) is fixed at one end, simply supported at the other, and subjected to a uniformly distributed load p/unit length as shown in Figure 11-2 rather than the point load originally shown. Initially, we will consider the beam to be made of, say, steel, capable of sustaining a plastic moment M_p in either a positive or a negative sense. We do this because later it will be necessary to consider beams to be made of reinforced concrete whose plastic moment will, in general, be different depending on the sense of the moment. This arises, of course, because the percentages of steel in the top and bottom of the section will usually be different.

With the application of load p, a negative maximum moment will be developed at the fixed support and a positive maximum moment will be developed at some point along the length of the beam. If p is increased, both moments will get larger until one of them reaches the value $\pm M_p$. But this gives only two hinge points for the structure, which cannot, therefore, collapse until the load p is further increased to bring the other maximum moment also to the value $\pm M_p$. We can begin with an elastic analysis of the loads and

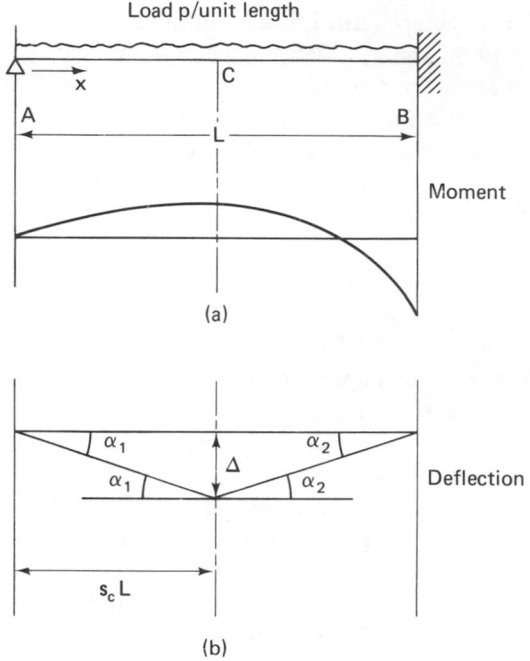

Figure 11-2. Uniformly distributed collapse load on beam simply supported at one end, fixed at the other: (a) moment; (b) deflection.

moments, in order to find the point in the beam where the maximum positive moment is developed.

For this case, the moment distribution in the beam is readily obtained.

$$M = \frac{1}{8}pL^2s(3 - 4s) \tag{11.4}$$

where $s = x/L$ and x is measured from the left end of the beam as shown in Figure 11-2. Differentiating M with respect to s and equating the result to zero give the result that the maximum positive moment in the beam occurs at $s = \frac{3}{8}$. Substituting this value in equation (11.4) gives M_{m+} to be $9pL^2/128$, whereas at B, of course, it is apparent that $M_{m-} = -pL^2/8$. At this stage, to proceed further, we need to make the assumption that the moment-rotation relation for the beam is elastic-perfectly plastic; that is, it is linear up to the plastic moment M_p and is flat thereafter. Since the moment at B is the bigger algebraically, it follows that it will reach the plastic value M_p first, at a value of p equal to $8M_p/L^2$. This can be taken as a first lower bound. However, at this value of p the maximum positive moment is $9M_p/16$. Consequently, p can still be increased, with the beam now supported by two hinges at A and at B, but with a fixed negative moment of $-M_p$ imposed at B. Eventually, p will reach a higher lower bound value p_l when M_p is achieved at the location of maximum internal bending moment. We can calculate this value p_l as follows.

The bending moment in the beam is still parabolic (and, of course, is equal to $-M_p$ at B) and may be expressed as follows:

$$M = \frac{pL^2}{2}s(1-s) - sM_p \tag{11.5}$$

We have to find the point at which M is a maximum inside the beam and then impose the value $+M_p$ at this location in order to determine p_l. By differentiation of M with respect to s again and equating the result to zero, we have

$$s = \frac{1}{2} - \frac{M_p}{pL^2} \tag{11.6}$$

Substituting this in equation (11.5), we find the value of the maximum internal bending moment, M_{m+}, to be

$$M_{m+} = \frac{pL^2}{8} + \frac{M_p^2}{2pL^2} - \frac{M_p}{2} \tag{11.7}$$

Now p has the value p_l, when M_{m+} is made equal to $+M_p$, so that we get a quadratic equation in p_l from which its value is determined:

$$p_l = (6 + 4\sqrt{2})\frac{M_p}{L^2} = 11.66\frac{M_p}{L^2} \tag{11.8}$$

This can be substituted in equation (11.6) to give $s = 0.414$ as the location of the maximum positive bending moment. By comparison with the previous elastic analysis it can be seen that the point of maximum internal moment has moved a small distance along the beam toward the fixed end as the final load increments were applied.

The upper bound analysis proceeds as follows. The material (beam) behavior is assumed to be rigid-plastic in moment-rotation space. The location C in Figure 11-2(a) of the maximum internal moment, again taken as M_p, is unknown, and we will take it at a distance $s_c L$ from the left end of the beam. At this point we assign an arbitrary vertical displacement Δ to the beam and proceed to establish the virtual work equation. When the upper bound load, p_u, is acting, the work done, E, by the applied load is given by the relation

$$E = \frac{\Delta}{2}p_u(1-s_c)L + \frac{\Delta}{2}p_u s_c L = p_u L\frac{\Delta}{2} \tag{11.9}$$

The rotation angle at the fixed-end hinge B is $\Delta/(1 - s_c)L$ and the rotation at the center hinge is the sum of this quantity plus the rotation at the pinned end A, Δ/s_cL. Thus, the energy dissipated, D, in the rotations is

$$D = \frac{M_p \Delta(1 + s_c)}{s_c L(1 - s_c)} \qquad (11.10)$$

Equating E and D gives

$$p_u = \frac{2M_p}{s_c L^2} \frac{(1 + s_c)}{(1 - s_c)} \qquad (11.11)$$

Once again, by differentiating p_u with respect to s_c and by equating to zero, we can find the location of the point of maximum internal moment

$$s_c = \sqrt{2} - 1 = 0.414 \qquad (11.12)$$

This is identical to the value obtained in the lower bound analysis, and, moreover, when we substitute it back in equation (11.11), we find that p_u also has the same value as before, $11.66\, M_p/L^2$.

Thus, in this analysis, we have again found that the lower and upper bounds are identical, so that the correct value of load to cause collapse has been calculated. This has occurred in these two illustrative examples because of the ideal elasto-plastic quality assigned to the beam, and it has been possible to calculate the worst-case collapse mechanism exactly by the differentiation process described. Although this is possible in a number of cases in practice, it will generally be found to be an extremely tedious process. A preferable technique used by engineers is to assume a collapse mechanism by inspection and then to evaluate the upper bound load required by this mechanism by the virtual work method. In general, in this approach the correct positions of the various hinges will not be obtained, and thus by the upper bound theorem, the calculated collapse load will be higher (and thus less conservative) than that which would have been obtained if the correct collapse mechanism had been found. It is not known how much higher the calculated loads are than the exact collapse load unless a lower bound analysis is performed. Frequently, in fact, the lower bound is not calculated. Instead, after several possible collapse mechanisms for a particular structure have been investigated and the lowest upper bound has been established for this set, a check is made of the moments at other critical points in the structure where plastic hinges have *not* been assumed to make sure that the limit moment has not been exceeded at any of these points. The lower bound conditions are also met by the results of the analysis if the moments at all these points are calculated to be below the limit moment M_p.

In the literature (5 and 11) there are a number of guides to help the designer select plausible collapse mechanisms in complicated cases. These include techniques of combining simpler mechanisms obtained by the subdivision of complicated structures into more elementary units. For steel structures, design methods have evolved to enable designers to arrive at an optimal structure (3 and 4) to carry their design loads (with a given load or safety factor). In this usage, an optimum structure is one that has a minimum weight for the task. Although the cost of steel structures is related to the weight of material used, other factors also influence the cost, so that, in practice, methods of optimal design have only been applied to minimize the cost (as well as the weight) of certain classes of structures.

If a designer does not wish to examine the collapse mechanism of the beam (Figure 11-2 discussed above) to the extent we have developed it, he may use experience or intuition to guess at the failure mechanism. In this problem the only thing to guess is the position of the point at which the second hinge will form in the beam. If, for example, the designer estimates that the hinge will form at the midpoint of the beam (not a very good guess, since the support conditions are asymmetrical), then he will arrive, via equation (11.11), at an upper bound value of p, p_u, equal to $12M_p/L^2$. This is only a few percent higher than the correct value. Exactly the same value for p would have been arrived at if a value of s_c had been selected equal to one-third for the hinge point. It is apparent that p_u varies only slowly with s_c in this problem, so that there are no severe penalties if the judgment exercised is not good. This is frequently the case.

11.3 BEAMS ON SOIL

Next, we will consider the effect of the presence of a soil subgrade. What happens when the beams of Figures 11-1 or 11-2 are resting on soil? In the form presented in the last section with the support indicated, this would be a rather artificial problem, and a more practical discussion can be centered on the infinite beam of Figure 11-3(a), which illustrates a number of points of interest. As the load P is applied and increased, the beam initially deflects elastically and presses into the soil; it is convenient to consider that the soil exhibits the Winkler behavior discussed previously, so that the solution given in Section 5.2.1 applies. The bending moment is shown in Figure 11-3(b) in which it is seen that, as before, there is a maximum positive moment under the load and that there are two symmetrical negative maxima some distance to each side. At a particular value of P the moment under the load will reach the yield value for the beam, M_p, but this does not constitute a failure mechanism for the beam. The load P can be increased still further.

It is convenient now to consider one-half of the beam only; the boundary conditions are those of a beam loaded at one end by a vertical (or lateral) force $P/2$ and a moment M_p as shown in Figure 11-3(c). As the load P goes up, the maximum negative moment will increase until it achieves the value

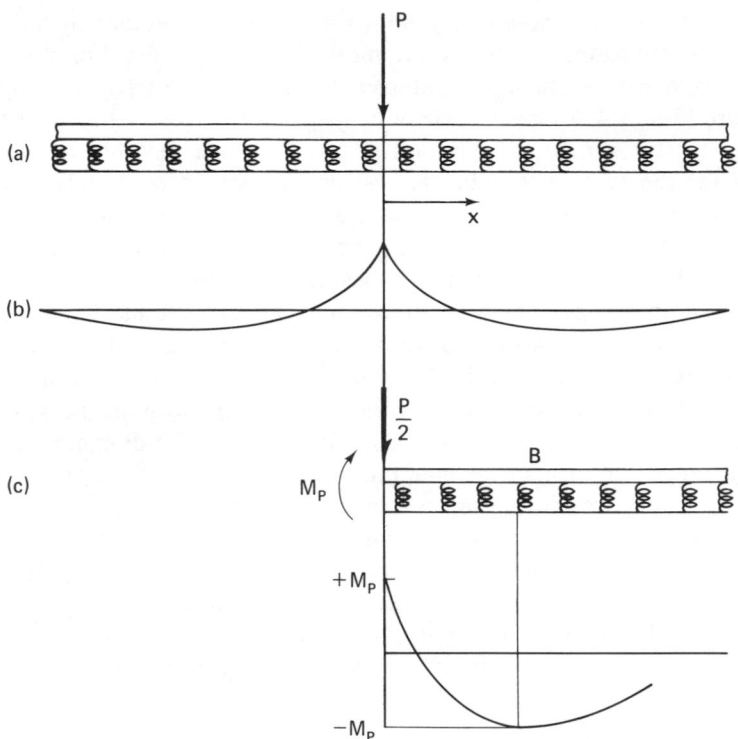

Figure 11-3. Collapse load on beam on elastic foundation: (a) load and beam arrangement; (b) deflection; (c) conversion of problem to semi-infinite beam.

$-M_p$ at the point B of Figure 11-3(c). (The beam is still taken to be homogeneous, so that M_p holds regardless of the sense of the moment.) At this stage we have to face a further difficulty. We have, according to the previous discussion, reached a collapse mechanism for the *beam*, but if we assume that the *soil* is still behaving as a Winkler material, a limit load has not yet been reached, so that we do not have a collapse mechanism for the *system*. This situation corresponds to that in the discussion at the beginning of Chapter 9. In fact, we cannot achieve a collapse load unless we terminate the Winkler soil behavior by a plastic yielding of the soil. From a practical point of view, however, the beam can be considered to have failed when the two hinges develop in it. Whether or not the soil has reached a yield state at this stage will depend on the deflections the beam has undergone and on the stress-deformation behavior of the soil. The problem is thus one of beam-soil interaction of a special kind.

We can distinguish two extreme cases. If the beam deflects a relatively large amount before the hinges develop, the soil may be considered to have yielded, and thus to exert a uniform (failure) loading along the underside of the beam. The failure soil reaction will depend on the soil properties and, in general, on the width of the beam. If, however, the deflection of the beam when the hinges have developed is small relative to the deflection required to fail the soil, a Winkler soil behavior may be assumed, and the limit load, for practical purposes, is that when the beam collapse mechanism is developed. The two cases will be considered in turn, since the first is the simpler of the two.

10.3.1   Soil yielding; uniform soil reaction

We examine the lower bound condition to begin with; the problem is illustrated in Figure 11-3. We are not concerned with conditions far to the right side. Once again, from equilibrium, the bending moment in the beam at section x is given by the expression

$$M = M_p + \frac{qx^2}{2} - \frac{Px}{2} \qquad (11.13)$$

where q is the soil reaction load (assumed to be the yield value) per unit length of the beam, and the beam is assumed to have unit width. Differentiating M with respect to x and setting equal to zero as before, we obtain the value of x, x_B, at point B, where the maximum negative moment occurs.

$$x_B = \frac{P}{2q} \qquad (11.14)$$

Substituting this back in equation (11.13) gives M at point B. Now we have to make this M equal to $-M_p$, the hinge moment for the beam, so that we obtain the lower bound value of P, P_l:

$$P_l = 4\sqrt{qM_p} \qquad (11.15)$$

Replacing P in equation (11.14) by P_l gives, finally,

$$x_B = 2\sqrt{\frac{M_p}{q}} = \frac{1}{2}\frac{P_l}{q} \qquad (11.16)$$

For the upper bound solution, we assume that a hinge has formed at a distance $x = x_B$ at present unknown, to the right of the load in Figure 11-3, and that the load has displaced the beam vertically downward a distance Δ at the point of application of the load. Considering the *half beam* again, we see that the work done by the load on this occasion has to overcome the energy dissipated both in the plastic hinges and in the underlying soil. In the hinges the dissipation is $2M_p\Delta/x_B$, and in the soil it is $qx_B\Delta/2$. The work done by the load is $P_u\Delta/2$. Equating these energy terms, we have

$$P_u = \frac{4M_p}{x_B} + qx_B \tag{11.17}$$

Differentiating, with respect to x_B, and equating to zero as before give the same result for x_B as that in equation (11.16). Substituting this value back in equation (11.17), we get

$$P_u = 4\sqrt{qM_p}$$

as in the lower bound solution, so that $P_u = P_l$ is the collapse load.

The square root dependence of the result for P is interesting compared to the linear relation for the simply supported beam examined earlier. A doubling of either the soil resistance or the plastic moment of the beam only increases the collapse load by a factor of $\sqrt{2}$. In addition, equation (11.16) for x_B indicates that an increase in soil resistance relative to the beam property M_p reduces the distance at which the second collapse hinge forms. The trend of these results parallels some of the earlier consequences of the linear elastic analysis of beams on a Winkler foundation.

It must be borne in mind in this analysis that, although the collapse mechanism assumes that the beam is rigid beyond the outer plastic hinge, elastic deformations are presumed to have occurred prior to formation of the hinge; these deformations bring the underlying soil to the assumed failure state with uniform soil pressure distribution.

If the beam has properties uniform throughout its width (perpendicular to the plane of the paper in the figures) and rests on a homogeneous bed of overconsolidated clay, an increase in the width of the beam by some factor (keeping the depth constant) increases both the hinge yield moment M_p *and* the soil resistance per unit length q by the same factor. This is true for the soil because the unit bearing capacity of the clay is independent of the foundation width. It follows from equations (11.15) and (11.16) that the collapse load for the beam is also increased by the same factor, while the distance x_B stays the same. If, however, the beam rests on cohesionless sand, the consequences are different, since the unit bearing capacity of a footing on sand varies approximately linearly with the width of the footing. Thus, an increase in the width of the beam by some factor r, say, raises the hinge moment M_p again by the same factor, but it increases the soil resistance q by r^2. In this circumstance, the collapse load increases by a factor $r^{3/2}$; that is, it increases faster than the width. The soil is effectively more resistant for the wider footing, so that the distance x_B decreases as $r^{-1/2}$, as shown by equation (11.16).

11.3.2 Soil behaving as a Winkler material

As we have done before, we examine the lower bound solution first. The basic equation to be used to describe the deflections of beam and soil is equation (5.5) and we are concerned with the moments in the semi-infinite beam loaded as in Figure 11-3(c). From equation (5.44) for load $P/2$ alone, we have the moment M at section x of the beam:

$$M = -\frac{P}{2\lambda}e^{-\lambda x}\sin \lambda x \qquad (11.18)$$

where λ is the reciprocal of the characteristic length, as defined in equation (5.12). When the moment $+M_p$ alone is applied at the end of the beam (origin of coordinates), equation (5.49) gives the result

$$M = M_p e^{-\lambda x}(\cos \lambda x + \sin \lambda x) \qquad (11.19)$$

Thus, as a result of the superposition of $P/2$ and M_p, since we are in the linear range of behavior as far as the semi-infinite beam is concerned, the total moment is found by adding equations (11.18) and (11.19) to get

$$M = M_p e^{-\lambda x}(\cos \lambda x + \sin \lambda x) - \frac{P}{2\lambda}e^{-\lambda x}\sin \lambda x \qquad (11.20)$$

Differentiating with respect to x as before and equating the result to zero [or alternatively, writing the expression for the superposed shears from equations (5.45) and (5.50) and equating it to zero] give the value of x at the point B, x_B, where the second hinge is imminent:

$$\tan \lambda x_B = \frac{P}{P - 4\lambda M_p} \qquad (11.21)$$

In this case, it is no longer possible to get an explicit expression for P_l, which must be found by trial-and-error substitution in equations (11.20) and (11.21), when M in the latter equation is replaced by $-M_p$. An example will be given following the upper bound calculation.

Once again for the upper bound, we assume a collapse mechanism with the outer hinge a distance x_B from the point of loading. Since we assume the beam to remain rigid between the hinge points, the soil reaction will vary linearly between a maximum under the loading point and a minimum below the outer hinge point. We generally assume with this mechanism that the deflection at the outer hinge is negligible compared with that under the load, so that we may take the soil reaction to vary from its peak value at the load point to zero at the outer hinge. In reality, the soil reaction will not be zero at the latter section.

With the deflection Δ_0 under the load, the work done by the load and the energy dissipated by the hinges remain the same as in the former case of the beam on the yielding soil. This time, the work done on the soil is different. At a section of width dx distant x from the origin, the work done against the soil resistance is $q(x) \, \Delta(x) \, dx$, and the total work done against the soil, W, is

$$W = \int_0^{x_B} q \, \Delta(x) \, dx \qquad (11.22)$$

However, for the Winkler material $q(x) = k \, \Delta(x)$, and because the deflection is linear, we have

$$\Delta(x) = \Delta_0 \left(1 - \frac{x}{x_B} \right)$$

Substituting in equation (11.22), we obtain

$$W = k \int_0^{x_B} \Delta^2 (x) \, dx = \frac{k \, \Delta_0^2 x_B}{3} \qquad (11.23)$$

The virtual work equation becomes

$$\frac{P_u \, \Delta_0}{2} = \frac{2M_p \, \Delta_0}{x_B} + \frac{k x_B \, \Delta_0^2}{3}$$

or

$$P_u = \frac{4M_p}{x_B} + \frac{2k x_B \, \Delta_0}{3} \qquad (11.24)$$

and we find that, in this case, the assumed deflection Δ_0 does not disappear from the equation. Proceeding with the usual minimization process, we get

$$x_B = \sqrt{\frac{6M_p}{k\Delta_0}} \qquad (11.25)$$

and, by substitution, we get

$$P_u = \frac{4}{3} \sqrt{6M_p k \Delta_0} \qquad (11.26)$$

Thus, to continue with the calculation of the collapse load, we need to estimate the deflection under the load. One way of doing this is to obtain an approximate value of the deflection under the load from equations (5.42) and (5.47), which refer to the point load and moment load cases on the semi-infinite beam, respectively. The deflection obtained in this way is approximate, since the case under study involves the deflection of a beam which is assumed to be rigid between the hinge points, so that the soil reaction under the beam has the triangular distribution described earlier. The deflection obtained from equations (5.42) and (5.47) applies to an elastic beam whose deflection is a continuous curve. However, we will proceed to obtain an estimate of the load P_u by this process. Superposing the deflections under the load point from the above equations and multiplying by k give

$$\Delta_0 = \lambda P_u - 2\lambda^2 M_p \tag{11.27}$$

which can be substituted into equation (11.26) to give a quadratic in P_u. Solving the quadratic, we find

$$P_u = \frac{8}{3}\lambda M_p \quad \text{or} \quad 8\lambda M_p$$

but, on examining equation (5.20) for the moment in an infinite beam, we find that it requires a load P equal to $4\lambda M_p$ to produce the plastic hinge moment under the load P. Thus, we take the upper value as our estimate for P_u:

$$P_u = 8\lambda M_p \tag{11.28}$$

Inserting this in equation (11.27) for $k\Delta_0$ gives the result that $k\Delta_0$ equals $6\lambda^2 M_p$, which, on substitution into equation (11.25), yields

$$\lambda x_B = 1 \tag{11.29}$$

If, as a check, we again refer to the *infinite* beam case, we find that the first peak negative moment occurs there at a distance of $\lambda x_B = \pi/2$ from the load point. It seems plausible to consider that the increase of load P required to bring the system to a state of imminent collapse should move the point of secondary maximum moment a short distance closer to the load point. It was observed in the case of the beam with one fixed end in Figure 11-2 that the section at which the maximum internal moment was developed also moved toward the fixed end as the load was increased to the collapse value.

At this stage we return to the lower bound calculation and perform the trial-and-error determination of the lower bound value of P_l by using the approximate upper bound value as a guide. It is necessary first to substitute different values of P_l in equation (11.21) to obtain λx_B. When this value and that of the P_l used is inserted in equation (11.20), the moment at λ_B, M_B is computed. We need to find the value of P_l that makes the moment M_B equal $-M_p$. On making this calculation it is found approximately that

$$P_l = 9.75\lambda M_p \tag{11.30}$$

and

$$\lambda x_B = 1.04 \tag{11.31}$$

The value of P_l is, in this case, higher than the upper bound; the latter is, therefore, incorrect, since we have seen from our previous calculations that a consistently performed upper bound determination coincides with the lower bound in these conditions of problem and method of analysis. The upper bound is lower than P_l because the soil reaction pressure at the outer hinge point is not zero as we have assumed it to be in the analysis. Using equations (5.46) and (5.51) for the semi-infinite beam with a value of $P_l/2$ and M_p, as obtained above, it can be calculated that the soil pressure under the loading point is $7.75\lambda^2 M_p$. Equations (5.42) and (5.47) with the same values and the appropriate λx_B give the pressure at the hinge B to be $2.00\lambda^2 M_p$. Thus, a better upper bound calculation in these circumstances would be to apply a soil reaction varying linearly from a maximum under the load to some other value, as indicated above, at the outer hinge. Since we have, however, established a value for the collapse load for this problem, we will not pursue the case further.

Because this second case involving the Winkler behavior of the subgrade soil was more difficult to analyze than the first, and since it is, perhaps, less rational in assuming yielding behavior of the beam, but not in the soil, we should determine whether or not, in fact, the beam deflections, just as it yields, are in the elastic range for soil behavior. We can do this by using the above result for the pressures under the center and outer hinge points of the beam and some of the properties of beams and soil in the usual range of proportions. The deflections at the center, w_0, and at the hinge point, w_B, are at least as great as the following values:

$$w_0 = 7.75\frac{\lambda^2 M_p}{k} \tag{11.32}$$

$$w_B = 2.00\frac{\lambda^2 M_p}{k} \tag{11.33}$$

since these obtain at the load when the beam is just about to collapse. For concrete beams, 12 in. (0.3 m) in width, reinforced top and bottom to correspond with the assumption of equal positive and negative plastic hinge moment capability, substitution of the appropriate beam and various soil properties into these equations shows that the deflections should typically be of the order of an inch or two (25 to 50 mm) when beam failure occurs.

These values are confirmed by examination of the results of tests on failures of steel and concrete beams and slabs (unsupported by soil). Generally, the collapse load, as studied here, is reached at maximum deflections of the order of 0.1 to 0.2 of the beam or slab depth or distance from the upper surface to the center of tensile reinforcement. Displacements of this magnitude in a beam supported on soil are sufficient to cause the underlying soil to yield over the whole span of the beam between the outer hinges. We can conclude, therefore, that it is reasonable (and also conservative) to employ the analysis developed for the case of uniform soil resistance to calculate the collapse load on a beam resting on soil.

TABLE 11.1

Values of Load Required to Collapse Beam on Yielding Soil

System	Collapse Load P_u, kips (kN)	Distance to Hinge x_B in. (m)
12-in. (0.3 m) deep beam		
Medium soft soil	120 (534)	70 (1.78)
Medium stiff soil	200 (890)	40 (1.02)
24-in. (0.61 m) deep beam		
Medium soft soil	240 (1068)	140 (3.56)
Medium stiff soil	400 (1780)	80 (2.03)

Table 11.1 indicates a range of values of collapse load and distance between the load and outer hinge for some typical beams 2 ft (0.61 m) wide and resting on soils of two different properties. Two beams were considered: one 12 in. (0.3 m) deep with $M_p = 10^6$ lb-in. (113 kN-m) [for the 2-ft (0.61 m) width]; the other 24 in. (0.61 m) deep with $M_p = 4 \times 10^6$ lb-in. (452 kN-m) ($M_p = \sigma_0 bh^2/4$). The soils are considered to be medium soft and medium stiff, respectively. They may be taken to be represented by clay with undrained shearing strengths of 1000 to 3000 psf (48 to 144 kN/m²), respectively, or by a dry sand with angles of shearing resistance of about 30° (medium loose) and 40° (medium dense), respectively. As pointed out before, in this table the bearing capacity of the sand only applies to the 2-ft (0.61 m) width of the beam. Wider or narrower beams would require a recalculation of the values used. The bearing capacity employed for the clay is independent of the width of beam.

11.4 SLABS OR PLATES

Limit or ultimate load design methods have also been applied to slabs. The application was begun in 1943 by Johansen (6), who deduced the *yield-line analysis* technique from a study of the collapse mechanisms evolving in concrete slabs (not supported by soil) during tests. He gave the name *yield-lines* to the extended plastic hinges that developed in failing slabs. The design method he originated was not recognized as deriving upper bound values of the loads until the theoretical bases of plasticity and limit design were established in the early 1950's. It has proved difficult to obtain lower bound solutions for slab problems in general, and only a few cases have been examined in the literature. Consequently, in general, yield-line or limit analysis has been widely employed in the design of slabs without concomitant checks of limit loads by lower bound investigations. The justification usually adduced is the close correspondence of calculated loads with experimental results when the calculated loads are obtained as the lowest values from a series of postulated collapse mechanisms.

The situation is analogous to that prevailing in slope stability analyses in soil mechanics; a slope stability analysis is an upper bound method; lower bounds are not obtained, and confidence in the values obtained is, instead, gained by favorable comparisons with field and laboratory experience of slope failures and landslides. For most problems of slab geometry and loading met in practice, various papers and textbooks (6, 7, and 12) give information on the collapse mechanisms that have been found to occur.

The technique of performing an upper bound analysis of the limit load which can act on a slab is similar to that described for beams. A failure or collapse mechanism is postulated and is varied by mathematical methods or by trial and error until the lowest load is found which will cause collapse. This is taken as the ultimate load for design purposes.

It is seen that the yield-line analysis method is based on an intuitive understanding of collapse mechanisms based on the study of mechanisms that have actually developed in practical situations. However, the understanding is based on failures occurring under certain rather restricted conditions that apply to the cases that have occurred in practice. The method supplies us with no guides for arriving at appropriate collapse mechanisms under other conditions. Specifically, studies of the behavior of plain or reinforced concrete slabs demonstrate that certain mechanisms develop for particular load and boundary conditions. In the application of this knowledge, essentially only the equilibrium condition, employed through differentiation of the energy relationships, is taken into account. The kinematics of the deformations are explicitly ignored, but they are implicit in choosing the mechanism. If a collapse load has to be calculated for a slab made of material other than concrete, the yield-line method does not tell us how to choose a collapse mechanism, except by carrying out tests with the new material. But the material properties and the kinematics are related, as we have learned, through the yield envelope of the material under combined stress conditions. If we know the shape of this envelope, the normality condition or flow rule dictates how the material must strain when it is stressed to failure. This is the kinematic requirement we need for the general problem. However, to begin with, we will examine the yield-line analysis method as it is usually treated in concrete works and we will defer a kinematic discussion until later.

11.4.1 Slabs without soil

A number of yield-line patterns for slabs under various conditions of loading and support is shown in Figure 11-4; these cases will be discussed briefly to describe the method.

We consider first the simply supported, square, flat slab of Figure 11-4(a) subjected to uniform load. It is found that when the load is gradually increased up to its ultimate value a cross-shaped pattern of hinge lines forms

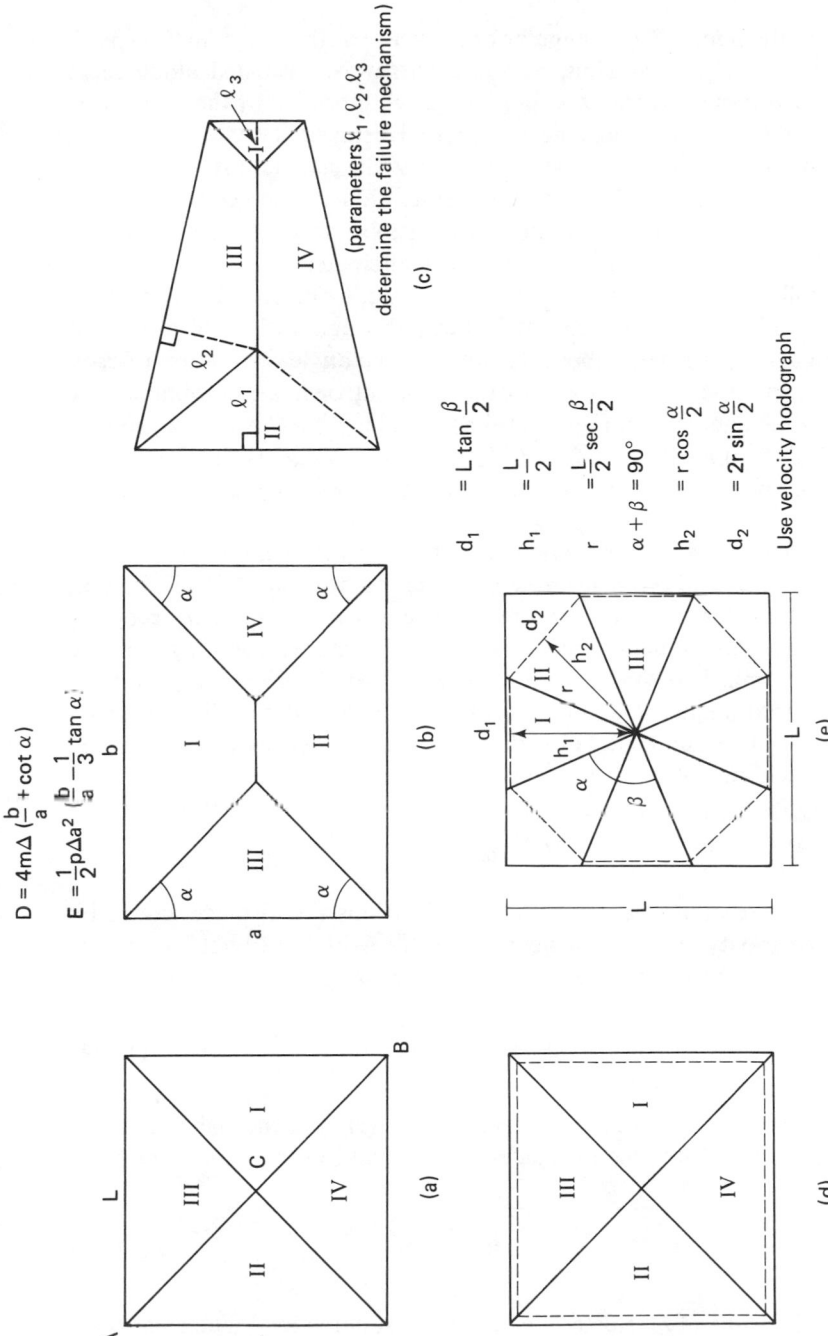

Figure 11-4. Collapse mechanisms for flat slabs. Solid lines are positive hinges and dashed lines are negative hinges: (a) simply supported square slab, uniformly loaded; (b) simply supported rectangular slab, uniformly loaded; here angle α is varied to minimize the upper bound load; (c) simply supported trapezoidal slab, uniformly loaded; parameters l_1, l_2, and l_3 determine the failure mechanism; (d) clamped-edge square slab, uniformly loaded; (e) clamped-edge square slab, uniformly loaded; alternate mechanism.

as shown in the figure. The triangular areas between the hinge lines displace essentially as rigid plates. Thus, at yield, energy is dissipated along these hinge or yield-lines only, and it is supplied, as with beams, by the work done as the applied load acts through the incremental distance that the slab deflects. The maximum deflection occurs at the center of the slab, and zero deflection occurs along the supported edges. A consideration of the geometry of the deformation process enables the angle changes between the rigid plates to be calculated for an evaluation of the energy dissipated.

If we follow this procedure in the present case, we can compute an upper bound load. If the center of the slab deflects through a distance Δ, so that each triangular area rotates about its support, the angle of rotation of each rigid triangular plate with respect to the adjacent one can be obtained by considering the line ACB in Figure 11-4(a). The length AC is $L/\sqrt{2}$, so that the slope of the line AC is $\sqrt{2}\,\Delta/L$. Since this also equals the slope of the line CB, it follows that the relative rotation across all the yield lines is $2\sqrt{2}\,\Delta/L$.

In a slab of this shape we would expect the reinforcement to be isotropic, so that the slab would have the same ultimate moment per unit width in all directions. We will denote this yield moment/unit width here as m, recognizing that, if the slab is made of concrete, the yield moment m for a positive moment (as applied to this case, assuming that the loading is a uniformly distributed dead weight on a horizontal slab) will generally be different from that for a negative moment. Thus, the energy dissipated by rotation of the section of yield-line AC will be $2m\Delta$, and, the total dissipation, E, for the four yield-lines is

$$E = 8m\Delta$$

It can be seen that the applied load p_uL^2 drops the distance moved by the center of gravity of each triangle $\Delta/3$, and the work done by it is therefore $p_uL^2\,\Delta/3$. Equating this to the energy dissipated, we get

$$p_u = \frac{24m}{L^2} \tag{11.34}$$

Since in design problems the load is generally given and the slab properties have to be found to bear the load safely, it is usual to write this equation in the form

$$m = \frac{p_uL^2}{24} \tag{11.35}$$

From this the value of m is determined, after a suitable load factor has been used to convert the design load to the limit or ultimate load p_u; the slab is then designed to have the hinge moment m.

For the simply supported slabs of other shapes shown in Figure 11-4(b) and (c), the yield displacements and yield-line rotations under a limit load are calculated from geometrical considerations once a deflection Δ has been assigned to some portion of the slab. If the slab has a clamped or built-in edge, as shown in Figure 11-4(d), it is necessary to include a term in the energy dissipation expression to account for the work done in forming hinges along the clamped edges. The geometry of the yielded configuration is the same as in Figure 11-4(a), but the rotations along the *clamped* edges are in this case $2\Delta/L$, and they must be multiplied by the hinge moments applying to rotation in the negative moment sense, since the triangular plates rotate downward with respect to the edge. The work done in rotation is, of course, positive. Assuming that the slab may be made of concrete, so that the negative hinge moment which it is capable of sustaining may be different from the positive moment, we will denote it by im, where i is a coefficient equal to or greater than zero. Thus, the total energy dissipated in these hinges (4 sides) is $8\,im\Delta$, and, with the work done by the load remaining the same as for the simply supported slab, we get the result

$$p_u = \frac{24m(1+i)}{L^2} \quad \text{or} \quad m = \frac{p_u L^2}{24(1+i)} \tag{11.36}$$

Thus, if the slab is concrete with equal amounts of reinforcing steel top and bottom, i will be equal to unity, and the slab with clamped edges will support twice the load of the similar simply supported slab.

This case of the slab with clamped edges can be used as an illustration of the effect of the choice of mechanisms on the upper bound result obtained. The mechanism shown in Figure 11-4(d) does not, in fact, give the lowest upper bound for the limit load. A lower bound can be obtained with the mechanism shown in Figure 11-4(e), in which the angle α can be varied to give a minimum load value at yield.

11.4.2   Slabs resting on soil

After this brief introduction to yield-line mechanisms for slabs, we can consider the problem of collapse of a slab resting on soil and subjected to various load conditions. For application to a variety of practical problems, the case of a vertical load uniformly distributed on a circular area of a slab infinite in extent, and resting on soil, is of interest and will be investigated first. By reference to the previous work on limit loads on beams resting on a Winkler foundation we can conclude that the additional labor involved in obtaining solutions for the case of slab deflections resisted by a Winkler subgrade reaction is not justified, and we will confine ourselves to the circumstance of a uniform soil reaction. We consider, therefore, that when the collapse load has been reached on the slab, the soil is also at the yield condition.

What is the failure configuration for this kind of structure? If the slab is isotropically reinforced (a reasonable assumption for a pavement but perhaps not for all foundation slabs), there would seem no reason to expect anything other than a symmetrical failure mechanism. In the initially elastic range of loading the slab deflects downward under the load, and a maximum moment and stresses are generated in it at the load point. The moment (positive under the load) decreases with distance, as in the beam-on-soil problem, and reaches a negative maximum at some radius from the load. In terms of the characteristic distance, $1/\beta$, expressed by equation (5.99), the negative maximum for a loaded area of small extent occurs at a radius of about $2/\beta$. The moment oscillates with radius, its magnitude dying out rapidly with distance from the loaded area. Thus, were we to increase the load, we would expect the yield or ultimate moment to develop first under the load and then a second yield moment end hinge to develop at some radius. In between there would be a dish-shaped area of deflection; it is to this region that we must apply our intuition to determine a collapse mechanism.

By analogy with the behavior of slabs supported from below by a column and loaded uniformly (6 and 7) it seems reasonable to guess that a conically shaped deflection region would develop as shown in Figure 11-5. This can also be deduced from the normality condition, as will be discussed later. As a consequence, between the loaded area and the outer hinge the moments in the θ-direction (m_θ) would be brought to a plastic hinge condition, with moment equal to m. The energy dissipated for this mechanism has been calculated (7) to be

$$D_1 = 2\pi m(1 + i)\Delta \tag{11.37}$$

where Δ is the central deflection, and we assume we are dealing with concrete in which the positive and negative moments required to form a hinge may be different. We will discuss later how this expression is obtained. The value of D_1 is seen to be independent of the radius, a, to the outer hinge. The soil under the failing portion of the slab to radius a exerts a uniform pressure q, and the additional energy required to do work against this soil is

$$D_2 = \int_0^a q\left(1 - \frac{r}{a}\right) \Delta 2\pi r \, dr = 2q\pi \Delta \left(\frac{r^2}{2} - \frac{r^3}{3a}\right)_0^a = \frac{q\pi \Delta a^2}{3} \tag{11.38}$$

If the uniformly distributed collapse load is p_u, as before, and it acts over a circular area of the slab of radius c, then the work done, E, by the load through the small virtual deflection Δ, is

$$E = \frac{2\pi p_u}{a}\left(\frac{ac^2}{2} - \frac{c^3}{3}\right)\Delta = P_u\left(1 - \frac{2}{3}\frac{c}{a}\right)\Delta \tag{11.39}$$

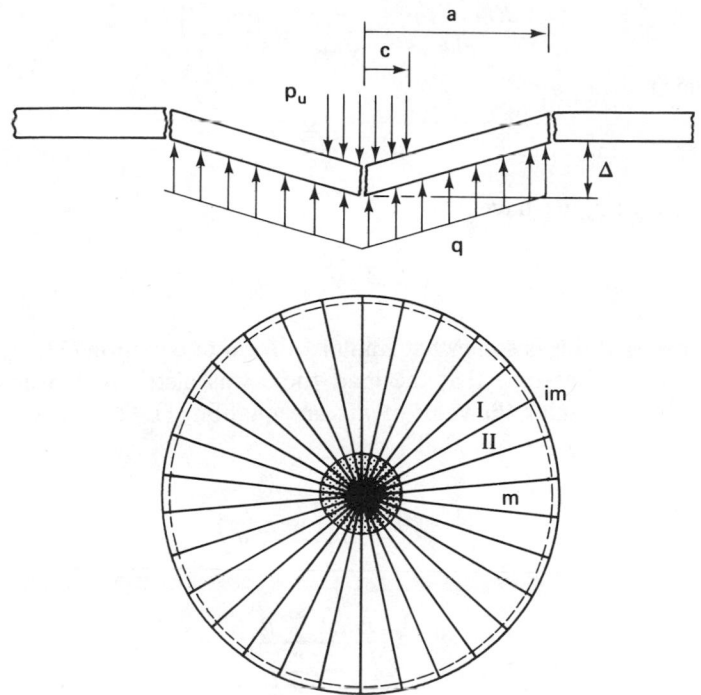

Figure 11-5. Load in interior of infinite flat slab resting on soil.

where P_u is the total load acting. This is obtained by the same integral employed to get equation (11.38), but with c as the upper limit and p_u for q. Equating the expressions for E and $D_1 + D_2$, we get

$$P_u\left(1 - \frac{2}{3}\frac{c}{a}\right) = 2\pi m(1 + i) + \frac{q\pi a^2}{3} \tag{11.40}$$

or, in terms of m,

$$m(1 + i) = \frac{P_u}{2\pi}\left(1 - \frac{2}{3}\frac{c}{a}\right) - \frac{qa^2}{6} \tag{11.41}$$

With this mechanism, we want to find the value of the radius a at which the P_u required is a minimum for a given value of m, or alternatively, at which m is maximized for a required P_u.[1] This can be calculated by differentiating either P_u or $m(1 + i)$ with respect to a and equating the result to zero. It is more convenient to perform the latter evaluation. We get

---

[1]Note that if we take $c = 0$, we run into trouble when minimizing; this is related to the fact that the moment is infinite beneath the point load in the simple elastic models.

$$\frac{d(m + im)}{da} = \frac{P_u c}{3\pi a^2} - \frac{qa}{3} = 0$$

from which

$$a^3 = \frac{cP_u}{\pi q} \tag{11.42}$$

or, in terms of p_u, we have

$$\frac{a}{c} = \sqrt[3]{\frac{p_u}{q}} \tag{11.43}$$

We observe that this is somewhat similar in form to equation (11.16) for the beam case, if we assume that the load there was distributed over a short length of beam. Taking the value of q from equation (11.42) and substituting in equation (11-41), we find that

$$m(1 + i) = \frac{P_u}{2\pi}\left(1 - \frac{c}{a}\right) \tag{11.44}$$

or

$$P_u = \frac{2\pi m(1 + i)}{\left(1 - \dfrac{c}{a}\right)} \tag{11.45}$$

in which the ratio c/a is given by equation (11.43) and $P_u = \pi c^2 p_u$.

A discussion of this result will be deferred until later. For a variety of practical problems, it is of interest to consider two further loading configurations: loads applied at right-angled corners and at edges of slabs supported on yielding soil. In the first case, the problem is shown in Figure 11-6, and it is convenient in the analysis to assume that the loaded area is the triangle at the tip of the corner. Although there is a variety of possible mechanisms by which collapse can be postulated in this problem, we will assume that failure will occur due to the plastic hinge developing across the corner along line BC distant a from the corner, as shown in the figure. This will not give the lowest upper bound, but it is satisfactory in the present context. The moment here is negative, so that the appropriate plastic hinge moment is im. In this case, for deflection Δ of the corner point A, the energy dissipated in the hinge is

$$D_1 = 2im\Delta \tag{11.46}$$

The work done against the soil pressure q is

$$D_2 = \frac{qa^2\,\Delta}{3} \tag{11.47}$$

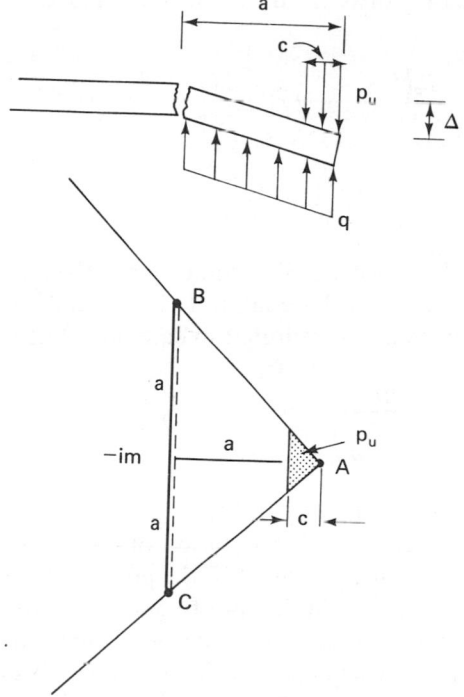

Figure 11-6.   Load at corner of flat slab on soil.

and the work done by the applied load is

$$E = p_u c^2 \left(1 - \frac{2}{3}\frac{c}{a}\right)\Delta = P_u\left(1 - \frac{2}{3}\frac{c}{a}\right)\Delta \qquad (11.48)$$

It is seen that there are similarities between this solution and that of the previous case of the infinite slab. In fact, the present approach suggests that a possible simple upper bound (but not, of course, the lowest one, since the conical deflection already considered gives that) mechanism for the infinite slab would be that of the square clamped plate of Figure 11-4(d) or (e). If, in fact, the load on the infinite plate was due to a square or rectangular column, the latter mechanisms would be more appropriate. In this case, we would get for a square of side $2c$,

$$P_u = \frac{8m(1+i)}{1 - \dfrac{c}{a}} \qquad (11.49)$$

$$a = \sqrt{\frac{P_u c}{4q}} \quad \text{or} \quad \frac{a}{c} = \sqrt{\frac{p_u}{q}} \qquad (11.50)$$

For the corner case, equating the work done and the dissipation, we have

$$P_u\left(1 - \frac{2}{3}\frac{c}{a}\right) = 2im + \frac{qa^2}{3} \tag{11.51}$$

or

$$im = \frac{P_u}{2}\left(1 - \frac{2}{3}\frac{c}{a}\right) - \frac{qa^2}{6} \tag{11.52}$$

After going through the process of determining a for the maximum m, we find that it is once again given by equation (11.42) and that, from it, the value of q can be obtained and substituted in equation (11.51) to give

$$P_u = \frac{2im}{\left(1 - \frac{c}{a}\right)} \quad \text{or} \quad im = \frac{P_u}{2}\left(1 - \frac{c}{a}\right) \tag{11.53}$$

with c/a again found from equation (11.43).

It can be seen in Figure 11-5 that the radii of the circular yield mechanism are positive hinges. Consequently, if in the present corner problem only top reinforcement were supplied in the slab to guard against the mechanism of failure we have examined here, and little or no bottom reinforcement, since no positive yield moments appear to be involved, it would be possible for a failure mechanism consisting of a fan and two rigid plates to develop, with positive hinges occurring because of the low positive (m) value of yield moment. In that case, the value of the constant appearing in equation (11.53) would not be 2 but $\pi/2$, and the quarter-circle would be a lower upper bound mechanism. Thus, as is pointed out in books on reinforced concrete slab design, steel must be supplied in the bottom of the slab to cause the assumed mechanism to develop.

The next case of interest is that in which the loaded area occurs on the edge of the slab supported by soil. The geometry of the problem is shown in Figure 11-7, which also illustrates the collapse mechanism that will be employed. Once again, it is not a mechanism that gives the *lowest* upper bound (the latter would incorporate a portion of a circular arc or fan in this case), but it does give an upper bound good enough for our purposes here, and which differs from a correct upper bound by only a few percent. We have an additional variable in this problem, in that we do not know *a priori* the value of the angle β, and we must determine it during the course of the solution. Alternatively, for a preliminary design estimate, the value of β could be taken as, say, 45°, and a reasonable estimate of an upper bound load could be obtained.

It is convenient in this problem to consider the loaded area to have the same shape as the postulated mechanism. This is not, of course, correct,

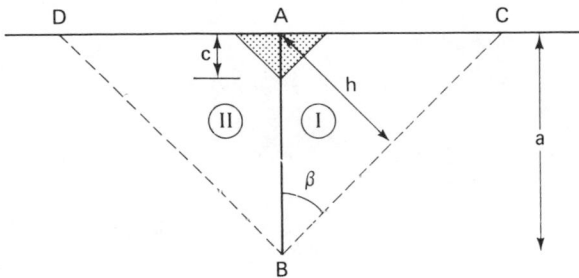

Figure 11-7. Load at edge of semi-infinite flat slab on soil.

because the real loaded area will have the definite shape of a rectangle for a
column base or of a circle or ellipse in the case of a wheel loading. In addition, the assumption also causes difficulty in extreme cases of the distribution
of reinforcement. It can be seen from Figure 11-7 that the assumed mechanism has one hinge, AB, of positive yield moment m and two hinges, BC and
BD, of negative moment im. If no top reinforcing is supplied in the concrete
slab, failure will occur in a mode that utilizes the weak negative moment
hinges as much as possible. Thus, it is intuitively obvious that as i tends
toward zero, the angle β will move toward 90°. With a loaded area of similar
shape, this leads to a rather unrealistic distribution of load. The load becomes, in fact, a uniformly distributed load along the entire edge when β
becomes 90°. However, for a realistic case that we will consider, β will not
be much greater than 45° and the loaded area has a reasonable shape. The
failure mechanism and collapse load would not change substantially were
the load to be applied over a rectangular or circular area centered over the
same region as the triangular area considered here.

The dissipation at the hinges can be calculated to be

$$D_1 = \frac{2\,im\,\Delta}{\sin\beta\cos\beta} + 2m\,\Delta\cot\beta \tag{11.54}$$

by the hodograph method to be described later. Work D_2 must be done
against the soil pressure q; this can be calculated from the product of the
total force on each triangular plate times the vertical distance the center of
gravity of the plate moves:

$$D_2 = \frac{qa^2\,\Delta}{3}\tan\beta \tag{11.55}$$

Finally, the collapse is generated by the load, which does work by moving
the virtual distance Δ:

$$E = p_u c^2 \left(1 - \frac{2}{3}\frac{c}{a}\right)\Delta\tan\beta$$

or

$$E = P_u\left(1 - \frac{2}{3}\frac{c}{a}\right)\Delta \qquad (11.56)$$

Equating the E and the D terms, we have

$$P_u\left(1 - \frac{2}{3}\frac{c}{a}\right) = \frac{2im}{\sin\beta\cos\beta} + 2m\cot\beta + \frac{qa^2}{3}\tan\beta \qquad (11.57)$$

or

$$m = \frac{\sin 2\beta}{4(i + \cos^2\beta)}\left[P_u\left(1 - \frac{2}{3}\frac{c}{a}\right) - \frac{qa^2}{3}\tan\beta\right] \qquad (11.58)$$

It will be seen that the condition which we have imposed of the loaded area being similar in shape to the collapse mechanism leads to an expression of the work done by the load identical in form to that obtained in the two previous cases of center and corner loading in which the shape of the loaded area was also similar to that of the deflected region.

First, as before, we differentiate equation (11.58) with respect to a and equate to zero to find the maximum value of m. This turns out to occur when

$$a^3 = \frac{P_u c}{q\tan\beta} \qquad (11.59)$$

or

$$\frac{a}{c} = \sqrt[3]{\frac{P_u}{q}} \qquad (11.60)$$

as in the previous examples. When q is obtained from equation (11.59) and substituted into equation (11.57), we get

$$P_u\left(1 - \frac{c}{a}\right) = \frac{4m(i + \cos^2\beta)}{\sin 2\beta} \qquad (11.61)$$

or

$$m = \frac{P_u\left(1 - \dfrac{c}{a}\right)\sin 2\beta}{4(i + \cos^2\beta)} \qquad (11.62)$$

However, the value of β which minimizes the load P_u or maximizes m still remains to be determined. It is obtained in the usual way when it is found that the minimum load occurs at

$$\cos 2\beta = -\frac{1}{2i + 1} \qquad (11.63)$$

The value of i can range from essentially zero, in the case of a slab possessing only bottom reinforcing, to some fairly large value if only top reinforcing is employed. In the first case, we see that β will tend toward the value 90°, and in the latter, toward 45° as a minimum. When equal top and bottom reinforcement is incorporated in the slab, $i = 1$, and β has the value 55°, approximately. In this case, which would be in the range usually encountered in practice, the assumed load distribution is quite satisfactory. Essentially, the same collapse mechanism and limit load would apply even if the loaded area were some other restricted shape.

The collapse mechanism assumed in this problem is again not that for which the lowest upper bound would be obtained but the limit load is only a few percent too high and may be considered a practical value. A lower upper bound would be found by employing a sector of a circular arc collapse pattern in place of the single positive line hinge AB in Figure 11-7.

It will be recognized that the technique described here of assuming yielding hinges to give collapse mechanisms and making a calculation of the work done and energy dissipated through hinge rotations is almost identical with the process of obtaining upper bounds in two-dimensional problems of plasticity in continua. The straight-line hinges correspond in plasticity configurations to the boundaries between sliding rigid blocks and the circular patterns correspond to the arcs of circular sliding surfaces. In yield-line theory, more complicated hinges incorporating logarithmic spirals have been employed. Such curves are also used in plasticity theory applied to soils possessing angles of friction.

From the exposition given here, it is possible to develop collapse mechanisms for multiple-load configurations. If an airfield pavement is stressed by the undercarriage of an aircraft in which each leg applies its load through two or four wheels in dual, tandem, or dual-tandem arrangements, collapse mechanisms which might suggest themselves are those shown in Figure 11-8(a) and (b) for interior loadings (9).

In Figure 11-8(a), for example, two semicircular areas of failure are shown around the loaded points, and these are connected by two negative hinge lines at the ends of the diameters. A positive hinge line joins the two load points. However, when an analysis is carried forward by the same technique as before to determine the radius a to the circular yield-line, it is found that the distance a is only slightly less than it is for a single-wheel loading, when the two load points are of the order of $2a_s$ apart, where a_s is the single-wheel value. Since we are representing the soil resistance as a uniform loading all over the deflected area, it follows that the soil resistance for the mechanism of Figure 11-8(a) [and for that of Figure 11-8(b)] is greater than it is for a circular failure area surrounding each load point as in the single-wheel case. The dissipation term for a separation of $2a_s$ is about the same in the two cases. This means that this mechanism gives a failure load for the two- (or four-) wheel configuration considerably greater than two (or four) times the single-wheel failure load. However, logically, the failure load must actually be less because of interaction. In consequence, Figure

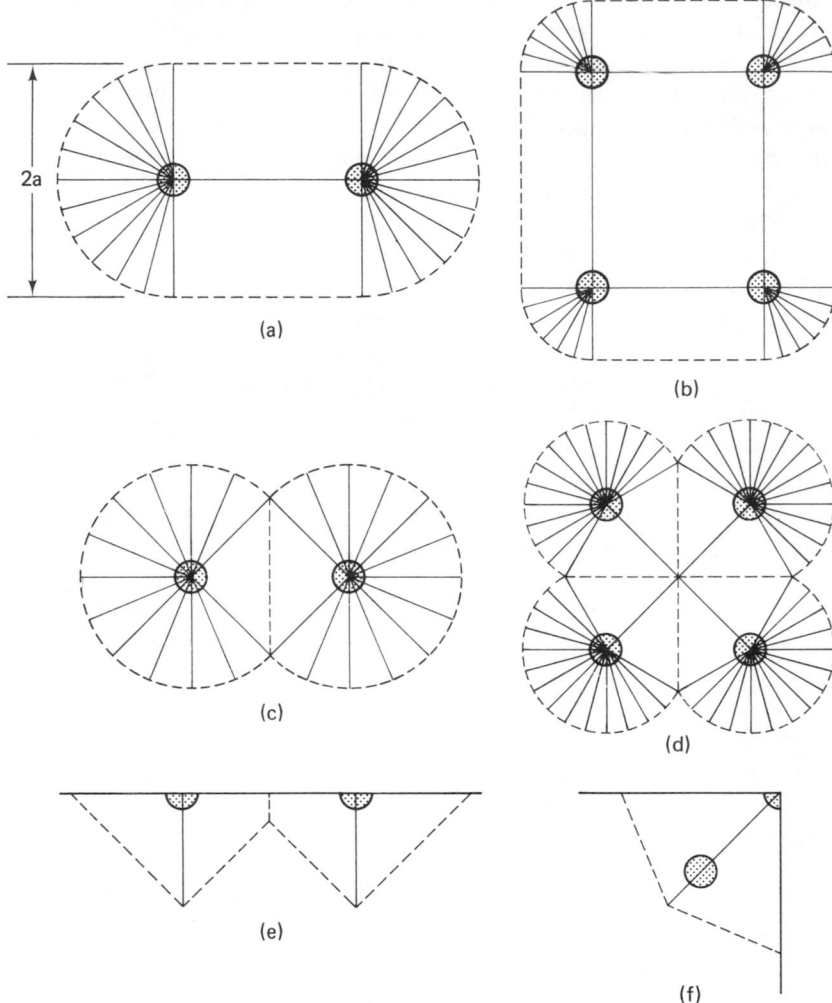

Figure 11-8. Combined collapse mechanisms for multiple loads on flat slabs on soil : (a) two loads closely spaced, interior (incorrect) ; (b) four loads closely spaced, interior (incorrect) ; (c) two loads closely spaced, interior (correct) ; (d) four loads closely spaced, interior (correct) ; (e) two loads, edge of semi-infinite slab ; (f) two loads, corner.

11-8(a) and (b) does not show the correct collapse mechanism. Instead, for the geometrical arrangement shown, and for uniform soil resistance at failure, the collapse mechanisms would be those of single-wheel loads. If the load centers are closer together than $2a_s$, the single-wheel collapse mechanisms will merge as shown in Figure 11-8(c) and (d). Possible collapse mechanisms for dual-wheel loads near an edge and a corner, respectively, are shown in Figure 11-8(e) and (f). For dual-tandem combinations, a check should be made of a separate combined failure mechanism for each pair of dual wheels.

In the discussion so far, the concrete has been assumed to be isotropic in its properties, which is frequently the case. Sometimes, however, it is convenient to include different amounts of reinforcement in the slab in two directions at right angles (orthotropic reinforcing) or at another angle. In such circumstances, the slab may be subjected to an affine geometrical trans-formation to convert it to an isotropic slab, which can then be analyzed by the use of a suitable collapse mechanism. These transformations are discussed in books on limit design (7 and 12), although the treatment of such a slab on a soil foundation does not seem to have been examined.

11.5 HODOGRAPH METHOD

In problems involving collapse in slabs in which the geometry of the hinges becomes complicated, it is a help in the calculation of the angular changes to draw an angular velocity (or angle change, since we have dropped the use of *rate* of dissipation, etc.) hodograph as shown in Figure 11-9. We do this by locating a pole O on the diagram from which rays are drawn to lengths proportional to the angular velocity, or angle changes of the different rigid plates. For any particular plate that has a hinge connecting it to the nonrotating body of the plate, the ray is drawn perpendicular to this hinge. The rays are numbered at their ends corresponding to the plate to which the rotation applies. Thus, for the edge-loading problem of Figure 11-7, we draw from the pole of Figure 11-9 a line OI at right angles to line BC with a length corresponding to the angular rotation of the hinge BC or plate I, Δ/h, since point A is taken to deflect the distance Δ. This establishes the point I. Similarly, the ray OII is drawn perpendicular to BD and, by symmetry, the same length. Now from this diagram the angle change across the hinge AB can be obtained as the distance $\overline{I\,II}$. This can be seen to be $2\Delta \cos \beta/h$, or since $h = a \sin \beta$, the angle change across AB is $2\Delta \cot \beta/a$. This can also be

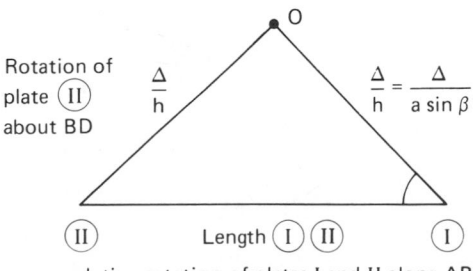

Figure 11-9. Angular velocity hodograph for edge load case (see Figure 11-7).

calculated, of course, from the angle change across CAD to be $2\Delta/AC$, which comes to the same value. The rotation hodograph demonstrated in Figure 11-9 for the case of an edge load on an infinite flat slab permits the relative rotations between component plates in complicated collapse mechanisms to be easily visualized. Figure 11-10 shows such hodographs for some of the mechanisms of Figure 11-4 and Figure 11-5. In each case, the lengths of the vectors $\overline{OI}$, $\overline{OII}$, etc., represent the rotation of each plate I, II, etc., with respect to the horizontal plane. The connecting lengths $\overline{I\,III}$, etc., indicate the relative rotation of each plate with respect to the adjacent one, such as that between plates I and III. Since each vector is perpendicular to the axis of rotation, the hodograph is easy to construct. For the trapezoidal collapse mechanism, the vertical deflection at either internal node can be chosen as the basic parameter. The circular hodograph of Figure 11-10(c) permits the relative rotations between adjacent sectors of the circular collapse mechanism to be discerned. The hodograph circle has radius Δ/a, so that the total (integrated) rotation between the sectors, where $\overline{I\,II}$ is the differential

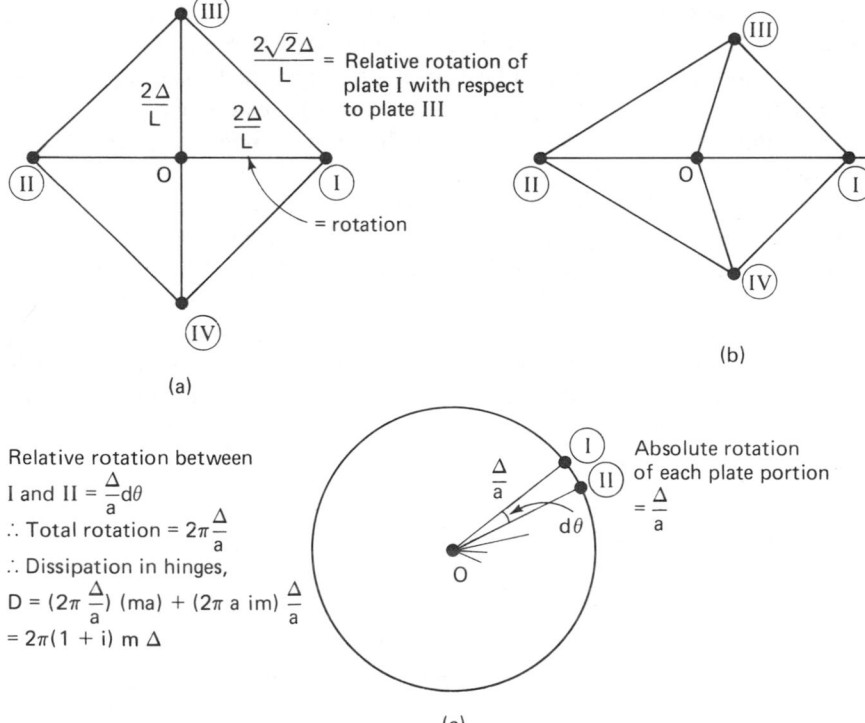

Figure 11-10. Rotation hodographs for cases shown in Figures 11-4 and 11-5: (a) square slab, Figure 11-4(a) and (d); (b) trapezoidal slab, Figure 11-4(c); (c) interior load in infinite slab, Figure 11-5.

rotation, is simply the circumference of the circle, $2\pi\Delta/a$. But the length of each positive hinge line is a (Figure 11-5), and the yield moment is m per unit of length, so that the total work dissipated by the positive yield moment is seen to be $2\pi\Delta/a \times a \times m = 2\pi\,\Delta m$. The rotation at the peripheral hinge is Δ/a, the distance around the hinge is $2\pi a$, the negative hinge moment is im, and therefore the work done by the negative hinge moments is $2\pi\,\Delta im$, for a total dissipation as given in equation (11.37).

11.6 KINEMATICS OF FAILURE

The collapse mechanisms described so far have been obtained basically by reference to the behavior of real slabs at failure in tests. We have not attempted to show whether or not the mechanisms are kinematically admissible in terms of the specific yield properties of the material involved. To take into account the kinematics, it is necessary to refer to the failure envelope of the material expressed as a function of generalized stresses, as discussed previously. The plastic strain increment vector, which must be couched in terms of generalized strains, must be normal to the failure envelope in appropriate regions of the generalized stress space (4).

For concrete slabs, the failure we are concerned with is in bending, which can occur in two orthogonal directions, with moments m_r and m_θ. Thus, the yield surface must be represented in the generalized m_r, m_θ stress space, as shown in Figure 11-11(a), since we are dealing in moments per unit length of slab. Generally, for an isotropically reinforced concrete slab with unequal reinforcement top and bottom (more at the bottom), the yield envelope is taken as the displaced square as shown in the figure. Here, the limiting positive and negative moments are m and im, respectively. If the same amount of reinforcement is present top and bottom, the center of the square will coincide with the origin of the coordinates, since $i = 1$. For other materials, the trace of the yield surface in m_r, m_θ space has a different shape. Were we concerned with a steel plate, for example, which might be employed for a temporary military airfield, the yield surface might be taken as the trace of the Tresca hexagon and would appear as shown in Figure 11-11(b). If the failure criterion were the one due to Von Mises, the trace on Figure 11-11 would be elliptical. The kinematic implication of the difference between the yield loci of Figure 11-11(a) and (b) is that the collapse mechanism would not be the same for two slabs, one formed from each material, and subjected to the same loading conditions. We will explore this briefly, as follows.

The generalized plastic strains that correspond to the moments of Figure 11-11 are, of course, the curvatures, or rotations of the slab in the two orthogonal directions. We refer to the curvature rates as κ_r and κ_θ, respectively, and they may be expressed in terms of the deflection w of the plate normal to its plane:

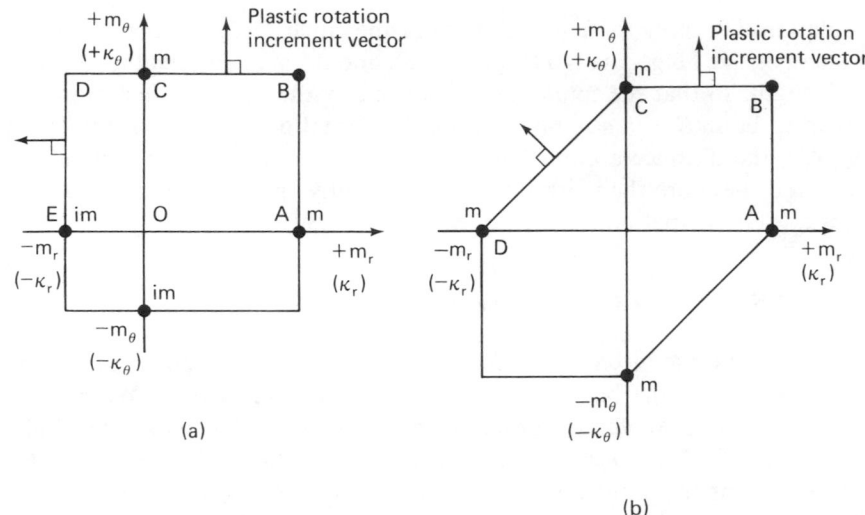

Figure 11-11. Generalized yield diagrams for slabs of different materials: (a) concrete, unequal reinforcement top and bottom; (b) steel plate.

$$\kappa_r = -\frac{d^2w}{dr^2}; \qquad \kappa_\theta = -\frac{1}{r}\frac{dw}{dr} \qquad (11.64)$$

As usual, the plane of plastic curvature space can be superimposed on the generalized stress space with coincidence of the principal axes of curvature and moment as indicated in Figure 11-11. The incremental plastic rate-of-curvature vectors are also indicated in the figure and direct, through equation (11-64), the kinematics of the collapse mode. If the moments acting on the plate are such that yielding occurs along section BC of the yield surface in either of Figure 11-11(a) or (b), then we can see that, to meet this condition, the rate of curvature κ_r must be zero, or

$$\frac{d^2w}{dr^2} = 0 \qquad (11.65)$$

Should the moments be such that yielding is described by section DE of the yield locus of Figure 11-11(a), the κ_θ is zero, or

$$\frac{1}{r}\frac{dw}{dr} = 0 \qquad (11.66)$$

Finally, yielding to meet the requirements of stresses combining to lie along section CD of Figure 11-11(b) implies that $+\kappa_\theta$ is equal to $-\kappa_r$ or

$$\frac{d^2w}{dr^2} = -\frac{1}{r}\frac{dw}{dr} \qquad (11.67)$$

We will examine the infinite plate on a yielding foundation loaded uniformly over a circular area from this point of view. Under the center of the loaded area, m_r and m_θ are both positive and equal to each other. At increasing radial distances m_r gets smaller, passes through zero at some radius, and then goes negative. We can conclude that, at the collapse load, $m_r = m_\theta =$ yield moment m at $r = 0$, and the stress condition must lie at point B in Figure 11-11. For $r > 0$, m_r decreases, so that to maintain the yield condition, m_θ must remain at its positive yield value, and the stress point moves along the path BC in Figure 11-11. At point C, m_r is zero. With increasing radius, m_r goes negative and to maintain the yield condition the stress point must move along path CD to D. At point D, m_r has reached its maximum negative value. At radii greater than this, both m_r and m_θ must decrease, so that the plate is not at the yield condition at these greater distances. We cannot, in this rigid-plastic analysis, say anything about the deflection in this region.

In both materials represented by Figure 11-11(a) and (b), equation (11.64) holds for the rate of curvature, and thus for the deflected shape of the plate, near the loaded area, where the stresses lie along the line BC. Equation (11.65) may be integrated to give

$$w = c_1 + c_2 r \tag{11.68}$$

in which c_1 and c_2 are constants that must be calculated to satisfy boundary and equilibrium conditions. From the previous intuitive analysis we can recognize that c_2 will be negative so that equation (11.68) represents a cone as the deformed shape of the plate. Beyond point C the difference in the properties of the two plate materials takes effect. For the concrete slab, the same kinematic requirement holds out to point D on Figure 11-11(a), and the deflected shape of the slab thus remains conical out to the circular hinge line of negative yield moment $(-im)$. However, for the other slab (possibly metallic), equation (11.67) holds for the relation between the curvature rates over the section of yield envelope CD. The solution of this equation is

$$w = c_3 + c_4 \log_e r \tag{11.69}$$

so that for this material, the conical shape near the loaded area changes into a logarithmic shape beyond the radius at which the radial moment m_r goes through zero. It does not seem likely that we would arrive at this conclusion intuitively for the case of the Tresca material. The logarithmic shape continues out to the radius at which the moments correspond to the conditions at point D of Figure 11-11(b), that is, where $m_r = -m$ and $m_\theta = 0$. The case of the Tresca material will not be pursued further here.

It is apparent that, by correctly taking into account the normality condition, we have confirmed the intuitively developed conical collapse mechanism in the case of the concrete slab.

Meyerhof (8) has adopted a different approach to the determination of the collapse load for a single loaded area. He uses the equations of equilibrium rather than the energy relation, and he takes the subgrade material to behave as a Winkler foundation. We present his results here, with some discussion. When the loaded radius c is very small, which for Meyerhof's analysis is a value of βc much less than 1 where β is defined in equation (5.99), he finds the collapse load to be

$$P_u = 2\pi m(1 + i) \tag{11.70}$$

This is the same expression as equation (11.45) with $c = 0$. For a larger loaded area ($\beta c > 0.2$), the collapse load is

$$P_u = \frac{4\pi m(1 + i)}{1 - \frac{4}{3}\frac{c}{a}} \tag{11.71}$$

In these equations, as before, the radius a is that of the circle where the maximum negative radial moment is reached. Meyerhof does not obtain a value for a at collapse by means of a minimization process; instead, he compares the plastic analysis with the elastic one. This requires the introduction of the characteristic length β into the solution. In the elastic solution the maximum negative radial moment occurs at a distance of about $\beta r = 1.9$,[2] and, as a load increases above the initial value required to cause yield immediately below it, it is likely that the radius at which the maximum negative radial bending moment occurs will decrease somewhat. Therefore, it would seem that a in equation (11-71) should be referred to a radius of about $\beta r = 1.9$. However, in developing the above equation for the collapse load, Meyerhof makes use of the equilibrium between the vertical load and the subgrade reaction. In so doing, he takes into account all the subgrade reaction out to the radius at which the plate deflection was zero. From the elastic solution, this radius is about $\beta r = 3.9$.[2] Meyerhof selects $\beta a = 3.9$ for substitution in equation (11.71) in order to refer the collapse load to the elastic characteristic β. The implication is that the maximum negative radial bending moment also occurs at $\beta a = 3.9$, which is not the case. The discussion of this point in the original paper (8) is not clear. However, the difference in collapse load in practice obtained by using either $\beta a = 2$ or 4 is not great.

On substituting $\beta a = 4$ into equation (11.71), we have

$$P_u = \frac{4\pi m(1 + i)}{1 - \frac{\beta c}{3}} \tag{11.72}$$

for the slab of infinite extent. In the calculation of β, Meyerhof points out that, although the flexural rigidity of *cracked* slabs of both plain and reinforced concrete is about one-half or less that of the uncracked section, the

---

[2]See Figure 5-9.

subgrade reaction coefficient is also smaller at large deflections because of the softening behavior of the soil in shear. Thus, the two effects tend to cancel and β to a first approximation can be assumed to be that of the uncracked slab resting on the soil whose reaction coefficient has a value appropriate to small deflections.

For load applied at the edge of a slab, Meyerhof's results are

$$P_u = \left(\frac{\pi}{2} + 2i\right)m \qquad (c \approx 0) \tag{11.73}$$

$$P_u = \frac{(\pi + 4i)m}{1 - \frac{2\beta c}{3}} \qquad (\beta c > 0.2) \tag{11.74}$$

and for a load at a corner, he gives

$$P_u = 2m \qquad (c \approx 0) \tag{11.75}$$

$$P_u = \frac{4im}{1 - \beta c} \qquad (\beta c > 0.2) \tag{11.76}$$

In using the equations for cases in which the loaded radius is very small, it is necessary to examine the other failure mechanism of punching shear, as suggested by Meyerhof, in the form

$$P_u = (2c + h)\pi h f_t \tag{11.77}$$

where f_t is the tensile strength of concrete and h is the slab thickness as employed previously.

For design purposes, a factor of safety in the range of from 2 to 3 should be applied to the collapse or punching load calculated from the above equations. Frequently, mechanical connections are made between highway and airfield pavement slabs to provide continuity across adjacent slabs and to inhibit the higher stresses developed by edge or corner loading conditions. Meyerhof (8) recommends that, depending on the degree of continuity provided by such a transfer device, the collapse value for a load applied across such a joint be calculated by linear interpolation (varying with the degree of load transfer capability) between the edge and interior collapse loads, or between the corner and edge collapse loads. The design of the transfer connection, for consistency, should also be based on the limit design approach.

11.7 FOOTINGS

The discussion to this point has concerned collapse loads on foundation slabs that were very large compared to the size of the loaded area. When the slab is of limited dimension and forms a footing supporting a single column, different considerations play a part. The first criterion applied to a footing design is soil failure, followed by allowable settlement. Since the latter controls the footing dimensions in most practical cases, it usually turns out that a footing has a substantial factor of safety (in the range from 4 to 6) against failure by soil yielding. If such a footing built on sand or clay is subjected to a steadily increasing central load, the footing will usually collapse before the soil failure load is reached. With adequate reinforcing around the column to prevent punching shear, slab collapse will take place by way of a mechanism similar to that of Figure 11-5. When the slab is relatively large (the usual case), failure will occur by the infinite slab mechanism, with negative moments around the perimeter; in a small slab the positive yield moments and cracking can extend to the slab edge where, of course, there is no moment restraint. In the latter case, the failure mechanism may have the cross-shaped appearance of Figure 11-4(a) or the hinge lines may have the shape of a plus, emerging at the centers of the edges.

With different soils, the pressure distribution under the footing at collapse changes; model experiments have shown (10) that clays exert a relatively uniform pressure, whereas sand pressure is a maximum under the center and less at the edges. The sand pressure distribution is parabolic to triangular. For square footings, centrally loaded, Meyerhof and Subba Rao show that model tests conform reasonably closely to collapse loads, P_u, indicated by the approximate expressions (10):

Uniform soil pressure (clay):

$$\frac{P_u}{m} = \frac{2.4\pi}{1 - 1.5\left(\frac{c}{b}\right)} \tag{11.78}$$

Triangular soil pressure (sand):

$$\frac{P_u}{m} = \frac{7.2\pi}{3 - 6\left(\frac{c}{b}\right)} \tag{11.79}$$

where m is the positive yield moment for the concrete slab, c is the column half-width, and b is the half-width of the footing.

When the footing is loaded eccentrically, the pressure distribution becomes different from these simple assumed variations, and the pressures at the edge, for a sufficiently large eccentricity, can reach the yield value for the soil. This then plays a part in affecting the collapse load.

The problem of the eccentrically loaded footing has also been examined by Meyerhof and Subba Rao (10), who present information by which the collapse load can be determined for specific footing dimensions and eccentricities.

It appears that more investigation is needed into the mechanism of collapse of simple footings and especially into combined footings.

PROBLEMS

1. A gridwork of reinforced concrete beams 4 ft (1.22 m) wide by 2 ft (0.61 m) deep is placed at ground surface at 20 ft (6.1 m) centers in both directions as shown in Figure P11-1. Columns apply loads of 250,000 lb (1.113 MN) to the beam intersections. If the soil is a uniform clay with a shearing strength of 1,000 psf (48 kN/m²), determine if the design is in trouble from the points of view of: (a) excessive deflection; (b) localized failure of beams by the development of hinges, on soil which yields; (c) overall bearing capacity. Take *EI* of beam to be 2.5 × 10¹¹ lb-in.² (718 MN-m²).

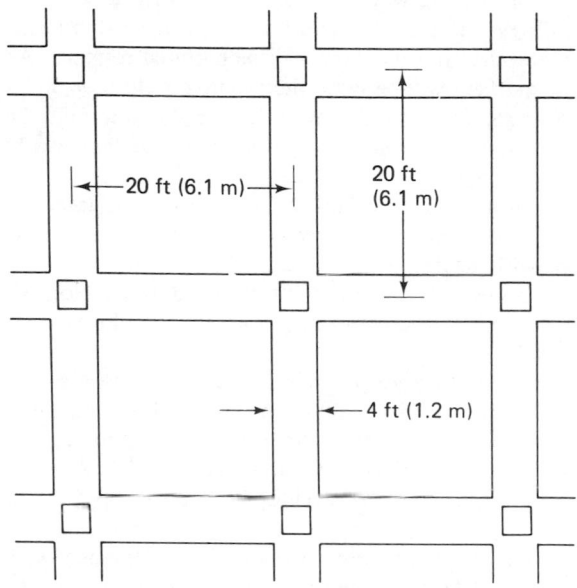

Figure P11-1.

2. In his original 1926 paper on pavements, Westergaard considered a concrete pavement 7 in. (178 mm) thick in the interior and 9 in. (229 mm) thick at the edges. From his equations he found that a 10,000 lb (45 kN) load acting over a circular area of radius 4 in. (102 mm) produced the following maximum tensile stresses:

$$\text{For interior load } \sigma_i = 319 \text{ psi (2.20 MN/m}^2)$$

$$\text{For edge load } \quad \sigma_e = 312 \text{ psi (2.15 MN/m}^2)$$

The subgrade coefficient, k, was assumed to be 50 pci (13.6 MN/m³), (E concrete = 3 × 10⁶ psi (2.07 × 10⁴ MN/m²), $\nu = 0.15$). Since these stresses both give a factor of safety of about 2 with respect to tensile cracking in concrete, Westergaard called the 7 and 9 in. thicknesses a "balanced design." (a) Analyze the problem by an upper bound method to obtain the *failure* load in each case; (b) Are the two factors of safety with respect to load still the same? Assume

467

that the subgrade is clay with strength appropriate to the k value used by Westergaard.

3. (a) What would be the effect on the pavement of Problem 2 if a load equal to $\frac{2}{3}$ of your calculated load at failure was repeatedly applied at the pavement edge? Field tests to determine k are discussed in Chapter 7. (b) What field test would you use to obtain the failure soil pressure in Problem 2? (c) In practice, would you use the same value of failure soil pressure for the internal load as for the edge load? Discuss.

4. A 2 ft (0.61 m) square concrete column of an industrial building carrying a load of 200 kips (890 kN) is to be supported at the center of a square concrete slab footing. The footing will rest on clay soil with a cohesion of 2,000 psf (96 kN/m²). Carry out a limit design for the concrete slab in order to arrive at the size of the slab and the moment it is required to resist. Assume that its positive yield moment will be twice the negative yield moment. How will you apply a factor of safety or load factor in this case? If you build the footing with an appropriate factor will you be concerned about settlements? Support your opinion with some approximate calculations.

5. Consider a long hollow tube steel pile 24 in. (0.61 m) in diameter, $\frac{3}{8}$ in. (9.5 mm) wall thickness, imbedded in a clay with a shear strength of 1,000 psf (48 kN/m²), uniform with depth. Calculate the ultimate lateral load that could be applied at ground level to cause a pile yield hinge to develop. What would be the approximate pile top displacement that you would estimate to occur at or near this load?

6. Design a reinforced concrete pavement to take the weight of a Boeing 747 fully-loaded aircraft (see Figure 6-5). The slab will rest on uniform clay with a shear strength of 1,000 psf (48 kN/m²). Design only for the collapse conditions in the slab interior and assume a load factor of 2.0. Assume also that the slab will have the same amount of steel top and bottom, and that the reinforcing is isotropic.

7. I notice that there are *re-entrant* corners on some of the new walkways around the campus (Figure P11-7). Suggest a collapse mechanism (or mechanisms) for this case and work out the collapse load if the loaded region is in the corner. Check your answer by comparison with the other cases we have looked at.

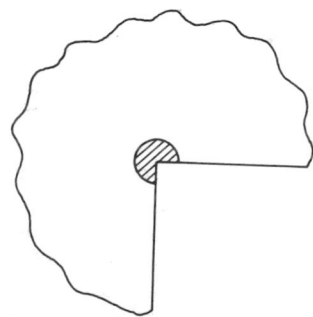

Figure P11-7.

8. Piles are frequently subjected to both lateral and vertical loads. Consider a square concrete pile embedded in the ground and loaded with a lateral load *T* at ground surface *and* a vertical load *P*. Construct and discuss possible yield surfaces in generalized *P-T* stress space for the pile/soil system. Describe the significance of the normality of the generalized plastic strain increment vector for the displacements of the top of the pile at failure under combined loadings.

REFERENCES

[1] CALLADINE, C. R. *Engineering Plasticity*. London: Pergamon Press, 1969.

[2] DRUCKER, D. C., PRAGER, W., AND H. J. GREENBERG. "Extended Limit Design Theorems for Continuous Media," *Quart. Appl. Math.*, *9*, 381–389 (1952).

[3] HEMP, W. S. *Optimum Structures*. Oxford, England: Clarendon Press, 1973.

[4] HOPKINS, H. G., AND W. PRAGER. "The Load Carrying Capacities of Circular Plates," *Journal of the Mechanics and Physics of Solids*, 2, 1–13 (1953).

[5] HORNE, M. R. *Plastic Theory of Structures*. Cambridge, Mass.: MIT Press, 1971.

[6] JOHANSEN, K. W. *Yield Line Theory*. London: Cement and Concrete Association, 1962.

[7] JONES, L. L., AND R. H. WOOD. *Yield Line Analysis of Slabs*. New York: Elsevier North-Holland, 1967.

[8] MEYERHOF, G. G. "Bearing Capacity of Floating Ice Sheets," *Proc. ASCE*, *86*, EM5, 113–145 (Oct. 1960).

[9] ——. "Load-Carrying Capacity of Concrete Pavements," *Proc. ASCE*, *88*, SM3, 89–116, (Jun, 1962).

[10] ——. AND K. S. SUBBA RAO. "Collapse Load of Reinforced Concrete Footings," *Proc. ASCE*, *100*, ST5, 1001–1018 (May 1974).

[11] NEAL, B. G. *The Plastic Methods of Structural Analysis*, 2nd ed. New York: John Wiley, 1963.

[12] REGAN, P. E., AND C. W. YU, *Limit State Design of Structural Concrete*. New York: John Wiley, 1973.

Appendices

Solid Mechanics $\quad$ *Appendix* A

A.1 INTRODUCTION

There are two results of the application of forces to a solid body: the development of *stresses* and *strains*, which give rise to *displacements* or deformations of the body. The majority of problems of interest to soil mechanics are *statically indeterminate*, that is, they cannot be resolved by consideration of the forces alone.

Three sets of conditions are required to be satisfied for the solution of a statically indeterminate problem. These are:

1. Equations of equilibrium.
2. Equations of compatibility of strains (the deformations must be consistent with each other and with applied geometrical constraints).
3. Equations of material behavior (*constitutive* equations).

A.1.1 Elementary example

A rigid, weightless beam (for example, a pier) is supported on one rigid (rock) and two spring supports (piles), as shown in Figure A-1. Free-body diagrams of the components are constructed in Figure A-2, where R_1, R_2,

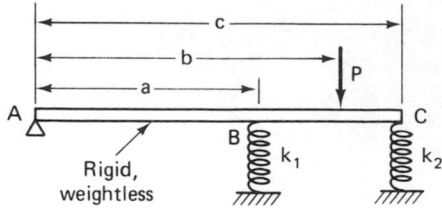

Figure A-1. Loaded beam resting on one simple support and two springs.

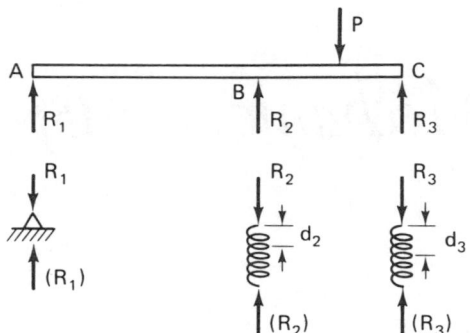

Figure A-2. Free-body diagrams of beam and supports.

and R_3 are reactions and d_2 and d_3 are deflections of the springs. Unknowns are $R_1, R_2, R_3, d_2,$ and d_3.

(1)   Equilibrium:

(a)   Vertical $P - R_1 - R_2 - R_3 = 0$ (A.1)

(b)   Moment about A $Pb - R_2a - R_3c = 0$ (A.2)

(2)   Compatibility (small rotation of beam):

$$\frac{d_2}{a} = \frac{d_3}{c}$$ (A.3)

(3)   Constitution:

$$R_2 = k_1 d_2$$ (A.4)

$$R_3 = k_2 d_3$$ (A.5)

Thus, we have five equations for five unknowns. Solve for $R_1, R_2,$ and R_3 and calculate d_2 and d_3 (and rotation of beam).

In this example, springs are considered *linear*, since equations (A.4) and

(A.5) involve only first powers, and *elastic*, since no restrictions have been placed on equations (A.4) and (A.5), which are equations describing the spring deflection-force property. This linear, elastic relation is shown in Figure A-3. It is seen that no limiting value of R (or d) is implied (no failure), that both positive and negative values of R and d are permissible, and that one value of d corresponds to one value of R. Therefore, it does not matter how a deflection d is arrived at, so the material is *stress-* and *strain-history independent*. The spring material in some cases might be better described by a *nonlinear*, but *elastic* relation, as in Figure A-4. Again, there is a one-to-one correspondence of a deflection to a force. This arises from the *elastic* property.

The material, again, might possess a limiting load R_u (yield) and be elastic and linear below that load, as in Figure A-5. This material is *elasto-plastic*; the behavior is *nonlinear*; in general, and for loads greater than R_u, a description must be given of the *unloading* path, say, as shown in Figure

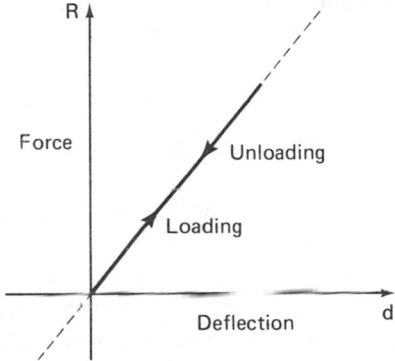

Figure A-3. Linear, reversible (elastic) relation between force and displacement of a spring or other deformable body.

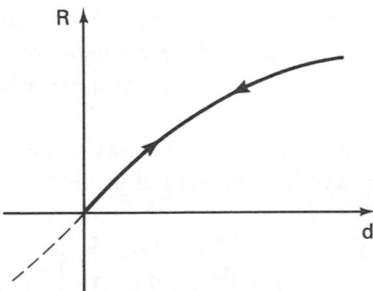

Figure A-4. Nonlinear, reversible (elastic) relation between force and displacement.

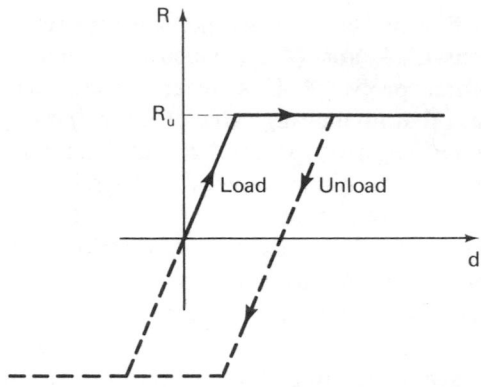

Figure A-5. Bilinear (elasto-plastic) irreversible relation between force and displacement.

A-5. In this case, the one-to-one correspondence of load and deflection is lost; the material becomes *load-history dependent*. More material behaviors will be given later. We will now return to treat stresses, strains, and displacements.

A.2 STRESSES IN CONTINUUM

The concept of stress as a limit of (force ÷ area) as the area shrinks to zero is assumed known. The possible existence of couple stresses (*Cosserat body*) is ignored. A *state of stress* at a point in a body is defined by its nine components, the stress tensor, with respect to a given coordinate system. The stress tensor is symmetric and of second order. Using a rectilinear Cartesian coordinate system, we see that the nine components consist of one normal and two shearing stresses acting on each of three mutually perpendicular faces of an elemental cube of material whose faces are aligned with the coordinate system. Strictly speaking, the stresses exist in the deformed body and should be referred to it. However, with restrictions on the strains to be considered, the deformed volume element may be taken to be indistinguishable from the original unstrained volume, whose coordinate system may then be employed.

The stresses in a right-handed x-, y-, z-axis system are shown in Figure A-6. The stress tensor may be represented in the following way:

$$\sigma_{ij} = \begin{bmatrix} \sigma_x & \tau_{xy} & \tau_{xz} \\ \tau_{yx} & \sigma_y & \tau_{yz} \\ \tau_{zx} & \tau_{zy} & \sigma_z \end{bmatrix} \tag{A.6}$$

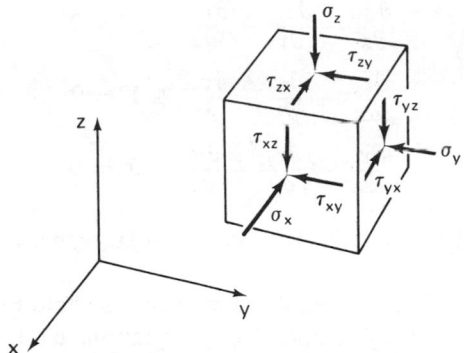

Figure A-6. Stresses acting on elemental body in right-handed Cartesian coordinate system.

and may be separated into parts as follows:

$$\sigma_{ij} = \begin{bmatrix} \bar{\sigma} & 0 & 0 \\ 0 & \bar{\sigma} & 0 \\ 0 & 0 & \bar{\sigma} \end{bmatrix} + \begin{bmatrix} \sigma'_x & \tau'_{xy} & \tau'_{xz} \\ \tau'_{yx} & \sigma'_y & \tau'_{yz} \\ \tau'_{zx} & \tau'_{zy} & \sigma'_z \end{bmatrix} \qquad (A.7)$$

in which $\bar{\sigma} = (\sigma_x + \sigma_y + \sigma_z)/3$ and σ'_x, etc., $= (\sigma_x - \bar{\sigma})$, etc. The first component of the right side of equation (A.7) is called the *hydrostatic component*, and the second component is called the *deviatoric component*. The stress $\bar{\sigma}$ is the hydrostatic component of the stress. It should be noted that the shearing stresses τ_{xy}, etc., are unaffected by the subtraction of the hydrostatic component, so that $\tau'_{xy} = \tau_{xy}$, etc. The reason for the form of equation (A.7) will appear in the section on constitutive equations.

A.2.1 Equilibrium

If the elemental volume in Figure A-6 is given dimensions dx, dy, and dz, the equilibrium of the element can be examined. Taking moment equilibrium first, it is found that

$$\tau_{xy} = \tau_{yx}; \qquad \tau_{xz} = \tau_{zx}; \qquad \tau_{yz} = \tau_{zy} \qquad (A.8)$$

so that there are, in fact, only six independent components of the state of stress.

Taking equilibrium in the x-, y-, and z-directions next and accounting for volumetric body forces X, Y, and Z in the appropriate directions give the *equilibrium* equations

$$\frac{\partial \sigma_x}{\partial x} + \frac{\partial \tau_{xy}}{\partial y} + \frac{\partial \tau_{xz}}{\partial z} + X = 0$$

$$\frac{\partial \tau_{yx}}{\partial x} + \frac{\partial \sigma_y}{\partial y} + \frac{\partial \tau_{yz}}{\partial z} + Y = 0 \qquad\qquad (A.9)$$

$$\frac{\partial \tau_{zx}}{\partial x} + \frac{\partial \tau_{zy}}{\partial y} + \frac{\partial \sigma_z}{\partial z} + Z = 0$$

The equations may be obtained in other (radial, spherical) coordinate systems.

At the surface of a body, equilibrium must also hold between the internal stresses and the boundary stresses. These equations have the form

$$\sigma_x \cos(x, n) + \tau_{xy} \cos(y, n) + \tau_{xz} \cos(z, n) = p_x \qquad (A.10)$$

where n is the direction of the normal to the boundary at the point and the x-component of the external boundary stress. A similar equation holds for p_y.

A.2.2 Transformation

If another set of orthogonal coordinate axes is selected (α, β, γ), the stresses $\sigma_\alpha, \tau_{\alpha\beta}$, etc., in this coordinate system may be obtained from the *transformation* equations:

$$\sigma_\alpha = \sigma_x \cos^2(x, \alpha) + 2\tau_{xy} \cos(x, \alpha) \cos(y, \alpha) + \sigma_y \cos^2(y, \alpha) + \cdots$$

$$\tau_{\alpha\beta} = \sigma_x \cos(x, \alpha) \cos(x, \beta) + \tau_{xy} [\cos(x, \alpha) \cos(y, \beta)$$

$$+ \cos(x, \beta) \cos(y, \alpha)] + \sigma_y \cos(y, \alpha) \cos(y, \beta) + \cdots \qquad (A.11)$$

etc. In two dimensions, these reduce to

$$\sigma_\alpha = \sigma_x \cos^2 \theta + \sigma_y \sin^2 \theta + 2\tau_{xy} \sin \theta \cos \theta$$

$$\sigma_\beta = \sigma_x \sin^2 \theta + \sigma_y \cos^2 \theta - 2\tau_{xy} \sin \theta \cos \theta \qquad (A.12)$$

$$\tau_{\alpha\beta} (= \tau_{\beta\alpha}) = (\sigma_y - \sigma_x) \sin \theta \cos \theta + \tau_{xy} (\cos^2 \theta - \sin^2 \theta)$$

when the α-axis makes angle θ with the x-axis. The relations between the stresses in equations (A.12) and (A.11) can be plotted in a σ-τ space diagram, and the stresses lie on the Mohr circle of stress. The stresses can be determined graphically from this diagram. The Mohr diagram can be drawn to represent three dimensions also.

A.2.3 Principal stresses and directions

For a given stress state at a point, it is possible to ask in which direction do the normal stresses have maximum and minimum values or in which planes through the point do the shearing stresses vanish. The two questions are interrelated. This can be approached as an extreme-value problem with constraints, since the direction cosines must satisfy certain conditions, but it is more easily approached in terms of the equilibrium of the elemental volume of Figure A-7.

In Figure A-7 it is anticipated that a face ABC of the element can be found on which the shearing stresses are zero, and σ (which turns out also

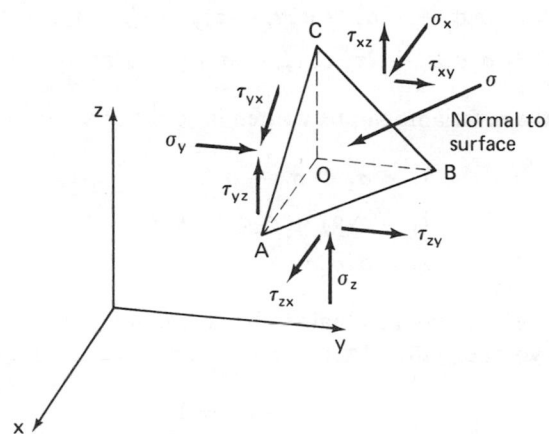

Figure A-7. Demonstration of existence of shear-free surfaces in a stressed element.

to be an extreme value of the normal stress) acts normal to the face. It is desired to find the direction cosines of this face (l, m, and n). Equations of equilibrium in the x-, y-, and z-directions are written to give

$$(\sigma_x - \sigma)l + \tau_{yx}m + \tau_{zx}n = 0$$
$$\tau_{xy}l + (\sigma_y - \sigma)m + \tau_{zy}n = 0 \qquad (A.13)$$
$$\tau_{xz}l + \tau_{yz}m + (\sigma_z - \sigma)n = 0$$

The necessary and sufficient condition that these equations in the unknowns l, m, and n have a solution is that the determinant of the coefficients vanish:

$$\begin{vmatrix} (\sigma_x - \sigma) & \tau_{yx} & \tau_{zx} \\ \tau_{xy} & (\sigma_y - \sigma) & \tau_{zy} \\ \tau_{xz} & \tau_{yz} & (\sigma_z - \sigma) \end{vmatrix} = 0 \qquad (A.14)$$

This can be expanded into an equation in σ:

$$-\sigma^3 + I_1\sigma^2 - I_2\sigma + I_3 = 0 \qquad (A.15)$$

Equation (A.15) is a cubic in σ, so that σ can be obtained in three values $\sigma_1, \sigma_2,$ and σ_3, the *principal stresses*. Because equation (A.15) contains real elements and is symmetric, the three roots are real. Since equation (A.15) does not depend on the original selection of a coordinate system, the coefficients $I_1, I_2,$ and I_3 are called *invariants*. From equation (A.14) they are given

$$I_1 = \sigma_x + \sigma_y + \sigma_z$$
$$I_2 = \sigma_x\sigma_y + \sigma_y\sigma_z + \sigma_x\sigma_z - \tau_{xy}^2 - \tau_{yz}^2 - \tau_{zx}^2 \tag{A.16}$$
$$I_3 = \sigma_x\sigma_y\sigma_z + 2\tau_{xy}\tau_{yz}\tau_{zx} - \sigma_x\tau_{yz}^2 - \sigma_y\tau_{xz}^2 - \sigma_z\tau_{xy}^2$$

In particular, the invariants can be written in terms of $\sigma_1, \sigma_2,$ and σ_3:

$$I_1 = \sigma_1 + \sigma_2 + \sigma_3;$$
$$I_2 = \sigma_1\sigma_2 + \sigma_2\sigma_3 + \sigma_3\sigma_1; \tag{A.17}$$
$$I_3 = \sigma_1\sigma_2\sigma_3$$

The direction cosines may be obtained by solving the set of equations composed of any two of equation (A.13) and the cosine constraint equation

$$l_1^2 + m_1^2 + n_1^2 = 1 \tag{A.18}$$

In equation (A.13) the value σ_1 is substituted for σ and $l_1, m_1,$ and n_1 for the direction cosines of the plane on which σ_1 acts. The other directions may be obtained similarly. It is found that the three principal stress directions are orthogonal.

It is seen that

$$I_1 = 3\bar{\sigma} \tag{A.19}$$

where $\bar{\sigma}$ is the hydrostatic component of stress. If the hydrostatic component is subtracted from the normal stresses, deviatoric invariants may be formed: I_1', I_2', I_3'

$$I_1' = (\sigma_x - \bar{\sigma}) + (\sigma_y - \bar{\sigma}) + (\sigma_z - \bar{\sigma}) = 0$$
$$I_2' = I_2 - \frac{1}{3}(I_1)^2 \tag{A.20}$$
$$I_3' = I_3 - \frac{1}{3}(I_1 I_2) + \frac{2}{27}(I_1)^3$$

The usual algebraic method of solving a cubic such as equation (A.15) is to substitute the deviatoric stress $(\sigma - \bar{\sigma})$ for the stress σ. The result is to form a new cubic equation

$$(\sigma')^3 - I_2'\sigma' - I_3' = 0 \tag{A.21}$$

which has the solution

$$\sigma_1 = \bar{\sigma} + \frac{2}{\sqrt{3}}\sqrt{I_2'}\,\sin\left(\alpha + \frac{2\pi}{3}\right)$$

$$\sigma_2 = \bar{\sigma} + \frac{2}{\sqrt{3}}\sqrt{I_2'}\,\sin\alpha \qquad\qquad (A.22)$$

$$\sigma_3 = \bar{\sigma} + \frac{2}{\sqrt{3}}\sqrt{I_2'}\,\sin\left(\alpha + \frac{4\pi}{3}\right)$$

where $-\pi/6 \le \alpha \le \pi/6$ and

$$\tan\alpha = \frac{1}{\sqrt{3}}\frac{(2\sigma_2 - \sigma_1 - \sigma_3)}{(\sigma_1 - \sigma_3)}$$

or $\qquad\qquad\qquad\qquad\qquad\qquad\qquad\qquad\qquad$ (A.23)

$$3\alpha = -\frac{3\sqrt{3}}{2}\frac{I_3'}{(I_2')^{3/2}}$$

The angle α is shown in Figure B-3(b). The group of principal stresses in equation (A.23) is called *Lode's parameter, μ,* where

$$\mu = -\frac{(2\sigma_2 - \sigma_3 - \sigma_1)}{(\sigma_1 - \sigma_3)} \qquad\qquad (A.24)$$

Cyclic permutation of the principal stresses in these equations results in the angle α being measured from reference axes 120° apart in the hydrostatic plane of principal stress space [see angle α, for example, in Figure B-3(b)].

A.2.4 Principal stress space

To demonstrate stress paths in stress tests, it is convenient to establish a principal stress space (compare with Mohr stress space) in which the coordinates are the three principal stresses, as in Figure A-8.

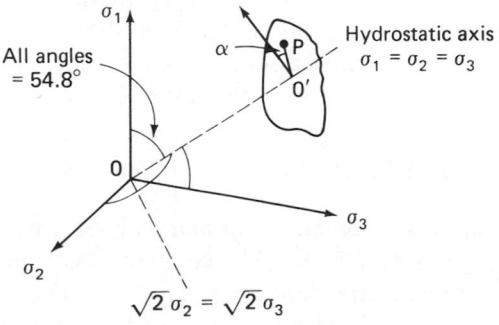

Figure A-8. Principal stress space.

A point in principal stress space represents a stress state at a point in (the physical space of) a solid. If the solid is *homogeneously* stressed, the point in stress space represents its entire stress state.

If the point in Figure A-8 lies on the hydrostatic axis, the stress condition is hydrostatic; off the hydrostatic axis, both hydrostatic and shearing stress states exist. The amount of the hydrostatic component is obtained by passing a plane through the point at right angles to the hydrostatic axis. The distance from the origin O, Figure A-8, to the intersection point O' is a measure of the *hydrostatic* stress, the distance $O'P$ to the stress point indicates the *shearing stress magnitude*, and an angle such as α on the figure gives the *distribution* of shearing stress.

The stress state could also be represented on an invariant space diagram (see Figure A-9). The use of the invariants will appear in the constitutive equations.

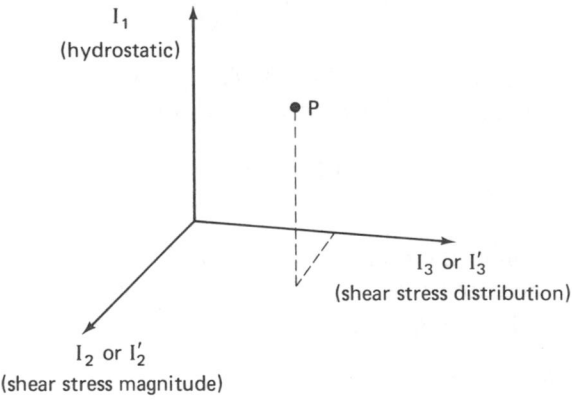

Figure A-9. Invariant space.

A similar development as for principal stresses can also be made with respect to the maximum values of the shearing stresses at a point. It is found that the maximum value is equal to one-half the difference between the largest and smallest principal stresses and acts on the plane whose normal bisects the angle between these principal planes.

A.3 STATE OF STRAIN IN A CONTINUUM

Figure A-10 shows a volume element of a body before and after straining. Points a, c, and p go to A, C, and P. The components of the displacement pP in the x-, y-, z-coordinate directions are u, v, and w. The relationship of the *change in length* of a line element such as pa ($PA–pa$) to its original length pa is called the *strain* of the element. Changes will also take place in the *angles*

Before deformation

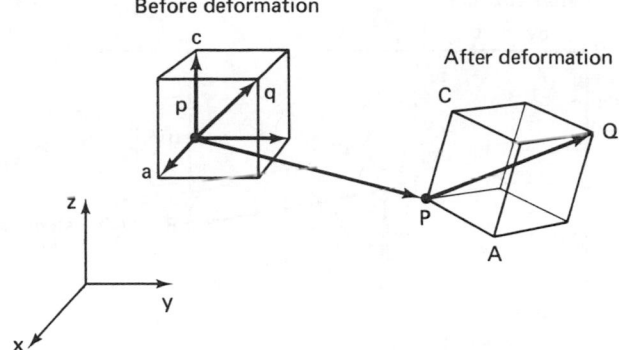

After deformation

Figure A-10. Displacement and straining of an element in three dimensions.

between the line elements as the elemental volume is strained. By considering the geometry of the deformation the following relations can be obtained between the linear strains and the angular changes and the displacements.

$$\epsilon_x = \frac{\partial u}{\partial x} + \frac{1}{2}\left[\left(\frac{\partial u}{\partial x}\right)^2 + \left(\frac{\partial v}{\partial x}\right)^2 + \left(\frac{\partial w}{\partial x}\right)^2\right]$$

$$\epsilon_y = \ldots$$

$$\epsilon_{xy} = \frac{1}{2}\left(\frac{\partial u}{\partial y} + \frac{\partial v}{\partial x} + \frac{\partial u}{\partial x}\cdot\frac{\partial u}{\partial y} + \frac{\partial v}{\partial x}\cdot\frac{\partial v}{\partial y} + \frac{\partial w}{\partial x}\cdot\frac{\partial w}{\partial y}\right)$$

$$\epsilon_{yz} = \text{etc.}$$

(A.25)

The ϵ_x, etc., are the relative elongations, or *normal strains*; the ϵ_{xy}, etc., are one-half the changes in the angles and are called the *shearing strains*.[1] Equation (A.25) is nonlinear; it is assumed that the strains are *small* in their derivation, although the rotations of elements are not so restricted. The equations may be linearized to the forms

$$\epsilon_x = \frac{\partial u}{\partial x}; \qquad \epsilon_y = \frac{\partial v}{\partial y}; \qquad \epsilon_z = \frac{\partial w}{\partial z}$$

$$\epsilon_{xy} = \frac{1}{2}\left(\frac{\partial u}{\partial y} + \frac{\partial v}{\partial x}\right); \qquad \epsilon_{yz} = \frac{1}{2}\left(\frac{\partial v}{\partial z} + \frac{\partial w}{\partial y}\right); \qquad \epsilon_{zx} = \frac{1}{2}\left(\frac{\partial w}{\partial x} + \frac{\partial u}{\partial z}\right)$$[1]

(A.26)

which can only be used if the *strains and rotations* are both *small*. Figure A-11 shows the displacements in two dimensions.

The rotations, which are *not* components of strain, are given by the equations

---

[1]Note: In engineering, the shearing strains are normally taken as

$$\gamma_{xy}, \gamma_{yz}, \gamma_{xz} = 2\epsilon_{xy}, \text{ etc.}$$

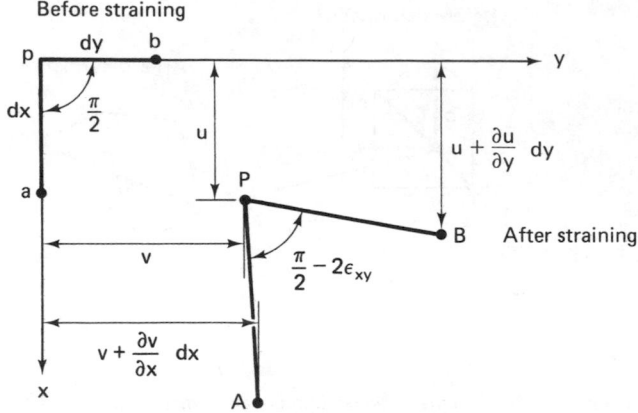

Figure A-11. Displacements, strains, and angle change in two dimensions.

$$\omega_x = \frac{1}{2}\left(\frac{\partial w}{\partial y} - \frac{\partial v}{\partial z}\right); \qquad \omega_y = \frac{1}{2}\left(\frac{\partial u}{\partial z} - \frac{\partial w}{\partial x}\right);$$

$$\omega_z = \frac{1}{2}\left(\frac{\partial v}{\partial x} - \frac{\partial u}{\partial y}\right) \tag{A.27}$$

Transformation equations for the strains can be written as for the stresses; principal strains, ϵ_1, ϵ_2, and ϵ_3, can be determined, as well as the orientation of their mutually othogonal planes, and strain invariants can be obtained analogous to the stress invariants in equations (A.16) and (A.20).

Since only three components of displacement serve to describe the movement of a point and there are six components of strain, it follows that there must be relationships among the strains. These are the *compatibility* equations, which may be defined physically by requiring that no cracks and no holes develop in the body upon straining.

$$\frac{\partial^2\epsilon_x}{\partial y^2} + \frac{\partial^2\epsilon_y}{\partial x^2} - 2\frac{\partial^2\epsilon_{xy}}{\partial x\,\partial y} = 0$$

etc. (two more)

$$\frac{\partial^2\epsilon_x}{\partial y\,\partial z} + \frac{\partial^2\epsilon_{yz}}{\partial x^2} - \frac{\partial^2\epsilon_{zx}}{\partial x\,\partial y} - \frac{\partial^2\epsilon_{xy}}{\partial x\,\partial z} = 0 \tag{A.28}$$

etc. (two more)

The compatibility equation (A.28) expresses the requirement that the deformations be consistent with the geometrical constraints, as was demonstrated by condition (2) in Section A.1.1.

A.3.1 Volume change

Since the orientation of the reference axes of a strained element of the material can be selected arbitrarily (the state of strain is invariant with respect to coordinate directions), the reference system will be aligned with the principal directions of strain. Choosing an element of original unit lengths of sides, forming a cube of unit volume, the new side lengths, after straining, become $(1 + \epsilon_1)$, $(1 + \epsilon_2)$, and $(1 + \epsilon_3)$. If the volumetric strain is defined as the ratio of the change in volume to the original volume and is given the symbol e, the equation follows:

$$e = \frac{(1 + \epsilon_1)(1 + \epsilon_2)(1 + \epsilon_3) - 1}{1} \qquad (A.29)$$

On expanding, equation (A.29) becomes

$$e = \overset{J_1}{(\epsilon_1 + \epsilon_2 + \epsilon_3)} + \overset{J_2}{(\epsilon_1\epsilon_2 + \epsilon_2\epsilon_3 + \epsilon_3\epsilon_1)} + \overset{J_3}{\epsilon_1\epsilon_2\epsilon_3} \qquad (A.30)$$

The first-, second-, and third-order groups of terms in this expression are the first, J_1, second, J_2, and third, J_3, normal strain invariants [see equation (A.17) for stress invariants in terms of principal values], respectively, so that

$$e = J_1 + J_2 + J_3 \qquad (A.31)$$

However, since the strains are considered to be *small*, J_2 and J_3 are small with respect to J_1 and, consequently,

$$e = J_1 \text{ approximately} \qquad (A.32)$$

A.4 SUMMARY

It is appropriate at this stage to summarize the situation and the limitations we have placed on the applicability of the equations.

Unknowns

We have nine components of stress; however, consideration of moment equilibrium showed that three of the shearing stress components were equal to the other three, so that the following are left:

6 components of stress
3 displacements
6 components of strain

{ There are, consequently, 15 quantities to be determined to describe the state of stress and deformation in a continuum.

These unknowns have to be determined from the equations we have established.

Equations

3 equilibrium, equation (A.9)
6 strain-displacement, equation (A.26)
[6 compatibility, equation (A.28)]

However, the 6 compatibility equations were obtained from the strain-displacement relations and are satisfied if the three displacements are continuous functions. Thus, in solving a problem *either* equation (A.26) *or* equation (A.28) may be counted, but not both. Therefore, we have available 9 equations with which to solve for 15 unknown quantities. Six additional equations are required; these are the *constitutive relations* to be discussed next. (Note that the 6 components of stress and the 3 of displacement can be taken as the unknowns; in this case, only the 3 equilibrum equtions so far serve to determine them.)

Some restrictions have already been placed on the equations devised. These are:

1. The time rate of change of quantities is small enough that inertia effects can be ignored. Consequently, problems in which relatively slow time effects occur are not excluded from analysis.

2. The displacement components are small in comparison with the dimensions of the strained body.

3. The strains and rotations are small.

The unknowns and equations may, of course, be expressed in terms of other (cylindrical, spherical, etc.) coordinate systems.

A.5 CONSTITUTIVE (STRESS-STRAIN) EQUATIONS

An element of a body that is strained under the action of some system of forces will possess a total strain energy in its strained state; on dividing by the volume of the element a quantity termed the *strain energy density, U*, is obtained. If a small change is made in one of the components of strain of the element, there will occur a change in the strain energy density. The total change dU in the strain energy density is equal to the sum of the changes in it arising from the contributions of changes in the various components of strain. The total differential dU is therefore given by the relation

$$dU = \frac{\partial U}{\partial \epsilon_x} d\epsilon_x + \frac{dU}{\partial \epsilon_y} d\epsilon_y + \ldots + \frac{\partial U}{\partial \gamma_{xy}} d\gamma_{xy} + \ldots \qquad (A.33)$$

Now consider that the element in question, in the strained state, is subjected to the stress system $\sigma_x, \sigma_y \ldots \tau_{xy}$, etc., which is causing its deformations. If a small increment $d\epsilon_x$, etc., is added to each of the strain components, without change in the stress state (or with only a small change), there will be a change dU in the strain energy density of the element. It can be shown that the change dU is given by the equation

$$dU = \sigma_x \, d\epsilon_x + \sigma_y \, d\epsilon_y + \ldots + \tau_{xy} \, d\gamma_{xy} + \ldots \qquad (A.34)$$

Comparing equation (A.33) and (A.34), we see that

$$\sigma_x = \frac{\partial U}{\partial \epsilon_x}; \qquad \sigma_y = \frac{\partial U}{\partial \epsilon_y}, \text{ etc.}; \qquad \tau_{xy} = \frac{\partial U}{\partial \gamma_{xy}}, \text{ etc.} \qquad (A.35)$$

Thus, the stress components are related to the strain components by equation (A.35).

Now, in a strained state, the strain energy density of the element, U, depends on the state of strain. In general, the state of strain at a point can always be related (by transformation equations) to the three principal strains ϵ_1, ϵ_2, and ϵ_3 and the direction cosines of the principal strain direction with respect to the selected x-, y-, z-coordinate system. Consequently, in general,

$$U = f(\epsilon_1, \epsilon_2, \epsilon_3, \text{ direction cosines}) \qquad (A.36)$$

Equation (A.36) holds for materials that are *not isotropic*, since the strain energy density is seen to depend on the orientation of the axes selected. If the material is *isotropic*, there can be no directional dependence of U, so that

$$U = f(\epsilon_1, \epsilon_2, \epsilon_3) \qquad (A.37)$$

However, the principal strains are obtained as the roots of a cubic equation, analogous to equation (A.15), but with strain invariant coefficients J_1, J_2, and J_3. Since the principal strain values are determined by these coefficients, it is therefore convenient to replace equation (A.37) by the alternative form

$$U = f(J_1, J_2, J_3) \qquad (A.38)$$

where the strain invariants are shown in equation (A.30).

When the functional form of U in equation (A.38) has been selected, equation (A.35) gives the desired *constitutive relations* between stress and strain. If these relations are to be *linear*, it follows that only quadratic functions of the strain invariants can be employed in equation (A.38). In this case, U can be represented, for example,

$$U = AJ_1^2 + BJ_2 \qquad (A.39)$$

Then the stress-strain relations are obtained by the use of equation (A.35). A similar expression for strain energy (or complementary strain energy) can be written in terms of the stress invariants and the strains can then be obtained in terms of the partial derivative of this energy function with respect to the corresponding stresses.

An alternative approach is to consider the strains as linear functions of all the stress components, in general, as follows:

$$\epsilon_x = s_{11}\sigma_x + s_{12}\sigma_y + \ldots s_{14}\tau_{xy} + \ldots$$
$$\epsilon_y = s_{21}\sigma_x + s_{22}\sigma_y + \ldots s_{24}\tau_{xy} + \ldots$$
$$\ldots \tag{A.40}$$
$$\gamma_{xy} = s_{41}\sigma_x + s_{42}\sigma_y + \ldots s_{44}\tau_{xy} + \ldots$$

In equation (A.40) all 36 material constants s_{mn} are not independent, for example, $s_{mn} = s_{nm}$ from symmetry, so that the number of independent constants reduces to 21. By considering the structural symmetry of the material, further reductions can be made, until the number of independent constants reduces to two for an isotropic material. These two constants are the familiar engineering ones (Young's modulus E and Poisson's ratio v) from which other convenient constants can be assembled, as shown below. Writing equation (A.40) for an isotropic material, we get

$$\epsilon_x = \frac{1}{E}\sigma_x - \frac{v}{E}\sigma_y - \frac{v}{E}\sigma_z$$

$$\epsilon_y = \frac{1}{E}\sigma_y - \frac{v}{E}\sigma_x - \frac{v}{E}\sigma_z \tag{A.41}$$

$$\epsilon_z = \frac{1}{E}\sigma_z - \frac{v}{E}\sigma_x - \frac{v}{E}\sigma_y$$

$$\gamma_{xy} = \frac{2(1+v)}{E}\tau_{xy}; \qquad \gamma_{yz} = \frac{2(1+v)}{E}\tau_{yz}; \qquad \gamma_{zx} = \frac{2(1+v)}{E}\tau_{zx}$$

from which we can see that

$$s_{11} = \frac{1}{E}, \qquad s_{12} = \frac{-v}{E}, \qquad s_{13} = \frac{-v}{E}, \qquad s_{14} = s_{15} = s_{16} = 0$$

and that

$$s_{22} = s_{33} = s_{11}; \qquad s_{21} = s_{23} = s_{41} = s_{42} \text{ etc.} = s_{12} = \frac{-v}{E};$$

$$\tag{A.42}$$

$$s_{14} = s_{24} = s_{34} = s_{41} = s_{42} \text{ etc.} = 0; \qquad s_{44} = s_{55} = \frac{2(1+v)}{E}$$

This type of linearly elastic behavior was first observed by Hooke, and equation (A.41) is a generalized form of Hooke's law.

From equation (A.30) the volumetric strain at a point is given, neglecting second- and third-order terms:

$$e = \epsilon_1 + \epsilon_2 + \epsilon_3 = \epsilon_x + \epsilon_y + \epsilon_z \tag{A.43}$$

so that the first three strains in equation (A.41) can be used to give an expression for e:

$$e = \frac{3(1 - 2v)(\sigma_x + \sigma_y + \sigma_z)}{E} = \frac{\bar{\sigma}}{K} \tag{A.44}$$

where $\bar{\sigma}$ is the hydrostatic stress, equation (A.7), and

$$K = \frac{E}{3(1 - 2v)} \tag{A.45}$$

is the *bulk* modulus of elasticity.

Since the equations for shearing strains [the last three of equation (A.41)] are given in terms of a combination of E and v, it is convenient to rewrite them in the form

$$\gamma_{xy} = \frac{\tau_{xy}}{G}; \qquad \gamma_{yz} = \frac{\tau_{yz}}{G}; \qquad \gamma_{zx} = \frac{\tau_{zx}}{G} \tag{A.46}$$

where

$$G = \frac{E}{2(1 + v)} \tag{A.47}$$

is the *shearing* modulus of elasticity. For the isotropic elastic material, K and G can be considered two alternate material constants. Equation (A.41) can be written in terms of K and G.

The first three of equation (A.41) can be solved for the stresses in terms of the strains, and the shearing stress expressions are obvious:

$$\sigma_x = \frac{E}{(1 + v)(1 - 2v)}[(1 - v)\epsilon_x + v(\epsilon_y + \epsilon_z)], \text{ etc.}$$

$$\tau_{xy} = \frac{E}{2(1 + v)}\gamma_{xy} = G\gamma_{xy}, \text{ etc.} \tag{A.48}$$

When both the state of stress and the state of strain are separated into their hydrostatic and deviatoric components in the manner of equation (A.7), each of the two components can be directly related. Equation (A.44) gives the relationship between volumetric strain and hydrostatic stress, and the deviatoric expressions are related as follows:

$$\sigma'_{ij} = 2G\epsilon'_{ij} \tag{A.49}$$

Although there are six deviatoric stresses and six deviatoric strains, the fact that $\sigma'_x = \sigma_x - \bar{\sigma}$, etc., means that

$$\sigma'_x + \sigma'_y + \sigma'_z = 0 \tag{A.50}$$

and $\epsilon'_x = \epsilon_x - e$ means that

$$\epsilon'_x + \epsilon'_y + \epsilon'_z = 0 \tag{A.51}$$

so that there are only five independent components of the deviatoric states of stress and strain. Consequently, equation (A.49) represents five equations. The sixth equation of constitution required is equation (A.44).

Thus, the six necessary constitutive relations in the linear theory are equation (A.40), in general, or equation (A.41) [or equation (A.44) and equation (A.49)] for an isotropic material in particular.

A.6 STRAIN ENERGY; COMPLEMENTARY STRAIN ENERGY

In Figure A-12 is shown a diagram of stress or force versus strain or displacement for a material or mechanism of an arbitrary nonlinear but elastic quality. The strain energy or total energy stored in the element or body up to the point $[\sigma_1(P_1); \epsilon_1(\delta_1)]$ is obtained by integration to give the area U:

$$U = \int_0^{\epsilon_1} \sigma \, d\epsilon \quad \text{or} \quad = \int_0^{\delta_1} P \, d\delta \tag{A.52}$$

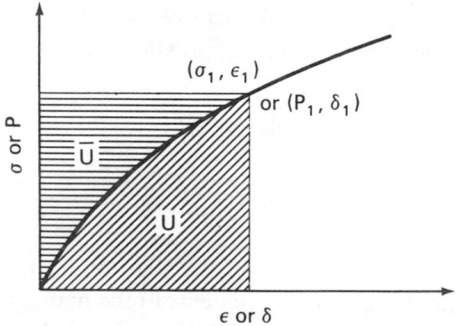

Figure A-12. Stress or force versus strain or displacement for nonlinearly elastic material; energy, complementary energy.

If the element has unit dimension, then U is the strain energy per unit volume or the strain energy density. In equation (A.52) the material property can be used to substitute strain, ϵ, for stress, σ, or displacement, δ, for force, P, so that the integration can be completed.

A related quantity is given by the area $\bar{U}$ between the curve of Figure A-12 and the $\sigma(P)$ axis; this is termed the *complementary* energy, where

$$\bar{U} = \int_0^{\sigma_1} \epsilon \, d\sigma \quad \text{or} \quad = \int_0^{P_1} \delta \, dP \qquad (A.53)$$

These integrals can be evaluated by again using the material property to substitute, in this case stress ϵ, for strain, σ, or force, P, for displacement, δ.

For a material with linearly elastic behavior, it is apparent that the strain energy and complementary energy are equal. However, the strain energy is formulated in terms of strains (displacements) and the complementary energy in terms of stresses (forces).

When an element of linearly elastic material has reached the state of strain $\epsilon_x, \epsilon_y \ldots \gamma_{xy}, \gamma_{xz}$, etc., as a result of gradual straining from an initially unstrained (and unstressed) state, integration of equation (A.34) gives the strain energy density in the final strained state as

$$U = \frac{1}{2}(\sigma_x \epsilon_x + \sigma_y \epsilon_y + \ldots \tau_{xy} \gamma_{xy} + \ldots) \qquad (A.54)$$

In the case of an isotropic material, substitution for the stresses $\sigma_x \ldots, \tau_{xy} \ldots$ from equation (A.48) gives the strain energy density in terms of strains only.

$$U = G\left[\epsilon_x^2 + \epsilon_y^2 + \epsilon_z^2 + \frac{v}{1 - 2v}e^2 + \frac{1}{2}(\gamma_{xy}^2 + \gamma_{yz}^2 + \gamma_{zx}^2)\right] \qquad (A.55)$$

The derivatives of this quantity with respect to each of the strains in turn give the successive stresses as indicated by equation (A.35).

When the expression for the complementary energy density is evaluated and expressed in terms of stresses only for an isotropic solid, the equation for $\bar{U}$ is

$$\bar{U} = \frac{1}{4G}\left[\sigma_x^2 + \sigma_y^2 + \sigma_z^2 - \frac{9v}{1 + v}\bar{\sigma}^2 + 2(\tau_{xy}^2 + \tau_{yz}^2 + \tau_{zx}^2)\right] \qquad (A.56)$$

Derivatives of $\bar{U}$ with respect to the stresses σ_x, σ_y, etc., in turn give the equivalent expressions for the strains, equation (A.41). Since the material is linear, in this case $U = \bar{U}$. Alternate formulations of equations (A.55) and (A.56) can be made by separating out deviatoric and volumetric components; this can lead to expressions of strain energy and complementary strain energy density in terms of the strain and stress invariants. For example,

$$\bar{U} = \frac{I_2'}{2G} + \frac{I_1^2}{18K} \tag{A.57}$$

If a body subjected both to body forces and external boundary forces is allowed to deform slightly under the forces to such a small extent that neither the external forces nor the stresses inside the body are considered to change during the deformation, the change in the strain energy of the body is given by the integral of equation (A.34) throughout the volume

$$\delta U_v = \int dU \, dV \tag{A.58}$$

It can be shown that this increment of total strain energy is equal to the work done δW by the external forces and the body forces during the increment of deformation. This is known as the *principle of virtual work* and is written

$$\delta W = \delta U_v \quad \text{or} \quad \delta U_v - \delta W = 0 \tag{A.59}$$

We now consider equation (A.59) as an expression of variation in a quantity π, which is the total potential energy of the body:

$$\pi = U_v - W \tag{A.60}$$

where U_v is the total strain energy stored in the body and W is the work done by surface and body forces in moving through the displacements to the final strained state of the body. The principle of minimum potential energy states that of all possible displacement states of a loaded body, that state of displacement which minimizes the potential energy is the correct one.

It is also possible to make use of the complementary strain energy, $\bar{U}_v$, in elastic materials in terms of a principle of minimum complementary energy. The principle is stated similarly to that of minimum potential energy but with stresses or forces exchanged for strains or displacements. It states that of all the systems of forces in equilibrium, those that minimize the complementary energy are also geometrically compatible. The complementary energy is written

$$\bar{\pi} = \bar{U}_v - \bar{W} \tag{A.61}$$

where $\bar{W}$ represents the complementary work done by a prescribed displacement. Both potential and complementary energies, π and $\bar{\pi}$, can be used in a calculation to place bounds on the stiffness of a system.

Both of these principles also apply to nonlinear elastic systems when the energies are defined as shown in Figure A-12. In this case, strain and complementary energies are not equal to each other. Similar considerations apply to electrical and hydraulic systems. They can also be developed for problems of thermal, chemical, and hydraulic diffusion.

Plasticity

Appendix B

B.1 PHENOMENOLOGY

The behavior of a cylinder of material in a compression test is shown in Figure B-1. When the test sample is unloaded from a point A on the loading curve, it exhibits some elastic, reversible behavior and some permanent deformation when it is completely unloaded to point B. A similar but not necessarily identical behavior is recorded in an extension test. On reloading from points B or B', a stress-strain path almost along the unloading curve BA ($B'A'$) is followed, so that the unloading-reloading process is essentially reversible. The original curve is encountered again near A, and is followed to C, if the load continues to increase. If, at C, unloading takes place again and is pursued into the extension range of stresses along path CDE, it will generally be found that the portion DE does not duplicate the original, extension path OA' which was followed by the material without a previous history of stressing. In particular, the stress levels along route DE are distinctly lower at corresponding strains than along either of the first paths OA or OA'. The weaker behavior exhibited by the material is called the *Bauschinger effect*. It is usually attributed to damage caused in the material (slips along crystal planes in a solid; slips between grains in a soil) by the initial compressive stressing.

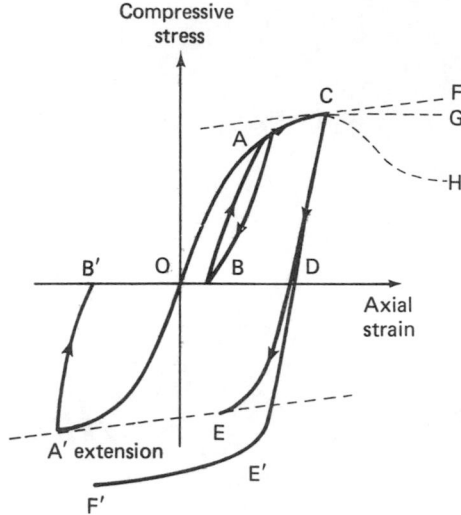

Figure B-1. Material behavior in axially symmetric tests.

In Figure B-1 it will be observed that, on loading, a continual increase in stress is required to produce increased strains in this material; this behavior is stable (that is, it does not lead to sudden collapse of the material at some stage in loading) and is referred to as *strain-hardening*.

Prager devised a simple mechanical model to demonstrate these effects; the model is a kinematic model, and the increased strength on straining and Bauschinger effect are therefore sometimes referred to as *kinematic hardening*.

The fact that the initial compression and extension test curves in Figure B-1 are, in some materials, almost identical is considered evidence of isotropic behavior. It follows that if this response were consistently followed on subsequent compression load/unload/extension load paths such as $OACDE'$, we would find that a path $E'F'$ would be defined in extension the same distance from the strain axis as the CF curve on the compression side. This behavior is called *isotropic hardening*. An assumed behavior of this kind is extensively used in plasticity analyses because of its mathematical convenience, although it is never, as far as I know, encountered in real materials. The Bauschinger effect turns up all the time, but it is difficult to take into account in analysis.

In many materials, and particularly in soils, some irreversible, plastic components of strain develop at essentially all levels of shearing stress. That is, unloading from any point along the curve OA of Figure B-1 results in some irrecoverable strain. This is also very inconvenient, and it is better to ignore it in preliminary theoretical developments. The generally curved nature of the stress-strain response in loading and unloading in Figure B-1 is also an

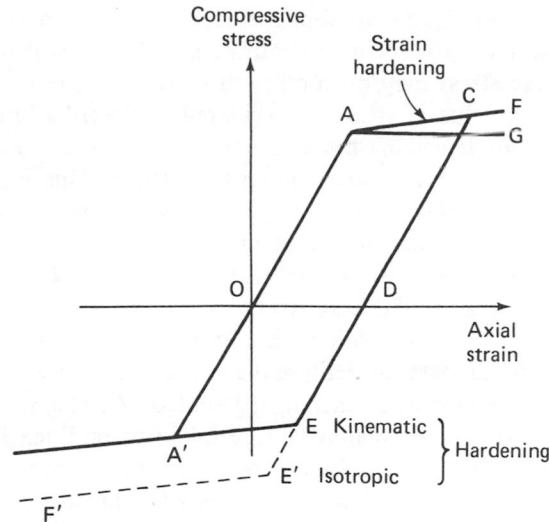

Figure B-2. Idealized material behavior.

obstacle that can be dispensed with in the next stage of our treatment. If these idealizations are made, a material behavior as shown in Figure B-2 is proposed. In this model it is considered that the initial behavior in the stress range $A'OA$ is linearly elastic. A load path to A or A' and unloading gives rise to no plastic strains, which are only incurred for stresses exceeding the levels at A and A' on first loading. The linearly elastic behavior on first loading is duplicated on subsequent unloading and reloading paths such as CDC in compression. Elastic behavior on unloading is encountered until the stresses reach their previous maximum level at C again, whereupon plastic straining is continued if the stresses are increased. In Figure B-2 the kinematically and isotropic hardening behavior of Figure B-1 is represented in linear form.

For stress paths lying between the bounding lines AF and EA' or $E'F'$ in Figure B-2, the material behavior is linearly elastic; the bounding lines represent a strain-hardening yield condition. The effect of strain-hardening can be considered to result from a yield surface being pushed or dragged to a higher stress level as the stresses increase. In its new position it is sometimes called a *loading* surface. In the case of the kinematic hardening model, the stress distance between the compression and extension yield boundaries can be thought of as constant, so that if a compressive stress raises the compressive boundary, the extensive boundary is pulled up with it an equal amount. Isotropic hardening can be associated with some mechanism, if one exists, whereby the material is strengthened for all paths to yield, by any path that raises the yield boundary.

In axially symmetrical tests on soils the axial stress-strain behavior is usually curved as in Figure B-1, but the curve reaches a maximum at some value of stress. Thereafter it may continue as a horizontal line CG (stable behavior), or the stress may drop off with increasing strain along CH (unstable). The peak value of stress is considered to describe failure in the soil and is employed in determining the failure envelope or surface of the soil as a function of, for example, the confining pressure. This envelope is also frequently called the *yield envelope*. However, since, as we have seen, irreversible or plastic deformations occur in the soil in the portion OA of the stress curve in Figure B-1, we will reserve the word *yield* for what happens along OA and *failure* for the peak stress conditions. In this way, successive yielding occurs on the way to failure, and in the region OA the soil is acting as a strain-hardening material. Failure then envelopes successive yield conditions. A choice of models is thereby presented. At the simpler level the behavior along OC approximately can be taken to be linearly elastic as in OA of Figure B-2, with a fixed failure condition (coinciding with yield here) running from A through G. This model describes an ideal linearly elastic, plastic behavior without strain-hardening. Alternatively, a model can be constructed in which AC of Figure B-2 is selected at a slope to match the unloading and reloading linear elastic behavior.

No general theories to describe the behavior of unstable elasto-plastic materials have been suggested.

B.2   GENERAL STRESS STATES

Although axial compression and extension tests serve to define some aspects of material behavior in the laboratory, they only give limited information applicable to the two- and three-dimensional situations encountered in practice. We need to know the deformational behavior in general, and the plastic and yield behavior in particular for this discussion, in the three-dimensional stress state. It will be initially assumed that the material is isotropic for the convenience of representing the stress states in principal stress space, as shown in Figure B-3. If the material is anisotropic, which must be the case in practice, the deformational behavior must be described in terms of a six-dimensional stress space with the six stresses of the stress tensor as coordinates. It is difficult to represent the latter space pictorially.

Using the stress space of Figure B-3, we will describe the various considerations that apply to yielding, failure, and plastic deformations. It should be borne in mind that all of the developments of plasticity theory came about through the study of metals, wherein the external plastic displacements arise from the movement of dislocations inside the crystals of a polycrystalline solid assemblage. Irreversible external deformations in sands are caused by the frictional sliding of grains at their points of contact. It is not clear what

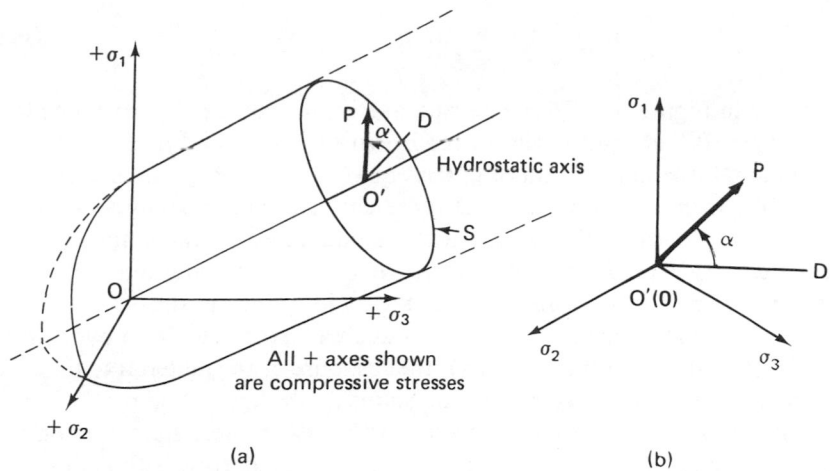

Figure B-3. Yield surface in stress space: (a) general perspective; (b) hydrostatic plane.

happens in clays, except that the forces and displacements at contact points between grains must be important. Whether or not the movement of one clay platelet at its contact with another is a frictional process is hard to say. Good correspondence between the predictions of a plasticity theory based on continuum metal physics and the performance of a sand under stress need not be expected. We will return to these points.

In metals subjected to high hydrostatic stresses in experiments it was found that (a) no irreversible volume changes occurred and (b) the shearing stress level required to cause plastic yielding was independent of the hydrostatic stress. The two results are presumably related and are a consequence of the lack of void space in the metal. Metals are composed of heterogeneous aggregates of crystals with apparently similar mechanical properties of stiffness, so that macroscopic external hydrostatic stresses do not cause microscopic internal shearing stresses. It is therefore concluded that in Figure B-3 the yield surface for a metal does not intersect the hydrostatic axis. (No permanent volume changes result from the application and removal of hydrostatic stress alone.) Since hydrostatic stress does not affect the yielding behavior, it may be subtracted from the total stress tensor, so that the yield condition can be described in terms of the five components of the deviatoric stress tensor $\sigma'_{ij} = \sigma_{ij} - \bar{\sigma}\,\delta_{ij}$ where $\bar{\sigma}$ is the mean or hydrostatic component of stress and δ_{ij} is the Kronecker delta. For isotropic behavior, this shear stress state can be represented by two of the principal deviatoric stresses $\sigma'_1, \sigma'_2,$ and σ'_3 (since their sum is zero), or more conveniently by the second and third invariants of the deviatoric stress tensor I'_2 and I'_3 (the first invariant is, of course, zero), or by I'_2 and the angle that incorporates Lode's stress parameter, μ, as given in Appendix A:

$$\alpha = \tan^{-1} \frac{-\mu}{\sqrt{3}} \quad \text{where} \quad \mu = -\frac{2\sigma_1 - \sigma_3 - \sigma_2}{\sigma_3 - \sigma_2} \tag{B.1}$$

as shown in Figure B-3. For a stress point P in an arbitrary hydrostatic plane S in Figure B-3 at right angles to the hydrostatic axis, and intersecting it in the point O', the radial distance $O'P$ is equal to $\sqrt{2I'_2}$. A value of α of $+90°$ (or $-30°$) from the direction $O'D$ represents an axially compressive stress state with σ_1 as the major principal stress and $\sigma_2 = \sigma_3$ the minor principal stress; $\alpha = +30°$ indicates an axially extensive stress condition with σ_2 as the minor principal stress and $\sigma_1 = \sigma_3$ the major principal stress.

Yield surfaces for various materials such as metals, with no hydrostatic effect, are cylinders in Figure B-3(a); the response of the material to different combined stress states determines the shape of the cross section. Since it is impossible to test a material in a hydrostatic tension stress state, the form of yield surface in the tension octant is a matter of conjecture although it presumably cuts the hydrostatic axis.

The stress conditions at yield in a metal can therefore be represented by an equation in the form

$$f(I'_2, I'_3) \quad \text{or} \quad f(I'_2, \alpha) = K \tag{B.2}$$

This equation describes a cylindrical surface in the principal stress space of Figure B-3. If the material yields when the maximum shear stress reaches a limiting value (Tresca's idea), the cross-sectional shape of the cylinder is a regular hexagon. When the strain energy of a linearly elastic solid is calculated, it can be divided into hydrostatic and shearing components; the shearing strain energy is a function of I'_2. Thus, if, as another yield hypothesis, yielding depends only on the shear strain energy in the solid reaching some maximum value, yield will occur at a fixed value of I'_2 which, we have seen, is the radial distance from the hydrostatic axis in Figure B-3. The cross section of this yield surface in Figure B-3 is therefore a circle. This was first suggested by von Mises on grounds of mathematical convenience, rather than for physical reasons. It actually turned out from experiments on metals that they generally yield according to the maximum shear strain energy or von Mises criterion, although real and consistent deviations from this criterion apparently occur.[1] Thus, from both the viewpoints of mathematical convenience and real material representation, the appropriate stress conditions at yield in an isotropic metal are described by the yield condition

---

[1] The metal tests are performed on tubes subjected to tensional and torsional combined stress states. It is difficult to manufacture such tubes without inducing anisotropy in the test material (tubes produced by drawing processes, for example, will be strongly anisotropic) so that some deviation from uniform behavior will always be present because of this anisotropy. This obscures the real behavior in yield.

$$\sqrt{2I_2'} = \text{constant} \tag{B.3}$$

which can be written more conveniently in the form

$$I_2' = k^2 \tag{B.4}$$

where k is a constant. In its most general form, when the ordering of the principal stresses is unknown, the Tresca criterion must be written as

$$4I_2'^3 - 27I_3'^2 - 36k^2I_2'^2 + 96k^4I_2' - 64k^6 = 0 \tag{B.5}$$

If the maximum (σ_1) and minimum (σ_3) principal stresses are known, then the Tresca criteria reduces to the simpler form

$$\sigma_1 - \sigma_3 = 2k \tag{B.6}$$

In a previously unstressed material which hardens on being strained, yield conditions such as given by equations (B-4), (B-5), or (B-6) indicate the stress conditions at which yield first develops and describe the yield surface in Figure B-3. If the stresses are increased beyond these values, then the yield surface is pushed out to a new position determined by the stress conditions developed before unloading is commenced. What happens to the yield surface depends practically on the material's response to yielding, or from a theoretical viewpoint on the type of hardening behavior assigned to it. If hardening is taken to be isotropic, the yield surface is considered to expand uniformly, as shown in Figure B-4 in which, for convenience, the von Mises yield criterion is represented. Kinematic hardening appears as a translation of the yield surface (Figure B-5), whereas, if hardening in some

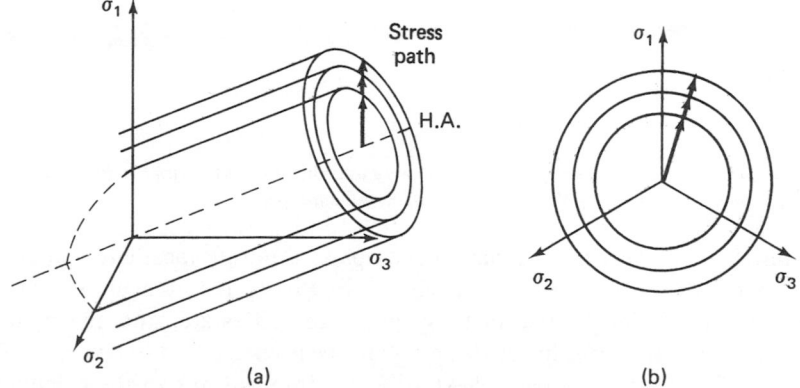

Figure B-4. Successive development of yield surfaces for isotropically hardening material (von Mises criterion shown): (a) general perspective; (b) hydrostatic plane.

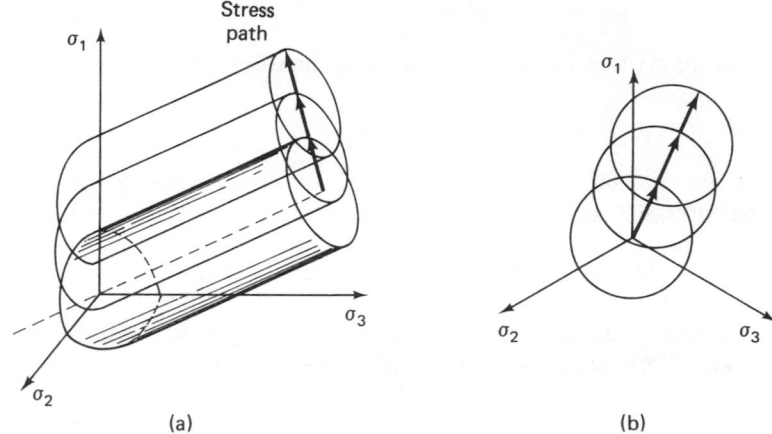

Figure B-5. Successive development of yield surfaces for kinematically hardening material (von Mises criterion): (a) general perspective; (b) hydrostatic plane.

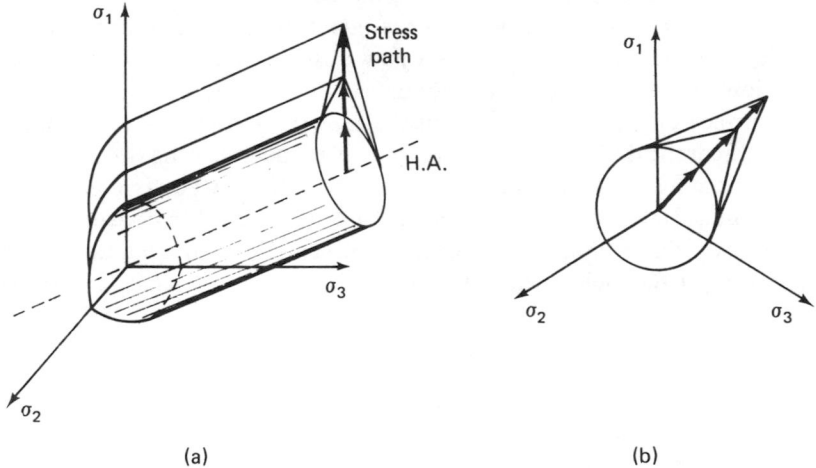

Figure B-6. Development of a corner or edge in yield surface for a hardening material: (a) general perspective; (b) hydrostatic plane.

materials (soils may be one class) develops as a purely local effect, a corner is formed on the yield surface as shown in Figure B-6. On unloading, the yield surface is left in its, so to speak, deformed state, and the material response to stress paths inside this new surface is considered to be elastic. The different hardening behaviors have different implications in the calculation of plastic deformations.

So far we have mentioned "loading" and "unloading" stress paths invoking an intuitive understanding of what was meant. Now we define them precisely. In general, the stresses in an element of material are represented by the stress tensor, σ_{ij}. When some function $f(\sigma_{ij})$ of the stresses reaches a critical value, K, as in equation (B-2), the material element is considered to yield. In particular, with the von Mises criterion, $f(\sigma_{ij})$ is taken equal to the second deviatoric invariant, I'_2. If the stresses in the material element are such that it is not yielding, then $f(\sigma_{ij}) < K$. At yield, whether the material is strain-hardening or not, $f(\sigma_{ij}) = K$. For a strain-hardening substance, K is not a constant, but it will depend on the previous history of stress and strain in the element and on the particular type of strain-hardening it exhibits or is assumed to follow. It should be noted that sometimes the yield criterion is taken in the form

$$f(\sigma_{ij}) - K = 0$$

or

$$F(\sigma_{ij}) = 0 \qquad\qquad (B.7)$$

and the stress conditions on a material element lie within or on the yield surface according to whether $F(\sigma_{ij})$ is less than or equal to zero, respectively. In either representation, if the element is at yield, so that $f(\sigma_{ij}) = K$ or $F(\sigma_{ij}) = 0$, a load increment df or dF may be applied to it. If the increment is positive, the yield criterion (initial or subsequent, depending on the value of K) will be exceeded, and this measure is considered to constitute *loading*. If the increment is negative, the stress point will retreat from the yield surface, so that the element is *unloaded*. A zero increment df or dF implies that the incremental stress path lies in the yield surface, and the loading is therefore *neutral*. However, the increment df or dF can be obtained from the product of the derivative of f or F with respect to the stress tensor times the incremental stress; thus,

$$df \quad \text{or} \quad dF = \frac{\partial f}{\partial \sigma_{ij}}\,d\sigma_{ij} \quad \text{or} \quad \frac{\partial F}{\partial \sigma_{ij}}\,d\sigma_{ij} \qquad\qquad (B.8)$$

so that we have, when $f(\sigma_{ij}) = K$ or $F(\sigma_{ij}) = 0$; loading, neutral loading, or unloading, if

$$\frac{\partial f}{\partial \sigma_{ij}}\,d\sigma_{ij} \quad \text{or} \quad \frac{\partial F}{\partial \sigma_{ij}}\,d\sigma_{ij} \lessgtr 0 \qquad\qquad (B.9)$$

respectively.

B.3 PLASTIC STRAINS

So far, yielding has been discussed only from the point of view of the stress states required to cause the material to develop some irreversible strains. The relationship of the strains generated to the stresses must be explored. It is not difficult to show that the relationship between plastic strains and the stresses is not clear-cut as in the case of elastic behavior. For example, consider a hollow cylinder of strain-hardening material. First, we stress it to a point somewhat beyond its initial yield surface in pure torsion. This will result in a plastic shear strain in the material, with no axial strains. In addition, a new yield surface is generated beyond the original surface. We unload the torque to some extent and apply an axial extensional load until the combined stress state lies just on the new yield surface. No new plastic deformations have been incurred along this load path, so that at the combined stress state the only plastic strain that exists is the one that was developed by the initial torsional loading.

Now we consider an identical fresh specimen of the same material. This time we load it in axial extension past the initial yield surface to a point at which we have developed in the material the same yield surface as in the first test. The result is an axial extension plastic strain. Again, we unload the axial force somewhat and now apply a torque in addition to the axial load, in such amount as to arrive at the same combined stress state as we obtained in the first test specimen, on the new yield surface. This stress path will cause no new plastic strains to develop in the material. Thus, we have two identical stress states but two widely differing conditions of plastic strain. The *elastic* components of strain are, of course, also identical. It follows that, in general, to determine the total strains or displacements in a final state of stress, the plastic strain increments along the stress path must be tracked and summed, unless a simple proportional stress or loading path is followed. This path is a straight line in stress space drawn through the origin. In order to establish some general relations between plastic strain increments and the stresses causing them, it is convenient to consider the work expended by an external agency on an element of work-hardening material in a cycle of loading and unloading and apply some postulates suggested by Drucker.

Drucker postulated (or defined) a *stable* plastic material to be one in which the work done by the external agency in its cycle of stress or load application and removal was positive. This led to two consequences: (a) The *yield surface* in principal stress space may be plane or curved; if it is curved, it must be *convex* outward. (b) The *plastic strain increment* vector must be at *right angles to* (*normal*) to the yield surface at the yield stress point when the surface varies smoothly in the vicinity [see Figure B-7(a)]. If the yield stress point occurs at a corner of the yield surface, for example, were the material to follow the Tresca criterion, then the plastic strain increment vector is defined only to the extent that it must lie between the outward normals of the two adjacent yield surfaces of the corner, as shown in Figure B-7(b).

So far, we have established certain restrictions on the nature of the yield surface and on the direction of the plastic strain increment vector with

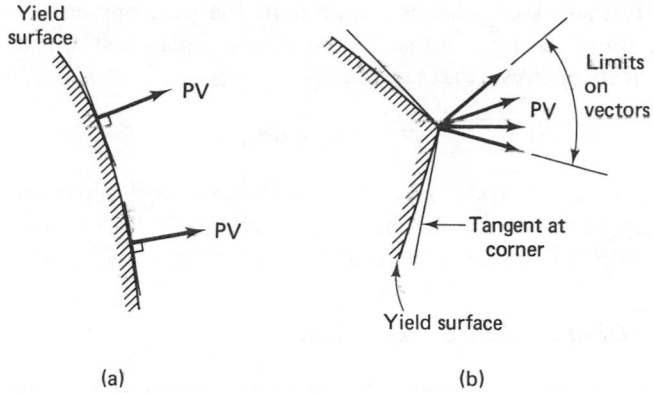

Figure B-7. Relation of stress and plastic strain increment (*PV*) vectors:
(a) *PV* at a continuous surface; (b) *PV* at a corner.

respect to it, but we have not obtained any stress-strain relations. We obtain these by observing that the vector gradient of the yield function f, $\partial f / \partial \sigma_{ij}$ is also normal to the yield surface, so that it must coincide with the plastic strain increment vector. Therefore, we take the incremental plastic strain to be proportional to the vector gradient of the yield function f:

$$d\epsilon_{ij}^p = d\lambda \, \frac{\partial f}{\partial \sigma_{ij}}$$

(B.10)

where $d\lambda$ is a constant of proportionality. It is not a material property such as Young's modulus in linearly elastic theory, since the plastic deformation may be unlimited at yield (for example, the ideal plastic material, as discussed below). The constant enables the relationships among the components of plastic strain to be established. For a work-hardening plastic material in general, $d\lambda$ depends on the stress, the stress increment, stress or load history, and the strain.

Equation (B-10) is generally referred to as the *associated flow rule* since it relates the plastic strain increments developed with the vector gradient of the yield (stress) function. In the flow of an ideal material without vorticity, it is customary to relate the velocity of flow with the gradient of a potential. Here we are dealing with plastic flow, and we have derived the plastic strain increment in terms of the gradient of the yield function, so that the yield function may be identified with a potential and the plastic strain increment vector may be identified with a velocity. Thus, the yield function $f(\sigma_{ij})$ is frequently referred to as the *plastic potential*, and the strain increment $d\epsilon_{ij}^p$ is often written in the form $\dot{\epsilon}_{ij}^p$ and is called *velocity*, although no quantitative description is applied to the implied time derivative.

One further assumption is required in the development of the stress-plastic strain relations; it is that the *incremental* plastic strains are linear in the *increments* of stress, that is to say,

$$d\epsilon_{ij}^p = C_{ijkl} \, d\sigma_{kl} \qquad (\text{B.11})$$

This means that if stress increments $d\sigma_{ij}^{(1)}$ and $d\sigma_{ij}^{(2)}$ separately produce, respectively, plastic strain increments $d\epsilon_{ij}^{(1)}$ and $d\epsilon_{ij}^{(2)}$, then a stress increment $(d\sigma_{ij}^{(1)} + d\sigma_{ij}^{(2)})$ will generate a plastic strain increment $d\epsilon_{ij}^{(1)} + d\epsilon_{ij}^{(2)}$.

B.4 IDEAL PLASTIC MATERIAL

The ideal plastic material is not work-hardening and is defined by the following conditions:

(a) $$d\sigma_{ij} \, d\epsilon_{ij}^p = 0 \qquad (\text{B.12})$$

(b) It possesses a yield function $f(\sigma_{ij})$, which is a function of stress only and not of plastic strain or stress history, so that the material exhibits no Bauschinger effect. Its yield surface is fixed in stress space.

(c) Plastic flow only occurs and is unlimited when the stress state is on the yield surface, i.e., $f(\sigma_{ij}) = K$; the stress state cannot lie outside the yield surface. Inside the yield surface $f(\sigma_{ij}) < K$, and the behavior is elastic.

(d) It follows from (c) that the incremental linearity described by equation (B.11) does not hold for an ideal plastic material. However, the incremental stress-strain relation of equation (B.10) still applies.

B.5 CONSEQUENCES OF PLASTICITY THEORY

We have seen that the yield surface is required to be convex outward. This condition is clearly satisfied by the von Mises and Tresca criteria. With soil materials, it has long been observed that they exhibit a frictional behavior; that is to say, failure occurs when the obliquity of the resultant stress on a failure or sliding surface reaches a maximum. One consequence of this is that the radial distance in the hydrostatic plane to yield in axial extension [see Figure B-3(b)] is less than the radial distance to yield in axial compression. In addition, the radial distance to yield increases with hydrostatic effective stress. The yield surface that satisfies these two basic observations is that known as the *Mohr-Coulomb yield criterion*. In principal stress space this yield surface is a cone with the apex at (or near, in the case of a material with cohesion) the origin; it has an irregular hexagonal shape in cross section, as

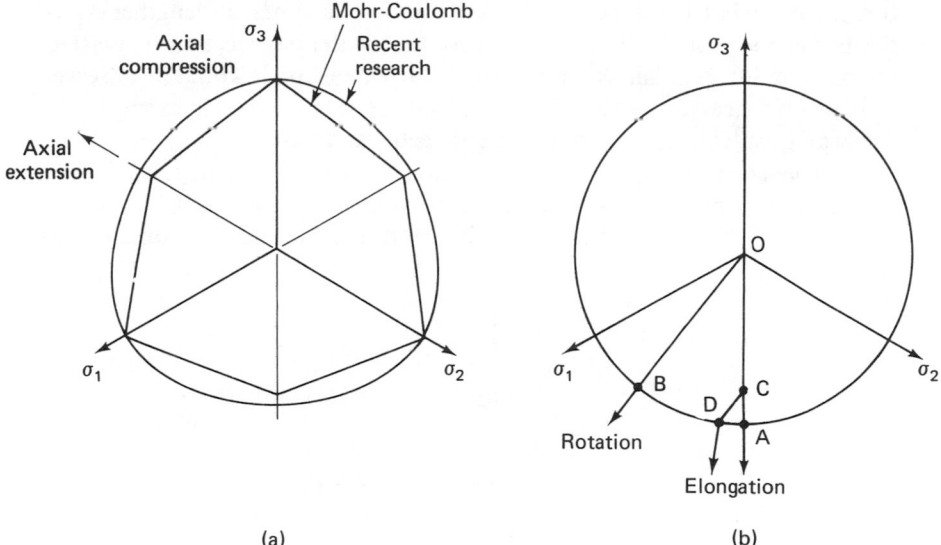

Figure B-8. Yield surfaces in hydrostatic plane of principal stress space: (a) Mohr-Coulomb and present concept; (b) *PV*'s for bar under different loading conditions.

shown in Figure B-8(a). In the last decade or so, refined three-dimensional testing of soils has shown that the yield surface has a more generally convex shape, as indicated in Figure B-8(a).

The normality requirement has interesting physical consequences that are not always clearly understood. The direction of the plastic strain increment vector (*PV*) at yield is directly related to the yielding or collapse mechanism in the specimen being tested. For example, suppose a metal bar (assumed ideally plastic) is taken to yield in extension. The *PV* is as shown at point *A* of Figure B-8(b) at the end of a load path *OA*, and physically the bar elongates as it yields. We can mark the *PV* *elongation* to denote the mechanism. If another identical bar were to be twisted to yield, the *PV* might appear as shown at point *B* in Figure B-8(b) with a load path *OB*. Physically one end of the bar would rotate plastically with respect to the other; *rotation* is marked on the *PV* at *B*.

Now suppose that yet another identical bar is loaded in extension to a point *C* just short of the yield surface at point *A* and is then twisted until its load path, now *CD*, parallel to *OB*, reaches yield at point *D*, close to *A*. Physically, what does the bar do? The normality requirement means that the *PV* at point *D* will be almost parallel to the *PV* at *A* and quite different in direction from the *PV* at *B*. Thus, the bar will yield in *elongation*, not by rotation. There will be a small *elastic* component of rotation due to the torque, but what the experimenter will observe is a plastic lengthening of the bar as he twists it. It is also the case that if the bar were to be twisted, almost to point B initially, and then loaded in extension, it would be observed to *rotate* plastically at yield. This would not be the case if, for example, the PV were parallel to the stress increment vector at yield.

A convenient example with soil is the following. A rectangular footing (Figure B-9) is placed on sand and loaded vertically until it fails and settles plastically. If the failure load P is maintained, the footing will come to rest

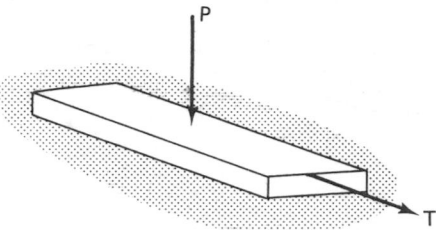

Figure B-9. Footing on soil subject to vertical and tangential loading.

a small depth below the surface, since a somewhat greater load will be required to continue failure under the small overburden pressure conditions. In effect, this is a strain-hardening type of behavior. Suppose that, with the vertical load that caused the initial failure still in place, a horizontal shearing force T directed along the long axis of the footing is applied. At a small value of T, the yield envelope of the footing-sand configuration is again reached, and the sand will yield. Which way will the footing go? Clearly, the normality requirement indicates that it will *settle* further, not slide in the direction of T.

That this actually happens is confirmed by the experience of anyone who has ever had a car stuck in sand. If, say, the car *coasts* to a stop on a sand beach, the wheels will come to rest and be embedded to a depth that depends on the load on the wheels and the yield condition for the sand. An increase in vertical load would cause the wheels to sink further. When the driver is ready to leave, he starts the engine and applies torque to the driving wheels. The driver thus develops the shearing force T in Figure B-9. The immediate consequence is that the driving wheels sink. The car may move only a small distance. With more torque, more settlement, and possibly some horizontal shearing or wheel slip occur. Of course, if excessive power is applied, the wheels spin and become sand pumps. At this stage, which usually occurs rapidly, the extraction process is hopeless. If the initially stationary car is towed or pushed, no torque is applied to the wheel footings and no sinkage of the vehicle is observed. From a practical point of view, lack of understanding of this point has led to incorrect interpretation of the causes of failure in a number of instances. For example, in an earthquake a building or a column-supported bridge collapses essentially vertically as a result of crushing of the columns. It has been reported that this was a consequence of increased vertical loads and that the vertical accelerations must have been high. Since the major part of the pre-earthquake column loads was axial due to the weight of the structure, it is more likely that the *horizontal* accelerations caused the stress increments that brought the columns to their yield value in compression, as in the case of the metal bar discussed above.

B.6 GENERALIZED YIELD COORDINATES; SURFACE

Instead of referring to material-based yield stresses and surface for the previous examples of metal bar and footing on sand, it is more convenient from an engineering point of view to construct a two-dimensional or multi-dimensional space in which the axes are the forces, loads, or moments in the system. It is apparent that, in the case of the footing, the stress conditions are different at every point in the underlying sand, so that when the footing is at yield under the vertical load, points representing yield in sand elements will be distributed all around the sand yield envelope.

The behavior of the footing is an integrated result of the elemental responses, which cannot be examined in detail. Thus, we may define a generalized load space for the bar or footing as shown in Figure B-10. Here, in two dimensions, the axes may be the extension force and torque in the bar or the vertical and shearing forces on the footing. The strain space to be superimposed for consideration of the PV's involves the bar extension and rotation or the footing settlement or horizontal movement. A yield surface for combined loadings can be drawn in the load space. Plastic displacement increment vectors occur normal to this surface to correlate with the direction of movement at yield.

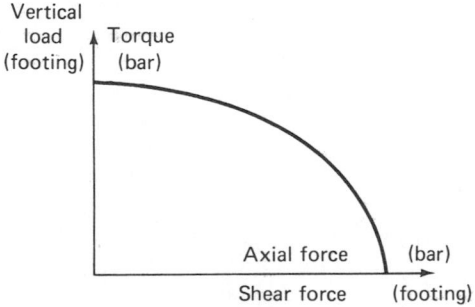

Figure B-10. Generalized failure surface for footing.

B.7 CAUTION

This has been an extremely simplified description of the complex world of plasticity, in particular with regard to soil behavior. Soils yield plastically under hydrostatic effective stresses and hydrostatic-shearing stress combinations, so that the yield surface for a soil closes on the hydrostatic (compressive) effective stress axis. Yield is distinguished from failure, so that it is currently considered that a soil passes through an infinite succession of strain-hardening yield surfaces on its way to a failure envelope. Several different theories exist to account for and describe this behavior.

Figures
and Tables

Appendix C

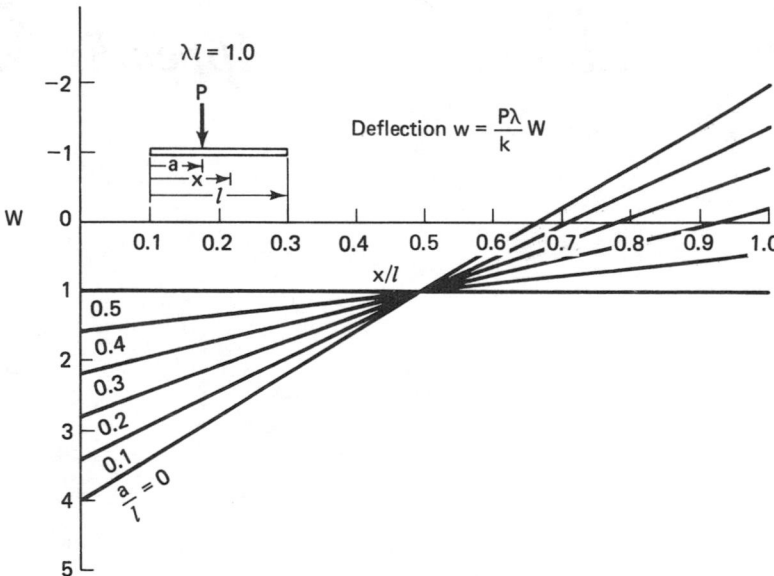

Figure C-1 Deflections of and moments in (influence functions) finite beams on elastic foundations. Beams are of various flexibilities with respect to subgrade and are loaded by a point load at different locations. (a) $\lambda \ell = 1.0$; deflections.

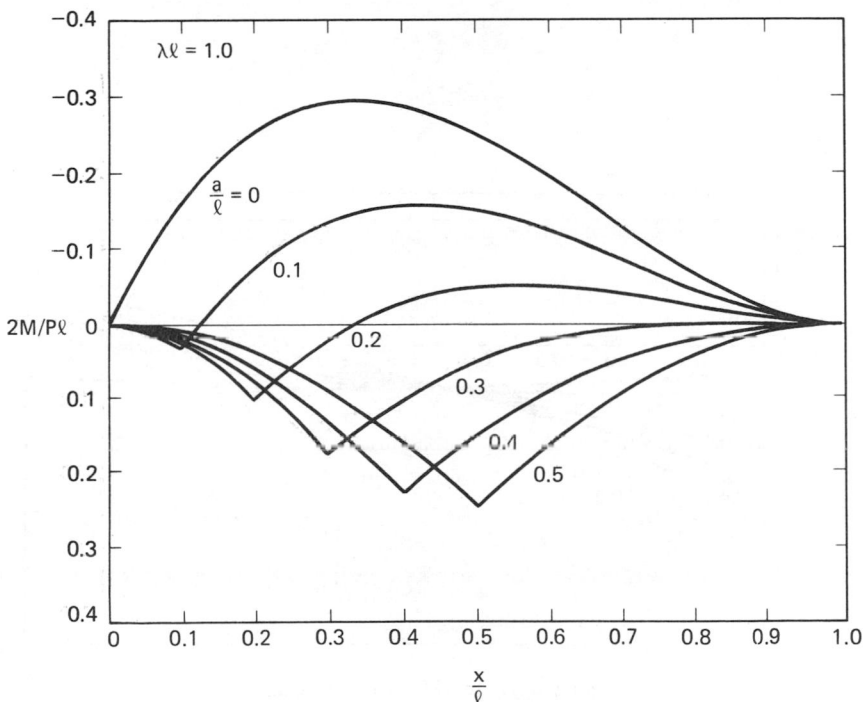

Figure C-1. (b) $\lambda \ell = 1.0$; moments.

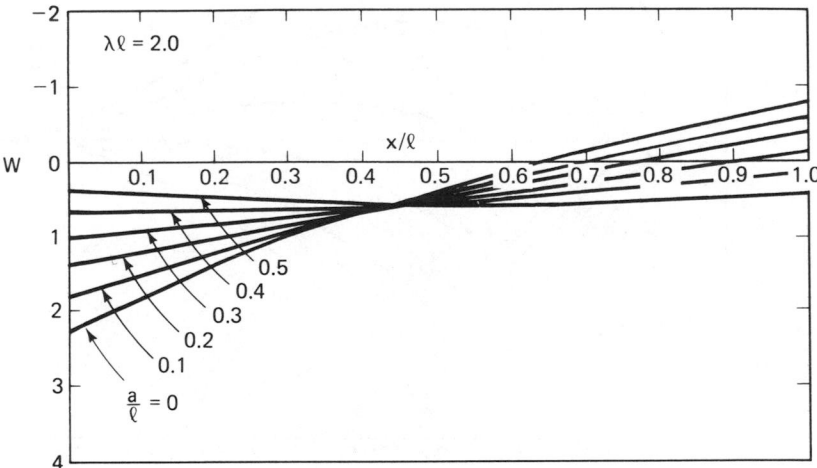

Figure C-1. (c) $\lambda \ell = 2.0$; deflections.

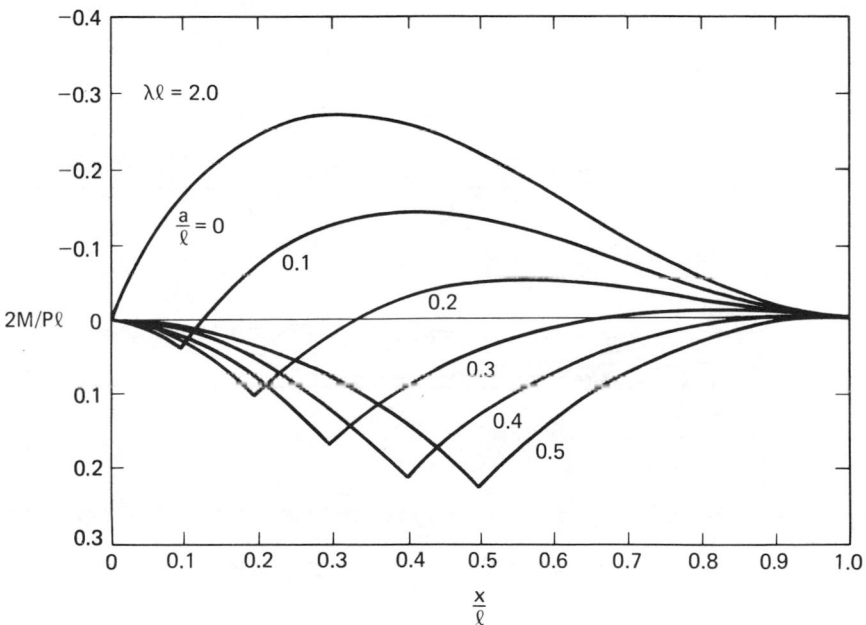

Figure C-1. (d) $\lambda\ell = 2.0$; moments.

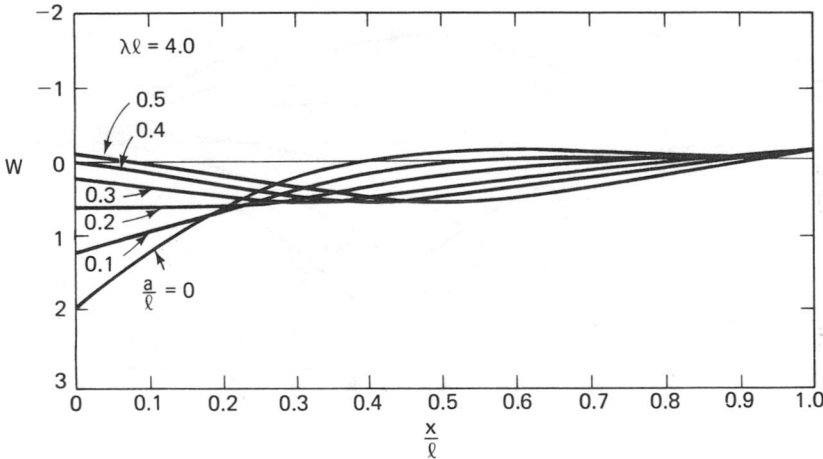

Figure C-1. (e) $\lambda\ell = 4.0$; deflections.

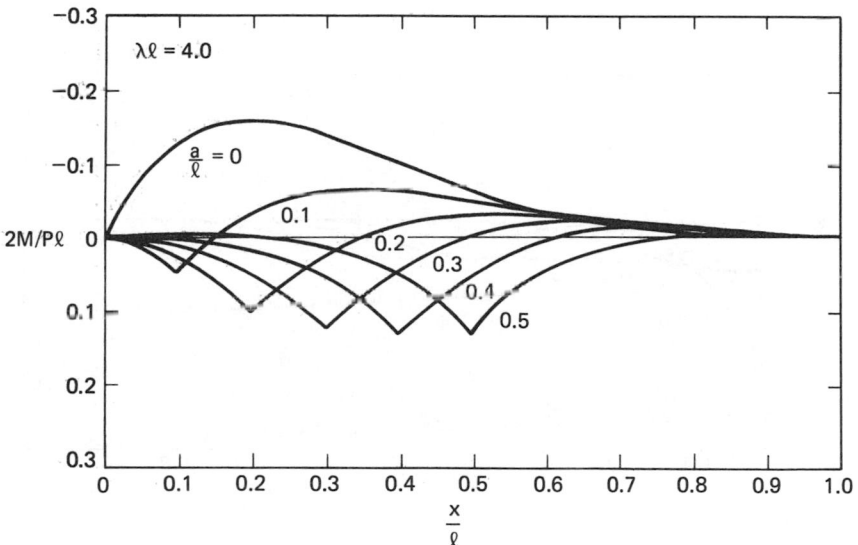

Figure C-1. (f) $\lambda\ell = 4.0$; moments.

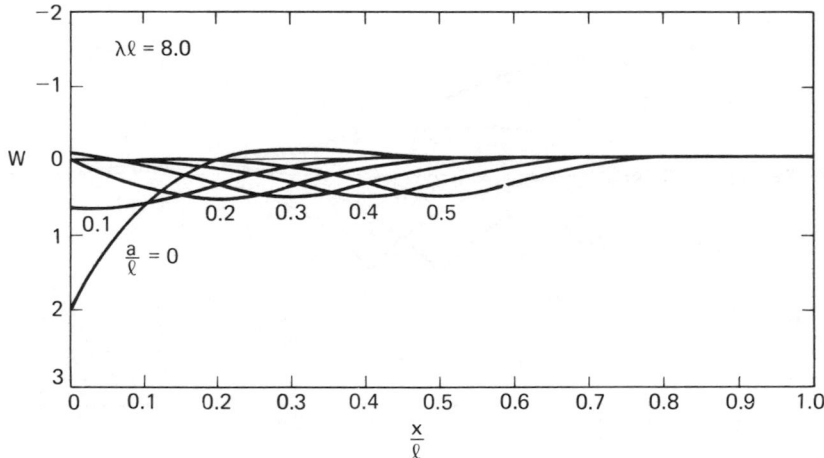

Figure C-1. (g) $\lambda \ell = 8.0$; deflections.

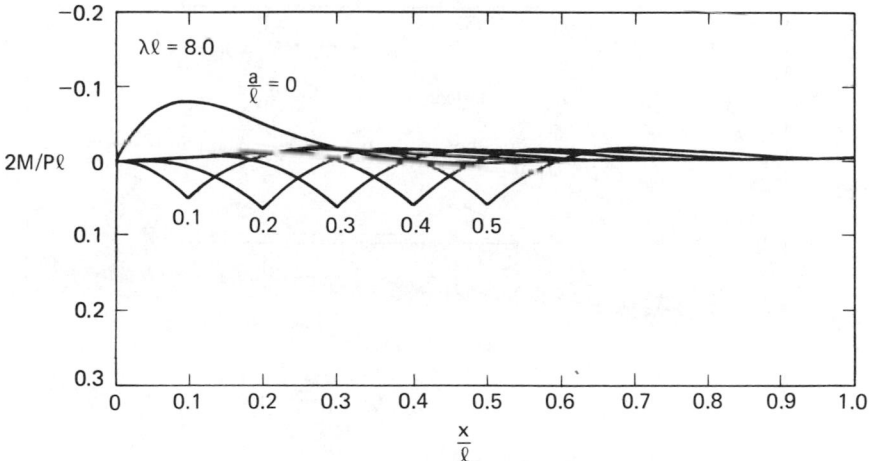

Figure C-1. (h) $\lambda\ell = 8.0$; moments.

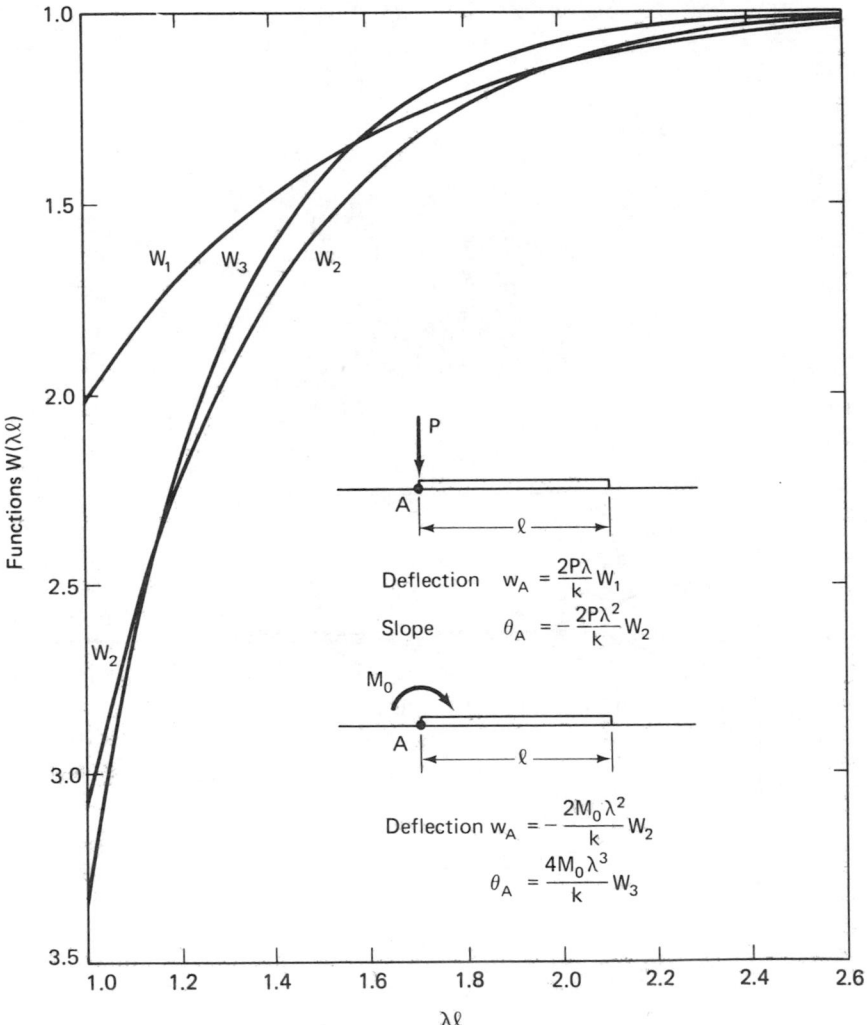

Figure C-2. Deflection and slope at end of finite beam as functions of beam/soil stiffness ratio. Beam is loaded with a point or moment load at end.

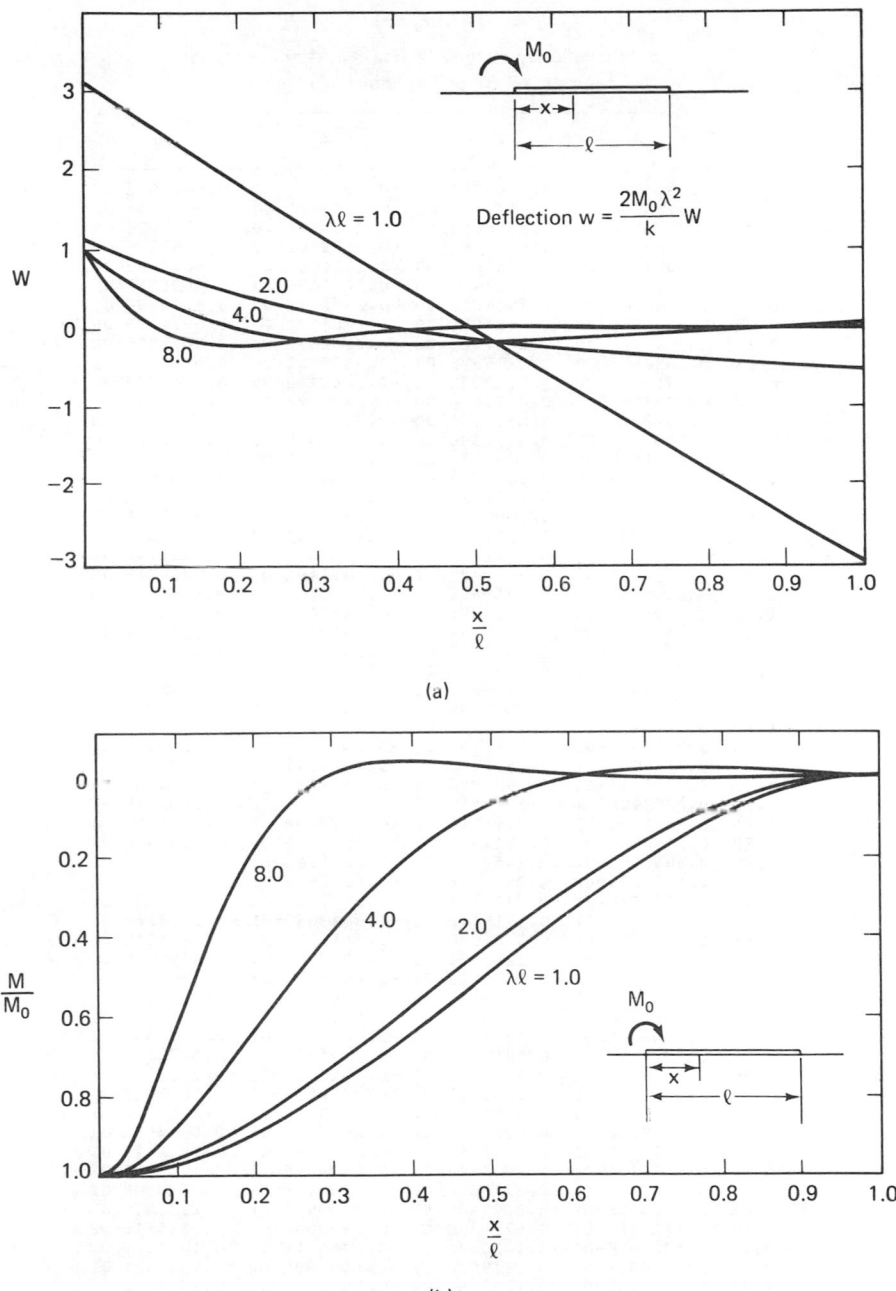

Figure C-3. Deflections and moments along finite beam as functions of beam/soil stiffness ratio. Beam loaded with end moment: (a) deflections; (b) moments.

TABLE C.1

Solutions to Equation $w'' - xw = 0$; Airy Functions:
Ai, Ai', Bi, Bi' as Functions of x.

x	Ai	Bi	Ai'	Bi'
0.0	0.355027E 00	0.614924E 00	−0.258818E 00	0.448286E 00
0.05	0.342094E 00	0.637353E 00	−0.258388E 00	0.449074E 00
0.10	0.329203E 00	0.659861E 00	−0.257131E 00	0.451512E 00
0.15	0.316394E 00	0.682534E 00	−0.255116E 00	0.455712E 00
0.20	0.303703E 00	0.705463E 00	−0.252406E 00	0.461788E 00
0.25	0.291164E 00	0.728746E 00	−0.249063E 00	0.469861E 00
0.30	0.278806E 00	0.752485E 00	−0.245147E 00	0.480049E 00
0.35	0.266658E 00	0.776788E 00	−0.240718E 00	0.492478E 00
0.40	0.254742E 00	0.801772E 00	−0.235833E 00	0.507281E 00
0.45	0.243081E 00	0.827559E 00	−0.230546E 00	0.524597E 00
0.50	0.231693E 00	0.854277E 00	−0.224911E 00	0.544573E 00
0.55	0.220595E 00	0.882063E 00	−0.218977E 00	0.567365E 00
0.60	0.209800E 00	0.911062E 00	−0.212793E 00	0.593144E 00
0.65	0.199319E 00	0.941430E 00	−0.206404E 00	0.622092E 00
0.70	0.189162E 00	0.973327E 00	−0.199851E 00	0.654406E 00
0.75	0.179336E 00	0.100693E 01	−0.193175E 00	0.690300E 00
0.80	0.169846E 00	0.104242E 01	−0.186413E 00	0.730007E 00
0.85	0.160696E 00	0.108000E 01	−0.179599E 00	0.773781E 00
0.90	0.151887E 00	0.111987E 01	−0.172764E 00	0.821904E 00
0.95	0.143419E 00	0.116227E 01	−0.165938E 00	0.874677E 00
1.00	0.135292E 00	0.120742E 01	−0.159148E 00	0.932436E 00
1.05	0.127504E 00	0.125560E 01	−0.152416E 00	0.995548E 00
1.10	0.120050E 00	0.130707E 01	−0.145767E 00	0.106441E 01
1.15	0.112925E 00	0.136214E 01	−0.139217E 00	0.113948E 01
1.20	0.106126E 00	0.142113E 01	−0.132785E 00	0.122123E 01
1.25	0.996445E-01	0.148439E 01	−0.126486E 00	0.131020E 01
1.30	0.934747E-01	0.155228E 01	−0.120334E 00	0.140699E 01
1.35	0.876086E-01	0.162523E 01	−0.114338E 00	0.151223E 01
1.40	0.820380E-01	0.170366E 01	−0.108510E 00	0.162664E 01
1.45	0.767546E-01	0.178806E 01	−0.102855E 00	0.175101E 01
1.50	0.717494E-01	0.187894E 01	−0.973819E-01	0.188621E 01
1.55	0.670133E-01	0.197687E 01	−0.920943E-01	0.203319E 01
1.60	0.625369E-01	0.208247E 01	−0.869959E-01	0.219299E 01
1.65	0.583106E-01	0.219641E 01	−0.820890E-01	0.236679E 01
1.70	0.543248E-01	0.231940E 01	−0.773749E-01	0.255585E 01
1.75	0.505699E-01	0.245227E 01	−0.728536E-01	0.276158E 01
1.80	0.470362E-01	0.259587E 01	−0.685248E-01	0.298554E 01
1.85	0.437142E-01	0.275115E 01	−0.643865E-01	0.322943E 01
1.90	0.405945E-01	0.291917E 01	−0.604368E-01	0.349516E 01
1.95	0.376674E-01	0.310107E 01	−0.566725E-01	0.378481E 01
2.00	0.349241E-01	0.329809E 01	−0.530904E-01	0.410068E 01
2.05	0.323555E-01	0.351161E 01	−0.496863E-01	0.444532E 01
2.10	0.299526E-01	0.374314E 01	−0.464560E-01	0.482154E 01
2.15	0.277070E-01	0.399435E 01	−0.433946E-01	0.523247E 01
2.20	0.256104E-01	0.426703E 01	−0.404972E-01	0.568154E 01
2.25	0.236546E-01	0.456321E 01	−0.377585E-01	0.617256E 01
2.30	0.218320E-01	0.488506E 01	−0.351731E-01	0.670973E 01
2.35	0.201349E-01	0.523502E 01	−0.327353E-01	0.729774E 01
2.40	0.185561E-01	0.561576E 01	−0.304395E-01	0.794177E 01
2.45	0.170887E-01	0.603023E 01	−0.282799E-01	0.864755E 01
2.50	0.157259E-01	0.648166E 01	−0.262509E-01	0.942142E 01
2.55	0.144615E-01	0.697362E 01	−0.243466E-01	0.102704E 02
2.60	0.132893E-01	0.751008E 01	−0.225613E-01	0.112024E 02
2.65	0.122035E-01	0.809540E 01	−0.208895E-01	0.122261E 02
2.70	0.111985E-01	0.873438E 01	−0.193256E-01	0.133511E 02
2.75	0.102692E-01	0.943237E 01	−0.178641E-01	0.145882E 02
2.80	0.941051E-02	0.101953E 02	−0.164598E-01	0.159492E 02
2.85	0.861770E-02	0.110296E 02	−0.152275E-01	0.174474E 02
2.90	0.788631E-02	0.119425E 02	−0.140421E-01	0.190978E 02
2.95	0.721212E-02	0.129422E 02	−0.129388E-01	0.209166E 02
3.00	0.659113E-02	0.140373E 02	−0.119130E-01	0.229222E 02

TABLE C.1 (Cont'd.)

x	Ai	Bi	Ai'	Bi'
3.00	0.659113E-02	0.140373E 02	-0.119130E-01	0.229222E 02
3.05	0.601961E-02	0.152378E 02	-0.109600E-01	0.251349E 02
3.10	0.549400E-02	0.165546E 02	-0.100756E-01	0.275777E 02
3.15	0.501098E-02	0.179999E 02	-0.925550E-02	0.302758E 02
3.20	0.456744E-02	0.195870E 02	-0.849582E-02	0.332577E 02
3.25	0.416045E-02	0.213309E 02	-0.779268E-02	0.365548E 02
3.30	0.378729E-02	0.232483E 02	-0.714249E-02	0.402026E 02
3.35	0.344539E-02	0.253577E 02	-0.654176E-02	0.442406E 02
3.40	0.313234E-02	0.276796E 02	-0.598722E-02	0.487130E 02
3.45	0.284594E-02	0.302370E 02	-0.547575E-02	0.536691E 02
3.50	0.258410E-02	0.330555E 02	-0.500441E-02	0.591643E 02
3.55	0.234488E-02	0.361634E 02	-0.457042E-02	0.652603E 02
3.60	0.212648E-02	0.395927E 02	-0.417113E-02	0.720266E 02
3.65	0.192723E-02	0.433786E 02	-0.380408E-02	0.795410E 02
3.70	0.174557E-02	0.475607E 02	-0.346694E-02	0.878906E 02
3.75	0.158007E-02	0.521832E 02	-0.315751E-02	0.971731E 02
3.80	0.142939E-02	0.572953E 02	-0.287375E-02	0.107498E 03
3.85	0.129230E-02	0.629524E 02	-0.261373E-02	0.118989E 03
3.90	0.116766E-02	0.692160E 02	-0.237564E-02	0.131783E 03
3.95	0.105440E-02	0.761551E 02	-0.215780E-02	0.146038E 03
4.00	0.951563E-03	0.838471E 02	-0.195864E-02	0.161927E 03
4.05	0.858251E-03	0.923781E 02	-0.177670E-02	0.179646E 03
4.10	0.773631E-03	0.101846E 03	-0.161061E-02	0.199417E 03
4.15	0.696946E-03	0.112358E 03	-0.145911E-02	0.221491E 03
4.20	0.627496E-03	0.124038E 03	-0.132100E-02	0.246146E 03
4.25	0.564640E-03	0.137021E 03	-0.119520E-02	0.273698E 03
4.30	0.507788E-03	0.151462E 03	-0.108070E-02	0.304505E 03
4.35	0.456398E-03	0.167532E 03	-0.976557E-03	0.338968E 03
4.40	0.409974E-03	0.185427E 03	-0.881893E-03	0.377542E 03
4.45	0.368063E-03	0.205364E 03	-0.795909E-03	0.420738E 03
4.50	0.330250E-03	0.227588E 03	-0.717866E-03	0.469135E 03
4.55	0.296166E-03	0.252375E 03	-0.647079E-03	0.523383E 03
4.60	0.265433E-03	0.280035E 03	-0.582915E-03	0.584226E 03
4.65	0.237764E-03	0.310921E 03	-0.524795E-03	0.652495E 03
4.70	0.212861E-03	0.345425E 03	-0.472184E-03	0.729139E 03
4.75	0.190461E-03	0.383993E 03	-0.424592E-03	0.815226E 03
4.80	0.170326E-03	0.427125E 03	-0.381571E-03	0.911964E 03
4.85	0.152236E-03	0.475389E 03	-0.342705E-03	0.102073E 04
4.90	0.135992E-03	0.529424E 03	-0.307616E-03	0.114308E 04
4.95	0.121417E-03	0.589952E 03	-0.275958E-03	0.128077E 04
5.00	0.108345E-03	0.657791E 03	-0.247414E-03	0.143582E 04
5.05	0.966281E-04	0.733861E 03	-0.221694E-03	0.161047E 04
5.10	0.861327E-04	0.819207E 03	-0.198533E-03	0.180733E 04
5.15	0.767362E-04	0.915003E 03	-0.177690E-03	0.202931E 04
5.20	0.683286E-04	0.102261E 04	-0.158944E-03	0.227975E 04
5.25	0.608102E-04	0.114352E 04	-0.142095E-03	0.256241E 04
5.30	0.540906E-04	0.127946E 04	-0.126960E-03	0.288162E 04
5.35	0.480884E-04	0.143238E 04	-0.113374E-03	0.324227E 04
5.40	0.427300E-04	0.160447E 04	-0.101185E-03	0.364992E 04
5.45	0.379489E-04	0.179826E 04	-0.902563E-04	0.411094E 04
5.50	0.336853E-04	0.201658E 04	-0.804634E-04	0.463255E 04
5.55	0.298855E-04	0.226265E 04	-0.716937E-04	0.522298E 04
5.60	0.265007E-04	0.254017E 04	-0.638447E-04	0.589165E 04
5.65	0.234872E-04	0.285331E 04	-0.568239E-04	0.664930E 04
5.70	0.208059E-04	0.320679E 04	-0.505479E-04	0.750813E 04
5.75	0.184213E-04	0.360604E 04	-0.449407E-04	0.848214E 04
5.80	0.163018E-04	0.405718E 04	-0.399341E-04	0.958726E 04
5.85	0.144189E-04	0.456724E 04	-0.354662E-04	0.108418E 05
5.90	0.127471E-04	0.514420E 04	-0.314814E-04	0.122665E 05
5.95	0.112635E-04	0.579714E 04	-0.279294E-04	0.138854E 05
6.00	0.994771E-05	0.653643E 04	-0.247652E-04	0.157256E 05

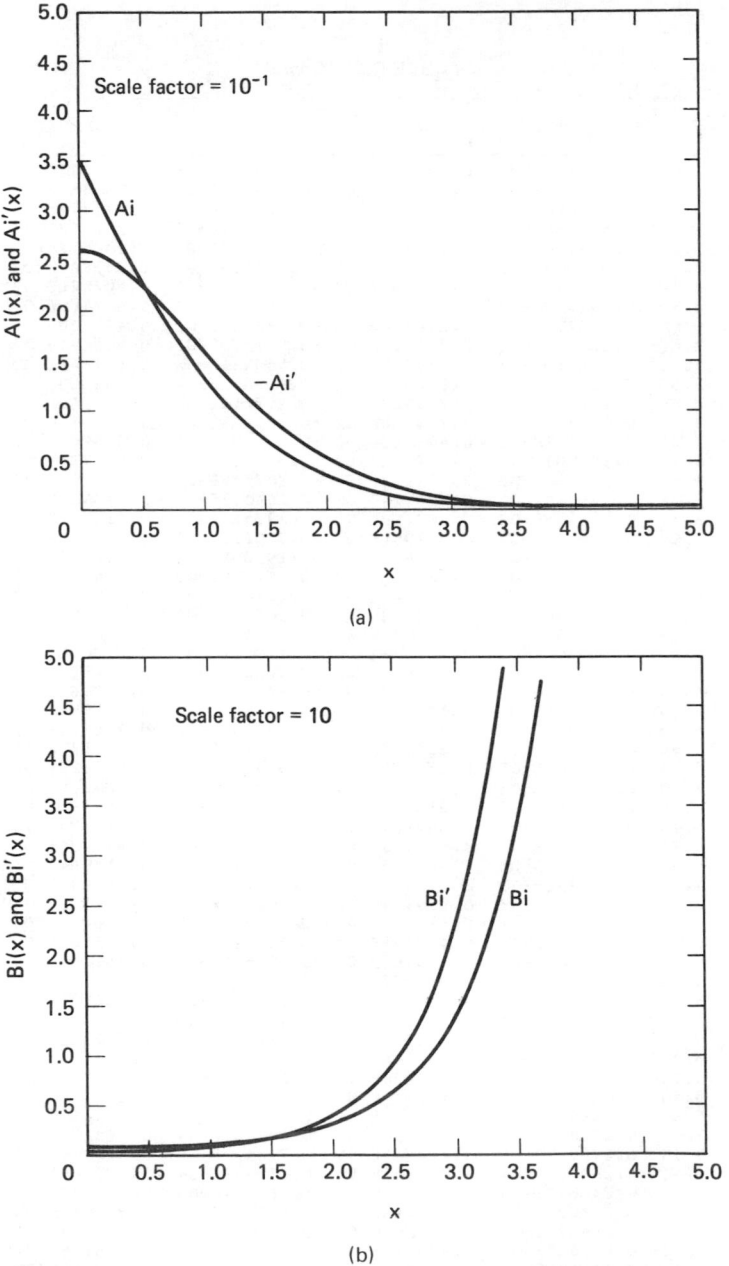

Figure C-4. Solutions to equation $w'' - xw = 0$. Airy functions: (a) *Ai*, *Ai'*; (b) *Bi*, *Bi'*.

TABLE C.2

Solutions to Equation $w'' - x^{1/2}w = 0$; Ri, Ri', Si, Si' as Functions of x.

x	Ri	Si	Ri'	Si'
0.00	0.968780E 00	0.968780E 00	-0.781216E 00	0.781216E 00
0.05	0.929861E 00	0.100799E 01	-0.774170E 00	0.788612E 00
0.10	0.891447E 0C	0.104775E 01	-0.761774E 00	0.802635E 00
0.15	0.853733E 00	0.108833E 01	-0.746387E 00	0.821454E 00
0.20	0.816843E 00	0.112997E 01	-0.728942E 0C	0.844682E 00
0.25	0.780864E 00	0.117287E 01	-0.710015E 00	0.871583E 00
0.30	0.745859E 00	0.121723E 01	-0.690015E 00	0.903312E 00
0.35	0.711875E 0C	0.126327E 01	-0.669253E 00	0.938660E 00
0.40	0.678942E 00	0.131117E 01	-0.647973E 00	0.978069E 00
0.45	0.647083E 00	0.136114E 01	-0.626372E 00	0.102162E 01
0.50	0.616308E 00	0.141340E 01	-0.604614E 00	0.106942E 01
0.55	0.586622E 00	0.146816E 01	-0.582834E 00	0.112161E 01
0.60	0.558023E 00	0.152564E 01	-0.561144E 00	0.117836E 01
0.65	0.530504E 00	0.158607E 01	-0.539638E 00	0.123985E 01
0.70	0.504055E 00	0.164970E 01	-0.518397E 00	0.130631E 01
0.75	0.478659E 00	0.171679E 01	-0.497486E 00	0.137796E 01
0.80	0.454300E 00	0.178759E 01	-0.476961E 00	0.145508E 01
0.85	0.430956E 00	0.186239E 01	-0.456866E 00	0.153795E 01
0.90	0.408605E 00	0.194149E 01	-0.437240E 00	0.162689E 01
0.95	0.387223E 0C	0.202519E 01	-0.418111E 0C	0.172226E 01
1.00	0.366785E 00	0.211383E 01	-0.399505E 00	0.182442E 01
1.05	0.347264E 00	0.220775E 01	-0.381438E 00	0.193378E 01
1.10	0.328632E 00	0.230733E 01	-0.363925E 00	0.205080E 01
1.15	0.310862E 00	0.241297E 01	-0.346973E 00	0.217559E 01
1.20	0.293926E 00	0.252507E 01	-0.330589E 00	0.230519E 01
1.25	0.277794E 00	0.264409E 01	-0.314775E 00	0.245274E 01
1.30	0.262438E 0C	0.277051E 01	-0.299531E 00	0.260557E 01
1.35	0.247831E 00	0.290483E 01	-0.284851E 00	0.276886E 01
1.40	0.233944E 00	0.304758E 01	-0.270733E 00	0.294332E 01
1.45	0.220749E 0C	0.319936E 01	-0.257168E 00	0.312971E 01
1.50	0.208218E 00	0.336077E 01	-0.244148E 0C	0.332886E 01
1.55	0.196325E 00	0.353247E 01	-0.231663E 00	0.354163E 01
1.60	0.185043E 00	0.371517E 01	-0.219702E 0C	0.376897E 01
1.65	0.174346E 00	0.390963E 01	-0.208253E 00	0.401192E 01
1.70	0.164209E 0C	0.411664E 01	-0.197303E 00	0.427155E 01
1.75	0.154608E 00	0.433708E 01	-0.186838E 00	0.454907E 01
1.80	0.145518E 00	0.457187E 01	-0.176845E 00	0.484574E 01
1.85	0.136916E 00	0.482199E 01	-0.167310E 0C	0.516292E 01
1.90	0.128779E 00	0.508853E 01	-0.158218E 00	0.550211E 01
1.95	0.121087E 00	0.537260E 01	-0.149554E 00	0.586488E 01
2.00	0.113817E 0C	0.567544E 01	-0.141304E 00	0.625294E 01
2.05	0.106950E 00	0.599335E 01	-0.133453E 0C	0.666814E 01
2.10	0.100465E 00	0.634273E 01	-0.125986E 0C	0.711246E 01
2.15	0.943450E-01	0.671011E 01	-0.118889E 0C	0.758803E 01
2.20	0.885705E-01	0.710210E 01	-0.112148E 00	0.809715E 01
2.25	0.831245E-01	0.752043E 01	-0.105747E 0C	0.864230E 01
2.30	0.779904E-01	0.796697E 01	-0.996745E-01	0.922614E 01
2.35	0.731519E-01	0.844373E 01	-0.939151E-01	0.985154E 01
2.40	0.685938E-01	0.895287E 01	-0.884561E-01	0.105216E 02
2.45	0.643015E-01	0.949669E 01	-0.832843E-01	0.112397E 02
2.50	0.602608E-01	0.100777E 02	-0.783871E-01	0.120093E 02
2.55	0.564584E-01	0.106986E 02	-0.737522E-01	0.128344E 02
2.60	0.528814E-01	0.113621E 02	-0.693675E-01	0.137192E 02
2.65	0.495177E-01	0.120715E 02	-0.652215E-01	0.146681E 02
2.70	0.463555E-01	0.128301E 02	-0.613030E-01	0.156860E 02
2.75	0.433838E-01	0.136414E 02	-0.576010E-01	0.167781E 02
2.80	0.405920E-01	0.145092E 02	-0.541052E-01	0.179501E 02
2.85	0.379700E-01	0.154378E 02	-0.508054E-01	0.192081E 02
2.90	0.355083E-01	0.164316E 02	-0.476919E-01	0.205586E 02
2.95	0.331979E-01	0.174953E 02	-0.447555E-01	0.220087E 02
3.00	0.310300E-01	0.186342E 02	-0.419871E-01	0.235661E 02

TABLE C.2 (*Cont'd.*)

x	Ri	Si	Ri'	Si'
3.00	0.310300E-01	0.186342E 02	-0.419871E-01	0.235661E 02
3.05	0.289965E-01	0.198539E 02	-0.393781E-01	0.252391E 02
3.10	0.270896E-01	0.211602E 02	-0.369204E-01	0.270365E 02
3.15	0.253021E-01	0.225598E 02	-0.346059E-01	0.289681E 02
3.20	0.236268E-01	0.240594E 02	-0.324272E-01	0.310441E 02
3.25	0.220572E-01	0.256668E 02	-0.303771E-01	0.332758E 02
3.30	0.205871E-01	0.273898E 02	-0.284486E-01	0.356754E 02
3.35	0.192104E-01	0.292373E 02	-0.266351E-01	0.382560E 02
3.40	0.179217E-01	0.312187E 02	-0.249304E-01	0.410317E 02
3.45	0.167157E-01	0.333440E 02	-0.233286E-01	0.440177E 02
3.50	0.155873E-01	0.356242E 02	-0.218238E-01	0.472308E 02
3.55	0.145318E-01	0.380712E 02	-0.204107E-01	0.506886E 02
3.60	0.135447E-01	0.406975E 02	-0.190841E-01	0.544106E 02
3.65	0.126220E-01	0.435170E 02	-0.178392E-01	0.584176E 02
3.70	0.117595E-01	0.465444E 02	-0.166712E-01	0.627323E 02
3.75	0.109537E-01	0.497957E 02	-0.155757E-01	0.673791E 02
3.80	0.102008E-01	0.532883E 02	-0.145486E-01	0.723844E 02
3.85	0.949774E-02	0.570406E 02	-0.135859E-01	0.777768E 02
3.90	0.884124E-02	0.610729E 02	-0.126838E-01	0.835873E 02
3.95	0.822840E-02	0.654069E 02	-0.118388E-01	0.898494E 02
4.00	0.765647E-02	0.700660E 02	-0.110474E-01	0.965992E 02
4.05	0.712283E-02	0.750756E 02	-0.103064E-01	0.103876E 03
4.10	0.662503E-02	0.804631E 02	-0.961294E-02	0.111723E 03
4.15	0.616079E-02	0.862581E 02	-0.896404E-02	0.120185E 03
4.20	0.572793E-02	0.924926E 02	-0.835702E-02	0.129312E 03
4.25	0.532443E-02	0.992013E 02	-0.778933E-02	0.139159E 03
4.30	0.494838E-02	0.106421E 03	-0.725858E-02	0.149783E 03
4.35	0.459800E-02	0.114194E 03	-0.676248E-02	0.161249E 03
4.40	0.427159E-02	0.122562E 03	-0.629890E-02	0.173624E 03
4.45	0.396760E-02	0.131572E 03	-0.586581E-02	0.186983E 03
4.50	0.368454E-02	0.141278E 03	-0.546132E-02	0.201407E 03
4.55	0.342102E-02	0.151732E 03	-0.508362E-02	0.216983E 03
4.60	0.317575E-02	0.162997E 03	-0.473104E-02	0.233806E 03
4.65	0.294752E-02	0.175135E 03	-0.440199E-02	0.251978E 03
4.70	0.273519E-02	0.188219E 03	-0.409496E-02	0.271610E 03
4.75	0.253768E-02	0.202323E 03	-0.380856E-02	0.292824E 03
4.80	0.235401E-02	0.217530E 03	-0.354146E-02	0.315751E 03
4.85	0.218324E-02	0.233929E 03	-0.329242E-02	0.340532E 03
4.90	0.202449E-02	0.251616E 03	-0.306027E-02	0.367322E 03
4.95	0.187695E-02	0.270697E 03	-0.284393E-02	0.396288E 03
5.00	0.173985E-02	0.291284E 03	-0.264235E-02	0.427611E 03
5.05	0.161248E-02	0.313501E 03	-0.245457E-02	0.461498E 03
5.10	0.149418E-02	0.337479E 03	-0.227969E-02	0.498136E 03
5.15	0.138431E-02	0.363364E 03	-0.211686E-02	0.537783E 03
5.20	0.128230E-02	0.391312E 03	-0.196529E-02	0.580684E 03
5.25	0.118761E-02	0.421491E 03	-0.182421E-02	0.627111E 03
5.30	0.109972E-02	0.454087E 03	-0.169295E-02	0.677363E 03
5.35	0.101816E-02	0.489297E 03	-0.157083E-02	0.731762E 03
5.40	0.942493E-03	0.527338E 03	-0.145725E-02	0.790660E 03
5.45	0.872303E-03	0.568444E 03	-0.135163E-02	0.854437E 03
5.50	0.807207E-03	0.612870E 03	-0.125343E-02	0.923508E 03
5.55	0.746845E-03	0.660891E 03	-0.116216E-02	0.998324E 03
5.60	0.690883E-03	0.712806E 03	-0.107734E-02	0.107937E 04
5.65	0.639011E-03	0.768941E 03	-0.998524E-03	0.116719E 04
5.70	0.590937E-03	0.829647E 03	-0.925312E-03	0.126235E 04
5.75	0.546393E-03	0.895309E 03	-0.857316E-03	0.136548E 04
5.80	0.505125E-03	0.966340E 03	-0.794176E-03	0.147728E 04
5.85	0.466900E-03	0.104319E 04	-0.735558E-03	0.159847E 04
5.90	0.431499E-03	0.112636E 04	-0.681147E-03	0.172987E 04
5.95	0.398720E-03	0.121636E 04	-0.630653E-03	0.187236E 04
6.00	0.368373E-03	0.131379E 04	-0.583801E-03	0.202691E 04

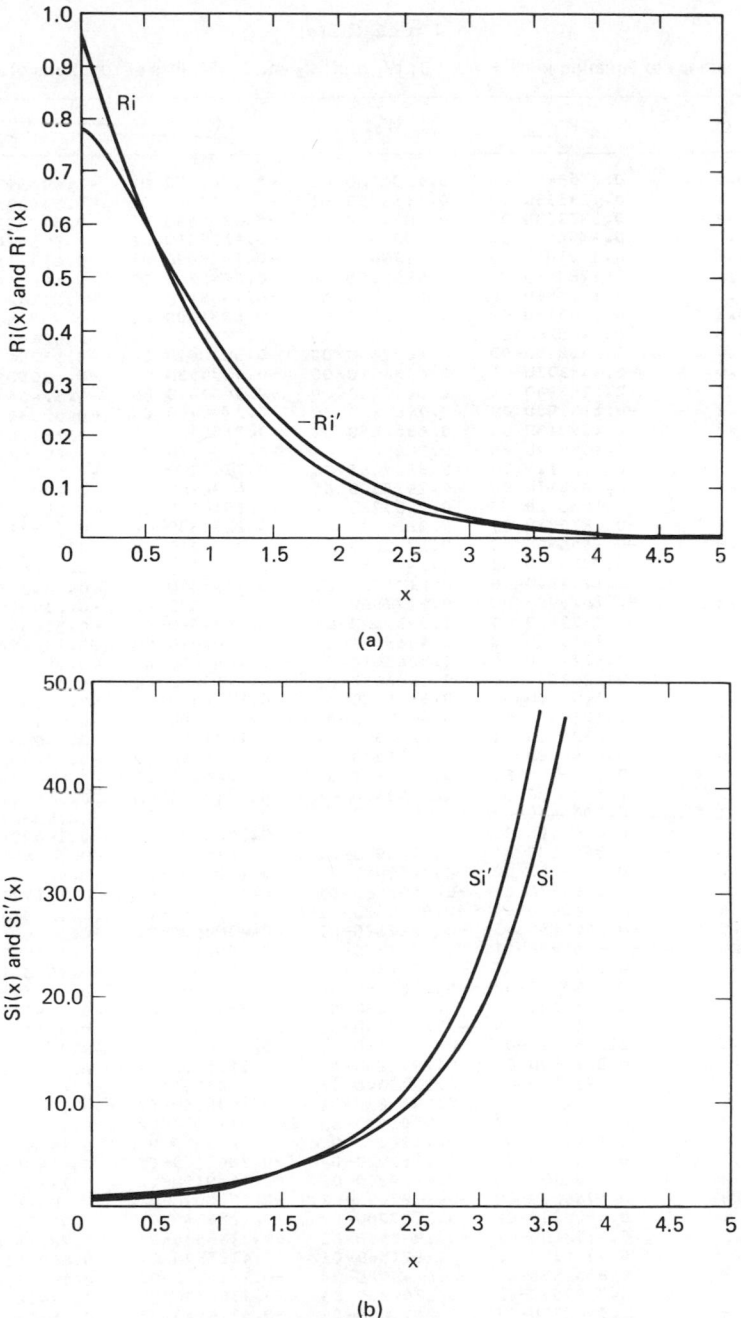

Figure C-5. Solutions to equation $w'' - x^{1/2}w = 0$. (a) *Ri, Ri'*; (b) *Si, Si'*.

TABLE C.3(a)

Solutions to Equation $w^{(iv)} + xw = 0$; W_1 and W_2 and Derivatives as functions of x.

x	W_1	W_2	W_1'	W_2'
0.0	0.875349D 00	0.120482D 01	−0.152775D 01	−0.496397D 00
0.1	0.729553D 00	0.115302D 01	−0.138915D 01	−0.538371D 00
0.2	0.597323D 00	0.109741D 01	−0.125643D 01	−0.572387D 00
0.3	0.478069D 00	0.103880D 01	−0.112968D 01	−0.598614D 00
0.4	0.371185D 00	0.977941D 00	−0.100903D 01	−0.617315D 00
0.5	0.276055D 00	0.915575D 00	−0.894624D 00	−0.628844D 00
0.6	0.192047D 00	0.852397D 00	−0.786613D 00	−0.633624D 00
0.7	0.118515D 00	0.789059D 00	−0.685120D 00	−0.632140D 00
0.8	0.548026D−01	0.726160D 00	−0.590243D 00	−0.624925D 00
0.9	0.243895D−03	0.664246D 00	−0.502047D 00	−0.612543D 00
1.0	−0.458303D−01	0.603804D 00	−0.420553D 00	−0.595586D 00
1.1	−0.840894D−01	0.545262D 00	−0.345744D 00	−0.574653D 00
1.2	−0.115198D 00	0.488986D 00	−0.277540D 00	−0.550347D 00
1.3	−0.139815D 00	0.435285D 00	−0.215832D 00	−0.523261D 00
1.4	−0.158577D 00	0.384407D 00	−0.160458D 00	−0.493971D 00
1.5	−0.172111D 00	0.336546D 00	−0.111217D 00	−0.463029D 00
1.6	−0.181017D 00	0.291839D 00	−0.678699D−01	−0.430957D 00
1.7	−0.185872D 00	0.250376D 00	−0.301415D−01	−0.398240D 00
1.8	−0.187223D 00	0.212198D 00	0.227049D−02	−0.365325D 00
1.9	−0.185585D 00	0.177304D 00	0.296925D−01	−0.332616D 00
2.0	−0.181439D 00	0.145655D 00	0.524696D−01	−0.300472D 00
2.1	−0.175234D 00	0.117180D 00	0.709585D−01	−0.269207D 00
2.2	−0.167378D 00	0.917756D−01	0.855263D−01	−0.239090D 00
2.3	−0.158246D 00	0.693160D−01	0.965404D−01	−0.210345D 00
2.4	−0.148176D 00	0.496548D−01	0.104367D 00	−0.183152D 00
2.5	−0.137467D 00	0.326291D−01	0.109365D 00	−0.157651D 00
2.6	−0.126385D 00	0.180647D−01	0.111886D 00	−0.133943D 00
2.7	−0.115161D 00	0.577868D−02	0.112264D 00	−0.112090D 00
2.8	−0.103993D 00	−0.441635D−02	0.110820D 00	−0.921259D−01
2.9	−0.930471D−01	−0.127095D−01	0.107855D 00	−0.740507D−01
3.0	−0.824646D−01	−0.192887D−01	0.103649D 00	−0.578403D−01
3.1	−0.723519D−01	−0.243381D−01	0.984612D−01	−0.434472D−01
3.2	−0.627972D−01	−0.280365D−01	0.925269D−01	−0.308049D−01
3.3	−0.538643D−01	−0.305547D−01	0.860590D−01	−0.198307D−01
3.4	−0.455968D−01	−0.320550D−01	0.792649D−01	−0.104294D−01
3.5	−0.382008D−01	−0.326895D−01	0.722568D−01	−0.249609D−02
3.6	−0.311466D−01	−0.325994D−01	0.652330D−01	0.408100D−02
3.7	−0.249712D−01	−0.319146D−01	0.582979D−01	0.941773D−02
3.8	−0.194806D−01	−0.307533D−01	0.515538D−01	0.136318D−01
3.9	−0.146513D−01	−0.292217D−01	0.450840D−01	0.168407D−01
4.0	−0.104525D−01	−0.274147D−01	0.389540D−01	0.191599D−01
4.1	−0.684754D−02	−0.254156D−01	0.332133D−01	0.207011D−01
4.2	−0.379569D−02	−0.232968D−01	0.278969D−01	0.215713D−01
4.3	−0.125332D−02	−0.211203D−01	0.230267D−01	0.218715D−01
4.4	0.824939D−03	−0.189386D−01	0.186134D−01	0.216963D−01
4.5	0.248469D−02	−0.167943D−01	0.146576D−01	0.211323D−01
4.6	0.377150D−02	−0.147225D−01	0.111525D−01	0.202592D−01
4.7	0.472975D−02	−0.127504D−01	0.808355D−02	0.191487D−01
4.8	0.540211D−02	−0.108985D−01	0.543095D−02	0.178644D−01
4.9	0.582904D−02	−0.918141D−02	0.317063D−02	0.164627D−01
5.0	0.604843D−02	−0.760829D−02	0.127534D−02	0.149921D−01
5.1	0.609532D−02	−0.618392D−02	−0.284315D−03	0.134941D−01
5.2	0.600175D−02	−0.490924D−02	−0.153919D−02	0.120035D−01
5.3	0.579667D−02	−0.378203D−02	−0.252009D−02	0.105488D−01
5.4	0.550579D−02	−0.279720D−02	−0.325813D−02	0.915297D−02
5.5	0.515210D−02	−0.194858D−02	−0.378353D−02	0.783343D−02
5.6	0.475529D−02	−0.122754D−02	−0.412532D−02	0.660311D−02
5.7	0.433233D−02	−0.624692D−03	−0.431098D−02	0.547083D−02
5.8	0.389754D−02	−0.129936D−03	−0.436617D−02	0.441800D−02
5.9	0.346275D−02	0.267182D−03	−0.431454D−02	0.351817D−02
6.0	0.303651D−02	0.577203D−03	−0.417759D−02	0.269955D−02

x	W_1''	W_2''	W_1'''	W_2'''
0.0	0.141516D 01	-0.459813D 00	-0.583056D 00	0.802511D 00
0.1	0.135672D 01	-0.379759D 00	-0.586941D 00	0.796658D 00
0.2	0.129757D 01	-0.300848D 00	-0.596766D 00	0.779822D 00
0.3	0.123725D 01	-0.224122D 00	-0.610083D 00	0.753162D 00
0.4	0.117551D 01	-0.150502D 00	-0.624821D 00	0.717915D 00
0.5	0.111229D 01	-0.807826D-01	-0.639262D 00	0.675358D 00
0.6	0.104771D 01	-0.156305D-01	-0.652016D 00	0.626790D 00
0.7	0.981986D 00	0.444179D-01	-0.661993D 00	0.573496D 00
0.8	0.915434D 00	0.989531D-01	-0.668380D 00	0.516732D 00
0.9	0.848448D 00	0.147689D 00	-0.670612D 00	0.457700D 00
1.0	0.781462D 00	0.190455D 00	-0.668344D 00	0.397531D 00
1.1	0.714935D 00	0.227192D 00	-0.661426D 00	0.337273D 00
1.2	0.649332D 00	0.257939D 00	-0.649875D 00	0.277874D 00
1.3	0.585109D 00	0.282824D 00	-0.633852D 00	0.220180D 00
1.4	0.522701D 00	0.302056D 00	-0.613633D 00	0.164926D 00
1.5	0.462510D 00	0.315911D 00	-0.589587D 00	0.112734D 00
1.6	0.404897D 00	0.324722D 00	-0.562156D 00	0.641128D-01
1.7	0.350175D 00	0.328866D 00	-0.531832D 00	0.194596D-01
1.8	0.298609D 00	0.328756D 00	-0.499137D 00	-0.209357D-01
1.9	0.250408D 00	0.324827D 00	-0.464612D 00	-0.568851D-01
2.0	0.205729D 00	0.317530D 00	-0.428793D 00	-0.882950D-01
2.1	0.164675D 00	0.307320D 00	-0.392208D 00	-0.115158D 00
2.2	0.127296D 00	0.294648D 00	-0.355357D 00	-0.137546D 00
2.3	0.935966D-01	0.279955D 00	-0.318712D 00	-0.155596D 00
2.4	0.635330D-01	0.263666D 00	-0.282700D 00	-0.169506D 00
2.5	0.370226D-01	0.246184D 00	-0.247707D 00	-0.179519D 00
2.6	0.139464D-01	0.227883D 00	-0.214069D 00	-0.185920D 00
2.7	-0.584596D-02	0.209110D 00	-0.182072D 00	-0.189021D 00
2.8	-0.225306D-01	0.190178D 00	-0.151951D 00	-0.189154D 00
2.9	-0.363048D-01	0.171366D 00	-0.123888D 00	-0.186663D 00
3.0	-0.473814D-01	0.152921D 00	-0.980193D-01	-0.181898D 00
3.1	-0.559845D-01	0.135051D 00	-0.744313D-01	-0.175204D 00
3.2	0.622450D-01	0.117933D 00	-0.531688D-01	-0.166919D 00
3.3	-0.666959D-01	0.101709D 00	-0.342362D-01	-0.157366D 00
3.4	-0.692688D-01	0.864917D-01	-0.176024D-01	-0.146851D 00
3.5	-0.702908D-01	0.723616D-01	-0.320467D-02	-0.135659D 00
3.6	-0.699813D-01	0.593736D-01	0.904609D-02	-0.124051D 00
3.7	-0.685500D-01	0.475577D-01	0.192614D-01	-0.112262D 00
3.8	-0.661931D-01	0.369205D-01	0.275705D-01	-0.100499D 00
3.9	-0.630946D-01	0.274508D-01	0.341167D-01	-0.889449D-01
4.0	-0.594234D-01	0.191194D-01	0.390512D-01	-0.777530D-01
4.1	-0.553328D-01	0.118838D-01	0.425321D-01	-0.670515D-01
4.2	-0.509601D-01	0.568943D-02	0.447198D-01	-0.569423D-01
4.3	-0.464266D-01	0.472988D-03	0.457735D-01	-0.475043D-01
4.4	-0.418379D-01	-0.383565D-02	0.458546D-01	-0.387939D-01
4.5	-0.372844D-01	-0.731121D-02	0.450969D-01	-0.308471D-01
4.6	-0.328415D-01	-0.100311D-01	0.436595D-01	-0.236821D-01
4.7	-0.285709D-01	-0.120737D-01	0.416705D-01	-0.173006D-01
4.8	-0.245217D-01	-0.135171D-01	0.392534D-01	-0.116906D-01
4.9	-0.207308D-01	-0.144370D-01	0.365206D-01	-0.682835D-02
5.0	-0.172247D-01	-0.149066D-01	0.335733D-01	-0.268028D-02
5.1	-0.140203D-01	-0.149955D-01	0.305007D-01	0.794756D-03
5.2	-0.111262D-01	-0.147685D-01	0.273807D-01	0.364378D-02
5.3	-0.854354D-02	-0.142859D-01	0.242799D-01	0.591793D-02
5.4	-0.626768D-02	-0.136023D-01	0.212538D-01	0.767090D-02
5.5	-0.428877D-02	-0.127672D-01	0.183478D-01	0.895758D-02
5.6	-0.259292D-02	-0.118245D-01	0.155976D-01	0.983283D-02
5.7	-0.116314D-02	-0.108126D-01	0.130303D-01	0.103504D-01
5.8	0.199067D-04	-0.976456D-02	0.106648D-01	0.105622D-01
5.9	0.976953D-03	-0.870866D-02	0.851304D-02	0.105174D-01
6.0	0.172978D-02	-0.766814D-02	0.658079D-02	0.102620D-01

TABLE C.3(b)

Solutions to Equation $w^{(iv)} + xw = 0$; W_3 and W_4 and Derivatives as Functions of x

x	W_3	W_4	W'_3	W'_4
0.0	0.141635D 01	−0.460201D 00	0.944204D 00	0.129959D 01
0.1000	0.150524D 01	−0.324274D 00	0.852019D 00	0.141843D 01
0.2000	0.158644D 01	−0.176616D 00	0.750313D 00	0.153424D 01
0.3000	0.165598D 01	−0.175270D−01	0.638852D 00	0.164703D 01
0.4000	0.171387D 01	0.152691D 00	0.517229D 00	0.175683D 01
0.5000	0.175907D 01	0.333739D 00	0.384852D 00	0.186362D 01
0.6000	0.179046D 01	0.525310C 00	0.240940D 00	0.196727D 01
0.7000	0.180684D 01	0.727080D 00	0.845143D−01	0.206754D 01
0.8000	0.180690D 01	0.938693D 00	−0.855934D−01	0.216403D 01
0.9000	0.178922D 01	0.115974D 01	−0.270739D 00	0.225611D 01
1.0000	0.175220D 01	0.138974D 01	−0.472449D 00	0.234287D 01
1.1000	0.169412D 01	0.162810D 01	−0.692406D 00	0.242310D 01
1.2000	0.161306D 01	0.187409D 01	−0.932418D 00	0.249521D 01
1.3000	0.150691D 01	0.212680D 01	−0.119438D 01	0.255718D 01
1.4000	0.137338D 01	0.238510D 01	−0.148025D 01	0.260650D 01
1.5000	0.120999D 01	0.264758D 01	−0.179196D 01	0.264012D 01
1.6000	0.101407D 01	0.291248D 01	−0.213138D 01	0.265439D 01
1.7000	0.782737D 00	0.317767D 01	−0.250023D 01	0.264499D 01
1.8000	0.512989D 00	0.344052D 01	−0.289999D 01	0.260688D 01
1.9000	0.201671D 00	0.369789D 01	−0.333178D 01	0.253428D 01
2.0000	−0.154457D 00	0.394601D 01	−0.379625D 01	0.242057D 01
2.1000	−0.558671D 00	0.418039D 01	−0.429346D 01	0.225830D 01
2.2000	−0.101421D 01	0.439577D 01	−0.482264D 01	0.203912D 01
2.3000	−0.152421D 01	0.458601D 01	−0.538210D 01	0.175381D 01
2.4000	−0.209155D 01	0.474399D 01	−0.596898D 01	0.139227D 01
2.5000	−0.271878D 01	0.486156D 01	−0.657900D 01	0.943521D 00
2.6000	−0.340793D 01	0.492940D 01	−0.720627D 01	0.395803D 00
2.7000	−0.416035D 01	0.493700D 01	−0.784295D 01	−0.263374D 00
2.8000	−0.497550D 01	0.487257D 01	−0.847899D 01	−0.104708D 01
2.9000	−0.585571D 01	0.472299D 01	−0.910182D 01	−0.196881D 01
3.0000	−0.679591D 01	0.447376D 01	−0.969597D 01	−0.304222D 01
3.1000	−0.779333D 01	0.410903D 01	−0.102428D 02	−0.428088D 01
3.2000	−0.884214D 01	0.361163D 01	−0.107199D 02	−0.569746D 01
3.3000	−0.993412D 01	0.296311D 01	−0.111012D 02	−0.730532D 01
3.4000	−0.110582D 02	0.214386D 01	−0.113561D 02	−0.911392D 01
3.5000	−0.122000D 02	0.113333D 01	−0.114494D 02	−0.111322D 02
3.6000	−0.133413D 02	−0.897579D−01	−0.113410D 02	−0.133658D 02
3.7000	−0.144599D 02	−0.154707D 01	−0.109854D 02	−0.158165D 02
3.8000	−0.155285D 02	−0.326020D 01	−0.103320D 02	−0.184813D 02
3.9000	−0.165145D 02	−0.525016D 01	−0.932450D 01	−0.213511D 02
4.0000	−0.173795D 02	−0.753671D 01	−0.790114D 01	−0.244096D 02
4.1000	−0.180787D 02	−0.101375D 02	−0.599499D 01	−0.276313D 02
4.2000	−0.185600D 02	−0.130672D 02	−0.353423D 01	−0.309805D 02
4.3000	−0.187645D 02	−0.163363D 02	−0.442818D 00	−0.344089D 02
4.4000	−0.186249D 02	−0.199496D 02	0.335856D 01	−0.378540D 02
4.5000	−0.180664D 02	−0.239050D 02	0.795122D 01	−0.412367D 02
4.6000	−0.170056D 02	−0.281916D 02	0.134168D 02	−0.444594D 02
4.7000	−0.153513D 02	−0.327876D 02	0.198350D 02	−0.474035D 02
4.8000	−0.130043D 02	−0.376583D 02	0.272811D 02	−0.499270D 02
4.9000	−0.985851D 01	−0.427535D 02	0.358224D 02	−0.518625D 02
5.0000	−0.580146D 01	−0.480047D 02	0.455141D 02	−0.530149D 02
5.1000	−0.716021D 00	−0.533228D 02	0.563950D 02	−0.531594D 02
5.2000	0.551776D 01	−0.585945D 02	0.684814D 02	−0.520401D 02
5.3000	0.130201D 02	−0.636791D 02	0.817608D 02	−0.493686D 02
5.4000	0.219081D 02	−0.684058D 02	0.961842D 02	−0.448239D 02
5.5000	0.322920D 02	−0.725697D 02	0.111658D 03	−0.380525D 02
5.6000	0.442699D 02	−0.759292D 02	0.128035D 03	−0.286701D 02
5.7000	0.579219D 02	−0.782029D 02	0.145101D 03	−0.162647D 02
5.8000	0.733034D 02	−0.790668D 02	0.162568C 03	−0.400583D 00
5.9000	0.904363D 02	−0.781526D 02	0.180058D 03	0.193743D 02
6.0000	0.109300D 03	−0.750462D 02	0.197091D 03	0.435197D 02

XYPLOT COMPLETED.

x	W_3''	W_4''	W_3'''	W_4'''
0.0	-0.874617D 00	0.120381D 01	-0.943408D 00	-0.306532D 00
0.1000	-0.969201D 00	0.117322D 01	-0.950753D 00	-0.304679D 00
0.2000	-0.106531D 01	0.114293D 01	-0.974068D 00	-0.301031D 00
0.3000	-0.116459D 01	0.111297D 01	-0.101468D 01	-0.298713D 00
0.4000	-0.126886D 01	0.108303D 01	-0.107374D 01	-0.301188D 00
0.5000	-0.137998D 01	0.105245D 01	-0.115196D 01	-0.312244D 00
0.6000	-0.149990D 01	0.102016D 01	-0.124967D 01	-0.335980D 00
0.7000	-0.163056D 01	0.984681D 00	-0.136668D 01	-0.376797D 00
0.8000	-0.177386D 01	0.944073D 00	-0.150230D 01	-0.439379D 00
0.9000	-0.193160D 01	0.895914D 00	-0.165525D 01	-0.528681D 00
1.0000	-0.210542D 01	0.837273D 00	-0.182360D 01	-0.649905D 00
1.1000	-0.229674D 01	0.764689D 00	-0.200468D 01	-0.808469D 00
1.2000	-0.250666D 01	0.674148D 00	-0.215500D 01	-0.100998D 01
1.3000	-0.273590D 01	0.561070D 00	-0.239018D 01	-0.126018D 01
1.4000	-0.298468D 01	0.420295D 00	-0.258481D 01	-0.156490D 01
1.5000	-0.325263D 01	0.246080D 00	-0.277235D 01	-0.192994D 01
1.6000	-0.353866D 01	0.321039D-01	-0.294499D 01	-0.236105D 01
1.7000	-0.384083D 01	-0.228516D 00	-0.309354D 01	-0.286372D 01
1.8000	-0.415621D 01	-0.543197D 00	-0.320727D 01	-0.344309D 01
1.9000	-0.448071D 01	-0.919843D 00	-0.327378D 01	-0.410372D 01
2.0000	-0.480892D 01	-0.136678D 01	-0.327885D 01	-0.484939D 01
2.1000	-0.513389D 01	-0.189265D 01	-0.320626D 01	-0.568282D 01
2.2000	-0.544697D 01	-0.250631D 01	-0.303775D 01	-0.660533D 01
2.3000	-0.573756D 01	-0.321667D 01	-0.275280D 01	-0.761647D 01
2.4000	-0.599290D 01	-0.403247D 01	-0.232862D 01	-0.871358D 01
2.5000	-0.619781D 01	-0.496208D 01	-0.174008D 01	-0.989128D 01
2.6000	-0.633452D 01	-0.601313D 01	-0.959684D 00	-0.111408D 02
2.7000	-0.638236D 01	-0.719222D 01	0.423338D-01	-0.124496D 02
2.8000	-0.631760D 01	-0.850445D 01	0.129787D 01	-0.138002D 02
2.9000	-0.611321D 01	-0.995289D 01	0.284072D 01	-0.151696D 02
3.0000	-0.573871D 01	-0.115380D 02	0.470615D 01	-0.165285D 02
3.1000	-0.516002D 01	-0.132570D 02	0.693045D 01	-0.178403D 02
3.2000	-0.433944D 01	-0.151030D 02	0.955016D 01	-0.190596D 02
3.3000	-0.323561D 01	-0.170641D 02	0.126012D 02	-0.201318D 02
3.4000	-0.180368D 01	-0.191223D 02	0.161176D 02	-0.209516D 02
3.5000	0.446041D-02	-0.212527D 02	0.201303D 02	-0.215618D 02
3.6000	0.223979D 01	-0.234220D 02	0.246652D 02	-0.217527D 02
3.7000	0.495553D 01	-0.255872D 02	0.297409D 02	-0.214602D 02
3.8000	0.820632D 01	-0.276940D 02	0.353667D 02	-0.205657D 02
3.9000	0.120471D 02	-0.296758D 02	0.415390D 02	-0.189350D 02
4.0000	0.165317D 02	-0.314515D 02	0.482385D 02	-0.164178D 02
4.1000	0.217111D 02	-0.329242D 02	0.554257D 02	-0.128474D 02
4.2000	0.276310D 02	-0.339798D 02	0.630372D 02	-0.804160D 01
4.3000	0.343296D 02	-0.344848D 02	0.709797D 02	-0.180277D 01
4.4000	0.418338D 02	-0.342855D 02	0.791256D 02	0.607992D 01
4.5000	0.501559D 02	-0.332066D 02	0.873060D 02	0.158283D 02
4.6000	0.592890D 02	-0.310500D 02	0.953047D 02	0.276717D 02
4.7000	0.692018D 02	-0.275947D 02	0.102851D 03	0.418417D 02
4.8000	0.798331D 02	-0.225965D 02	0.109613D 03	0.585667D 02
4.9000	0.910849D 02	-0.157890D 02	0.115189D 03	0.780630D 02
5.0000	0.102816D 03	-0.688501D 01	0.119102D 03	0.100525D 03
5.1000	0.114831D 03	0.442035D 01	0.120789D 03	0.126114D 03
5.2000	0.126877D 03	0.184459D 02	0.119599D 03	0.154942D 03
5.3000	0.138630D 03	0.355183D 02	0.114785D 03	0.187055D 03
5.4000	0.149685D 03	0.559650D 02	0.105498D 03	0.222412D 03
5.5000	0.159549D 03	0.801039D 02	0.907903D 02	0.260861D 03
5.6000	0.167628D 03	0.108231D 03	0.696101D 02	0.302116D 03
5.7000	0.173217D 03	0.140605D 03	0.408099D 02	0.345719D 03
5.8000	0.175494D 03	0.177431D 03	0.315351D 01	0.391010D 03
5.9000	0.173509D 03	0.218834D 03	-0.446693D 02	0.437092D 03
6.0000	0.166176D 03	0.264837D 03	-0.104022D 03	0.482786D 03

XYPLOT COMPLETED.

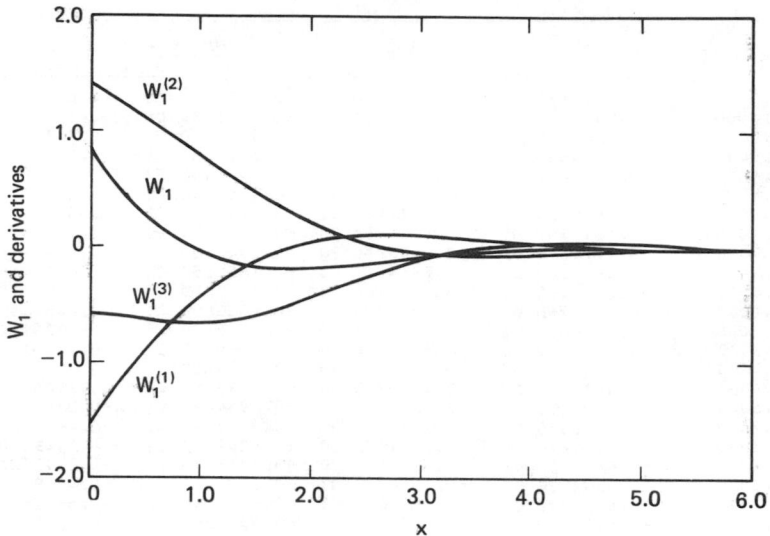

Figure C-6. Solutions to equation $w^{(iv)} + xw = 0$: (a) W_1, W_1', W_1'', W_1'''.

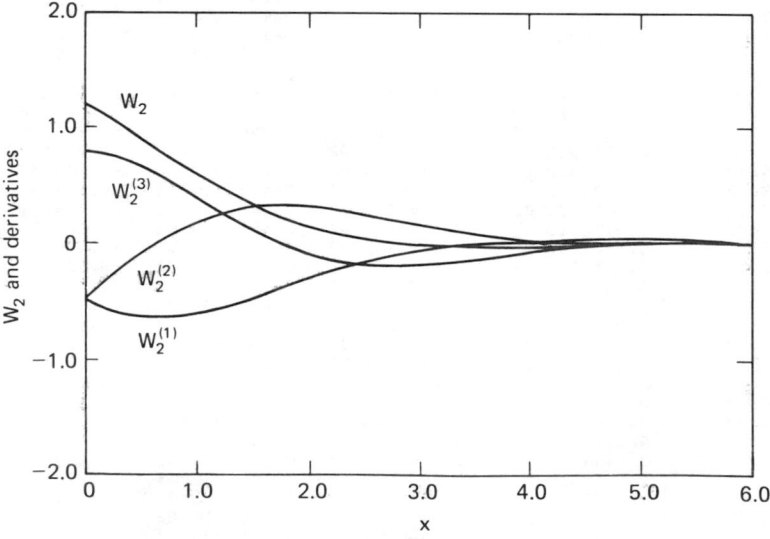

Figure C-6. (b) W_2, W_2', W_2'', W_2'''.

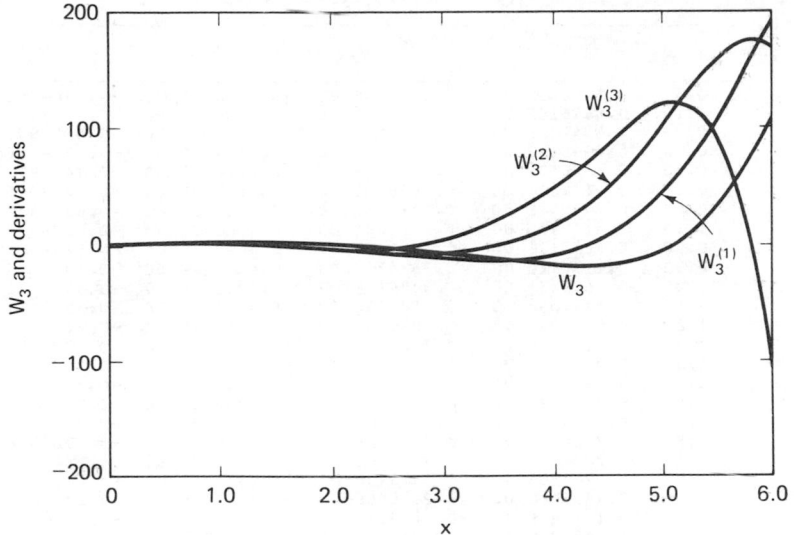

Figure C-6. (c) W_3, W'_3, W''_3, W'''_3.

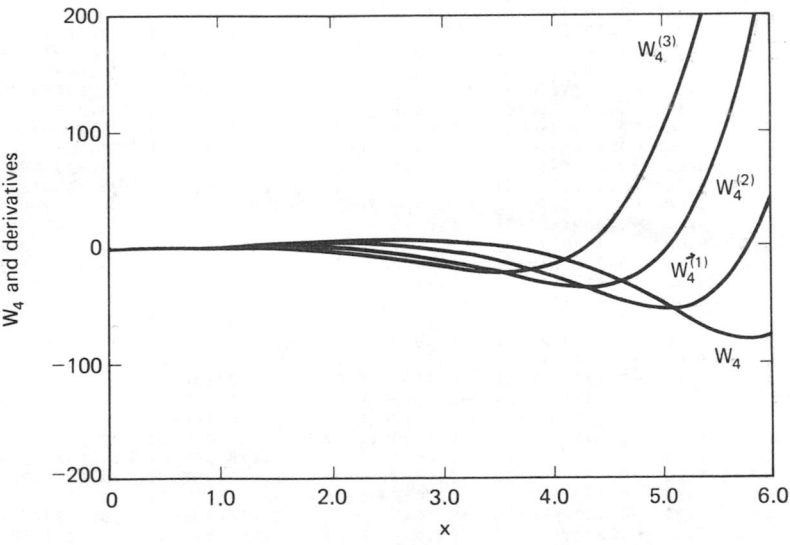

Figure C-6. (d) W_4, W'_4, W''_4, W'''_4.

TABLE C.4

Two Bounded Solutions to Equation $w^{(iv)} + x^{1/2}w = 0$;
W_1, W_2, and Derivatives as Functions of x

x	W_1	W_2	W'_1	W'_2
0.00	0.100000E 01	0.100000E 01	-0.276070E 00	-0.886498E 00
0.10	0.969618E 00	0.912914E 00	-0.330091E 00	-0.854746E 00
0.20	0.934272E 00	0.829141E 00	-0.375405E 00	-0.820297E 00
0.30	0.894815E 00	0.748534E 00	-0.412374E 00	-0.783481E 00
0.40	0.852060E 00	0.672511E 00	-0.441452E 00	-0.744687E 00
0.50	0.806771E 00	0.600048E 00	-0.463153E 00	-0.704334E 00
0.60	0.759657E 00	0.531681E 00	-0.478032E 00	-0.662847E 00
0.70	0.711373E 00	0.467503E 00	-0.486668E 00	-0.620643E 00
0.80	0.662512E 00	0.407563E 00	-0.489653E 00	-0.578125E 00
0.90	0.613611E 00	0.351876E 00	-0.487581E 00	-0.535669E 00
1.00	0.565145E 00	0.300416E 00	-0.481040E 00	-0.493625E 00
1.10	0.517533E 00	0.253127E 00	-0.470602E 00	-0.452311E 00
1.20	0.471136E 00	0.209920E 00	-0.456823E 00	-0.412008E 00
1.30	0.426262E 00	0.170663E 00	-0.440230E 00	-0.372965E 00
1.40	0.383167E 00	0.135278E 00	-0.421326E 00	-0.335395E 00
1.50	0.342058E 00	0.103549E 00	-0.400581E 00	-0.299473E 00
1.60	0.303097E 00	0.753236E-01	-0.378429E 00	-0.265345E 00
1.70	0.266405E 00	0.504166E-01	-0.355273E 00	-0.233121E 00
1.80	0.232064E 00	0.286333E-01	-0.331478E 00	-0.202881E 00
1.90	0.200120E 00	0.977254E-02	-0.307373E 00	-0.174677E 00
2.00	0.170590E 00	-0.637090E-02	-0.283252E 00	-0.148536E 00
2.10	0.143461E 00	-0.200036E-01	-0.259373E 00	-0.124460E 00
2.20	0.118699E 00	-0.313311E-01	-0.235960E 00	-0.102429E 00
2.30	0.962473E-01	-0.405564E-01	-0.213207E 00	-0.824068E-01
2.40	0.760307E-01	-0.478776E-01	-0.191274E 00	-0.643380E-01
2.50	0.579609E-01	-0.534869E-01	-0.170292E 00	-0.481555E-01
2.60	0.419372E-01	-0.575690E-01	-0.150366E 00	-0.337797E-01
2.70	0.278497E-01	-0.603001E-01	-0.131577E 00	-0.211219E-01
2.80	0.155819E-01	-0.618474E-01	-0.113981E 00	-0.100863E-01
2.90	0.501256E-02	-0.623681E-01	-0.976132E-01	-0.571404E-03
3.00	-0.398226E-02	-0.620089E-01	-0.824914E-01	0.752749E-02
3.10	-0.115273E-01	-0.609062E-01	-0.686164E-01	0.143172E-01
3.20	-0.177467E-01	-0.591855E-01	-0.559748E-01	0.199051E-01
3.30	-0.227625E-01	-0.569617E-01	-0.445403E-01	0.243978E-01
3.40	-0.266938E-01	-0.543390E-01	-0.342764E-01	0.278998E-01
3.50	-0.296553E-01	-0.514114E-01	-0.251379E-01	0.305130E-01
3.60	-0.317571E-01	-0.482627E-01	-0.170724E-01	0.323357E-01
3.70	-0.331036E-01	-0.449674E-01	-0.100221E-01	0.334616E-01
3.80	-0.337933E-01	-0.415907E-01	-0.392489E-02	0.339799E-01
3.90	-0.339182E-01	-0.381889E-01	0.128387E-02	0.339743E-01
4.00	-0.335639E-01	-0.348107E-01	0.567066E-02	0.335230E-01
4.10	-0.328093E-01	-0.314968E-01	0.930254E-02	0.326986E-01
4.20	-0.317264E-01	-0.282811E-01	0.122463E-01	0.315678E-01
4.30	-0.303808E-01	-0.251914E-01	0.145676E-01	0.301915E-01
4.40	-0.288315E-01	-0.222492E-01	0.163305E-01	0.286247E-01
4.50	-0.271312E-01	-0.194711E-01	0.175968E-01	0.269169E-01
4.60	-0.253267E-01	-0.168690E-01	0.184254E-01	0.251121E-01
4.70	-0.234588E-01	-0.144506E-01	0.188723E-01	0.232490E-01
4.80	-0.215632E-01	-0.122201E-01	0.189901E-01	0.213614E-01
4.90	-0.196702E-01	-0.101782E-01	0.188278E-01	0.194783E-01
5.00	-0.178254E-01	-0.832342E-02	0.184308E-01	0.176244E-01
5.10	-0.159904E-01	-0.665168E-02	0.178409E-01	0.158202E-01
5.20	-0.142424E-01	-0.515715E-02	0.170958E-01	0.140827E-01
5.30	-0.125753E-01	-0.383247E-02	0.162301E-01	0.124251E-01
5.40	-0.109994E-01	-0.266912E-02	0.152743E-01	0.108577E-01
5.50	-0.952250E-02	-0.165767E-02	0.142557E-01	0.938814E-02
5.60	-0.814956E-02	-0.788663E-03	0.131983E-01	0.802156E-02
5.70	-0.688343E-02	-0.498687E-04	0.121229E-01	0.676028E-02
5.80	-0.572499E-02	0.567550E-03	0.110473E-01	0.560587E-02
5.90	-0.467346E-02	0.107484E-02	0.998669E-02	0.455752E-02
6.00	-0.372673E-02	0.148253E-02	0.895359E-02	0.361324E-02

TABLE C.4 (Cont'd)

Two Bounded Solutions to Equation $w^{(iv)} + x^{1/2}w = 0$

x	W_1''	W_2''	W_1'''	W_2'''
0.00	−0.584606E 00	0.303220E 00	0.852356E 00	0.290394E 00
0.10	−0.496174E 00	0.331476E 00	0.871928E 00	0.270699E 00
0.20	−0.410723E 00	0.356940E 00	0.835258E 00	0.237226E 00
0.30	−0.329419E 00	0.378731E 00	0.789625E 00	0.197923E 00
0.40	−0.252996E 00	0.396438E 00	0.738012E 00	0.155985E 00
0.50	−0.181949E 00	0.409906E 00	0.682417E 00	0.113392E 00
0.60	−0.116594E 00	0.419141E 00	0.624370E 00	0.715048E−01
0.70	−0.571150E−01	0.424264E 00	0.565105E 00	0.312979E−01
0.80	−0.357967E−02	0.425480E 00	0.505646E 00	−0.652702E−02
0.90	0.440365E−01	0.423054E 00	0.446852E 00	−0.414726E−01
1.00	0.858369E−01	0.417293E 00	0.389439E 00	−0.732015E−01
1.10	0.121990E 00	0.408528E 00	0.334002E 00	−0.101505E 00
1.20	0.152720E 00	0.397110E 00	0.281025E 00	−0.126278E 00
1.30	0.178290E 00	0.383391E 00	0.230895E 00	−0.147502E 00
1.40	0.199004E 00	0.367726E 00	0.183908E 00	−0.165227E 00
1.50	0.215184E 00	0.350459E 00	0.140281E 00	−0.179558E 00
1.60	0.227177E 00	0.331922E 00	0.100158E 00	−0.190648E 00
1.70	0.235335E 00	0.312431E 00	0.636188E−01	−0.198681E 00
1.80	0.240021E 00	0.292281E 00	0.306864E−01	−0.203869E 00
1.90	0.241592E 00	0.271745E 00	0.133275E−02	−0.206443E 00
2.00	0.240404E 00	0.251071E 00	−0.245130E−01	−0.206645E 00
2.10	0.236803E 00	0.230486E 00	−0.469587E−01	−0.204725E 00
2.20	0.231121E 00	0.210189E 00	−0.661427E−01	−0.200931E 00
2.30	0.223677E 00	0.190354E 00	−0.822287E−01	−0.195512E 00
2.40	0.214772E 00	0.171133E 00	−0.953999E−01	−0.188709E 00
2.50	0.204688E 00	0.152651E 00	−0.105854E 00	−0.180753E 00
2.60	0.193685E 00	0.135013E 00	−0.113799E 00	−0.171866E 00
2.70	0.182005E 00	0.118302E 00	−0.119450E 00	−0.162254E 00
2.80	0.169864E 00	0.102580E 00	−0.123024E 00	−0.152111E 00
2.90	0.157462E 00	0.878919E−01	−0.124737E 00	−0.141612E 00
3.00	0.144972E 00	0.742642E−01	−0.124801E 00	−0.130920E 00
3.10	0.132550E 00	0.617096E−01	−0.123425E 00	−0.120177E 00
3.20	0.120328E 00	0.502262E−01	−0.120807E 00	−0.109513E 00
3.30	0.108423E 00	0.398007E−01	−0.117137E 00	−0.990369E−01
3.40	0.969298E−01	0.304093E−01	−0.112595E 00	−0.888466E−01
3.50	0.859276E−01	0.220192E−01	−0.107347E 00	−0.790222E−01
3.60	0.754789E−01	0.145904E−01	−0.101548E 00	−0.696301E−01
3.70	0.656315E−01	0.807698E−02	−0.953411E−01	−0.607233E−01
3.80	0.564200E−01	0.242822E−02	−0.888539E−01	−0.523423E−01
3.90	0.478662E−01	−0.240999E−02	−0.822024E−01	−0.445162E−01
4.00	0.399815E−01	−0.649414E−02	−0.754893E−01	−0.372635E−01
4.10	0.327674E−01	−0.988212E−02	−0.688048E−01	−0.305935E−01
4.20	0.262170E−01	−0.126323E−01	−0.622266E−01	−0.245073E−01
4.30	0.203163E−01	−0.148028E−01	−0.558211E−01	−0.189985E−01
4.40	0.150452E−01	−0.164509E−01	−0.496436E−01	−0.140546E−01
4.50	0.103785E−01	−0.176320E−01	−0.437390E−01	−0.965772E−02
4.60	0.628707E−02	−0.183999E−01	−0.381431E−01	−0.578564E−02
4.70	0.273867E−02	−0.188058E−01	−0.328827E−01	−0.241261E−02
4.80	−0.301287E−03	−0.188981E−01	−0.279767E−01	0.489892E−03
4.90	−0.286889E−02	−0.187225E−01	−0.234370E−01	0.295239E−02
5.00	−0.500110E−02	−0.183212E−01	−0.192693E−01	0.500677E−02
5.10	−0.673514E−02	−0.177336E−01	−0.154734E−01	0.668569E−02
5.20	−0.810801E−02	−0.169955E−01	−0.120447E−01	0.802205E−02
5.30	−0.915602E−02	−0.161395E−01	−0.897427E−02	0.904853E−02
5.40	−0.991441E−02	−0.151951E−01	−0.624999E−02	0.979721E−02
5.50	−0.104171E−01	−0.141883E−01	−0.385681E−02	0.102992E−01
5.60	−0.106962E−01	−0.131424E−01	−0.177750E−02	0.105845E−01
5.70	−0.107824E−01	−0.120777E−01	0.678971E−05	0.106814E−01
5.80	−0.107040E−01	−0.110115E−01	0.151613E−02	0.106169E−01
5.90	−0.104876E−01	−0.995884E−02	0.277133E−02	0.104161E−01
6.00	−0.101575E−01	−0.893207E−02	0.379358E−02	0.101022E−01

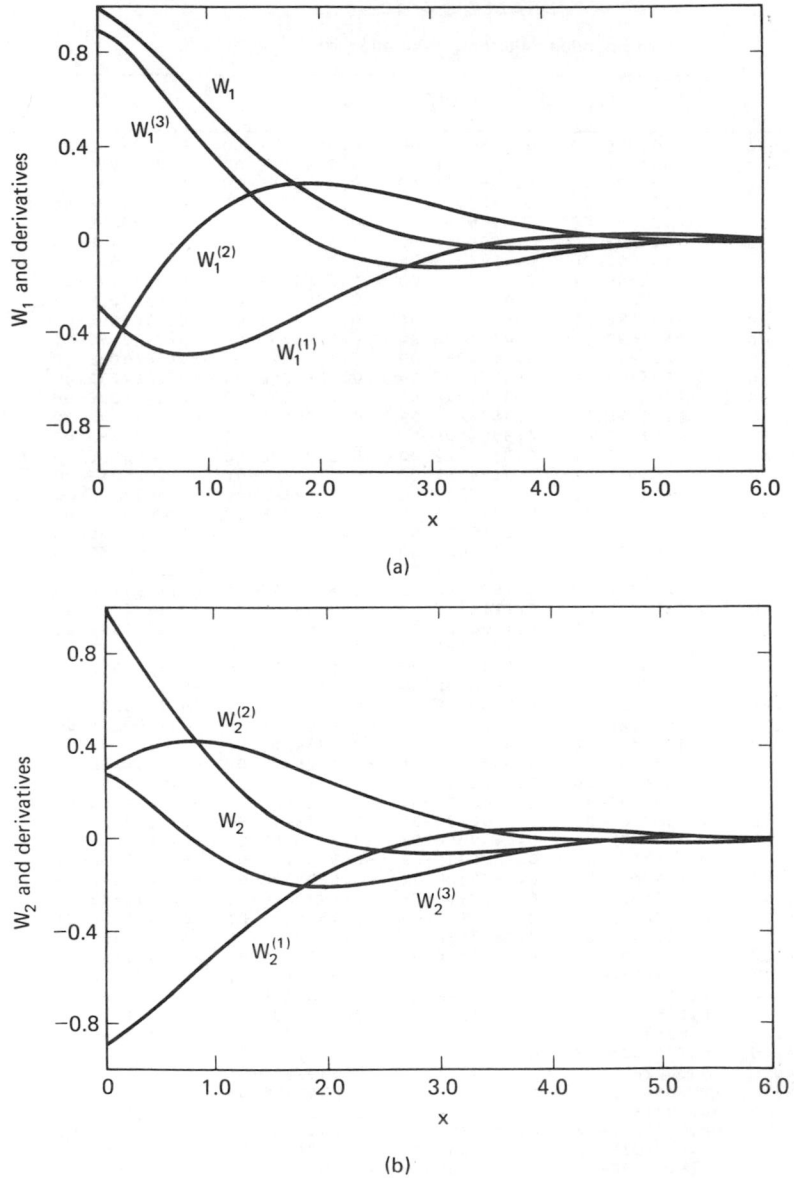

Figure C-7. Two bounded solutions to equation $w^{(\text{iv})} + x^{1/2}w = 0$: (a) W_1, W_1', W_1'', W_1'''; (b) W_2, W_2', W_2'', W_2'''.

Name Index

Subject Index